Trends in Mathematics

Trends in Mathematics is a series devoted to the publication of volumes arising from conferences and lecture series focusing on a particular topic from any area of mathematics. Its aim is to make current developments available to the community as rapidly as possible without compromise to quality and to archive these for reference.

Proposals for volumes can be submitted using the Online Book Project Submission Form at our website www.birkhauser-science.com.

Material submitted for publication must be screened and prepared as follows:

All contributions should undergo a reviewing process similar to that carried out by journals and be checked for correct use of language which, as a rule, is English. Articles without proofs, or which do not contain any significantly new results, should be rejected. High quality survey papers, however, are welcome.

We expect the organizers to deliver manuscripts in a form that is essentially ready for direct reproduction. Any version of TeX is acceptable, but the entire collection of files must be in one particular dialect of TeX and unified according to simple instructions available from Birkhäuser.

More information about this series at http://www.springer.com/series/4961

Birkhäuser

Trends in Mathematics

Trends in Mathematics is a series devoted to the publication of volumes arising from conferences and lecture series focusing on a particular topic from any area of mathematics. Its aim is to make current developments available to the community as rapidly as possible without compromise to quality and to archive these for reference.

Proposals for volumes can be submitted using the Online Book Project Submission Form at our website www.birkhauser-science.com.

Material submitted for publication must be screened and prepared as follows:

All contributions should undergo a reviewing process similar to that carried out by journals and be checked for correct use of language which, as a rule, is English. Articles without proofs, or which do not contain any significantly new results, should be rejected. High quality survey papers, however, are welcome.

We expect the organizers to deliver manuscripts in a form that is essentially ready for direct reproduction. Any version of TEX is acceptable, but the entire collection of files must be in one particular dialect of TEX and unified according to simple instructions available from Birkhäuser.

Furthermore, in order to guarantee the timely appearance of the proceedings it is essential that the final version of the entire material be submitted no later than one year after the conference.

More information about this series at http://www.springer.com/series/4961

V. Madhu • A. Manimaran • D. Easwaramoorthy
D. Kalpanapriya • M. Mubashir Unnissa
Editors

Advances in Algebra and Analysis

International Conference on Advances in Mathematical Sciences, Vellore, India, December 2017 - Volume I

Birkhäuser

Editors
V. Madhu
Department of Mathematics
School of Advanced Sciences
Vellore Institute of Technology
Vellore, Tamil Nadu, India

A. Manimaran
Department of Mathematics
School of Advanced Sciences
Vellore Institute of Technology
Vellore, Tamil Nadu, India

D. Easwaramoorthy
Department of Mathematics
School of Advanced Sciences
Vellore Institute of Technology
Vellore, Tamil Nadu, India

D. Kalpanapriya
Department of Mathematics
School of Advanced Sciences
Vellore Institute of Technology
Vellore, Tamil Nadu, India

M. Mubashir Unnissa
Department of Mathematics
School of Advanced Sciences
Vellore Institute of Technology
Vellore, Tamil Nadu, India

ISSN 2297-0215 ISSN 2297-024X (electronic)
Trends in Mathematics
ISBN 978-3-030-01119-2 ISBN 978-3-030-01120-8 (eBook)
https://doi.org/10.1007/978-3-030-01120-8

Library of Congress Control Number: 2018966815

This book is published under the imprint Birkhäuser, www.birkhauser-science.com by the registered company Springer Nature Switzerland AG
The registered company address is: Gewerbestrasse 11, 6330 Cham, Switzerland

Preface

The Department of Mathematics, School of Advanced Sciences, Vellore Institute of Technology (Deemed to be University), Vellore, Tamil Nadu, India, organized the International Conference on Advances in Mathematical Sciences—2017 (ICAMS 2017) in association with the Society for Industrial and Applied Mathematics VIT Chapter from December 1, 2017 to December 3, 2017. The major objective of ICAMS 2017 was to promote scientific and educational activities toward the advancement of common man's life by improving the theory and practice of various disciplines of Mathematics. This prestigious conference was partially financially supported by the Council of Scientific and Industrial Research (CSIR), India. The Department of Mathematics has 90 qualified faculty members and 30 research scholars, and all were delicately involved in organizing ICAMS 2017 grandly. In addition, 30 leading researchers from around the world served as an advisory committee for this conference. Overall, more than 450 participants (professors/scholars/students) enriched their knowledge in the wings of Mathematics.

There were 9 eminent speakers from overseas and 33 experts from various states of India who delivered the keynote address and invited talks in this conference. Many leading scientists and researchers worldwide submitted their quality research articles to ICAMS. Moreover, 305 original research articles were shortlisted for ICAMS 2017 oral presentations that were authored by dynamic researchers from 25 states in India and 20 countries around the world. We hope that ICAMS will further stimulate research in Mathematics, share research interest and information, and create a forum of collaboration and build a trust relationship. We feel honored and privileged to serve the best of recent developments in the field of Mathematics to the reader.

A basic premise of this book is that quality assurance is effectively achieved through the selection of quality research articles by a scientific committee consisting of more than 100 reviewers from all over the world. This book comprises the contribution of several dynamic researchers in 52 chapters. Each chapter identifies the existing challenges in the areas of Algebra, Analysis, Operations Research, and Statistics and emphasizes the importance of establishing new methods and algorithms to address the challenges. Each chapter presents a research problem, the technique suitable for solving the problem with sufficient mathematical background, and discussions on the obtained results with physical interruptions to understand the domain of applicability. This book also provides a comprehensive literature survey which reveals the challenges, outcomes, and developments of higher level mathematics in this decade. The theoretical coverage of this book is relatively at a higher level to meet the global orientation of mathematics and its applications in science and engineering.

The target audience of this book is postgraduate students, researchers, and industrialists. This book promotes a vision of pure and applied mathematics as integral to modern science and engineering. Each chapter contains important information emphasizing core Mathematics, intended for the professional who already possesses a basic understanding. In this book, theoretically oriented readers will find an overview of Mathematics and its applications. Industrialists will find a variety of techniques with sufficient discussion in terms of physical point of view to adapt for solving the particular application based on mathematical models. The reader can make use of the literature survey of this book to identify the current trends in Mathematics. It is our hope and expectation that this book will provide an effective learning experience and referenced resource for all young mathematicians.

As Editors, we would like to express our sincere thanks to all the administrative authorities of Vellore Institute of Technology, Vellore, for their motivation and support. We also extend our profound thanks to all faculty members and research scholars of the Department of Mathematics and all staff members of our institute. We especially thank all the members of the organizing committee of ICAMS 2017 who worked as a team by investing their time to make the conference a great success. We thank the national funding agency, Council of Scientific and Industrial Research (CSIR), Government of India, for the financial support they contributed toward the successful completion of this international conference. We express our sincere gratitude to all the referees for spending their valuable time to review the manuscripts, which led to substantial improvements and selection of the research papers for publication. The organizing committee is grateful to Mr. Christopher Tominich, Editor at Birkhäuser/Springer, for his continuous encouragement and support toward the publication of this book.

Vellore, India V. Madhu
Vellore, India A. Manimaran
Vellore, India D. Easwaramoorthy
Vellore, India D. Kalpanapriya
Vellore, India M. Mubashir Unnissa

Contents

Volume II Contents

Part VI Fluid Dynamics

Part I
Algebra

IT-2 Fuzzy Automata and IT-2 Fuzzy Languages

M. K. Dubey, Priyanka Pal, and S. P. Tiwari

Abstract The objective of this work is to give certain determinization and algebraic studies for an interval type-2 (IT-2) fuzzy automaton and language. We introduce a deterministic IT-2 fuzzy automaton and prove that it is behavioural equivalent to an IT-2 fuzzy automaton. Also, for a given IT-2 fuzzy language, we give certain recipe for constructions of deterministic IT-2 fuzzy automata.

1 Introduction

The notion of type-2 fuzzy sets was introduced by Zadeh [21], who gives the substructure to model and abbreviate the impact of uncertainty in fuzzy logic rule-based systems. The author in [9] has pointed out that the membership function of type-1 fuzzy sets is totally crisp and hence not able to model certain uncertainty involved in the model, whereas in case of type-2 fuzzy sets, it is capable to model such uncertainty because of their fuzzy membership functions. Also, the membership function of type-2 fuzzy sets is three dimensional which gives additional degrees of freedom to model the uncertainty directly in comparison to type-1 fuzzy sets which have two-dimensional membership function. However, it is not easy to understand and use the concept of type-2 fuzzy sets, which can be seen by the fact that almost all applications use interval type-2 fuzzy set for the sake of all computations to perform easily [10].

From the commencement of the theory of fuzzy sets, Santos [12], Wee [17] and Wee and Fu [18] introduced and studied fuzzy automata and languages, and after Malik, Mordeson and Sen [11] have further studied and developed. In the last few decades, many works on fuzzy automata and languages have been done (cf., [1, 2, 4, 5, 7, 8, 13–16, 20]). During the decades, it has been observed that fuzzy automata

M. K. Dubey (✉) · P. Pal · S. P. Tiwari
Department of Applied Mathematics, Indian Institute of Technology (ISM), Dhanbad, India
e-mail: maheshdubey6@gmail.com; mahesh@am.ism.ac.in; priyankapal2192@gmail.com; priyanka.2015DR0217@am.ac.in; sptiwarimaths@gmail.com; sptiwari@iitism.ac.in

V. Madhu et al. (eds.), *Advances in Algebra and Analysis*, Trends in Mathematics,
https://doi.org/10.1007/978-3-030-01120-8_1

and fuzzy languages have obtained not only conversion of classical automata to fuzzy automata but also a broad field of applications[2].

Fuzzy automata and fuzzy languages referred above are either based upon type-1 fuzzy sets or on certain lattice structures (cf., [4, 5, 8, 20]). Since we know, type-1 fuzzy sets cannot be able to minimize the uncertainty involved in the model, and Mendel [10] suggested to use an IT-2 fuzzy set model of a word in the concept of computing with words. Recently, Jiang and Tang [6] introduced and studied the concepts of IT-2 fuzzy automata and languages and give the platform to develop the above model of nonclassical computations. In this note, we give a brief look at certain studies for IT-2 fuzzy automata and languages, which may be carried out in details. In particular, we begin by introducing a deterministic IT-2 fuzzy automaton and prove that it is behavioural equivalent to an IT-2 fuzzy automaton. Further, for a given IT-2 fuzzy language, we give the certain recipe for constructions of deterministic IT-2 fuzzy automata. Finally, we give a brief look at an algebraic study of an IT-2 fuzzy automaton.

2 IT-2 Fuzzy Sets

In this section, we memorize certain notions allied with an IT-2 fuzzy set. We initiate with the following notion of a type-2 fuzzy set. For more description, we refer to [9, 10, 19, 21].

Definition 2.1 ([9]) A **type-2 fuzzy set** $\widetilde{F}$ in a nonempty set Y is characterized by a type-2 membership function $\mu_{\widetilde{F}}(y, v)$, where $y \in Y$ and $v \in J_y \subseteq [0, 1]$, i.e.:

$$\widetilde{F} = \int_{y\in Y} \int_{v\in J_y} \mu_{\widetilde{F}}(y, v)/(y, v),\ \ J_y \subseteq [0, 1],\ in\ which\ 0 \leq \mu_{\widetilde{F}}(y, v) \leq 1.$$

From Definition 2.1, it has been observed that when uncertainties disappear, a type-2 membership function must reduce to a type-1 membership function, and in this case, the variable v equals $\mu_F(y)$ and $0 \leq \mu_{\widetilde{F}}(y) \leq 1$.

Definition 2.2 ([10]) A type-2 fuzzy set $\widetilde{F}$ in Y is called an **IT-2 fuzzy set** if $\mu_{\widetilde{F}}(y, v) = 1, \forall y \in Y$ and $\forall v \in J_y$. An IT-2 fuzzy set $\widetilde{F}$ can be expressed as $\widetilde{F} = \int_{y\in Y} \int_{v\in J_y} 1/(y, v),\ J_y \subseteq [0, 1]$.

For an IT-2 fuzzy set, we consider $J_y = [\underline{\mu}_{\widetilde{F}}(y), \overline{\mu}_{\widetilde{F}}(y)]$ for all $y \in Y$, where $\underline{\mu}_{\widetilde{F}}(y)$ and $\overline{\mu}_{\widetilde{F}}(y)$ are, respectively, called the lower membership function (LPF) and upper membership function (UMF) of $\widetilde{F}$ which are two type-1 membership functions that bound the footprint of uncertainty. We shall denote by $IT2F(Y)$, the set of all IT-2 fuzzy sets in Y. For more details on IT-2 fuzzy sets and their operations, we refer to [10].

3 IT-2 Fuzzy Automata and IT-2 Fuzzy Languages

In this section, we give a brief look to determinization of an IT-2 fuzzy automaton. In particular, we introduce a deterministic IT-2 fuzzy automaton and prove that it is behavioural equivalent to an IT-2 fuzzy automaton. We initiate with the following concept of an IT-2 fuzzy automaton.

Definition 3.1 ([6]) An **IT-2 fuzzy automaton (IT2FA)** is a five-tuple $\widetilde{\mathscr{M}} = (S, X, \widetilde{\lambda}, \widetilde{i}, \widetilde{f})$, where S, X are nonempty sets called **set of states** and **set of inputs** and the characterization of $\widetilde{\lambda}$, $\widetilde{i}$ and $\widetilde{f}$ is as follows:

(i) $\widetilde{\lambda} : S \times X \to IT2F(S)$, called the **transition map**, such that for a given $s \in S$ and $x \in X$, $\widetilde{\lambda}(s, x)$ is an IT-2 fuzzy subset of S, and it may be seen as the possibility distribution of the states that the automaton in state s and with input x can enter.

(ii) $\widetilde{i}$ and $\widetilde{f}$ are IT-2 fuzzy subsets of S, called the **IT-2 fuzzy set of initial states** and **IT-2 fuzzy set of final states**, respectively.

Now, we need to extend the transition function for defining the notion of the degree to which a string of input symbols is accepted by an IT-2 fuzzy automaton, which is given below.

Definition 3.2 Let $\widetilde{\mathscr{M}} = (S, X, \widetilde{\lambda}, \widetilde{i}, \widetilde{f})$ be an IT-2 fuzzy automaton. The transition map $\widetilde{\lambda}$ can be extended to $\widetilde{\lambda^*} : S \times X^* \to IT2F(S)$, where

$$\widetilde{\lambda^*}(s, e) = 1/[1, 1]/s,$$

$$\widetilde{\lambda^*}(s, wx) = \bigcup_{s' \in S} \left[\widetilde{\lambda^*}(s, w)(s') \cdot \widetilde{\lambda}(s', x)\right],$$

$\forall w \in X^*$ and $\forall x \in X$, where $1/[1, 1]/s$ is an IT-2 fuzzy subset of S with membership 1. Also, $\widetilde{\lambda^*}(s, w)(s') \cdot \widetilde{\lambda}(s', x)$ stands for the scalar product of IT-2 fuzzy set $\widetilde{\lambda}(s', x)$ with the scalar quantity $\widetilde{\lambda^*}(s, w)(s')$.

Definition 3.3 An IT-2 fuzzy language $\widetilde{\rho} \in IT2F(X^*)$ is said to be accepted by an IT-2 fuzzy automaton $\widetilde{\mathscr{M}} = (S, X, \widetilde{\lambda}, \widetilde{i}, \widetilde{f})$, if $\forall w \in X^*$

$$\begin{aligned}\widetilde{\rho}(w) &= 1/[\vee\{\underline{\mu}_{\widetilde{i}}(s) \wedge \underline{\mu}_{\widetilde{\lambda^*}(s,w)}(s') \wedge \underline{\mu}_{\widetilde{f}}(s') : s, s' \in S\},\\ &\quad \vee\{\overline{\mu}_{\widetilde{i}}(s) \wedge \overline{\mu}_{\widetilde{\lambda^*}(s,w)}(s') \wedge \overline{\mu}_{\widetilde{f}}(s') : s, s' \in S\}].\end{aligned}$$

The notion of a deterministic IT-2 fuzzy automaton is introduced as follows.

Definition 3.4 A **deterministic IT-2 fuzzy automaton (DIT2FA)** is a five-tuple $\widetilde{\mathscr{M}} = (S, X, \lambda, s_0, \widetilde{f})$, where S and X are as in an IT-2 fuzzy automaton; s_0 is the **initial state**; $\lambda : S \times X \to S$ is a map, called **state transition map**; and $\widetilde{f}$ is an IT-2 fuzzy set in S, called the **IT-2 fuzzy set of final states**.

Definition 3.5 The transition map λ can be extended to $\lambda^* : S \times X^* \to S$, such that $\lambda^*(s, e) = s$ and $\lambda^*(s, wa) = \lambda(\lambda^*(s, w), a), \forall w \in X^*$ and $a \in X$.

Definition 3.6 An IT-2 fuzzy language $\widetilde{\rho} \in IT2F(X^*)$ is said to be accepted by a deterministic IT-2 fuzzy automaton $\widetilde{\mathscr{M}} = (S, X, \lambda, s_0, \widetilde{f})$, if for all $w \in X^*$,

$$\widetilde{\rho}(w) = 1/[\underline{\mu}_{\widetilde{f}}(\lambda^*(s_0, w)), \overline{\mu}_{\widetilde{f}}(\lambda^*(s_0, w))].$$

We shall denote an IT-2 fuzzy language $\widetilde{\rho}$ by $\widetilde{\rho}_{\widetilde{\mathscr{M}}}$, if $\widetilde{\rho}$ is accepted by a deterministic IT-2 fuzzy automaton $\widetilde{\mathscr{M}}$.

Now, the following result is towards the behavioural equivalent between an IT-2 fuzzy automaton and a deterministic IT-2 fuzzy automaton.

Proposition 3.1 *A $\widetilde{\rho} \in IT2F(X^*)$ is accepted by an IT-2 fuzzy automaton if and only if it is accepted by a deterministic IT-2 fuzzy automaton.*

Proof Let $\widetilde{\mathscr{M}} = (S, X, \widetilde{\lambda}, \widetilde{i}, \widetilde{f})$ be an IT-2 fuzzy automaton. Then for all $w \in X^*$ and for all $s \in S$, define an IT-2 fuzzy subset of S as under:

$$\widetilde{i}_w(s) = 1/[\vee_{s' \in S}\{\underline{\mu}_{\widetilde{i}}(s') \wedge \underline{\mu}_{\widetilde{\lambda}^*(s', w)}(s)\}, \vee_{s' \in S}\{\overline{\mu}_{\widetilde{i}}(s') \wedge \overline{\mu}_{\widetilde{\lambda}^*(s', w)}(s)\}],$$

or that $\widetilde{i}_w(s) = \bigcup_{s' \in S}\left[\widetilde{i}(s') \cdot \widetilde{\lambda}^*(s', w)(s)\right]$. Now, let $S' = \{\widetilde{i}_w : w \in X^*\}$ and the map $\lambda^{*'} : S' \times X^* \to S'$ such that $\lambda^{*'}(\widetilde{i}_w, w') = \widetilde{i}_{ww'}, \ \forall w, w' \in X^*$. It is clear that $\lambda^{*'}$ is well-defined. Now, $\widetilde{\mathscr{M}'} = (S', X, \lambda^{*'}, \widetilde{i}_e, \widetilde{f}')$ is a DIT2FA, where the IT-2 fuzzy subset of final states $\widetilde{f}' \in IT2F(S')$ is defined as under:

$$\begin{aligned}
\widetilde{f}'(\widetilde{i}_w) &= \bigcup_{s \in S}\left[\widetilde{i}_w(s) \cdot \widetilde{f}(s)\right] \\
&= \bigcup_{s \in S}\left[\bigcup_{s' \in S}\{i(s') \cdot \widetilde{\lambda}^*(s', w)(s)\} \cdot \widetilde{f}(s)\right] \\
&= \bigcup_{s' \in S}\left[\widetilde{i}(s') \cdot \widetilde{\lambda}^*(s', w)(s) \cdot \widetilde{f}(s)\right],
\end{aligned}$$

or that

$$\begin{aligned}
\widetilde{f}'(\widetilde{i}_w)(s) = 1/[&\vee_{s' \in S}\{\underline{\mu}_{\widetilde{i}}(s') \wedge \underline{\mu}_{\widetilde{\lambda}^*(s', w)}(s) \wedge \underline{\mu}_{\widetilde{f}}(s)\}, \\
&\vee_{s' \in S}\{\overline{\mu}_{\widetilde{i}}(s') \wedge \overline{\mu}_{\widetilde{\lambda}^*(s', w)}(s) \wedge \overline{\mu}_{\widetilde{f}}(s)\}].
\end{aligned}$$

Finally, let $\widetilde{\rho} \in IT2F(X^*)$ be accepted by $\widetilde{\mathscr{M}}$. Then for all $w \in X^*$,

$$\begin{aligned}
\widetilde{\rho}(w) = 1/[&\vee_{s, s' \in S}\{\underline{\mu}_{\widetilde{i}}(s) \wedge \underline{\mu}_{\widetilde{\lambda}^*(s, w)}(s') \wedge \underline{\mu}_{\widetilde{f}}(s')\}, \\
&\vee_{s, s' \in S}\{\overline{\mu}_{\widetilde{i}}(s) \wedge \overline{\mu}_{\widetilde{\lambda}^*(s, w)}(s') \wedge \overline{\mu}_{\widetilde{f}}(s')\}]
\end{aligned}$$

$$= 1/[\vee_{s' \in S}\{\vee_{s \in S}\left\{\underline{\mu}_{\widetilde{i}}(s) \wedge \underline{\mu}_{\widetilde{\lambda}^*(s,w)}(s')\right\} \wedge \underline{\mu}_{\widetilde{f}}(s')\},$$

$$\vee_{s' \in S}\{\vee_{s \in S}\left\{\overline{\mu}_{\widetilde{i}}(s) \wedge \overline{\mu}_{\widetilde{\lambda}^*(s,w)}(s')\right\} \wedge \overline{\mu}_{\widetilde{f}}(s')\}]$$

$$= 1/[\vee_{s' \in S}\{\underline{\mu}_{\widetilde{i}_w}(s') \wedge \underline{\mu}_{\widetilde{f}}(s')\}, \vee_{s' \in S}\{\overline{\mu}_{\widetilde{i}_w}(s') \wedge \overline{\mu}_{\widetilde{f}'}(s')\}]$$

$$= 1/[\vee_{s' \in S}\{\underline{\mu}_{\widetilde{f}}(\widetilde{i}_w)(s'), \vee_{s \in S}\{\overline{\mu}_{\widetilde{f}}(\widetilde{i}_w)(s')\}]$$

$$= 1/[\underline{\mu}_{\widetilde{f}}(\lambda^{*'}(\widetilde{i}_e, w)), \overline{\mu}_{\widetilde{f}}(\lambda^{*'}(\widetilde{i}_e, w)] = \widetilde{\rho}_{\widetilde{\mathscr{A}}'}(w).$$

Thus $\widetilde{\rho}$ is accepted by a DIT2FA $\widetilde{\mathscr{M}}'$.

Similarly, we can show that converse is also true.

4 Construction of Deterministic IT-2 Fuzzy Automata for IT-2 Fuzzy Languages

In this section, we give the recipe to constructions of a DIT2FA for a given IT-2 fuzzy language. In particular, we give two recipes for such constructions and prove that both the DIT2FA are homomorphic. The first such recipe is based on right congruence relation (Myhill-Nerode relation), while the other is based on the derivative of given IT-2 fuzzy language. We initiate with the following construction based on right congruence relation.

Proposition 4.1 *Let $\widetilde{\rho} \in IT2F(X^*)$. Then there exists a deterministic IT-2 fuzzy automaton, which accepts $\widetilde{\rho}$.*

Proof Let us define a relation $R_{\widetilde{\rho}}$ on X^* such that $uR_{\widetilde{\rho}}v \Leftrightarrow \widetilde{\rho}(uw) = \widetilde{\rho}(vw), \forall w \in X^*$. Then $R_{\widetilde{\rho}}$ is an equivalence relation on X^*. Now, let $S_{R_{\widetilde{\rho}}} = X^*/R_{\widetilde{\rho}} = \{[u]_{R_{\widetilde{\rho}}} : u \in X^*\}$, where $[u]_{R_{\widetilde{\rho}}} = \{v \in X^* : uR_{\widetilde{\rho}}v\}$. Define the maps $\lambda^*_{R_{\widetilde{\rho}}} : S_{R_{\widetilde{\rho}}} \times X^* \to S_{R_{\widetilde{\rho}}}$ such that $\lambda^*_{R_{\widetilde{\rho}}}([u]_{R_{\widetilde{\rho}}}, v) = [uv]_{R_{\widetilde{\rho}}}$ and $\widetilde{f}_{R_{\widetilde{\rho}}} \in IT2F(S_{R_{\widetilde{\rho}}})$ such that $\widetilde{f}_{R_{\widetilde{\rho}}}([u]_{R_{\widetilde{\rho}}}) = \widetilde{\rho}(u)$. Now, it is easy to check that both the maps $\lambda^*_{R_{\widetilde{\rho}}}$ and $\widetilde{f}_{R_{\widetilde{\rho}}}$ are well-defined. Thus $\widetilde{\mathscr{M}_{R_{\widetilde{\rho}}}} = (S_{R_{\widetilde{\rho}}}, X, \lambda^*_{R_{\widetilde{\rho}}}, [e]_{R_{\widetilde{\rho}}}, \widetilde{f}_{R_{\widetilde{\rho}}})$ is a deterministic IT-2 fuzzy automaton. Finally, for all $u \in X^*$,

$$\widetilde{\rho}_{\mathscr{M}_{R_{\widetilde{\rho}}}}(u) = \widetilde{f}_{R_{\widetilde{\rho}}}(\lambda^*_{R_{\widetilde{\rho}}}([e]_{R_{\widetilde{\rho}}}, u))$$

$$= 1/[\underline{\mu}_{\widetilde{f}_{R_{\widetilde{\rho}}}}(\lambda^*_{R_{\widetilde{\rho}}}([e]_{R_{\widetilde{\rho}}}, u)), \overline{\mu}_{\widetilde{f}_{R_{\widetilde{\rho}}}}(\lambda^*_{R_{\widetilde{\rho}}}([e]_{R_{\widetilde{\rho}}}, u))]$$

$$= 1/[\underline{\mu}_{\widetilde{f}_{R_{\widetilde{\rho}}}}([u]_{R_{\widetilde{\rho}}}), \overline{\mu}_{\widetilde{f}_{R_{\widetilde{\rho}}}}([u]_{R_{\widetilde{\rho}}})] = \widetilde{f}_{R_{\widetilde{\rho}}}([u]_{R_{\widetilde{\rho}}}) = \widetilde{\rho}(u).$$

Hence DIT2FA $\mathscr{M}_{R_{\widetilde{\rho}}}$ accepts $\widetilde{\rho}$.

Now, we introduce the following concept of derivative of an IT-2 fuzzy language.

Definition 4.1 Let $\widetilde{\rho} \in IT2F(X^*)$ and $u \in X^*$. An IT-2 fuzzy language $\widetilde{\rho}^u$, defined by $\widetilde{\rho}^u(v) = \widetilde{\rho}(uv), \forall v \in X^*$ is called a **derivative** of $\widetilde{\rho}$ with respect to u.

The following recipe is construction of a DIT2FA with the help of derivative of given IT-2 fuzzy language.

Let $\widetilde{\rho} \in IT2F(X^*)$. Now, assume $S^{\widetilde{\rho}} = \{\widetilde{\rho}^u : u \in X^*\}$, and define $\lambda^{*\widetilde{\rho}}$ and $f^{\widetilde{\rho}}$ as under:

$$\lambda^{*\widetilde{\rho}} : S^{\widetilde{\rho}} \times X^* \to S^{\widetilde{\rho}} \text{ such that } \lambda^{*\widetilde{\rho}}(\widetilde{\rho}^u, v) = \widetilde{\rho}^{uv}, \forall \widetilde{\rho}^u \in S^{\widetilde{\rho}}, \forall v \in X^*, \text{ and}$$

$$f^{\widetilde{\rho}} \in IT2F(S^{\widetilde{\rho}}) \text{ such that } f^{\widetilde{\rho}}(\widetilde{\rho}^u) = \widetilde{\rho}^u(e), \forall \widetilde{\rho}^u \in S^{\widetilde{\rho}}.$$

Then it can be easily seen that the maps $\lambda^{*\widetilde{\rho}}$ and $f^{\widetilde{\rho}}$ are well-defined. Thus $\widetilde{\mathscr{M}^{\rho}} = (S^{\widetilde{\rho}}, X, \lambda^{*\widetilde{\rho}}, \widetilde{\rho}^e, f^{\widetilde{\rho}})$ is a deterministic IT-2 fuzzy automaton. Now, for all $w \in X^*$ $\widetilde{\rho}_{\widetilde{\mathscr{A}^{\rho}}}(w) = f^{\widetilde{\rho}}(\lambda^{*\widetilde{\rho}}(\widetilde{\rho}^e, w)) = f^{\widetilde{\rho}}(\widetilde{\rho}^{ew}) = f^{\widetilde{\rho}}(\widetilde{\rho}^w) = \widetilde{\rho}^w(e) = \widetilde{\rho}(we) = \widetilde{\rho}(w)$, it shows that $\widetilde{\mathscr{M}^{\rho}}$ accepts $\widetilde{\rho}$.

Before starting next, we familiarize the following concept of homomorphism between two DIT2FA.

Definition 4.2 Let $\widetilde{\mathscr{M}} = \left(S, X, \lambda, s_0, \widetilde{f}\right)$ and $\widetilde{\mathscr{M}'} = \left(S', X, \lambda', s_0', \widetilde{f'}\right)$ be two deterministic IT-2 fuzzy automata. A map $\phi : S \to S'$ is called a **homomorphism** from $\widetilde{\mathscr{M}}$ to $\widetilde{\mathscr{M}'}$ if

(i) $\phi(s_0) = s_0'$;
(ii) $\phi(\lambda(s, u)) = \lambda'(\phi(s), u)$; and
(iii) $\widetilde{f}(s) = \widetilde{f'}(\phi(s)), \forall s \in S \text{ and } \forall u \in X^*$.

$\widetilde{\mathscr{M}'}$ is called the **homomorphic image** of $\widetilde{\mathscr{M}}$ if ϕ is an onto map.

Proposition 4.2 *Let $\widetilde{\rho} \in IT2F(X^*)$. Then DIT2FA $\mathscr{M}_{R_{\widetilde{\rho}}} = \Big(S_{R_{\widetilde{\rho}}}, X, \lambda^*_{R_{\widetilde{\rho}}}, [e]_{R_{\widetilde{\rho}}}, \widetilde{f}_{R_{\widetilde{\rho}}}\Big)$ is a homomorphic image of DIT2FA $\widetilde{\mathscr{M}^{\rho}} = (S^{\widetilde{\rho}}, X, \lambda^{*\widetilde{\rho}}, \widetilde{\rho}^e, f^{\widetilde{\rho}})$.*

Proof Define a map $\phi : \widetilde{\mathscr{M}^{\rho}} \to \widetilde{\mathscr{M}}_{R_{\widetilde{\rho}}}$ such that $\phi(\widetilde{\rho}^u) = [u]_{R_{\widetilde{\rho}}}, \forall\, \widetilde{\rho}^u \in S^{\rho}$ and $u \in X^*$. Then it is easy to check that ϕ is well-defined onto map. Now, $\phi(\lambda^{*\widetilde{\rho}}(\widetilde{\rho}^u, w)) = \phi(\lambda^{*\widetilde{\rho}}(\lambda^{*\widetilde{\rho}}(\widetilde{\rho}^e, u), w)) = \phi(\lambda^{*\widetilde{\rho}}(\widetilde{\rho}^e, uw)) = [uw]_{R_{\widetilde{\rho}}} = \lambda^*_{R_{\widetilde{\rho}}}([u]_{R_{\widetilde{\rho}}}, w) = \lambda^*_{R_{\widetilde{\rho}}}(\phi(\widetilde{\rho}^u), w)$. Also, for all $\widetilde{\rho}^u \in S^{\widetilde{\rho}}$, $f^{\widetilde{\rho}}(\widetilde{\rho}^u) = f^{\widetilde{\rho}}(\lambda^{*\widetilde{\rho}}(\widetilde{\rho}^e, u)) = f^{\widetilde{\rho}}(\widetilde{\rho}^u) = \widetilde{\rho}^u(e) = \widetilde{\rho}(u) = \widetilde{f}_{R_{\widetilde{\rho}}}([u]_{R_{\widetilde{\rho}}}) = \widetilde{f}_{R_{\widetilde{\rho}}}(\phi(\widetilde{\rho}^u))$. Hence $\phi : \widetilde{\mathscr{M}^{\rho}} \to \widetilde{\mathscr{M}}_{R_{\widetilde{\rho}}}$ is a homomorphism, and $\mathscr{M}_{R_{\widetilde{\rho}}}$ is a homomorphic image of $\widetilde{\mathscr{M}^{\rho}}$.

5 Algebraic Aspects of an IT-2 Fuzzy Automaton

This section is towards the study of algebraic aspects of an IT2FA. For a better understanding of the structure of automata and their applications, such type of studies has been done for both classical and nonclassical automata (cf., [3, 11, 14]). Here, we consider an IT-2 fuzzy automaton having no IT-2 fuzzy sets of initial and final states. In particular, an IT-2 fuzzy automaton is a three-tuple $\widetilde{\mathscr{M}} = (S, X, \widetilde{\lambda})$, where S is nonempty finite set of states, X is an input set and $\widetilde{\lambda}$ is transition map having usual meaning as in Definition 3.1. We begin with the following.

Definition 5.1 Let $\widetilde{\mathscr{M}} = (S, X, \widetilde{\lambda})$ be an IT2FFA and $\widetilde{U} \in IT2F(S)$. The **IT-2 fuzzy source** and the **IT-2 fuzzy successor** of $\widetilde{U}$ are, respectively, defined as,

$$\begin{aligned} IT2FSO(\widetilde{U})(s) = 1/[&\vee\{\underline{\mu}_{\widetilde{\lambda}^*(s,w)}(s') \wedge \underline{\mu}_{\widetilde{U}}(s')\}, \\ &\vee\{\overline{\mu}_{\widetilde{\lambda}^*(s,w)}(s') \wedge \overline{\mu}_{\widetilde{U}}(s')\} : s' \in S, w \in X^*], \textit{and} \\ IT2FSU(\widetilde{U})(s) = 1/[&\vee\{\underline{\mu}_{\widetilde{U}}(s') \wedge \underline{\mu}_{\widetilde{\lambda}^*(s',w)}(s)\}, \\ &\vee \overline{\mu}_{\widetilde{U}}(s') \wedge \overline{\mu}_{\widetilde{\lambda}^*(s',w)}(s)\} : s' \in S, w \in X^*]. \end{aligned}$$

Proposition 5.1 *Let $\widetilde{\mathscr{M}} = (S, X, \widetilde{\lambda})$ be an IT2FFA. Then for all $\widetilde{U}, \widetilde{U'}, \widetilde{U}_i \in IT2F(S)$*

(i) if $\widetilde{U} \subseteq \widetilde{U'}$, then $IT2FSO(\widetilde{U}) \subseteq IT2FSO(\widetilde{U'})$ and $IT2FSU(\widetilde{U}) \subseteq IT2FSU(\widetilde{U'})$;
(ii) $\widetilde{U} \subseteq IT2FSO(\widetilde{U})$ and $\widetilde{U} \subseteq IT2FSU(\widetilde{U})$;
(iii) $IT2FSO(\bigcup \widetilde{U}_i : i \in I) = \bigcup IT2FSO(\widetilde{U}_i : i \in I)$ and $IT2FSU(\bigcup \widetilde{U}_i : i \in I) = \bigcup IT2FSU(\widetilde{U}_i : i \in I)$;
(iv) $IT2FSO(IT2FSO(\widetilde{U})) = IT2FSO(\widetilde{U})$ and $IT2FSU(IT2FSU(\widetilde{U})) = IT2FSU(\widetilde{U})$.

Definition 5.2 Let $\widetilde{\mathscr{M}} = (S, X, \widetilde{\lambda})$ be an IT2FFA and $\widetilde{U} \in IT2F(S)$. Then $\widetilde{U}$ is called an **IT-2 fuzzy subsystem** of $\widetilde{\mathscr{M}}$ if for all $s \in S$, $IT2FSU(\widetilde{U}) \subseteq \widetilde{U}$, i.e.:

$$\begin{aligned} \underline{\mu}_{\widetilde{U}}(s) &\geq \underline{\mu}_{\widetilde{U}}(s') \wedge \underline{\mu}_{\widetilde{\lambda}^*(s',w)}(s) \textit{and} \\ \overline{\mu}_{\widetilde{U}}(s) &\leq \overline{\mu}_{\widetilde{U}}(s') \wedge \overline{\mu}_{\widetilde{\lambda}^*(s',w)}(s), \end{aligned}$$

$\forall s' \in S$ and $\forall w \in X^*$.

Remark 5.1 From the above, it can be observed that for any IT2FFA, $\widetilde{\mathscr{M}} = (S, X, \widetilde{\lambda})$ and $\widetilde{U} \in IT2F(S)$, IT2FSU($\widetilde{U}$) is always an IT-2 fuzzy subsystem of $\widetilde{\mathscr{M}}$.

6 Conclusion

In this note, we have tried to introduce deterministic IT-2 fuzzy automata and shown that it is behavioural equivalent to IT-2 fuzzy automata. Also, we have provided certain recipes to constructions of DTI2FA for a given IT-2 fuzzy language and prove that they are homomorphic. Finally, we have given a brief look on algebraic aspects of IT2FFA. These studies are beginning to develop the theory of automata and languages based on IT2 fuzzy sets.

References

1. Bělohlávek, R.: Determinism and fuzzy automata. Information Sciences, **143**, 205–209 (2002)
2. Ćirić, M., Ignjatović, J.: Fuzziness in automata theory: why? how?. Studies in Fuzziness and Soft Computing, **298**, 109–114 (2013)
3. Holcombe, W. M. L.: Algebraic Automata Theory. Cambridge University Press, (1987)
4. Ignjatović, J., Ćirić, M., Bogdanović, S.: Determinization of fuzzy automata with membership values in complete residuated lattices. Information Sciences, **178**, 164–180 (2008)
5. Ignjatović, J., Ćirić, M., Bogdanović, S., Petković, T.: Myhill-Nerode type theory for fuzzy languages and automata. Fuzzy Sets and Systems, **161**, 1288–1324 (2010)
6. Jiang, Y., Tang, Y.: An interval type-2 fuzzy model of computing with words. Information Sciences, **281**, 418–442 (2014)
7. Jun, Y. B.: Intuitionistic fuzzy finite state machines. Journal of Applied Mathematics and Computing, **17**, 109–120 (2005)
8. Li, Y., Pedrycz, W.: Fuzzy finite automata and fuzzy regular expressions with membership values in lattice-ordered monoids. Fuzzy Sets and Systems, **156**, 68–92 (2005)
9. Mendel, J. M., John, R. I.: Type-2 fuzzy sets made Simple. IEEE Transaction on Fuzzy Systems, **10**, 117–127 (2002)
10. Mendel, J. M., John, R. I., Liu, F.: Interval Type-2 fuzzy logic Systems made Simple. IEEE Transaction on Fuzzy Systems, **14**, 808–821 (2006)
11. Mordeson, J. N., Malik, D. S.: Fuzzy Automata and Languages, Theory and Applications. Chapman and Hall/CRC. London/Boca Raton, (2000)
12. Santos, E. S.: Max-product machines. Journal of Mathematical Analysis and Applications, **37**, 677–686 (1972)
13. Tiwari, S.P., Yadav, V. K., Singh, A.K.: Construction of a minimal realization and monoid for a fuzzy language: a categorical approach. Journal of Applied Mathematics and Computing, **47**, 401–416 (2015)
14. Tiwari, S.P., Yadav, V. K., Singh, A. K.: On algebraic study of fuzzy automata. International Journal of Machine Learning and Cybernetics, **6**, 479–485 (2015)
15. Tiwari, S.P., Yadav, V. K., Dubey, M.K.: Minimal realization for fuzzy behaviour: A bicategory-theoretic approach. Journal of Intelligent & Fuzzy Systems, **30**, 1057–1065 (2016)
16. Tiwari, S.P., Gautam, V., Dubey, M.K.: On fuzzy multiset automata. Journal of Applied Mathematics and Computing, **51**, 643–657 (2016)
17. Wee, W. G.: On generalizations of adaptive algorithm and application of the fuzzy sets concept to pattern classification. Ph. D. Thesis, Purdue University, Lafayette, IN, (1967)
18. Wee, W. G., Fu, K.S.: A formulation of fuzzy automata and its application as a model of learning systems. IEEE Transactions on Systems, Man and Cybernetics, **5**, 215–223 (1969)

19. Wu, D., Mendel, J. M.: Uncertainty measures for interval type-2 fuzzy Sets. Information Sciences, **177**, 5378–5393 (2007)
20. Wu, L., Qiu, D.: Automata theory based on complete residuated lattice-valued logic: Reduction and minimization. Fuzzy Sets and Systems, **161**, 1635–1656 (2010)
21. Zadeh, L. A.: The concept of a linguistic variable and its application to approximate reasoning -1. Information Sciences, **8**, 199–249 (1975)

Level Sets of i_v_Fuzzy β-Subalgebras

P. Hemavathi and K. Palanivel

Abstract This paper explores the new idea of level sets of i_v_fuzzy β-subalgebras and discussed some of their properties.

1 Introduction

The study of two class of algebras, B-algebras and β-algebras, was initiated by Neggers and Kim in 2002 [6, 7]. In 1965, the notion of fuzzy sets has been introduced by Zadeh [8]. In 1994 Biswas [2] proposed Rosenfeld's fuzzy subgroups with i-v membership functions. Borumand Saeid et al. [3] investigated about the i-v-f B-algebras in 2006. The concept of fuzzy β-subalgebras in β-algebras was introduced by Ayub Anasri et al. [1] in 2013. Inspired by this, Hemavathi et al. [4, 5] introduced the notion of i-v-f β-subalgebra and i-v-f translation and multiplication of i-v-f β-subalgebras. This paper aspires to define the level subset of i-v-f β-subalgebras with the help of i-v β-subalgebra in β-algebras, and it deals some of their properties and elegant results.

2 Preliminaries

In this section some primary definitions and results are reviewed which are required in the sequel.

P. Hemavathi
Research scholar, School of Advanced Sciences, Vellore institute of Technology, Vellore, India
e-mail: happy.hema85@gmail.com

K. Palanivel (✉)
Vellore Institute of Technology, Vellore, India
e-mail: drkpalanivel@gmail.com; palanivel.k@vit.ac.in

V. Madhu et al. (eds.), *Advances in Algebra and Analysis*, Trends in Mathematics,
https://doi.org/10.1007/978-3-030-01120-8_2

Definition 2.1 A β-algebra is a nonempty set X with a constant 0 and two binary operations $(+)$ and $(-)$ satisfying the following axioms:

(*i*) x-0 = x
(*ii*) (0-x)+x=0
(*iii*) $(x - y) - z = x - (z + y)\ \forall\, x, y, z \in X$

Example 2.2 Let $X = \{0, u, v, w\}$ be a set with constant 0: from the following Cayley's table, the binary operations $(+)$ and $(-)$ are defined on X:

+	0	u	v	w
0	0	u	v	w
u	u	v	w	0
v	v	w	0	u
w	w	0	u	v

−	0	u	v	w
0	0	w	v	u
u	u	0	w	v
v	v	u	0	w
w	w	v	u	0

Then $(X, +, -, 0)$ is a β-algebra.

Definition 2.3 Let S be a nonempty subset of a β-algebra $(X, +, -, 0)$ which is said to be a β-sub algebra of X, if

(*i*) $x + y \in S$
(*ii*) $x - y \in S\ \forall\ x, y \in S$

Definition 2.4 Let A be a fuzzy set of X, for $\alpha \in [0, 1]$. Then $A_\alpha = \{x \in X : \sigma(x) \geq \alpha\}$ is known as a level set of A.

Proposition 2.5 *Let A be a fuzzy set of a set X. For $\alpha_1, \alpha_2 \in [0, 1]$, if $\alpha_1 \leq \alpha_2$, then $A_{\alpha_2} \leq_{\alpha_1}$ where A_{α_1} and A_{α_2} are the corresponding level sets of A.*

Definition 2.6 Let A be a fuzzy set of X. For $\alpha \in [0, 1]$, the set $A^\alpha = \{x \in X : \sigma(x) \leq \alpha\}$ is called as a lower level set of A.

Definition 2.7 Let $A = \{x, \overline{\alpha}_A(x) : x \in X\}$ be an interval-valued fuzzy subset in X. Then $\overline{\sigma}_A$ is said to be an interval-valued fuzzy(i_v_fuzzy) β-sub algebra of X if

(*i*) $\overline{\sigma}_A(x + y) \geq rmin\{\overline{\sigma}_A(x), \overline{\sigma}_A(y)\}\ \forall\, x, y \in X.$
(*ii*) $\overline{\sigma}_A(x - y) \geq rmin\{\overline{\sigma}_A(x), \overline{\sigma}_A(y)\}\ \forall\, x, y \in X.$

Definition 2.8 Let A be an i_v_ fuzzy subset of $X, \overline{\alpha} \in D[0, 1]$. Then $A_{\overline{\alpha}} = \{x \in X : \overline{A_{\overline{\alpha}}}(x) \geq \overline{\alpha}\}$ is called a i_v_ level set of A.

3 Level Sets of i_v_ Fuzzy β-Subalgebras

This section classifies the β-subalgebras by their family of level sets of interval-valued fuzzy(i_v_ fuzzy) β-subalgebras of a β-algebra.

Definition 3.1 Consider A be an i_v_ fuzzy β-subalgebra of $X, \overline{\alpha} \in D[0, 1]$. Then $A_{\overline{\alpha}} = \{x \in X : \overline{A_{\overline{\alpha}}}(x) \geq \overline{\alpha}\}$ is called a i_v_ level β-subalgebra of A.

Theorem 3.2 *If $A = \{x, \overline{\sigma}_A(x) : x \in X\}$ is an i_v_fuzzy set in X, then $A_{\overline{\alpha}}$ is a subalgebra of X, for every $\overline{\alpha} \in D[0, 1]$.*

Proof For $x, y \in \overline{\sigma}_{A_{\overline{\alpha}}}$ & $\overline{\sigma}_A(x) \geq \overline{\alpha}$ & $\overline{\sigma}_A(y) \geq \overline{\alpha}$

$$\begin{aligned}\overline{\sigma}_A(x+y) &\geq rmin\{\overline{\sigma}_A(x), \overline{\sigma}_A(y)\}\\ &\geq rmin\{\overline{\alpha}, \overline{\alpha}\}\\ &\geq \overline{\alpha}\\ \Rightarrow x + y &\in \overline{A_{\overline{\alpha}}}\end{aligned}$$

Similarly, t $x - y \in \overline{A_{\overline{\alpha}}}$.

Hence $A_{\overline{\alpha}}$ is subalgebra of X.

Theorem 3.3 *If $A = \{\langle x, \overline{\sigma}_A(x)\rangle : x \in X\}$ is an i_v_fuzzy set in X such that $A_{\overline{\alpha}}$ is a subalgebra of X for every $\overline{\alpha} \in D[0, 1]$. Then A is an i_v_fuzzy β-subalgebra of X.*

Proof Consider $A = \{\langle x, \overline{\sigma}_A(x)\rangle : x \in X\}$ is an i_v_ fuzzy set in X.
Assume that $A_{\overline{\alpha}}$ is a subalgebra of X for $\overline{\alpha} \in D[0, 1]$.
Now, $\overline{\alpha} = rmin\{\overline{\sigma}_A(x), \overline{\sigma}_A(y)\}$
For $x + y$ *and* $x - y \in \overline{A_{\overline{\alpha}}}$,
$\Rightarrow x + y \in \overline{A_{\overline{\alpha}}} \geq rmin\{\overline{\sigma}_A(x), \overline{\sigma}_A(y)\} = \overline{\alpha}$
Similarly, $x - y$.
$\therefore A$ is an i_v_ fuzzy β-subalgebra of X.

Theorem 3.4 *Any β-subalgebra of X can be realized as a level β-subalgebra for some i_v_fuzzy β-subalgebra of X.*

Proof Consider A be an i_v_ fuzzy β-subalgebra of X.
We define

$$\overline{\sigma}_A(x) = \begin{cases}\overline{\alpha} & x \in X\\ [0, 0], & \text{elsewhere}\end{cases}$$

Case (i)
Both $x, y \in A$.

$$\begin{aligned}\overline{\sigma}_A(x+y) &\geq rmin\{\overline{\sigma}_A(x), \overline{\sigma}_A(y)\}\\ &\geq rmin\{\overline{\alpha}, \overline{\alpha}\}\\ &= \overline{\alpha}\end{aligned}$$

Similarly $\overline{\sigma}_A(x - y) \geq \overline{\alpha}$

Case (ii)
Both $x, y \notin A$.

$$\begin{aligned}\overline{\sigma}_A(x+y) &\geq rmin\{\overline{\sigma}_A(x), \overline{\mu}_A(y)\}\\ &\geq rmin\{[0,0],[0,0]\}\\ &= [0,0]\end{aligned}$$

Similarly $\overline{\sigma}_A(x-y) \geq [0,0]$.

Case (iii)
$x \in A$ & $y \notin A$.

$$\begin{aligned}\overline{\sigma}_A(x+y) &\geq rmin\{\overline{\sigma}_A(x), \overline{\sigma}_A(y)\}\\ &\geq rmin\{\overline{\alpha},[0,0]\}\\ &= [0,0]\end{aligned}$$

Similarly, $\overline{\sigma}_A(x-y) \geq [0,0]$.

Case(iv) Interchanging the character of x and y in case (iii), then A is an i_v_ fuzzy β-subalgebra of X.

Lemma 3.5 *If A and B be two-level set of i_v_fuzzy β-subalgebra of X, $\overline{\sigma}_A(x) \leq \overline{\sigma}_B(x)$, then $A \subseteq B$.*

Proof By definition of i_v_ level β-subalgebra,
$A_{\overline{\alpha}} = \{\langle x, \overline{\sigma}_A(x) \geq \overline{\alpha}\rangle\}$ &
$B_{\overline{\alpha}_1} = \{\langle x, \overline{\sigma}_A(x) \geq \overline{\alpha}_1\rangle\}$ where $\overline{\alpha} \leq \overline{\alpha}_1$
If $x \in \overline{\sigma}_B(\overline{\alpha}_1)$ then $\overline{\sigma}_B \geq \overline{\alpha}_1 \geq \overline{\alpha} \Rightarrow x \in \overline{\sigma}_A(\overline{\alpha})$
$\therefore \overline{\sigma}_B(x) \geq \overline{\sigma}_A(x)$
Hence $A \subseteq B$.

Theorem 3.6 *Let X be any β-algebra. If $\{A_i\}$ any sequence of β-subalgebra of X such that $A_0 \subset A_1 \subset \ldots \subset A_n = X$, then $\exists$ an i_v_fuzzy β-subalgebras $\overline{\sigma}$ of X whose i_v_ level β-subalgebras are exactly the β-subalgebra $\{A_i\}$.*

Proof Consider a set of numbers $\overline{\alpha}_0 > \overline{\alpha}_1 > \ldots > \overline{\alpha}_n$, where each $\alpha_i \in D[0,1]$.
Let $\overline{\sigma}$ be a fuzzy set represented by
$\overline{\sigma}_{A_0} = \overline{\alpha_0}$ & $\overline{\sigma}(A_i \sim A_{i-1}) = \overline{\alpha_i}, 0 < i \leq n$
to prove $\overline{\sigma}$ is i_v_ fuzzy β-subalgebra of X.
For this $x, y \in X$.

Case (i)

Let $x, y \in A_i \sim A_{i-1}$.

Then $x, y \in A_i \Rightarrow x + y \in A_i \& x - y \in A_i$.Also $x, y \in A_i \sim A_{i-1} \Rightarrow \overline{\sigma}(x) = s_i \Rightarrow \overline{\sigma}(y) \Rightarrow rmin\{\overline{\sigma}(x), \overline{\sigma}(y)\} = \overline{\alpha}_i$.

Now since A_i is subalgebra $x+y$ & $x-y \in A_i \Rightarrow x+y \,\&\, x-y \in A_i \sim A_{i-1}$ *or* $x + y, x - y \in A_{i-1}$. $\overline{\sigma}(x + y) = \overline{\sigma}(x - y) \geq \overline{\alpha_i}$

Thus $\overline{\sigma}(x - y) \geq \overline{\alpha_i} = rmin\{\overline{\sigma}(x), \overline{\sigma}_(y)\}$ & $\overline{\sigma}(x + y) \geq \overline{\alpha_i} = rmin\{\overline{\sigma}(x), \overline{\sigma}_(y)\}$.

Case (ii) For $i > j \Rightarrow \overline{\alpha_j} > \overline{\alpha_i} \Rightarrow A_j \subset A_i$:

Let $x \in A_i \sim A_{i-1}$ & $y \in A_j \sim A_{j-1}$.

Then $\overline{\sigma}(x) = \overline{\alpha_i}$ & $\overline{\sigma}(y) = \overline{\alpha_j} > \alpha_i$.

Hence $rmin\{\overline{\sigma}(x), \overline{\sigma}_(y)\} = rmin\{\overline{\alpha}_i, \overline{\alpha}_j\} = \overline{\alpha_i}$.

Further, $y \in A_i \sim A_{j-1} \Rightarrow y \in A_j \subset A_i \Rightarrow x, y \in A_i$.

Since A_i is β-subalgebra of X,$x + y$ & $x - y \in A_i$,

$\therefore \overline{\sigma}_{A_i}(x - y) \geq \overline{s_i} = rmin\{\overline{\sigma}(x), \overline{\sigma}(y)\}$ &
$\overline{\sigma}_{A_i}(x + y) \geq \overline{\alpha_i} = rmin\{\overline{\sigma}(x), \overline{\sigma}(y)\}$.

Thus in both cases, $\overline{\sigma}$ is a i_v_ fuzzy β-subalgebra of X.

From the definition of $\overline{\sigma}$, it follows that $Im(\overline{\sigma}) = \{\overline{\alpha_0}, \overline{\alpha_1} \ldots \overline{s_n}\}$.

Hence $\overline{\sigma}_{\overline{\alpha_i}} = \{x \in X : \overline{\sigma}(x) \geq \overline{\alpha_i}\}$, for $0 \leq i \leq n$ are the i_v_ level β-subalgebras of X by Theorem 3.3.

Then the sequence $\{\overline{\sigma}_{\overline{t_i}}\}$ of i_v_level β-subalgebras of $\overline{\sigma}$ is in the form of $\overline{\sigma}_{\overline{\alpha_0}} \subset \overline{\sigma}_{\overline{\alpha_1}} \subset \ldots \subset \overline{\sigma}_{\overline{t_n}} = X$

Now, $\overline{\sigma}_{\overline{\alpha_0}} = \{x \in X : \overline{\sigma}(x) \leq \overline{\alpha}_0\} = A_0$.

Finally, to prove $\overline{\sigma}_{\overline{t_i}} = A_i$ for $0 \leq i \leq n$

clearly,$A_i \subseteq \overline{\sigma}_{\overline{\alpha}_i}$. If $x \in \overline{\sigma}_{\overline{\alpha}_i}$, then $\overline{\sigma}(x) \geq \overline{\alpha}_i$ which implies

$\overline{\sigma}(x) \in \{\overline{\alpha}_1, \overline{\alpha}_2, \ldots \overline{\alpha}_n\}$.

Here $x \in A_0 or A_1 \ldots or A_i$. It follows that $x \in A_i$;

$\therefore \overline{\sigma}_{\overline{\alpha}_i} = A_i$ for $0 \leq i \leq n$.

Thus the i_v_ level β-subalgebras of $\overline{\sigma}$ are exactly the β-subalgebras of X.

Theorem 3.7 *Let $A = \{x, \overline{\sigma}_A(x) : x \in X\}$ be an i_v_fuzzy β-subalgebra of X. If $Im(A)$ is finite $\overline{\alpha}_0 < \overline{\alpha}_1 < \ldots < \overline{\alpha}_n$, then any $\overline{\alpha}_i, \overline{\alpha}_j \in Im(\overline{\sigma}_A)$,$\overline{\sigma}_{\alpha_i} = \overline{\sigma}_{\alpha_j}$ implies $\alpha_i = \alpha_j$.*

Proof Assume that $\overline{\alpha}_i \neq \overline{\alpha}_j$.

If $x \in \overline{\sigma}_{\alpha_j}$, then $\overline{\sigma}_A(x) \geq \overline{\alpha}_j > \overline{\alpha}_i \Rightarrow x \in \overline{\sigma}_{\alpha_i}$; there exists $x \in X$ such that $\overline{\alpha}_i \leq \overline{\sigma}(x) < \overline{\alpha}_j \Rightarrow x \in \overline{\sigma}_{\overline{\alpha}_i}$ but $x \in \overline{\sigma}_{\overline{\alpha}_j}$.

Hence $\overline{\sigma}_{\alpha_j} \subset \overline{\sigma}_{\alpha_i}$ and $\overline{\sigma}_{\alpha_j} \neq \overline{\sigma}_{\alpha_i}$

which is a contradiction.

Theorem 3.8 *Let $A = \{x, \overline{\sigma}_A(x) : x \in X\}$ be an i_v_fuzzy β-subalgebra of X. Two-level subalgebras $A_{\overline{\alpha}_1}$ & $A_{\overline{\alpha}_2}$(with $\overline{\alpha}_1 < \overline{\alpha}_2$) of A are equal if and only if there is no $x \in X$ such that $\overline{\alpha}_1 \leq \overline{\sigma}_A(x) < \overline{\alpha}_2$.*

Proof Assume that$A_{\overline{\alpha}_1} = A_{\overline{\alpha}_2}$ for $\overline{\alpha}_1 < \overline{\alpha}_2$.

Then there exists $x \in X$ such that the membership function $\overline{\alpha}_1 < \overline{\sigma}_A(x) < \overline{\alpha}_2$.

Hence $\overline{\sigma}_{\overline{\alpha}_2}$ is a proper subset of $\overline{\sigma}_{\overline{s}_1}$

which is a contradiction.

Conversely, suppose that there is no $x \in X$ such that the membership function $\overline{\alpha}_1 < \overline{\sigma}_A(x) < \overline{\alpha}_2$.

$\because$ $\overline{\alpha}_1 < \overline{\alpha}_2$ then $\overline{\sigma}_{\overline{\alpha}_2} \subseteq \overline{\sigma}_{\overline{\alpha}_1}$

If $x \in \overline{\sigma}_{\overline{\alpha}_1}$, then $\overline{\sigma}(x) \geq \overline{\alpha}_1$ & $\overline{\sigma}(x) \geq \overline{\alpha}_2$ because $\overline{\sigma}(x)$ does not lie between $\overline{\alpha}_1$ & $\overline{\alpha}_2$.

Hence $x \in \overline{\sigma}_{\overline{\alpha}_2} \Rightarrow \overline{\sigma}_{\overline{\alpha}_1} \subseteq \overline{\sigma}_{\overline{\alpha}_2}$;

$\therefore \overline{\sigma}_{\overline{\alpha}_1} = \overline{\sigma}_{\overline{\alpha}_2}$

References

1. Aub Ayub Anasri, M., Chandramouleeswaran, M.: Fuzzy β-subalgebras of β-algebras. International journal of mathematical sciences and engineering applications. **5**(7), 239–249 (2013)
2. Biswas, R.: Rosenfeld's fuzzy subgroups with Interval valued membership functions. Fuzzy sets and systems. **63**(1), 87–90 (1994)
3. Borumand saeid, A.: Interval valued fuzzy B-algebras. Iranian Journal of fuzzy systems. **3**(2),63–73 (2006)
4. Hemavathi, P., Muralikrishna, P., Palanivel, K.: A note on interval valued fuzzy β-subalgebras. Global Journal of Pure and Applied Mathematics. **11**(4), 2553–2560 (2015)
5. Hemavathi, P., Muralikrishna, P., Palanivel, K.:Study on i-v fuzzy translation and multiplication of i-v fuzzy β-subalgebras, International Journal of Pure and Applied Mathematics, **109**(2), 245–256 (2016)
6. Neggers, J., Kim Hee Sik.: On B-algebras. Math. Vesnik. **54**(1–2),21–29 (2002)
7. Neggers, J., Kim Hee Sik.: On β-algebras. Math. Solvaca. **52**(5), 517–530 (2002)
8. Zadeh, L.A.: Fuzzy sets. Inform. Control. **8**(3), 338–353 (1965)

Interval-Valued Fuzzy Subalgebra and Fuzzy INK-Ideal in INK-Algebra

M. Kaviyarasu, K. Indhira, V. M. Chandrasekaran, and Jacob Kavikumar

Abstract In this paper we examine IVF INK-ideal in INK-algebras by giving a couple of definitions and relative hypotheses. The image and preimage of IVF INK-ideal become i-v fuzzy INK-ideals in INK-algebras.

1 Introduction

In 1966, Imai and Iséki introduced two classes of abstract algebras: BCK-algebra and BCI-algebra [2]. It is known that the class of BCK-algebra is proper subclass of the class of BCI-algebra [3, 4]. Neggers et al. [5] introduced a notion called Q-algebra, which is a generalization of BCH/ BCI/BCK-algebras, and generalized some theorems discussed in BCI-algebra. The concept of a fuzzy set was introduced by Zadeh [7]. Xi [6] applied the concept of fuzzy set to BCK-algebra and gave some of its properties. In [8], Zadeh made an extension of the concept of fuzzy set by an interval-valued fuzzy set (i.e., a fuzzy set with an interval-valued membership function). This interval-valued fuzzy set is referred to as i-v fuzzy set. Zadeh also constructed a method of approximate inference using his i-v fuzzy sets. In [1], Biswas defined interval-valued fuzzy subgroups and investigated some elementary properties.

In this paper, using the notion of interval-valued fuzzy set by Zadeh, we introduce the concept of interval-valued fuzzy INK-ideals in INK-algebra (briefly i-v fuzzy INK-ideals in INK-algebra) and study some of their properties. We prove that every INK-ideals of INK-algebra X can be realized as i-v level INK-ideals of a INK-algebra X, and then we obtain some related results which have been mentioned in the abstract.

M. Kaviyarasu · K. Indhira (✉) · V. M. Chandrasekaran
Department of Mathematics, VIT University, Vellore, India
e-mail: kavitamilm@gmail.com; kaviyarasu.m2015@vit.ac.in; kindhira@vit.ac.in; vmcsn@vit.ac.in

J. Kavikumar
Faculty of Science, Universiti Tun Hussein Onn Malaysia, Malaysia
e-mail: kavi@uthm.edu.my

V. Madhu et al. (eds.), *Advances in Algebra and Analysis*, Trends in Mathematics,
https://doi.org/10.1007/978-3-030-01120-8_3

2 Preliminaries

Definition 1 An algebra $(A, \odot, 0)$ is approached INK-algebra on the off chance that it satisfies the accompanying conditions:

INK-I. $a \odot a = 0$
INK-II. $a \odot 0 = a$
INK-III. $0 \odot a = a$
INK-IV. $(b \odot a) \odot (b \odot c) = (a \odot c)$, for all $0, a, b, c \in A$.

Definition 2 Let B nonvoid set be a subset of a INK-algebra A. Then B is called a INK-ideal of A if, for all $a, b, c \in A$.

INKI(1). $0 \in B$
INKI(2). $(c * a) * (c * b) \in B$, and $b \in B$ imply $a \in B$.

Definition 3 Let A be a INK-algebra and $T \subseteq A$. Then T is said to be INK-subalgebra, if $a \odot b \in T$, for all $a, b \in A$.

Definition 4 An ideal C of a INK-algebra X is said to be closed if $0 * a \in C$ for all $a \in C$.

Definition 5 Let A be a nonvoid set. A mapping $\xi : a \to [0, 1]$. ξ is called a fuzzy set in X. The complement of ξ denoted by $\xi^-(a) = 1 - \xi(a)$, for all $a \in A$.

Definition 6 A fuzzy set ξ in a INK-algebra A is called a fuzzy subalgebra of A if, $\xi(a \odot b) \geq min\{\xi(a), \xi(b)\}$, for all $a, b \in A$.

Definition 7 Let ξ be a fuzzy set of a set A. For a fixed $t \in [0, 1]$, the set $\xi_t = \{a \in A/\xi(a)\}$ is called an upper level of ξ.

Definition 8 Let ξ be a fuzzy set of INK-algebra A which is said to be fuzzy ideal of A, if

FI1. $\xi(0) \geq \xi(a)$
FI2. $\xi(a) \geq min\{\xi(a \odot b), \xi(b)\}$, for all $a, b \in A$.

Definition 9 A fuzzy subset ξ in a INK-algebra A is called a fuzzy INK-ideal of A, if for all $a, b, c \in A$,

FINKI-1. $\xi(0) \geq \xi(a)$
FINKI-2. $\xi(a) \geq min\{\xi(c \odot a) \odot (c \odot b), \xi(y)\}$.

3 Interval-Valued Fuzzy INK-Ideal in INK-Algebra

An i-v fuzzy set τ on A is given by $\tau = \{(a, [\xi_\tau^L(a), \xi_\tau^U(a)]), a \in A\}$ and is denoted by $\tau = [\xi_\tau^L, \xi_\tau^U]$, where ξ_τ^L and ξ_τ^U are any two fuzzy sets in A such that $\xi_\tau^L \leq \xi_\tau^U$. Let $\xi_\tau^-(a) \geq [\xi_\tau^L, \xi_\tau^U]$, and let $\theta[0, 1]$ be the family of all closed sub-intervals of [0, 1]. It is clear that if $\xi_\tau^L(a) = \xi_\tau^U(a) = c$, where $0 \leq c \leq 1$, then $\xi_\tau^-(a) = [c, c]$ is in $\theta[0, 1]$. Thus $\xi_\tau^-(a) \in [0, 1]$ for all $a \in A$. Then the i-v fuzzy set τ is given by

$\tau = \{(a, \xi_\tau^-(a)), a \in A\}$, where $\xi_\tau^- : A \to \theta[0, 1]$.

Definition 10 An i-v fuzzy set τ in A is called an interval-valued fuzzy (IVF) INK-algebra of A if $\xi_\tau^-(a * b) \geq r\ min\{\xi_\tau^-(a), \xi_\tau^-(b)\}$ for all $a, b \in A$.

Proposition 1 *If τ is a IVF (INK)-subalgebra of A, then for all $a \in A$, $\xi_\tau^-(0) \geq \xi_\tau^-(a)$.*

Proof For all $a \in A$, we have

$$\begin{aligned}
\xi_\tau^-(0) &= \xi_\tau^-(a \odot a)\\
&= r\ min\{\xi_\tau^-(a), \xi_\tau^-(a)\}\\
&= r\ min\{[\xi_\tau^L(a), \xi_\tau^U(a)], [\xi_\tau^L(a), \xi_\tau^U(a)]\}\\
&= [\xi_\tau^L(a), \xi_\tau^U(a)]\\
\xi_\tau^-(0) &= \xi_\tau^-(a)
\end{aligned}$$

Theorem 1 *Let τ be a IVF (INK)-subalgebra of A. If there exists a sequence $\{a_n\}$ in A such that $lim_{n\to\infty}\xi_\tau^-(a_n) = [1, 1]$, then $\xi_\tau^-(0) = [1, 1]$.*

Proof By proposition, we have $\xi_\tau^-(0) \geq \xi_\tau^-(a)$.
Then $\xi_\tau^-(0) \geq \xi_\tau^-(a_n)$; so $[1, 1] \geq \xi_\tau^-(0) \geq lim_{n\to\infty}\xi_\tau^-(a_n)$.
Hence $\xi_\tau^-(0) = [1, 1]$.

Theorem 2 *A IVF set $\tau = [\xi_\tau^L, \xi_\tau^U]$ in A is a IVF INK-algebra of A if and only if ξ_τ^L and ξ_τ^U are fuzzy INK-subalgebra of A.*

Proof Let ξ_τ^L and ξ_τ^U are fuzzy INK-subalgebra of A and $a, b \in A$.
Then

$$\begin{aligned}
&= [\xi_\tau^-(a \odot a)]\\
&\geq [min\{\xi_\tau^L(a), \xi_\tau^L(b)\},\ min\{\xi_\tau^U(a), \xi_\tau^U(b)\}]\\
&= r\ min\{[\xi_\tau^L(a), \xi_\tau^U(a)], [\xi_\tau^L(b), \xi_\tau^U(b)]\}\\
\xi_\tau^-(a * b) &= r\ min\{\xi_\tau^-(a), \xi_\tau^-(b)\}.
\end{aligned}$$

Conversely, suppose that τ is a INF (INK)-subalgebra of X. Then for all $a \in A$,

$$\begin{aligned}
[\xi_\tau^L(a * b), \xi_\tau^U(a * b)] &= \xi_\tau^-(a * b)\\
&\leq r\ min\{\xi_\tau^-(a), \xi_A^-(\tau)\}\\
&= r\ min\{[\xi_\tau^L(a), \xi_\tau^U(a)], [\xi_\tau^L(b), \xi_\tau^U(b)]\}\\
&= [min\{\xi_\tau^-(a), \xi_\tau^-(b)\}]\\
\xi_\tau^L(a * b) &\geq [min\{\xi_\tau^L(a), \xi_\tau^U(a)\}]\\
\xi_\tau^U(a * b) &\geq [min\{\xi_\tau^U(a), \xi_\tau^U(a)\}].
\end{aligned}$$

Hence ξ_A^L and ξ_A^U are fuzzy INK-subalgebras of A.

Theorem 3 *A IVFS* $\tau = \left[\xi_\tau^L, \xi_\tau^U\right]$ *in A is a IVF(INK)-ideal of A if and only if* ξ_τ^L *and* ξ_τ^U *are fuzzy INK-ideals of A.*

Proof Let ξ_τ^L and ξ_τ^U are fuzzy INK-ideals of A and $a, b, c \in A$.

$$\begin{aligned}
\xi_\tau^-(a) &= \left[\xi_\tau^L(a), \xi_\tau^U(a)\right] \\
&\geq \left[min\left\{\xi_\tau^L(c \odot a) \odot (z \odot y), \xi_\tau^L(b)\right\}, min\left\{\xi_\tau^U(c \odot a) \odot (c \odot b), \xi_\tau^U(b)\right\}\right] \\
&= r\ min\left\{\left[\xi_\tau^L(c \odot a) \odot (c \odot b), \xi_\tau^U(c \odot a) \odot (c \odot b)\right], \left[\xi_\tau^L(b), \xi_\tau^U(b)\right]\right\} \\
\xi_\tau^-(a) &= r\ min\left\{\xi_\tau^-(c \odot a) \odot (c \odot b), \xi_\tau^-(b)\right\}.
\end{aligned}$$

τ is a IVF (INK)-ideal of A. Conversely, suppose that τ is a IVF (INK)-ideal of A. For all $a, b, c \in A$

$$\begin{aligned}
\left[\xi_\tau^L(c \odot a) \odot (c \odot b), \xi_\tau^U(b)\right] &= \xi_\tau^-(a) \\
&\leq r\ min\left\{\xi_\tau^-(c \odot a) \odot (c \odot b), \xi_\tau^-(b)\right\} \\
&= r\ min\left\{\left[\xi_\tau^L(c \odot a) \odot (c \odot b), \xi_\tau^U(c \odot a)\right.\right. \\
&\quad \left.\left.\odot (c \odot b)\right], \left[\xi_\tau^L(b), \xi_\tau^U(b)\right]\right\} \\
&= \left[min\left\{\xi_\tau^L(c \odot a) \odot (c \odot b), \xi_\tau^L(b)\right\},\right. \\
&\quad \left. min\left\{\xi_\tau^U(c \odot a) \odot (c \odot b), \xi_\tau^U(b)\right\}\right] \\
\xi_\tau^L(a) &\geq min\left\{\xi_\tau^L(c \odot a) \odot (c \odot b), \xi_\tau^L(b)\right\} \\
\xi_\tau^U(a) &\geq min\left\{\xi_\tau^U(c \odot a) \odot (c \odot b), \xi_\tau^U(b)\right\}
\end{aligned}$$

Hence ξ_τ^L and ξ_τ^U are fuzzy INK-ideals of A.

Theorem 4 *Let* τ_1 *and* τ_2 *be IVF (INK)-ideal of a INK-algebra A. Then* $\tau_1 \cap \tau_2$ *is a IVF (INK)-ideal of A.*

Proof Letτ_1 and τ_2 be i-v fuzzy INK-ideal of a INK-algebra A. Then

$$\begin{aligned}
\xi_{\tau_1 \cap \tau_2}^-(0) &= \left[\xi_{\tau_1 \cap \tau_2}^L(0), \xi_{\tau_1 \cap \tau_2}^L(0)\right] \\
\xi_{\tau_1 \cap \tau_2}^-(0) &= \left[\xi_{\tau_1 \cap A_2}^L(a), \xi_{\tau_1 \cap \tau_2}^L(a)\right] \\
\xi_{\tau_1 \cap \tau_2}^-(0) &= \xi^-\tau_1 \cap \tau_2(a)
\end{aligned}$$

Suppose $a, b, c \in A$ such that $(c \odot a) \odot (a \odot b) \in \tau_1 \cap \tau_2$. Since τ_1 and τ_2 are i-v fuzzy INK-ideals of a INK-algebra A by Theorem 3, we get

$$\begin{aligned}
\xi_{\tau_1 \cap \tau_2}^-(a) &= \left[\xi_{\tau_1 \cap \tau_2}^L(a), \xi_{\tau_1 \cap \tau_2}^U(a)\right] \\
&= \left[min\left\{\xi_{\tau_1 \cap \tau_2}^L(c \odot a) \odot (c \odot b), \xi_{\tau_1 \cap \tau_2}^L(b)\right\},\right. \\
&\quad \left. min\left\{\xi_{\tau_1 \cap A_2}^U(c \odot a) \odot (c \odot b), \xi_{\tau_1 \cap \tau_2}^U(b)\right\}\right] \\
&= \left[min\left\{\xi_{\tau_1 \cap A_2}^L(c \odot a) \odot (c \odot b), \xi_{\tau_1 \cap \tau_2}^U(c \odot a) \odot (c \odot b)\right\},\right. \\
&\quad \left. min\left\{\xi_{\tau_1 \cap \tau_2}^L(b), \xi_{\tau_1 \cap \tau_2}^U(b)\right\}\right] \\
\xi_{\tau_1 \cap A_2}^-(a) &= \left[r\ min\left\{\xi_{\tau_1 \cap \tau_2}^-(c \odot a) \odot (c \odot b), \xi_{\tau_1 \cap \tau_2}^-(b)\right\}\right]
\end{aligned}$$

Theorem 5 *Let A be a INK-algebra and τ be an i-v fuzzy subset in A. Then τ is an i-v fuzzy INK-ideal of A, if and only if $U^-(\tau; [\vartheta_1, \vartheta_2]) = \{a \in A/\xi_\tau^-(a) \geq [\vartheta_1, \vartheta_2]\}$ is a INK-ideal of τ, for every $[\vartheta_1, \vartheta_2] \in D(0, 1)$. We call $U^-(\tau; [\vartheta_1, \vartheta_2])$ the i-v level INK-ideal of τ.*

Proof Assume that τ is an i-v fuzzy INK-ideal of A.

Let $[\vartheta_1, \vartheta_2]) \in \theta(0, 1)$ such that $(c \odot b) \odot (c \odot b), b \in U^-(\tau; [\vartheta_1, \vartheta_2])$ $\xi_\tau^-(a) \geq r\ min\{\xi_\tau^-(c \odot a) \odot (c \odot b), \xi_\tau^-(b)\} \geq r\ min\{[\vartheta_1, \vartheta_2], [\vartheta_1, \vartheta_2]\} = [\vartheta_1, \vartheta_2]$. Therefore $a \in U^-(\tau; [\vartheta_1, \vartheta_2])$ and then $U^-(\tau; [\vartheta_1, \vartheta_2])$ is i-v level INK-ideal of τ.

Conversely, assume that $U^-(\tau; [\vartheta_1, \vartheta_2]) \neq \phi$ is a INK-ideal of A, for every $[\vartheta_1, \vartheta_2] \in \theta(0, 1)$. On the contrary, suppose that there exist $a_0, b_0, c_0 \in A$ such that

$$\begin{aligned}
\xi_\tau^-(a_0) &\geq r\ min\{\xi_\tau^-(c_0 \odot a_0) \odot (c_0 \odot b_0), \xi_\tau^-(b_0)\} \\
\xi_\tau^-(a_0) &= [\vartheta_1, \vartheta_2], \xi_\tau^-(c_0 \odot a_0) \odot (c_0 \odot b_0) \\
&= [\eta_1, \eta_2], \xi_\tau^-(b_0) \\
&= [\vartheta_3, \vartheta_4].
\end{aligned}$$

If $[\vartheta_1, \vartheta_2] < r\ min\{[\eta_1, \eta_2], [\eta_3, \eta_4]\}$

$= min\{min[\eta_1, \eta_2], min[\eta_3, \eta_4]\}$.

So $\vartheta_1 < min[\eta_1, \eta_2]$ and $\vartheta_1 < min[\eta_3, \eta_4]$.

Consider $[\tau_3, \tau_4] = \frac{1}{2}\{\xi_\tau^-(a_0) + \xi_\tau^-(c_0 \odot a_0) \odot (c_0 \odot b_0), \xi_\tau^-(b_0)\}$.

Then we get $[\tau_1, \tau_2] = \frac{1}{2}\{[\eta_1, \eta_2] + r\ min\{[\eta_1, \eta_2], [\eta_3, \eta_4]\}\}$

$= \frac{1}{2}\{(\vartheta_1 + min\{[\eta_1, \eta_3]\}), (\vartheta_2 + min\{[\eta_2, \eta_4]\})\}$

$min\{\eta_1, \eta_3\} > \tau_1 = \frac{1}{2}([\eta_1, \eta_3]) > \eta_1$

$min\{\eta_2, \eta_4\} > \tau_2 = \frac{1}{2}([\eta_2, \eta_4]) > \eta_2$.

Hence $[min\{\eta_1, \eta_3\}, min\{\eta_2, \eta_4\}] > [\tau_1, \tau_2] > [\vartheta_1, \vartheta_2].(c_0 \odot a_0) \odot (c_0 \odot b_0) \in U^-(\tau; [\tau_1, \tau_2])$. Then $\xi_\tau^-(a) \geq r\ min\{\xi_\tau^-(c \odot a) \odot (c \odot b), \xi_\tau^-(b)\}$, for all $a, b, c \in A$.

4 Homomorphism of INK-Algebra

Definition 11 Let $g : (X; \odot, 0) \to (Y; , \odot', 0)$ be a mapping from set X into a set Y. Let τ be an i-v fuzzy subset in Y. Then the inverse image of τ, denoted by $g^{-1}(\tau)$, is an i-v fuzzy subset in X with the membership function given $\xi_{g_\tau^{-1}}(a) = \xi^-(g(a))$, for all $x \in X$.

Theorem 6 *An into homomorphic preimage of a fuzzy INK-ideal is also fuzzy INK-ideal.*

Proof Let $f : X \to X'$ be an into homomorphism of INK-algebra, B a fuzzy

INK-ideal of X', and ξ the preimage of B under f; then $B(f(x)) = \xi(x)$, for all $x \in X$.

Then

$$\xi(0) = B(f(0))$$
$$\xi(0) \geq B(f(x))$$
$$\xi(0) = B(f(x)).$$

Let $x, y, z \in X$; then

$$\xi(x) = B(f(x))$$
$$\xi(x) \geq min\{B((f(z) \odot f(x)) \odot (f(z) \odot f(y))), B(f(y))\}$$
$$\xi(x) = min\{B(f(z \odot x) \odot (z \odot y)), B(f(y))\}$$
$$\xi(x) = min\{\xi((z \odot x) \odot (z \odot y)), \xi(y)\}$$

Hence $\xi(x) = B(f(x)) = B \circ f(x)$ is a fuzzy INK-ideal of X.

Proposition 2 *Let $g : X \to Y$. Let $n = [n^L, n^U]$ and $m = [m^L, m^U]$ be i-v fuzzy subset in X and Y. Then*

1. $g^{-1}(m) = [g^{-1}(m^L), g^{-1}(m^U)]$
2. $g(n) = [g(n^L), f(n^U)]$.

Theorem 7 *Let $g : X \to Y$ be a homomorphism from a INK-algebra X into a INK-algebra Y. If B is an i-v fuzzy INK-ideal of Y, then the inverse image $f^{-1}(B)$ of B is an i-v fuzzy INK-ideal of X.*

Proof Since $B = [\xi_B^L, \xi_B^U]$ is an i-v fuzzy INK-ideal of Y, it follows from Theorem 3 that (ξ_B^L) and (ξ_B^U) are fuzzy INK-ideals of Y. Using Theorem 6, we know that $f^{-1}(\xi_B^L)$ and $f^{-1}(\xi_B^U)$ are fuzzy INK-ideals of X. Hence by Proposition 2, $f^{-1}(B) = [f^{-1}(\xi_B^L), f^{-1}(\xi_B^U)]$ is an i-v fuzzy INK-ideal of X.

Theorem 8 *Let $f : X \to Y$ be a homomorphism. If A is an i-v fuzzy INK-ideal of X, then f[A] of A is an i-v fuzzy INK-ideal of Y.*

Proof Assume that A is an i-v fuzzy INK-ideal of X. Note that $A = [\xi_A^L, \xi_A^U]$ is an i-v fuzzy INK-ideal of X. Let $f : X \to Y$ be a homomorphism between INK-algebra X and Y. For every fuzzy INK-ideal ξ in X, $f(\xi)$ is a fuzzy INK-ideal of Y. Then the image $f(\xi_A^L)$ and $f(\xi_A^U)$ are fuzzy INK-ideals of Y. Combining Theorem 3 and Proposition 2, we conclude that $f[A] = [f(\xi_A^L), f(\xi_A^U)]$ is an i-v fuzzy INK-ideal of Y.

References

1. Biswas, R.: Rosenfeld's fuzz subgroups with interval valued membership function. Fuzzy Sets and Systems. **63**, 87–90 (1994)
2. Imai, Y., Iséki, K.: On axiom systems of propositional calculi, X I V. proc. Japan Academy (1996) Available via DIALOG. https://projecteuclid.org/euclid.pja/1195522169. Cited 12 Jan 1966

3. Iséki, K., Tanaka, T.: An introduction to the theory of BCK-algebras. Math. Japonica. **23**, 1–26 (1978)
4. Mostafa, S.M.: Fuzzy implicative ideal in BCK-algebras. Fuzzy Sets and Systems. (1997) https://doi.org/10.1016/S0165-0114(96)00017-6
5. Neggers, J., Ahn, S.S., Kim, H.S.: On Q-algebras. Int. J. Math. Math. Sci. **27**, 749–757 (2001)
6. Xi, O.G.: Fuzzy BCK-algebra. Math. Japon. **36**, 935–942 (1991)
7. Zadeh, L.A: Fuzzy sets, Information and Control 8 (1965) Available via DIALOG. https://doi.org/10.1016/S0019-9958(65)90241-X. Cited 30 June 1965.
8. Zadeh, L.A.: The concept of a linguistic variable and its application to approximate reasoning-I, Information Sciences 8 (1975) Available via DIALOG. https://doi.org/10.1016/0020-0255(75)90036-5

On Dendrites Generated by Symmetric Polygonal Systems: The Case of Regular Polygons

Mary Samuel, Dmitry Mekhontsev, and Andrey Tetenov

Abstract We define G-symmetric polygonal systems of similarities and study the properties of symmetric dendrites, which appear as their attractors. This allows us to find the conditions under which the attractor of a zipper becomes a dendrite.

1 Introduction

Though the study of dendrites from the viewpoint of general topology proceeded for more than 75 years [3, 7], the attempts to study the geometrical properties of self-similar dendrites were rather fragmentary. Hata [5] studied the connectedness properties of self-similar sets and proved that if a dendrite is an attractor of a system of weak contractions in a complete metric space, then the set of its endpoints is infinite. Bandt [2] showed that the Jordan arcs connecting pairs of points of a post-critically finite self-similar dendrite are self-similar, and the set of their possible dimensions is finite. Kigami [6] applied the methods of harmonic calculus on fractals to dendrites and developed new approaches to the study of their structure. Croydon [4] obtained heat kernel estimates for continuum random trees.

In [8, 10, 11] we considered contractible P-polyhedral systems $\mathcal{S}$ of contraction similarities in $\mathbb{R}^d$ defined by some polyhedron $P \subset \mathbb{R}^d$. We proved that their

---

M. Samuel (✉)
Department of Mathematics, Bharata Mata College, Kochi, India
e-mail: marysamuel2000@gmail.com

D. Mekhontsev
Sobolev Mathematics Institute, Novosibirsk, Russia
e-mail: mekhontsev@gmail.com

A. Tetenov
Gorno-Altaisk State University, Altay Republits, Russia

Novosibirsk State University, Novosibirsk, Russia

Sobolev Mathematics Institute, Novosibirsk, Russia
e-mail: atet@mail.ru

V. Madhu et al. (eds.), *Advances in Algebra and Analysis*, Trends in Mathematics,
https://doi.org/10.1007/978-3-030-01120-8_4

attractors are dendrites K in $\mathbb{R}^d$, and that the upper bound for the orders of points $x \in K$ depend only on P, while Hausdorff dimension of the set of the cut points of K is strictly smaller than the one of the set of its end points unless K is a Jordan arc.

Now we extend our approach to the case of symmetric P-polygonal systems $\mathcal{S}$ and show that their attractors K are symmetric dendrites K, whose main trees are symmetric n-pods (Proposition 2); all the vertices of the polygon P are the end points of K; that for $n > 5$ each ramification point of K has the order n (Proposition 3); that the augmented system $\tilde{\mathcal{S}}$ contain subsystems $\mathcal{Z}$ which are zippers whose attractors are subdendrites of the dendrite K (Theorem 5).

1.1 Dendrites

Definition 1 A *dendrite* is a locally connected continuum containing no simple closed curve.

In the case of dendrites, the order $Ord(p, X)$ of a point p with respect to X is equal to the number of components of $X \setminus \{p\}$. $EP(X)$ denotes the set of points of order 1 or *end points* of X. $CP(X)$ is the set of all points of order ≥ 2 or *cut points* of X. $RP(X)$ is the set of points of order at least 3, or *ramification points* of X.

According to [3, Theorem 1.1], for a continuum X, the following conditions are equivalent: X is dendrite; every two distinct points of X are separated by a third point; the intersection of every two connected subsets of X is connected; X is locally connected and uniquely arcwise connected.

1.2 Self-similar Sets

Let (X, d) be a complete metric space. A mapping $F : X \to X$ is a contraction if $\mathrm{Lip}\, F < 1$ and a similarity if $d(S(x), S(y)) = rd(x, y)$ for all $x, y \in X$ and fixed r.

Definition 2 Let $\mathcal{S} = \{S_1, S_2, \ldots, S_m\}$ be a system of contractions of c.m.s. (X, d). A nonempty compact set $K \subset X$ is the *attractor* of the system $\mathcal{S}$, if $K = \bigcup_{i=1}^{m} S_i(K)$.

The system $\mathcal{S}$ defines the Hutchinson operator T by the equation $T(A) = \bigcup_{i=1}^{m} S_i(A)$. By Hutchinson's Theorem, the attractor K is uniquely defined by $\mathcal{S}$, and for any compact $A \subset X$, the sequence $T^n(A)$ converges to K.

We also call the subset $K \subset X$ self-similar with respect to $\mathcal{S}$. In our paper, the maps $S_i \in \mathcal{S}$ are similarities and X is $\mathbb{R}^2$.

Notation. $I = \{1, 2, \ldots, m\}$ is the set of indices, and $I^* = \bigcup_{n=1}^{\infty} I^n$ is the set of all multiindices $\mathbf{j} = j_1 j_2 \ldots j_n$. By $\mathbf{ij}$ we denote the concatenation of $\mathbf{i}$ and $\mathbf{j}$; $\mathbf{i} \sqsubset \mathbf{j}$, if $\mathbf{j} = \mathbf{ik}$ for some $\mathbf{k} \in I^*$; if $\mathbf{i} \not\sqsubset \mathbf{j}$ and $\mathbf{j} \not\sqsubset \mathbf{i}$, $\mathbf{i}$ and $\mathbf{j}$ are *incomparable*; we write

$S_{\mathbf{j}} = S_{j_1 j_2 \dots j_n} = S_{j_1} S_{j_2} \dots S_{j_n}$ and for the set $A \subset X$ we denote $S_{\mathbf{j}}(A)$ by $A_{\mathbf{j}}$; $G_{\mathcal{S}} = \{S_{\mathbf{j}}, \mathbf{j} \in I^*\}$ denotes the semigroup, generated by $\mathcal{S}$. The set of all infinite sequences $I^\infty = \{\boldsymbol{\alpha} = \alpha_1\alpha_2 \dots, \quad \alpha_i \in I\}$ is the *index space*; and $\pi : I^\infty \to K$ is the *index map*, which sends a sequence $\boldsymbol{\alpha}$ to the point $\bigcap_{n=1}^{\infty} K_{\alpha_1 \dots \alpha_n}$.

1.3 Zippers

The simplest way to construct a self-similar curve is to take a polygonal line and then replace each of its segments by a smaller copy of the same polygonal line; this construction is called zipper and was studied in [1, 9].

Definition 3 A system $\mathcal{S} = \{S_1, \dots, S_m\}$ of contractions of X is called a *zipper* with vertices $\{z_0, \dots, z_m\}$ and signature $\varepsilon = (\varepsilon_1, \dots, \varepsilon_m)$, $\varepsilon_i \in \{0, 1\}$, if for $i = 1, \dots m$, $S_i(z_0) = z_{i-1+\varepsilon_i}$ and $S_i(z_m) = z_{i-\varepsilon_i}$.

A zipper $\mathcal{S}$ is a *Jordan zipper* if and only if one (and hence every) of the structural parametrizations of its attractor establishes a homeomorphism of the interval $J = [0, 1]$ onto $K(\mathcal{S})$.

Theorem 1 ([1]) *Let $\mathcal{S} = \{S_1, \dots, S_m\}$ be a zipper with vertices $\{z_0, \dots, z_m\}$ in X such that all $S_j : X \to X$ are injective. If for any $i, j \in I$, the set $K_i \cap K_j$ is empty for $|i - j| > 1$ and is a singleton for $|i - j| = 1$, then $\mathcal{S}$ is a Jordan zipper and $K(\mathcal{S})$ is a Jordan arc with endpoints z_0 and z_m.*

2 Contractible P-Polygonal Systems

Let P be a polygon in $\mathbb{R}^2$ and $V_P = \{A_1, \dots, A_{n_P}\}$, $n_P = \#V_P$, be the set of its vertices.

Let $\mathcal{S} = \{S_1, \dots, S_m\}$ be a system of contracting similarities, such that:

(D1) For any $k \in I$, the set $P_k = S_k(P)$ is contained in P;
(D2) For any $i \neq j$, $i, j \in I$, $P_i \bigcap P_j$ is either empty or a common vertex of P_i and P_j;
(D3) For any $A_k \in V_P$ there is a map $S_i \in \mathcal{S}$ and a vertex $A_l \in V_P$ such that $S_i(A_l) = A_k$;
(D4) The set $\widetilde{P} = \bigcup_{i=1}^{m} P_i$ is contractible.

Definition 4 The system $(P, \mathcal{S})$ satisfying the conditions **D1–D4** is called a contractible P-polygonal system of similarities (CPS).

Applying Hutchinson operator $T(A) = \bigcup_{i \in I} S_i(A)$ of $\mathcal{S}$ to the polygon P, we define $\widetilde{P}^{(1)} = \bigcup_{i \in I} P_i$ and consequently $\widetilde{P}^{(n+1)} = T(\widetilde{P}^{(n)})$, obtaining a nested family of contractible compact sets $\widetilde{P}^{(1)} \supset \widetilde{P}^{(2)} \supset \ldots \supset \widetilde{P}^{(n)} \supset \ldots$, whose intersection by Hutchinson's theorem is the attractor K.

The following Theorem was proved by the authors in [8, 10, 11]:

Theorem 2 *Let K be the attractor of CPS $\mathcal{S}$. Then K is a dendrite.*

Since K is a dendrite, for any vertices $A_i, A_j \in V_P$, there is a unique Jordan arc $\gamma_{ij} \subset K$ connecting A_i, A_j. The set $\hat{\gamma} = \bigcup_{i \neq j} \gamma_{ij}$ is a subcontinuum of the dendrite K, all of whose end points are contained in V_P, so $\hat{\gamma}$ is a topological tree [3, **A.17**].

Definition 5 The union $\hat{\gamma} = \bigcup_{i \neq j} \gamma_{ij}$ is called *the main tree* of the dendrite K. The ramification points of $\hat{\gamma}$ are called *main ramification points* of the dendrite K.

We consider $\hat{\gamma}$ as a topological graph whose vertex set $V_{\hat{\gamma}}$ is the union of V_P and the ramification set $RP(\hat{\gamma})$, while the edges of $\hat{\gamma}$ are the components of $\hat{\gamma} \setminus V_{\hat{\gamma}}$.

We proved in [8] the following relation between the vertices of P and end points, cut points and ramification points of $\hat{\gamma}$:

Proposition 1

a) For any $x \in \hat{\gamma}$, $\hat{\gamma} = \bigcup_{j=1}^{n} \gamma_{A_j x}$;
b) A_i is a cut point of $\hat{\gamma}$, if there are j_1, j_2 such that $\gamma_{j_1 i} \cap \gamma_{j_2 i} = \{A_i\}$;
c) the only end points of $\hat{\gamma}$ are the vertices A_j such that $A_j \notin CP(\hat{\gamma})$;
d) $Ord(A_i, K) \leq (n-1)(\left\lceil \frac{\theta_{max}}{\theta_{min}} \right\rceil - 1)$, where $\theta_{max}, \theta_{min}$ are maximal and minimal values of vertex angles of P; if $\#\pi^{-1}(A_i) = 1$, $Ord(A_i, K) \leq n - 1$.

As it was proved in [8, 10, 11], each cut point y of K lies in some image $S_{\mathbf{j}}(\hat{\gamma})$ of the main tree and if $\#\pi^{-1}(y) = 1$, for some $\mathbf{j} \in I^*$, $Ord(y, K) = Ord(y, S_{\mathbf{j}}(\hat{\gamma}))$. If $y \in G_{\mathcal{S}}(V_P)$, then $Ord(y, K) \leq (n_P - 1)\left(\left\lceil \frac{2\pi}{\theta_{min}} \right\rceil - 1\right)$.

The dimension of $EP(K)$ is always greater then the one of $CP(K)$ [10, 11]:

Theorem 3 *Let $\mathcal{S}$ be CPS and K be its attractor. (i) $\dim_H(CP(K)) = \dim_H(\hat{\gamma}) \leq \dim_H EP(K) = \dim_H(K)$; (ii) $\dim_H(CP(K)) = \dim_H(K)$ iff K is a Jordan arc.*

3 Symmetric Polygonal Systems

Definition 6 Let P be a polygon and G be a nontrivial symmetry group of P. Let $\mathcal{S}$ be a CPS such that for any $g \in G$ and any $S_i \in \mathcal{S}$, there are such $g' \in G$ and $S_j \in \mathcal{S}$ that $g \cdot S_i = S_j \cdot g'$. Then the system of mappings $\mathcal{S} = \{S_i, i = 1, 2, \ldots, m\}$ is called a contractible G-symmetric P-polygonal system.

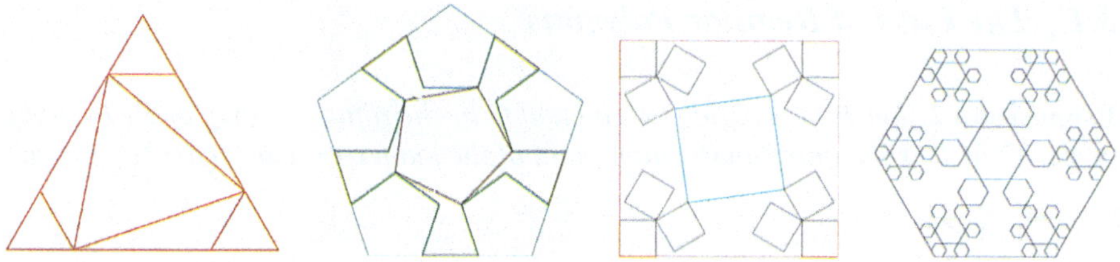

For convenience we will call such systems symmetric polygonal systems or SPS, if this does not cause ambiguity in choice of P and G.

Theorem 4 *The attractor K of SPS and its main tree $\hat{\gamma}$ are symmetric with respect to the group G.*

Proof Let $\mathcal{S} = \{S_1, \ldots, S_m\}$ and $g \in G$. The map $g^* : \mathcal{S} \to \mathcal{S}$, sending each S_i to respective S_j is a permutation of $\mathcal{S}$, therefore $g(\bigcup_{i=1}^m S_i(P)) = \bigcup_{i=1}^m S_i(P)$, or $g(\widetilde{P}) = \widetilde{P}$. The Definition 6 implies that for any $\mathbf{i} = i_1 \ldots i_k$ there is such $\mathbf{j} = j_1 \ldots j_k$ and such $g' \in G$ that $g \cdot S_\mathbf{i} = S_\mathbf{j} \cdot g'$. Therefore for any g, $g(\widetilde{P}^k) = \widetilde{P}^k$. Since $K = \bigcap\limits_{k=1}^{\infty} \widetilde{P}^k$, $g(K) = K$. Since g preserves the set V_P, $g(\hat{\gamma}) = \hat{\gamma}$.

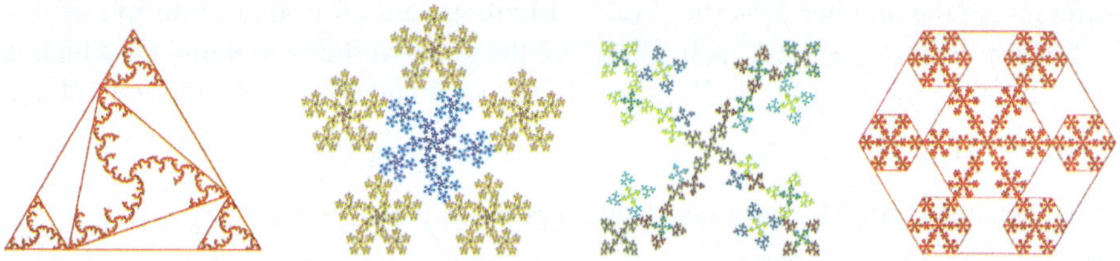

Corollary 1 *If $\mathcal{S}$ is SPS then $\mathcal{S}^{(n)} = \{S_\mathbf{j}, \mathbf{j} \in I^n\}$ is SPS with the same G and P.*

Corollary 2 *Let $\mathcal{S} = \{S_1, \ldots, S_m\}$ be SPS with the attractor K, $g_1, \ldots, g_m \in G$ and $\mathcal{S}' = \{S_1 g_1, \ldots, S_m g_m\}$. Then K is the attractor of the system $\mathcal{S}'$.*

Proof Let K' be the attractor of $\mathcal{S}'$ and put $\widetilde{P}' = \bigcup_{i=1}^m (S_i \circ g_i(P))$. For any i, $g_i(P) = P$, therefore $\widetilde{P}' = \widetilde{P}$ and $\widetilde{P}'^{(k)} = \widetilde{P}^{(k)}$. Then $K' = \bigcap\limits_{k=1}^{\infty} \widetilde{P}'^{(k)} = K$.

Definition 7 Let $\mathcal{S} = \{S_1, \ldots, S_m\}$ be a G-symmetric P-polygonal system. The system $\widetilde{\mathcal{S}} = \{S_i \cdot g, S_i \in \mathcal{S}, g \in G\}$ is called the *augmented system* for $\mathcal{S}$.

The system $\widetilde{\mathcal{S}}$ has the same attractor K as $\mathcal{S}$ and generates the augmented semigroup $G(\widetilde{\mathcal{S}})$ consisting of all maps of the form $S_\mathbf{j} \circ g_i$, where $g_i \in G$.

3.1 The Case of Regular Polygons

Proposition 2 *Let P be a regular n-gon and G be the rotation group of P. Then the center O of P is the only ramification point of the main tree and $Ord(O, \hat{\gamma}) = n$.*

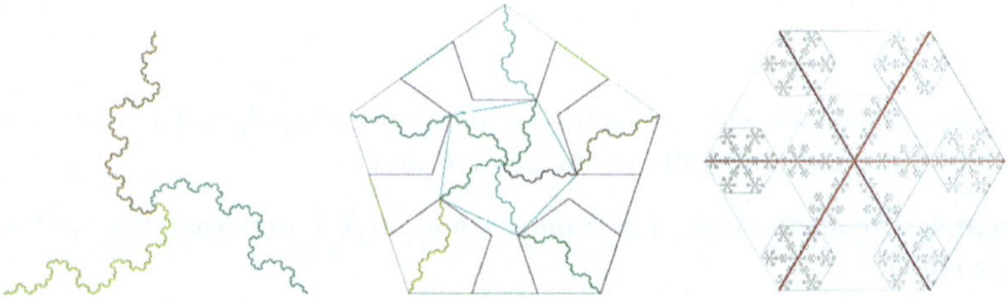

Proof Consider the main tree $\hat{\gamma}$. It is a fine finite system [6], which is invariant with respect to G. Let f be the rotation of P in the angle $2\pi/n$.

Let V and E be the numbers of vertices and edges of $\hat{\gamma}$. For any edge $\lambda \subset \hat{\gamma}$, $f(\lambda) \cap \lambda$ is either empty or is equal to $\{O\}$, and then O is the endpoint of λ and $f(\lambda)$. In each case all the edges $f^k(\lambda)$ are different. Therefore E is a multiple of n.

If A' is a vertex of $\hat{\gamma}$ and $A' \neq O$, then all the points $f^k(A'), k = 1, \ldots, n$ are different, so the number of vertices of $\hat{\gamma}$, different from O, is also a multiple of n.

Since $\hat{\gamma}$ is a tree, $V = E + 1$. Therefore the set of vertices contains O, which is the only invariant point for f. Denote the unique subarc of $\hat{\gamma}$ with endpoints O and A_k by γ_k. Then for any $k = 1, \ldots, n$, $\gamma_k = f^k(\gamma_n)$. By Proposition 1 $\bigcup\limits_{k=1}^{n} \gamma_k = \hat{\gamma}$. Thus the center O is the only ramification point of $\hat{\gamma}$ and $Ord(O, \hat{\gamma}) = n$.

Corollary 3 *All vertices of the polygon P are the end points of the main tree.*

Proof For any $k = 1, \ldots, n$, there is a unique arc γ_k of the main tree meeting the vertex A_k of the polygon P, so $Ord(A_k, \hat{\gamma}) = 1$ by Proposition 1. Since all the vertex angles of P are equal, for each $A_k \in V_P$, there is unique $S_k \in \mathcal{S}$ such that $P_k = S_k(P) \ni A_k$, so $\#\pi^{-1}(A_k) = 1$ and therefore $Ord(A_k, K) = Ord(A_k, \hat{\gamma}) = 1$.

Lemma 1 *Each arc γ_k is the attractor of a Jordan zipper.*

Proof We prove the statement for the arc γ_n, because $\gamma_k = f^k(\gamma_n)$.

If $n > 3$, there is $S_0 \in \mathcal{S}$, whose fixed point is O. Indeed, there is some $S_0 \in \mathcal{S}$ for which $P_0 = S_0(P) \ni O$. The point O cannot be the vertex of P_0, otherwise $f(P_0)$ and P_0 would intersect more than in one point. Therefore $f(P_0) = P_0$ and $S_0(O) = O$.

Observe that for any $A_i, A_j \in V_P$, the arc $\gamma_{A_i A_j} = \gamma_i \cup \gamma_j$.

There is a unique chain of polygons $P_{l_k} = S_{l_k}(P), k = 0, \ldots, s$ connecting P_0 and P_n and containing γ_n, where $S_{l_0} = S_0$ and $S_{l_s} = S_n$. For each $k = 1, \ldots, s$, there are i_k, j_k such that $\gamma_n \cap P_{l_k} = S_{l_k}(f^{i_k}(\gamma_n) \cup f^{j_k}(\gamma_n))$, so

$\gamma_n = \bigcup_{k=1}^{s} S_{l_k}\left(f^{i_k}(\gamma_n) \cup f^{j_k}(\gamma_n)\right) \cup S_0(\gamma_n)$. The arcs on the right hand satisfy the conditions of Theorem 1, so the system $\{S_0, S_{l_1} f^{i_1}, S_{l_1} f^{j_1}, \ldots, S_{l_s} f^{i_s}, S_{l_s} f^{j_s}\}$ is a Jordan zipper whose attractor is a Jordan arc with endpoints O and A_n.

If $n = 3$, it is possible that for some l_1, O is a vertex of a triangle $S_{l_1}(P)$, and there is a unique chain of subpolygons $P_{l_k} = S_{l_k}(P), k = 1, \ldots, s$, where $S_{l_s} = S_3$. By the same argument, a system $\{S_{l_1} f^{i_1}, S_{l_1} f^{j_1}, \ldots, S_{l_s} f^{i_s}, S_{l_s} f^{j_s}\}$ is a Jordan zipper whose attractor is a Jordan arc with endpoints O and A_3.

Corollary 4 *If P is a regular n-gon and the symmetry group G of the system $\mathcal{S}$ is the dihedral group D_n, then γ_{OA_i} is the line segment and the set of cut points of K has dimension 1.*

Proof D_n contains a symmetry with respect to the line OA_n, so γ_n is a line segment.

Thus we see that Proposition 1 implies the following:

Proposition 3 *Let $\mathcal{S}$ be SPS, where P is a regular n-gon and G contains the rotation group of P. Then:*

a) $V_p \subset EP(\hat{\gamma}) \subset EP(K)$;
b) For each cut point $y \in K \setminus \bigcup_{\mathbf{j} \in I^*} S_{\mathbf{j}}(V_P)$, either $y = S_{\mathbf{i}}(O)$ for some $\mathbf{i} \in I^*$ and $Ord(y, K) = n$ or $Ord(y, K) = 2$.
c) For any $y \in \bigcup_{\mathbf{j} \in I^*} S_{\mathbf{j}}(V_P)$, there is unique $x \in \bigcup_{i \in I} S_i(V_P)$, such that

$$Ord(y, K) = Ord(x, K) = \#\pi^{-1}(y) = \#\pi^{-1}(x)$$
$$= \#\{i \in I : x \in S_i(V_P)\} \le 1 + \left\lceil \frac{4}{n-2} \right\rceil$$

Proof All vertex angles of P are $\theta = \dfrac{(n-2)\pi}{n}$, so $\left\lceil \dfrac{2\pi}{\theta_{min}} \right\rceil - 1 = 1 + \left\lceil \dfrac{4}{n-2} \right\rceil$.

a) Take a vertex $A_i \in V_P$. There is unique $j \in I$ such that $A_i \in S_j(V_P)$. For that reason $\#\pi^{-1}(A_i) = 1$. Since $S_j(P)$ cannot contain the center O, $\#(S_j(V_P) \cap \hat{\gamma}) = 2$, therefore by Theorem 1, $Ord(A_i, \hat{\gamma}) = 1$ and $Ord(A_i, K) = 1$, so $A_i \in EP(K)$.
b) If for some $\mathbf{j} \in I^*$, $y = S_{\mathbf{j}}(O)$, then $Ord(y, K) = n$. Since for any point $x \in CP(\hat{\gamma}) \setminus \{O\}$, $Ord(x, \hat{\gamma}) = 2$, the same is true for $y = S_{\mathbf{j}}(x)$ for any $y \in I^*$.
c) Now let $\mathcal{C} = \{C_1, \ldots, C_N\}$ be the full collection of those points $C_k \in \bigcup_{i \in I} S_i(V_P)$ for which $s_k := \#\{j \in I : S_j(V_P) \ni C_k\} \ge 3$. By Theorem 1, $\#\pi^{-1}(C_k) = s_k$ and $Ord(C_k, K) = s_k$, while $s_k \le 1 + \left\lceil \dfrac{4}{n-2} \right\rceil$

Then, if $y \in S_{\mathbf{j}}(C_k)$ for some $\mathbf{j} \in I^*$ and $C_k \in \mathcal{C}$, then $\#\pi^{-1}(y) = s_k = Ord(y, K)$.

Thus we get all possible ramification orders for regular n-gons:

1. If $n \geq 6$ then all ramification points of K are the images $S_{\mathbf{j}}(O)$ of the centre O and have the order n.
2. If $n = 4$ or 5 then there is a finite set of ramification points $x_1, \ldots, x_r$, whose order is equal to 3 such that each x_k is a common vertex of polygons $S_{k1}(P), S_{k2}(P), S_{k3}(P)$. Then each ramification point is represented either as $S_{\mathbf{j}}(O)$ and has the order n or as $S_{\mathbf{j}}(x_k)$ and has the order 3.
3. If $n = 3$ the center is a ramification point of order 3 and those ramification points which are not images of O will have an order less than or equal to 5.

3.2 Self-similar Zippers, Whose Attractors Are Dendrites

Theorem 5 *Let $(\mathcal{S}, P)$ be a G-symmetric P-polygonal system of similarities. Let A, B be two vertices of the polygon P and L be the line segment $[A, B]$. If $\mathcal{Z} = \{S'_1, \ldots, S'_k\}$ is such family of maps from $\tilde{\mathcal{S}}$ that $\tilde{L} = \bigcup_{i=1}^{k} S'_i(L)$ is a polygonal line connecting A and B, then the attractor $K_{\mathcal{Z}}$ of $\mathcal{Z}$ is a subcontinuum of K. If for some subpolygon P_j, $\tilde{L} \cap P_j$ contains more than one segment, then $K_{\mathcal{Z}}$ is a dendrite.*

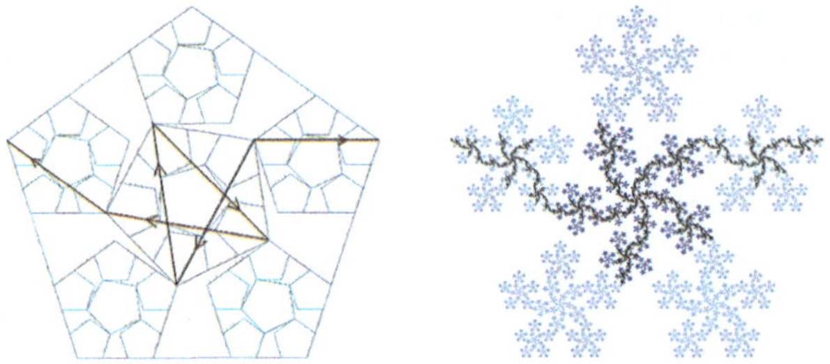

Proof Since $\mathcal{Z} \subset \tilde{\mathcal{S}}$, the attractor $K_{\mathcal{Z}}$ is a subset of K. The system $\mathcal{Z}$ is a zipper with vertices A, B, therefore $K_{\mathcal{Z}}$ is a continuum, and therefore is a subdendrite of the dendrite K. Let γ_{AB} be the Jordan arc connecting A and B in $K_{\mathcal{Z}}$, and, therefore, in K. By the proof of Lemma 1, $\gamma_{AB} = \gamma_{OA} \cup \gamma_{OB}$. If the maps S'_{i_1}, S'_{i_2} send L to two segments belonging to the same subpolygon P_{i_0}, then $S'_{i_1}(\gamma_{AB}) \bigcup S'_{i_2}(\gamma_{AB})$ is equal to $S'_{i_1}(\gamma_{OA} \bigcup \gamma_{OB}) \bigcup S'_{i_2}(\gamma_{OA} \bigcup \gamma_{OB})$. At least 3 points in $\{S'_{i_1}(A), S'_{i_1}(B), S'_{i_2}(A), S'_{i_2}(B)\}$ are different, therefore $S'_{i_1}(O)$ is a ramification point of $K_{\mathcal{Z}}$ of order at least 3.

Corollary 5 *Let u_i be the number of segments of the intersection $\tilde{L} \cap P_i$ and $u = \max u_i$. Then maximal order of ramification points of $K_{\mathcal{Z}}$ is greater or equal to $\min(u + 1, n)$.*

Proof Suppose $\widetilde{L} \bigcap P_i$ contains u segments of $\widetilde{L}$. Then the set $K_{\mathbb{Z}} \cap P_i$ contains at least $u + 1$ vertices of P_i if $u < n - 1$ and contains n vertices otherwise, so it contains at least $u + 1$ (resp. exactly n) different images of the arc γ_{OA}.

Acknowledgements Supported by Russian Foundation of Basic Research projects 16-01-00414 and 18-501-51021

References

1. Aseev, V. V., Tetenov, A. V., Kravchenko, A. S.: On Self-Similar Jordan Curves on the Plane. Sib. Math. J. **44**(3), 379–386 (2003).
2. Bandt, C., Stahnke, J.: Self-similar sets 6. Interior distance on deterministic fractals. preprint, Greifswald 1990.
3. Charatonik, J., Charatonik, W.: Dendrites. Aport. Math. Comun. **22** 227–253(1998).
4. Croydon, D.: Random fractal dendrites, Ph.D. thesis. St. Cross College, University of Oxford, Trinity(2006)
5. Hata, M.: On the structure of self-similar sets. Japan. J. Appl. Math. **3**, 381–414.(1985)
6. Kigami, J.: Harmonic calculus on limits of networks and its application to dendrites. J. Funct. Anal. **128**(1) 48–86, (1995)
7. Kuratowski, K.: Topology. Vols. 1 and 2. Academic Press and PWN, New York(1966)
8. Samuel, M., Tetenov, A. V. , Vaulin, D.A.: Self-Similar Dendrites Generated by Polygonal Systems in the Plane. Sib. Electron. Math. Rep. **14**, 737–751(2017)
9. Tetenov, A. V.: Self-similar Jordan arcs and graph-oriented IFS. Sib. Math.J. **47**(5), 1147–1153 (2006).
10. Tetenov, A. V., Samuel, M., Vaulin, D.A.: On dendrites generated by polyhedral systems and their ramification points. Proc. Krasovskii Inst. Math. Mech. UB RAS **23**(4), 281–291 (2017) doi: 10.21538/0134-4889-2017-23-4-281-291 (in Russian)
11. Tetenov A. V., Samuel, M., Vaulin, D.A.: On dendrites, generated by polyhedral systems and their ramification points. arXiv:1707.02875v1 [math.MG], 7 Jul 2017.

Efficient Authentication Scheme Based on the Twisted Near-Ring Root Extraction Problem

V. Muthukumaran, D. Ezhilmaran, and G. S. G. N. Anjaneyulu

Abstract An authentication protocol is a type of computer communication protocol or cryptography protocol precisely constructed for transferring authentication data between two entities. The aim of this chapter is to propose two new entity authentication schemes that work in the center of the near-ring. The security of the proposed schemes is dependent on the intractability of the twisted near-ring root extraction problem over the near-ring.

1 Introduction

Several digital signatures have recently been proposed based on the non-abelian structure given in [1]. In 2007, Chowdhury [2] described an authenticated scheme established in a non-abelian semi-group. Sibert et al. [3] discovered an entity authentication scheme based on the root extraction problem (REP). In 2017, Muthukumaran and Ezhilmaran [4] described an authentication protocol based on the REP in a near-ring structure. In 2005, Shpilrain and Ushakov [5] proposed new authentication based on the twisted conjugacy problem in non-commutative groups; in 2007, Ferrero [7] suggested a near-ring link with groups and semigroups; and in 2009, Wang and Hu [8] described a signature scheme based on the root extraction problem over braid groups. Further back, in 1988, Guillou and Quisquater [9] discovered a zero knowledge protocol fitted to security microprocessing and transmission, and more recently, in 2015, Ranjan and Om [10] cryptanalyzed the authentication schemes based on braid groups. In this chapter, we introduce the twisted near-ring root extraction problem (TNREP) and describe two authentication scheme established on a near-ring.

The rest of the chapter is organized as follows: in Sect. 2 we describe some basic definitions of a near-ring, the center of a near-ring, the near-ring root extraction

V. Muthukumaran (✉) · D. Ezhilmaran · G. S. G. N. Anjaneyulu
VIT, Vellore, India
e-mail: muthu.v2404@gmail.com; muthu.kumaranv2014@vit.ac.in;
ezhil.devarasan@yahoo.com; ezhilmaran.d@vit.ac.in; anjaneyulu.gsgn@vit.ac.in

V. Madhu et al. (eds.), *Advances in Algebra and Analysis*, Trends in Mathematics,
https://doi.org/10.1007/978-3-030-01120-8_5

problem, and the TNREP. In Sect. 3, we suggest two authentication schemes based on the center of near-ring. In Sect. 4, we conclude this article.

2 Preliminaries [6]

Definition 1 A near-ring is a set N together with binary operations "+" and "·" such that

1. (N, +) is a group (not necessarily abelian)
2. (N, ·) is a semigroup
3. For all $n_1, n_2, n_3 \epsilon N$; $(n_1 + n_2) \cdot n_3 = n_1 \cdot n_3 + n_2 \cdot n_3$ (right distributive law)

This near-ring is termed a right near-ring. If the set N satisfies $n_1(n_2 + n_3) = n_1 \cdot n_2 + n_1 \cdot n_3$ instead of the last condition, then we call N a left near-ring.

Definition 2 For a near-ring $(N, +, \bullet)$, let $C(N) = \{a \in N | ab = baforallb \in N\}$ be its multiplicative center. Let N_1 and N_2 be two subnear-rings of the near-ring N, which satisfies the following conditions:

i. Both N_1 and N_2 are large.
ii. $C(N_1) \cap C(N_2) = \{1_N\}$, where $\{1_N\}$ is the identity of N.

2.1 *Cryptography Assumptions in a Near-Ring [6]*

In this subsection, we describe two cryptography assumptions, which are related to the classical root extraction problem.

Near-Ring Root Extraction Problems

- Instance: For the given $z \in N$ and an integer $a \geq 2$
- Objective: Find $x \in N$ such that $z = x^a$ if such an x exists

TNREP

- Instance: For the given $\varphi \in End(N)$, $z \in N$ and an integer $a \geq 2$
- Objective: Find $x \in N$ such that $z = \varphi x^a$ if such an x exists

3 Proposed Entity Authentication Schemes

Initial Setup Let N be a non-abelian near-ring with two subnear-rings N_1, N_2. The elements of the above subnear-rings satisfy the non-abelian condition $ab = ba$ and we also take H as a fixed collision-free hash function on N.

Scheme I

Phase I. Entity Authentication

1. Alice (A) chooses two arbitrary integers $m \geq 2$ and $n \geq 2$)
2. Alice chooses $a_1 \in C(N_1)$ and $a_2 \in C(N_2)$such that TNREP for a_1, a_2 is hard enough
3. Alice computes $y = \phi(a_1)^m \phi(a_2)^n$
4. Alice's public key is (y, m, n) and the secret key is the pair
5. Alice sends the y value to the Trusted Authority (TA)(X) through a secure channel.

Phase II. Entity Authentication

1. Alice chooses $b_1 \in C(N_1)$ and $b_2 \in C(N_2)$ and sends the challenge $f = \phi(b_1)^m \phi(b_2)^n$
2. Alice sends the response $\omega = H(\phi(a_1)^m f \phi(a_2)^n)$ to Bob. Bob gets the value of y from a TA through a secure channel and checks if $\omega' = H(\phi(b_1)^m y \phi(b_2)^n)$. If they match, authentication is successful.

Proof $\omega' = H(\phi(b_1)^m (\phi(a_1)^m \phi(a_2)^n) \phi(b_2)^n)$
$= H(\phi(a_1)^m (\phi(b_1)^m \phi(b_2)^n) \phi(a_2)^n)$
$= H(\phi(a_1)^m f \phi(a_2)^n)$
$= \omega$

3.1 Proposition

Our entity authentication scheme I is a perfectly honest verifier zero-knowledge interactive proof of knowledge of a_1 and a_2

Proof

Completeness
Assume that, at phase II (ii), Alice sent ω' through a TA. Then, Bob accepts Alice's key if we have $\omega' = H(\phi(b_1)^m y \phi(b_2)^n)$, which is equivalent to

$$= H(\phi(a_1)^m (\phi(b_1)^m \phi(b_2)^n) \phi(a_2)^n) \quad (3.1.1)$$

According to the hypothesis, $a_1, b_1 \in C(N_1)$, while $a_2, b_2 \in C(N_2)$, so that $a_1 b_1 = b_1 a_1 and a_2 b_2 = b_2 a_2$. Therefore, Eq. 3.1.1 is equivalent to $\omega = \omega'$.

Soundness
Assume a cheater (C) is accepted as non-negligible. This means that C can compute $H(\phi(b_1)^m y \phi(b_2)^n$ with non-negligible probability. As H is supposed to be an ideal hash function, this means that C can compute a m satisfying $H(\omega) = H(\phi(b_1)^m y \phi(b_2)^n$ with non-negligible probability. There are two possibilities: Either we have $\omega = H(\phi(b_1)^m y \phi(b_2)^n$, which contradicts the hypothesis that the

TNREP for $b_1 and b_2$ is hard, or $\omega \neq H(\phi(b_1)^m y\phi(b_2)^n$, which means that C and Bob are able to find a collision for H, contradicting the hypothesis that H might be collision-free.

Honest Verifier Zero-Knowledge
Consider the probabilistic Turing machine defined as follows: It chooses random elements $b_1 and b_2$ using the same drawing as the honest verifier and outputs the instance $(b_1, b_2, H(\phi(b_1)^m y\phi(b_2)^n)$. Then, the instance generated by this simulator follows the same probability distribution as those generated by the interactive pair (A, B).

Scheme II

Phase I. Entity Authentication

1. Alice chooses a sufficiently complicated s in N, two-integer $m \geq 2 and n \geq 2)$
2. Alice chooses $a_1 \in C(N_1),)$ such that TNREP for a_1 is hard enough
3. Alice computes $y = \phi(a_1)^m \phi(a_1)^n$
4. Alice's public key is (y, m, n, s) and the secret key is a_1
5. Alice sends the y value to the Trusted Authority (TA)(X) through a secure channel.

Phase II. Entity Authentication

1. Alice chooses $b_1 \in C(N_2)$, and sends the challenge $f = \phi(b_1)^m \phi(b_1)^n$
2. Alice sends the response $\omega = H(\phi(a_1)^m f\phi(a_1)^n)$ to Bob. Bob gets the value of y from a TA through a secure channel and checks if $\omega' = H(\phi(b_1)^m y\phi(b_1)^n)$. If they match, authentication is successful.

Proof $\omega' = H(\phi(b_1)^m (\phi(a_1)^m \phi(a_1)^n)\phi(b_1)^n)$
$= H(\phi(a_1)^m (\phi(b_1)^m \phi(b_1)^n)\phi(a_1)^n)$
$= H(\phi(a_1)^m f\phi(a_1)^n)$
$= \omega$

3.2 Proposition

Our entity authentication scheme I is a perfect honest verifier zero-knowledge interactive proof of knowledge of a_1.

Proof

Completeness
Assume that, at phase II (ii), Alice sent ω' through a TA. Then, Bob accepts Alice's key if we have $\omega' = H(\phi(b_1)^m y\phi(b_1)^n)$, which is equivalent to

$$= H(\phi(a_1)^m (\phi(b_1)^m \phi(b_1)^n)\phi(a_1)^n) \quad (3.2.1)$$

According to the hypothesis, $a_1 \in C(N_1)$, while $b_1 \in C(N_2)$, so that $a_1 b_1 = b_1 a_1$. Therefore, Eq. 3.2.1 is equivalent to $\omega = \omega^{'}$

Soundness
Assume that a cheater (C) is accepted as non-negligible. This means that C can compute $H(\phi(b_1)^m y\phi(b_1)^n$ with non-negligible probability. As H is supposed to be an ideal hash function, this means that C can compute a m satisfying $H(\omega) = H(\phi(b_1)^m y\phi(b_1)^n$ with non-negligible probability. There are two possibilities: Either we have $\omega = H(\phi(b_1)^m y\phi(b_1)^n$, which contradicts the hypothesis that the TNREP for b_1 is hard, or $\omega \neq H(\phi(b_1)^m y\phi(b_1)^n$, which means that C and Bob are able to find a collision for H, contradicting the hypothesis that H might be collision-free.

Honest Verifier Zero-Knowledge
Consider the probabilistic Turing machine defined as follows: It chooses random elements b_1 using the same drawing as the honest verifier and outputs the instance $(b_1, H(\phi(b_1)^m y\phi(b_1)^n)$. Then, the instance generated by this simulator follows the same probability distribution as those generated by the interactive pair (A, B).

4 Conclusion

In this article, we have designed entity authentication schemes based on the center of a near-ring. The security of the schemes relies on the hardness of the TNREP over the near-ring structure. The proposed schemes have been secure against insider attack, replay attack, and stolen verifier attack. We wish to continue our research in the field of secure key generation and mutual authentication schemes using the near-ring. We will also widen our research to develop multi-server authentication schemes using the near-ring.

References

1. Anshel, I., Anshel, M., and Goldfeld D.: An algebraic method for public-key cryptography, Mathematical Research Letters. **6**,287–292 (1999)
2. Chowdhury, M.M.: Key agreement and authentication schemes using non-commutative semi-groups, arXiv preprint arXiv:0708.239 (2007)
3. Sibert, H., Dehornoy P., and Girault M.: Entity authentication schemes using braid word reduction, Discrete Applied Mathematics. **154**,420–436 (2006)
4. Muthukumaran, V., Ezhilmaran, D.: Symmetric decomposition problem in zero-knowledge authentication schemes using near- ring structure, International Journal of Applied Engineering Research. vol **11**,36–40 (2016)
5. Shpilrain, V., Ushakov, A.: An authentication scheme based on the twisted conjugacy problem, In International Conference on Applied Cryptography and Network Security. Springer Berlin Heidelberg, 366–372 (2008)

6. Muthukumaran, V., Ezhilmaran D.: Efficient authentication scheme based on near-ring root extraction problem, In IOP Conference Series: Materials Science and Engineering. **263**, 042137(2017)
7. Ferrero, Giovanni: Near-rings: Some developments linked to semigroups and groups, Springer Science and Business Media (2013)
8. Wang, B.C., Hu, Y.P.: Signature scheme based on the root extraction problem over braid groups, IET Information Security, **3**, 53–59 (2009)
9. Guillou, L.C., Quisquater J.J.: A practical zero knowledge protocol fitted to security microprocessor minimizing both transmission and memory, Advances in Cryptology-Encrypt '88, Proceeding: Springer Verlag, 1088,123–128 (1988)
10. Pratik Ranjan, Hari Om: Cryptanalysis of braid groups based authentication schemes, NGCT (2015)

Dimensionality Reduction Technique to Solve E-Crime Motives

R. Aarthee and D. Ezhilmaran

Abstract The dimensionality reduction technique is a great way of math or statistics to minimize the size of data as much as possible, as little information is possible. With a large number of variables, the dispersed matrix may be too large to be studied and interpreted correctly. There will be too much correlation between the variables to be considered. Graphics data is also not particularly useful because the dataset is large. To interpret the more meaningful data, it is essential to reduce the number of variables to a few linear combinations. Each linear combination will correspond to one major component. Dimensionality reduction technique used to transform dataset onto a lower dimensional subspace for visualization and exploration. This technique is also called as principal component analysis. In this article, we are developing an analysis of the essential elements of the cybercrime motives database and discovering some of the high reasons for increasing cybercriminals.

1 Introduction

In many applications, PCA comprises of examining factors estimated on people. At the point when n and p are huge, the point is to blend the colossal amount of data into a simple and justifiable form [1]. PCA is a variable reduction methodology. It is valuable when we have gotten information on various factors, and trust that there is some excess in those factors [2]. PCA is concerned with explaining the variance-covariance structure of the data through a few linear combinations of the original variables. Its general objectives are data reduction and interpretation. A PCA can show relationships that were not previously suspected, and it allows interpretations that would not ordinarily result. The PCA is the statistical technique that changes the original variable set to a smaller set, the irrelevant variable that represents most of the information in the original set of variables [3]. PCA is a way of identifying

R. Aarthee (✉) · D. Ezhilmaran
VIT, Vellore, India
e-mail: aarsi229@gmail.com; aarthee.r@vit.ac.in; ezhil.devarasan@gmail.com; ezhilmaran.d@vit.ac.in

V. Madhu et al. (eds.), *Advances in Algebra and Analysis*, Trends in Mathematics,
https://doi.org/10.1007/978-3-030-01120-8_6

patterns in data. The data is displayed in a way that identifies similarities and differences. When samples are in the database they can be compressed without losing much information. The main idea of PCA is to reduce the dimensions of a dataset with a large number of relative variables [4]. In this work, using data reduction techniques, we find the reasons that are strongly impacted by cybercrimes.

2 Preliminaries

PCA will require a small background of matrix algebra. So we will discuss some basic concepts of matrix algebra.

Eigenvectors and Eigenvalues

Let A be a square matrix. If λ is a scalar and X is a non-zero column vector satisfying

$$AX = \lambda X \tag{1}$$

X is an eigenvector of A; λ is an eigenvalue of A.
Eigenvectors are possible only for square matrices.
Eigenvectors of a matrix are orthogonal.
λ is an eigenvalue of a $n\ x\ n$ matrix A, with the corresponding eigenvector X.
$(A - \lambda I)X = 0$, with $X \neq 0$ leads to $|A - \lambda I| = 0$.
There are at most n distinct eigenvalues of A

3 Principal Component Analysis

The explanations of PCA is the variance-covariance structure of the data through several linear combinations of the original variables. Its overall goal is to reduce and interpret data. PCA is the backbone of modern data analysis, the black box widely used, but sometimes misunderstood [5]. PCA is a coherent technique that analyzes a data table, where a survey is described by a variable that can identify multiple correlations [6]. PCA is used to explain the structure of dispersed by several linear combinations of the initial variables called essential components. The analysis is used to reduce and interpret data [7]. PCA has become the most useful tool for compression, viewing, and data viewing [8]. PCA is a statistical technique that is widely used in mixed data analysis [9]. PCA is a well-developed tool for data analysis and size reduction [10]. PCA's goal is to find diagonal factors that represent the largest variant direction. PCA is used in many programs, including machine learning [11], image processing [12], neuro computing of engineering, and computer systems, especially for a large database.

4 Dataset Information

For our analysis we collected a database from the National Crime Records Bureau (NCRB), which is an Indian government agency responsible for collecting and analyzing crime data as defined by the Indian Penal Code. The dataset are as follows. U: States V1: Personal revenge/settling scores V2: Emotional motives like anger, revenge, etc. V3: Greed/financial gain V4: Extortion V5: Causing disrepute V6: Prank/satisfaction of gaining control V7: Fraud/illegal gain V8: Insult to modesty of women V9: Sexual exploitation V10: Political motives V11: Inciting hate crimes against community V12: Inciting hate crimes against country V13: Disrupt public services V14: Sale/purchase of illegal drugs/items V15: For developing own business/interest V16: For spreading piracy V17: Serious psychiatric illness, viz., perversion, etc. V18: Steal information for espionage V19: Motives of blackmailing V20: Others Now we examine the eigenvalues to determine how many principal components should be considered:

On the off chance that we take these eigenvalues and include them up, we will get the total variance of 20.1872. The proportion of variation explained by each eigenvalue is given in the third column. For example, 8.8467 divided by 20.1872 equals 0.443. The cumulative percentage clarified is acquired by including the progressive extents of variety disclosed to get the running aggregate. For instance, 0.442 plus 0.171 equals 0.614, and so forth. Therefore, about 61% of the variation is explained by the first two eigenvalues together. Next we need to look at successive differences between the eigenvalues. Subtracting the second eigenvalues, we get a difference of 5.42. The difference between the second and third eigenvalues is 1.41; the next difference is 0.46. Subsequent differences are even smaller. A sharp drop starting with one eigenvalue and then onto the next may fill in as another marker of what number of eigenvalues to consider. The first three principal components explain 71% of the variation which is represent in Table 1. This is an acceptably moderate percentage. We can also determine the number of principal components to look at a scree plot. With the eigenvalues ordered from largest to the smallest, a scree plot is the plot of versus. The number of component is determined at the point, beyond which the remaining eigenvalues are all relatively small and of comparable size. The following plot is made in Minitab.

As we see, we could have stopped at the second principal component, but we continued till the third component. Relatively speaking, contribution of the third component is small compared to the second component which is shown in Fig. 1.

Table 1 Eigenvalues and proportion of variation explained by the principal components

Component	Eigen value	Proportion	Cumulative
1	8.8467	0.442	0.442
2	3.4283	0.171	0.614
3	2.0203	0.101	0.715
4	1.5591	0.078	0.793
5	1.2407	0.062	0.855
6	0.8503	0.043	0.897
7	0.5987	0.030	0.927
8	0.5183	0.026	0.953
9	0.3188	0.016	0.969
10	0.2521	0.013	0.982
11	0.1267	0.006	0.988
12	0.0962	0.005	0.993
13	0.0579	0.003	0.996
14	0.038	0.002	0.998
15	0.208	0.001	0.999
16	0.0139	0.001	0.999
17	0.0074	0	1
18	0.0046	0	1
19	0.0009	0	1
20	0.0003	0	1
	20.1872		

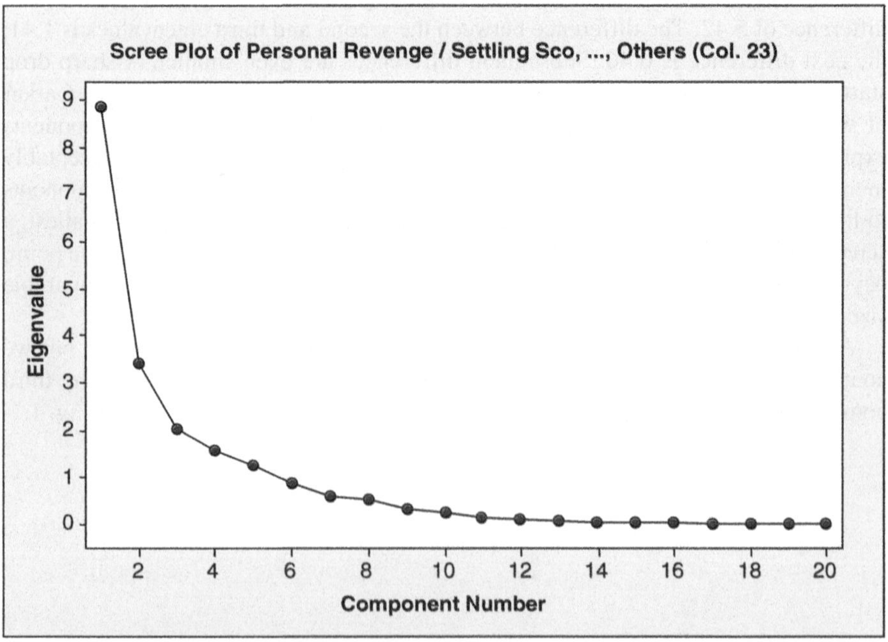

Fig. 1 The scree plot for the variables

5 Interpretation of the Principal Components

To interpret each component, we must compute the correlations procedure. In the variable statement, we will include the first three principal components, "Principal Component 1, Principal Component2, and Principal Component3", in addition to all twenty of the original variables. We will use these correlations between the principal components and the original variables to interpret these principal components. The first principal component is strongly correlated with only two of the values financial gain and sexual exploitation. This component can be viewed as a measure of how financial gain leads to cheapest sexual exploitation. The second principal component is strongly correlated with six of the original variables. The second principal component increases with increasing the prank, fraud, insult to modesty of women, purchase of illegal drugs, serious psychiatric illness, and other reasons. The details of the values mentioned in Table 2. Like this, we can analyze each principal component, and we can get an idea which reason is most important to do cybercrime by an individual.

Table 2 Some samples of principal components

	Principal component				
Variable	1	2	3	4	5
Personal revenge/settling scores	−0.195	−0.01	−0.042	−0.402	0.405
Emotional motives like anger, revenge, etc.	−0.229	0.034	0.07	−0.301	0.413
Greed/financial gain	−0.313	0.034	0.098	0.184	0.063
Extortion	−0.259	0.267	0.244	−0.05	−0.061
Causing disrepute	−0.286	0.144	0.13	0.182	0.099
Prank/satisfaction of gaining control	−0.23	0.332	0.229	−0.039	−0.162
Fraud/illegal gain	−0.236	−0.316	0.003	−0.202	−0.137
Insult to modesty of women	−0.17	−0.401	0.012	−0.164	−0.113
Sexual exploitation	−0.305	−0.031	0.196	−0.129	−0.065
Political motives	−0.245	0.065	−0.189	−0.008	0.07
Inciting hate crimes against community	−0.283	0.182	0.243	0.002	−0.15
Inciting hate crimes against country	−0.122	0.128	−0.05	0.521	0.236
Disrupt public services	−0.177	−0.012	−0.563	0.151	−0.039
Sale/purchase of illegal drugs/items	−0.162	−0.314	0.071	0.175	0.275
For developing own business/interest	−0.239	0.018	−0.018	0.204	−0.359
For spreading piracy	−0.176	0.218	−0.417	−0.205	−0.224
Serious psychiatric illness, viz., perversion, etc.	−0.122	−0.369	0.106	0.044	−0.449
Steal information for espionage	−0.2	−0.243	−0.089	0.4	0.191
Motives of blackmailing	−0.238	0.085	−0.449	−0.135	−0.036
Others	−0.139	−0.361	0.449	−0.062	0.083

6 Summary

Using PCA, we decide on how many components should include only the resulting results. The main purpose of this analysis is descriptive that it is not testing hypotheses. So our decision in many respects must be made on the basis of what gives us a good, brief description of the data. We have to decide how important it is. The report is not necessary from the standpoint of statistical tests, but in this case. From the point of view of the city and sociology, we must decide what is important in the context of the problem. This resolution may differ from discipline. The analysis of principal components has been widely applied in the field of sociology and environment as well as marketing research. As we know, PCA reduces data size. But the individual PCA does not show any specific physical variables we've seen. By using PCA techniques we obtained a result that anger, revenge etc, extortion; and inciting hate crimes against the community plays a major role to do cyber crime in the society. Therefore, PCA techniques are useful for reducing data. In the future, by analyzing the principal component and then implementing predictive regression, those variables come from the core component itself. So we can use PCA as a predictive or criterion variable in the next analysis.

References

1. Saporta,G., Niang,N.,: Principal component analysis: application to statistical process control. Data analysis, 1–23(2009)
2. O'Rourke,N., Psych,R., Hatcher,L.,: A step-by-step approach to using SAS for factor analysis and structural equation modeling. Sas Institute,(2013).
3. Dunteman,G.H., Principal components analysis. Sage,**69**,(1989)
4. Jolliffe,I.T., Principal component analysis and factor analysis. In Principal component analysis.Springer, New York, NY. 115–128(1986)
5. Shlens,J., A tutorial on principal component analysis. arXiv preprint, 1404–1100 (2014)
6. Abdi,H., Williams,L.J., Principal component analysis. Wiley interdisciplinary reviews: computational statistics, **2(4)**. 433–459 (2010)
7. Olive,D.J., Robust multivariate analysis. Springer.(2018)
8. Elhamifar,E., Vidal,R.,Sparse subspace clustering. In Computer Vision and Pattern Recognition, 2009. CVPR 2009, 2790–2797(2009)
9. Liu,W., Zhang,H., Tao,D., Wang,Y., Lu.K., Large-scale paralleled sparse principal component analysis. Multimedia Tools and Applications, **75(3)**. 1481–1493 (2016)
10. Guan,N., Tao,D., Luo,Z., Yuan,B.,Online nonnegative matrix factorization with robust stochastic approximation. IEEE Transactions on Neural Networks and Learning Systems, **23(7)**. 1087–1099(2012)
11. Xu,C., Tao,D., Large-margin multi-view information bottleneck. IEEE Transactions on Pattern Analysis and Machine Intelligence, **36(8)**.1559–1572(2014)
12. Tao,D., Li,X., Wu,X., Maybank,S.J., General tensor discriminant analysis and gabor features for gait recognition. IEEE Transactions on Pattern Analysis and Machine Intelligence, **29(10)**(2007)

Partially Ordered Gamma Near-Rings

T. Nagaiah

Abstract The notions of partially ordered Γ-near-ring(POGN), T-fuzzy ideal of POGN(TFIPOGN), T-fuzzy K-ideal of a POGN(TFKIPOGN), quotient POGN, and normal TFKIPOGN are introduced and investigated the basic properties. I also propose some necessary sufficient conditions on POGN under the T-norm.

1 Introduction

In 1965 Zadeh [20] introduced the fuzzy set. The notion of total graphs of a commutative rings were studied by Anderson and Badawi [2]. Fuzzy groups were considered by Rosenfeld [18]. In Meldrum [9] and Pilz [16] are introduced the notion of near-rings. Nobusawa [15] recently introduced the notion of a Gamma-ring, and Barnes [3] study the Gamma-homomorphism. Satyanarayana[19] study the concept of Γ-near-ring. Nagaiah et al. [11, 12, 14] study the concept of fuzzy ideal partially ordered semigroups and T-fuzzy ideals of Gamma near-rings.

In 1975, Radhakrishna [17] defined the partially ordered, fully ordered near-rings and non-associative rings. We proposed the new concept fuzzy subΓ-near-ring, fuzzy ideal of POGN, TFIPOGN, and TFKIPOGN and study their properties. Also, we have studied some properties and their results discussed lucid manner. For some other recent papers on fuzzy ideals of near-rings, see [1, 4–8, 10, 13, 21].

2 Preliminaries

Definition 1 ([1]) A function $T : [0, 1] \times [0, 1] \to [0, 1]$ is said to be t-norm if it satisfies the following:

T. Nagaiah (✉)
Department of Mathematics, Kakatiya Univeristy, Warangal, Telangana, India
e-mail: nagaiahku@gmail.com; nagaiah@uascku.ac.in

V. Madhu et al. (eds.), *Advances in Algebra and Analysis*, Trends in Mathematics,
https://doi.org/10.1007/978-3-030-01120-8_7

$(i)\ T(\alpha, 1) = \alpha$
$(ii)\ T(\alpha, \beta) = T(\beta, \alpha)\ (commutativity)$
$(iii)\ T(\alpha, T(\beta, \gamma)) = T(T(\alpha, \beta), \gamma)\ (assosiativity)$
$(iv)\ T(\alpha, \beta) \leq T(\alpha, \gamma)\ when\ ever \beta \leq \gamma\ (monotonicity), for\ all\ \alpha, \beta, \gamma \in [0, 1].$

Definition 2 ([11]) Let $\hbar$ be a fuzzy subset of X and $\alpha \in [0,1 - \sup\{\hbar(x)/\ x \in X\}]$, $\beta \in [0, 1]$. The mappings $\hbar_\alpha^T : X \to [0, 1]$, $\hbar_\beta^M : X \to [0, 1]$, and $\hbar_{\beta,\alpha}^P : X \to [0, 1]$ are called fuzzy translation, fuzzy multiplication, and fuzzy magnified translation of $\hbar$, respectively, if $\hbar_\alpha^T(x) = \hbar(x) + \alpha$, $\hbar_\beta^M(x) = \beta.\hbar(x)$, and $\hbar_{\beta,\alpha}^P(x) = \beta.\hbar(x) + \alpha$ for all $x \in X$, respectively.

3 T-Fuzzy Ideal of Partially Ordered Γ-Near-Ring

In this section, we propose fuzzy ideal, TFIPOGN and TFKIPOGN.

Definition 3 A near-ring M is called POGN if it satisfies the following:

(i) $r \leq s$ then $r + g \leq s + g$ for all $r, s, g \in M$
(ii) $r \leq s$ and $g \geq 0$ then $r\alpha g \leq s\alpha g$ and $g\alpha r \leq g\alpha s$ for all $r, s, g \in M$ and $\alpha \in \Gamma$.

Definition 4 A fuzzy subset δ of POGN M is said to be a fuzzy sub Γ-near-ring of M if it satisfies the following:

(i) $\delta(r - s) \geq \wedge[\delta(r), \delta(s)]$
(ii) $\delta(r\alpha s) \geq \wedge[\delta(r), \delta(s)]$
(iii) $r \leq s \Rightarrow \delta(r) \geq \delta(s)$ for all $r, s \in M$ and $\alpha \in \Gamma$.

Definition 5 Let δ be a non-empty fuzzy subset of POGN M. Then δ is called a fuzzy right (resp. left) ideal POGN if
(P_1) δ(r-s) $\geq \wedge[\delta(r), \delta(s)]$
(P_2) $\delta((r + t)\alpha s - r\alpha s) \geq \delta(t)(\delta(r\alpha s) \geq \delta(s))$
$(P_3) r \leq s$ then $\delta(r) \geq \delta(s)$, for all $r, s, t \in M$ and $\alpha \in \Gamma$

Definition 6 A fuzzy subset δ of POGN M is called T-fuzzy right (resp. left) ideal of POGN if it satisfies P_2, P_3, and $(P_4) : \delta(r - s) \geq T(\delta(r), \delta(s))$ for all $r, s, t \in M, \alpha \in \Gamma$.

Definition 7 A T-fuzzy ideal is said to be a T-fuzzy K-ideal of POGN M if it satisfies $(P_5) : \delta(r) \geq T(\delta(r - s), \delta(y))$ for all $r, s \in M$ and $\alpha \in \Gamma$.

4 Main Results

Theorem 1 *Every fuzzy ideal of M is a TFIPOGN M.*

Proof From the definition, we have $\delta(r - s) \geq \min\{\delta(r), \delta(s)\} \geq T\{\delta(r), \delta(s)\}$, $r, s \in M$.

Theorem 2 *A fuzzy subset δ of a M is a T-fuzzy left ideal (TFLI) of M if and only if it satisfies the following:*

$$(i)\chi_M o\delta \subseteq \delta\,(ii)\delta(r+s) \geq T(\delta(r), \delta(s)) \textit{for all } r, s \in M.$$

Proof Suppose δ is a TFLI of M and $r \in$ M.

Case (i) If$(\chi_M o\delta)$(r)=0 then $(\chi_M o\delta)$(r)=0 $\leq \delta$(r) and hence $\chi_M o\delta \subseteq \delta$
Case (ii) If$(\chi_M o\delta)(r) \neq 0$ so there exist $s, t \in M$ and $\alpha_1 \in \Gamma$ such that $r \leq s\alpha_1 t$. $(\chi_M o\delta)(r) = \sup\limits_{r \leq s\alpha_1 t} T(\chi_M(s), \delta(t)) = \sup\limits_{r \leq s\alpha_1 t} \delta(t) \leq \delta(s\alpha_1 t) \leq \delta(r)$. This implies $\chi_M o\delta \subseteq \delta$. And also $\delta(r+s) = \delta(r-(-s)) \geq T(\delta(r), \delta(-s)) \geq T(\delta(r), \delta(s))$. Therefore $\delta(r+s) \geq T(\delta(r), \delta(s))$.

Conversely suppose that (i) and (ii) conditions are holds for any fuzzy subset δ. Let r,s$\in M, \alpha_1 \in \Gamma$. Since $\chi_M o\delta \subseteq \delta$ then

$$(i)\delta(r-s) = \sup_{r-s \leq c-d} T(\chi_M(c), \delta(-d)) \geq T(\chi_M(r), \delta(s)) \geq T(\delta(r), \delta(s))$$

$$(ii)\delta(r\alpha_1 s) \geq (\chi_M o\delta)(r\alpha_1 s) = \sup_{r\alpha_1 s \leq p\alpha_1 q} T(\chi_M(p), \delta(q)) \geq T(\chi_M(r), \delta(s)) = \delta(s)$$

$$(iii) \textit{Let } r, s \in M \textit{ such that } r \leq s. \textit{ Then} \delta(r) \geq (\chi_M o\delta)(r\alpha_1 s) = \sup_{r\alpha_1 s \leq s\alpha_1 s} T(\chi_M(s), \delta(s)) \geq \delta(s).$$

Therefore $\delta(r) \geq \delta(s)$. Hence δ is a TFLI of M.

Theorem 3 *If Ω_1 and Ω_2 are TFKI of M, then $\Omega_1 \wedge \Omega_2$ is also TFKI of M.*

Proof Let Ω_1 and Ω_2 be TFLKI of M and $x_1, y_1 \in M, \alpha_1 \in \Gamma$. Then

$$\begin{aligned}(i)\ (\Omega_1 \wedge \Omega_2)(x_1 - y_1) &= T(\Omega_1(x_1 - y_1), \Omega_2(x_1 - y_1)) \\ &\geq T(T(\Omega_1(x_1), \Omega_1(y_1)), T(\Omega_2(x_1), \Omega_2(y_1))) \\ &= T(T(\Omega_1(x_1), \Omega_2(x_1)), T(\Omega_2(y_1), \Omega_1(y_1))) \\ &= T((\Omega_1 \wedge \Omega_2)(x_1), (\Omega_1 \wedge \Omega_2)(y_1))\end{aligned}$$

As Ω_1 and Ω_2 are TFLKI of M. Then we have $\Omega_1(x_1\alpha_1 y_1) \geq \Omega_1(y_1)$ and $\Omega_2(x_1\alpha_1 y_1) \geq \Omega_2(y_1)$

$$\begin{aligned}(ii)\ (\Omega_1 \wedge \Omega_2)(x_1\alpha_1 y_1) &= T(\Omega_1(x_1\alpha_1 y_1), \Omega_2(x_1\alpha_1 y_1)) \\ &\geq T(\Omega_1(y_1), \Omega_2(y_1)) \\ &= T(\Omega_1(y_1), \Omega_2(y_1)) \\ &= (\Omega_1 \wedge \Omega_2)(y_1)\end{aligned}$$

(iii) Suppose $x_1, y_1 \in M$ and $x_1 \leq y_1$, then $\Omega_1(x_1) \geq \Omega_1(y_1)$ and $\Omega_2(x_1) \geq \Omega_2(y_1)$.

$$\begin{aligned} Now\ (\Omega_1 \wedge \Omega_2)(x_1) &= T(\Omega_1(x_1), \Omega_2(x_1)) \\ &\geq T(\Omega_1(y_1), \Omega_1(y_1)) \\ &\geq (\Omega_1 \wedge \Omega_2)(y_1) \end{aligned}$$

Hence $\Omega_1 \wedge \Omega_2$ is a TFLI. Since Ω_1 and Ω_2 are TFLKI of M, then we have $\Omega_1(x_1) \geq$T$(\Omega_1(x_1 - y_1), \Omega_1(y_1))$ and $\Omega_2(x_1) \geq T(\Omega_2(x_1 - y_1), \Omega_2(y_1))$ for all $x_1, y_1 \in M$.

$$\begin{aligned} (\Omega_1 \wedge \Omega_2)(x_1) &= T(\Omega_1(x_1), \Omega_2(x_1)) \\ &\geq T(T(\Omega_1(x_1 - y_1), \Omega_1(y_1)), T(\Omega_2(x_1 - y_1), \Omega_2(y_1)) \\ &\geq T((\Omega_1 \wedge \Omega_2(x_1 - y_1), (\Omega_1 \wedge \Omega_2(y_1))\ for\ all\ x_1, y_1 \in M \end{aligned}$$

We also prove that $\Omega_1 \wedge \Omega_2((y_1 + z_1)\alpha x_1 - y_1\alpha x_1) \geq \Omega_1 \wedge \Omega_2(z_1)$ for all $x_1, y_1, z_1 \in M$ and $\alpha \in \Gamma$. Hence $\Omega_1 \wedge \Omega_2$ is a TFKI of M.

Theorem 4 *A fuzzy subset Ω is a TFKI of M if and only if Ω_α^T is a TFKI of M, provided t-norm holds for combined translations.*

Proof Let $r, s, i \in M$ and $\alpha_1 \in \Gamma$. Then from definition, we have

$$\begin{aligned} &(i)\ \Omega_\alpha^T(r - s) = T(\Omega_\alpha^T(r), \Omega_\alpha^T(s)) \\ &(ii)\ \Omega_\alpha^T(r\alpha_1 s) = \Omega(r\alpha_1 s) + \alpha \geq \Omega(s) + \alpha = \Omega_\alpha^T(s) \\ &(iii)\ \Omega_\alpha^T(r) = \Omega(r) + \alpha \geq T(\Omega(r - s), \Omega(s)) + \alpha \\ &\qquad = T(\Omega(r - s) + \alpha, \Omega(s) + \alpha) = T(\Omega_\alpha^T(r - s), \Omega_\alpha^T(s)) \end{aligned}$$

(iv) Let $r \leq s$ for all $r, s \in M$. This implies $\Omega(r) + \alpha \geq \Omega(s) + \alpha$ and hence $\Omega_\alpha^T(r) \geq \Omega_\alpha^T(s)$. (v) $\Omega_\alpha^T((r + i)\alpha_1 s - r\alpha_1 s) = \Omega(r + i)\alpha_1 s - r\alpha_1 s) + \alpha \geq \Omega(i) + \alpha = \Omega_\alpha^T(i)$. Also $\Omega_\alpha^T(r) \geq \Omega_\alpha^T(s)$, for all $r, s \in M$. Hence $\Omega_\alpha^T(r)$ is TFKI of M. Conversely suppose $\Omega_\alpha^T$is a TFKI of M. Then obviously Ω is a TFIPOGN M. Let $\Omega(s) = p_1$ and $\Omega(r - s) = p_2$, $p = \min(p_1, p_2) \geq T(p_1, p_2)$. Then $s \in \Omega_p$,and $r - s \in \Omega_p$. Since Ω_p is a K-ideal, $r \in \Omega_p$ which implies that $\Omega(r) \geq p = \min(p_1, p_2) \geq T(p_1, p_2) = T(\Omega(s), \Omega(r - s))$. Hence Ω is a TFKI of M.

Theorem 5 *A fuzzy subset Ω is a TFKIPOGN M if and only if Ω_β^M, the fuzzy multiplication of Ω is a TFKIPOGN M, where $\beta \in [0, 1]$.*

Proof The proof of this theorem similar to Theorem 4.

Theorem 6 *If Ω is a TFKI of M if and only if $\Omega_{\beta,\alpha}^P$ is a TFKI of M, provided t-norm holds for combined translation and $\beta \in [0, 1]$.*

Proof Let $x_1, y_1, z_1 \in M$ and $\alpha_1 \in \Gamma$. Then we have

$$
\begin{aligned}
(i)\ \Omega^P_{\beta,\alpha}(x_1 - y_1) &\geq \beta.T(\Omega(x_1), \Omega(y_1)) + \alpha \\
&= T(\beta.\Omega(x_1) + \alpha, \beta.\Omega(y_1) + \alpha) \\
&= T(\Omega^P_{\beta,\alpha}(x_1), \Omega^P_{\beta,\alpha}(y_1)) \\
(ii)\ \Omega^P_{\beta,\alpha}(x_1\alpha_1 y_1) &= \beta.\Omega(x_1\alpha_1 y_1) + \alpha \geq \beta.\Omega(y_1) + \alpha \\
&= \Omega^P_{\beta,\alpha}(y_1) \\
(iii)\ \Omega^P_{\beta,\alpha}(x_1) &= \beta.\Omega(x_1) + \alpha \geq \beta.T(\Omega(x_1 - y_1), \Omega(y_1)) + \alpha \\
&\geq T(\beta.\Omega(x_1 - y_1) + \alpha, \beta.\Omega(y_1) + \alpha) \\
&\geq T(\Omega^P_{\beta,\alpha}(x_1 - y_1), \Omega^P_{\beta,\alpha}(y_1)) \\
(iv) Let\ x_1, y_1 \in M\ with x_1 &\leq y_1 \\
\Rightarrow \Omega(x_1) &\geq \Omega(y_1)\ and\ hence \beta.\Omega(x_1) + \alpha \geq \beta.\Omega(y_1) + \alpha \\
This\ implies\ \Omega^P_{\beta,\alpha}(x_1) &\geq \Omega^P_{\beta,\alpha}(y_1).
\end{aligned}
$$

Also $\Omega^P_{\beta,\alpha}(x_1+z)\alpha_1 y_1 - x_1\alpha_1 y_1) = \beta.\Omega(x_1+z)\alpha_1 y_1 - x_1\alpha_1 y_1) + \alpha \geq \beta.\Omega(z) + \alpha = \Omega^P_{\beta,\alpha}(z)$. Hence $\Omega^P_{\beta,\alpha}$ is a TFKI of M.

Conversely suppose $\Omega^P_{\beta,\alpha}$ is a TFKI of M. Then obviously Ω is a TFI of M. From the definition it is clear that Ω is a TFKI of M.

Theorem 7 *If $\widetilde{\hbar}$ is a TFI of M, and fuzzy set $\hbar^*$ of M/K defined by $\widetilde{\hbar}^\star(a_1 + K) = \bigvee_{x_1 \in K} \widetilde{\hbar}(a_1 + x_1)$. Then $\widetilde{\hbar}^\star$ is TFI of quotient Γ-near-ring M/K of M with respect to K where K is an ideal of M.*

Proof Let $a_1, b_1 \in M$ be such that $a_1 + K = b_1 + K$,then $b_1 = a_1 + y_1$ for some $y_1 \in K$. Then

$$
\widetilde{\hbar}^\star(b_1 + K) = \bigvee_{x_1 \in K} \widetilde{\hbar}(b_1 + x_1) = \bigvee_{x_1 + y_1 = z \in K} \widetilde{\hbar}(a_1 + z) = \widetilde{\hbar}^*(a_1 + K)
$$

Therefore $\widetilde{\hbar}^\star$ is well-defined.

$$
\begin{aligned}
&Let\ x_1 + K, y_1 + K \in N/K\ then \\
&\widetilde{\hbar}^*[(x_1 + K) - (y_1 + K)] = \widetilde{\hbar}^*[(x_1 - y_1) + K] \\
&= \bigvee_{z_1 = u_1 - v_1 \in K} \widetilde{\hbar}[(x_1 - y_1) + (u_1 - v_1)] \\
&= \bigvee_{z_1 = u_1 - v_1 \in K} \widetilde{\hbar}[(x_1 + u_1) - (y_1 + v_1)] \\
&\geq \bigvee_{u_1, v_1 \in K}, \widetilde{T}[\widetilde{\hbar}(x_1 + u_1) - \widetilde{\hbar}(y_1 + v_1)] \\
&= T(\widetilde{\hbar}^*(x_1 + K), \widetilde{\hbar}^*(y_1 + K)).
\end{aligned}
$$

$$
\begin{aligned}
Now\ \widetilde{\hbar}^*((x_1+K)\alpha_1(y_1+K)) &= \widetilde{\hbar}^*(x_1\alpha_1 y_1+K)\\
&= \bigvee_{z_1\in K} \widetilde{\hbar}(x_1\alpha_1 y_1+z_1))\\
&= \bigvee_{z_1=u_1\alpha_1 v_1\in K} \widetilde{\hbar}(x_1\alpha_1 y_1+u_1\alpha_1 v_1))\\
&= \bigvee_{u_1\alpha_1 v_1\in K} \widetilde{\hbar}((x_1+u_1)\alpha_1(y_1+v_1))\\
&\geq \bigvee_{u_1,v_1\in K} \widetilde{\hbar}((y_1+v_1))\\
&= \widetilde{\hbar}^*(y_1+K)
\end{aligned}
$$

Take $x_1, y_1 \in M$ such that $x_1 \leq y_1$. This implies $\hbar(x_1) \geq \hbar(y_1)$. From this inequality we get $\bigvee_{z_1\in K} \widetilde{\hbar}(x_1+z_1) \geq \bigvee_{z_1\in K} \widetilde{\hbar}(y_1+z_1)$. This implies $\widetilde{\hbar}^*(x_1+K) \geq \widetilde{\hbar}^*(y_1+K)$. Also we prove that $\widetilde{\hbar}^*(((y_1+K)+(z_1+K)\alpha_1(x_1+K)-(y_1+K)\alpha_1(x_1+K)) \geq \widetilde{\hbar}^*(z_1)$ for all $x_1, y_1, z_1 \in M$ and $\alpha_1 \in \Gamma$.

Theorem 8 *The imaginable fuzzy subset Ω of M is a TFLKI of M if and only if the strongest fuzzy relation $\hbar_\Omega$ on M is an imaginable TFLKI of $M \times M$.*

Proof Suppose Ω is an imaginable TFLKI of M. Then obviously $\hbar_\Omega$ is TFLKI of $M \times M$, for any $(\xi_1, \xi_2), (\zeta_1, \zeta_2) \in M \times M$.

$$
\begin{aligned}
\hbar_\Omega(\xi_1,\xi_2) &= T(\Omega(\xi_1),\Omega(\xi_2)) \geq T(T(\Omega(\xi_1-\zeta_1),\Omega(\zeta_1)),\\
&\quad T(\Omega(\xi_2-\zeta_2),\Omega(\zeta_2)))\\
&= T(\hbar_\Omega(\xi_1-\zeta_1,\xi_2-\zeta_2),\hbar_\Omega(\zeta_1-\zeta_2))\\
T(\hbar_\Omega(\xi_1,\xi_2),\hbar_\Omega(\xi_1,\xi_2)) &= T(T(\Omega(\xi_1),\Omega(\xi_2),T(\Omega(\xi_1),\Omega(\xi_2))\\
&= T(T(\Omega(\xi_1),\Omega(\xi_1),T(\Omega(\xi_2),\Omega(\xi_2)))\\
&= T(\Omega(\xi_1),\Omega(\xi_2)) = \hbar_\Omega(\xi_1,\xi_2)
\end{aligned}
$$

Suppose $(\xi_1, \xi_2), (\zeta_1, \zeta_2) \in$ M×M and $(\xi_1, \xi_2) \leq (\zeta_1, \zeta_2)$, then $\xi_1 \leq \zeta_1 and \xi_2 \leq \zeta_2$. Therefore $T(\Omega(\xi_1), \Omega(\xi_2)) \geq T(\Omega(\zeta_1), \Omega(\zeta_2))$. Hence $\hbar_\Omega(\xi_1, \xi_2) \geq \hbar_\Omega(\zeta_1, \zeta_2)$. Thus $\hbar_\Omega$ is an imaginable TFLKI of M. Let $\xi_1, \zeta_1 \in$ M and $\alpha_1 \in \Gamma$. Then

$$
\begin{aligned}
\Omega(\xi_1-\zeta_1) &= T(\Omega(\xi_1-\zeta_1),\Omega(\xi_1-\zeta_1))\\
&= \hbar_\Omega(\xi_1-\zeta_1,\xi_1-\zeta_1) = \hbar_\Omega((\xi_1,\xi_1)-(\zeta_1,\zeta_1))\\
&\geq T(\hbar_\Omega((\xi_1,\xi_1),(\zeta_1,\zeta_1)) = T(T(\Omega(\xi_1),\Omega(\xi_1)),T(\Omega(\zeta_1),\Omega(\zeta_1)))\\
&= T(\hbar_\Omega((\xi_1,\zeta_1),\hbar_\Omega((\xi_1,\zeta_1)) = T(\hbar_\Omega(\xi_1,\zeta_1)) = T(\Omega(\xi_1),\Omega(\zeta_1))
\end{aligned}
$$

$$\begin{aligned}
\Omega(\xi_1\alpha_1\zeta_1) &= T(\Omega(\xi_1\alpha_1\zeta_1), \Omega(\xi_1\alpha_1\zeta_1)) = \hbar_\Omega(\xi_1\alpha_1\zeta_1, \xi_1\alpha_1\zeta_1) \\
&= \hbar_\Omega((\xi_1, \xi_1)\alpha_1(\zeta_1, \zeta_1)) \geq T(\hbar_\Omega(\zeta_1, \zeta_1)) \\
&= T(\Omega(\zeta_1), \Omega(\zeta_1)) = \Omega(\zeta_1) \\
\Omega(\xi_1) &= T(\Omega(\xi_1), \Omega(\xi_1)) = \hbar_\Omega(\xi_1,) \\
&\geq T(\hbar_\Omega(\xi_1 - \zeta_1, \xi_1 - \zeta_1), \hbar_\Omega(\zeta_1, \zeta_1)) \\
&= T(\Omega(\xi_1 - \zeta_1), \Omega(\zeta_1))
\end{aligned}$$

If $\xi_1, \zeta_1 \in M$ *and* $\xi_1 \leq \zeta_1$, then certainly we get $T(\Omega(\xi_1), \Omega(\xi_1)) \geq T(\Omega(\zeta_1), \Omega(\zeta_1))$. This implies $\Omega(\xi_1) \geq \Omega(\zeta_1)$.

5 Conclusion

In study of the structure of an algebraic system, ideals with special properties play an important role. This paper studies the notions of partially ordered gamma near-rings, T-fuzzy ideals of partially ordered gamma near-rings, and T-fuzzy K-ideals of partially ordered gamma near-rings and investigated a relationship between these. I also propose some necessary sufficient conditions on POGN with respect to T-norm. Finally, I study T-fuzzy ideal of a quotient gamma near-rings. I hope that this work will serve as a foundation for further study of the theory of partially ordered gamma near-algebras, Smarandache fuzzy gamma near-rings, and Smarandache fuzzy partially ordered gamma near-algebras.

Acknowledgements The author is grateful to the referees for their careful reading and valuable suggestions which helped in improving this paper. I am also thankful to the editors for their valuable comments and suggestions.

References

1. Akram, M.: On T-Fuzzy Ideals in Near-rings, International Journal of Mathematics and Mathematical Sciences., 1–14 (2007).
2. Anderson, David F., Badawi, A.: The total graph of a commutative ring, Journal of Algebra., **320**, 2706–2719 (2008).
3. Barnes, W. E.: On the Gamma-rings of Nobusawa, Pacific Journal of Maths., **18** , 411–422 (1966).
4. Davvaz, B.: Fuzzy Ideals of Near-rings with interval valued membership functions, J. Sci.I.R.Iran., **12(2)**, 171–175 (2001).
5. Jun, Y. B., Hong, S. M and Kim, HS.: Fuzzy ideals in near-rings, Bulletin of the Korean Mathematical Society., **35(3)**, 455–464 (1998).
6. Jun, Y. B., Kim, K. H., Ozturk, M. A.: Fuzzy maximal ideals of Gamma near-rings, Turkish Journal of Mathematics., **25(4)**, 457–464 (2001).
7. Jun, Y. B., Sapanci, M and Ozturk, M. A.: Fuzzy ideal in Gamma near-ring, Tr. J. Of mathematics., **22**, 449–459 (1998).

8. Kim, S. D and kim, H. S.: On fuzzy ideal of near-rings, Bull Korean Math.Soc., **33(4)**, 593–601 (1996).
9. Meldrum, J. D. P.: Near-rings and their links with groups, Pitman London, (1985).
10. Muralikrishna Rao, M.: T-Fuzzy ideals in ordered Γ-semirings, Annals of fuzzy Mathematics and informatics., **13(2)**, 253–276 (2017).
11. Nagaiah, T., Vijay kumar, K., Iampan, A and Srinivas, T.: A study of fuzzy ideals in PO-Γ-Semigroups, Palestine journal of mathematics., **6(2)**, 591–597 (2017).
12. Nagaiah, T., Vijay Kumar, K., Narasimha Swamy, P and Srinivas, T.: A Study of fuzzy ideals in PO-Γ-Semigroups in terms of anti fuzzy ideals, Annals of fuzzy mathematics and informatics., **14(3)**, 225–236 (2017).
13. Nagaiah, T., Bhaskar, L.: Fuzzy ideals of partially ordered Gamma near-rings, International journal scientific and innavative mathematical research (IJSIMR), **5(8)**, 8–14 (2017).
14. Nagaiah, T.: Contribution to near-ring theory and fuzzy ideals in near-rings and semirings, Doctoral thesis, Kakatiya University (2012).
15. Nobusawa, N.: On a generalization of the ring theory, Osaka J.Math., **1**, 81–89 (1964)
16. Pilz, G.: Near-rings, North-Holland mathematics studies, (1977).
17. Radhakrishna, A.: On lattice ordered near-rings and Nonassociative Rings, Doctoral thesis, Indian Istitute of Technology, (1975).
18. Rosenfeld, A., Fuzzy group, Journal of Mathematics., **35**, 512–517 (1971).
19. Satyanarayana, Bh.: Contribution to near-ring theory, Doctoral thesis, Nagarjuna University (1984).
20. Zadeh, L. A.: Fuzzy sets, Information and control., **8(3)**, 338–353 (1965).
21. Zhan. J.: On properties of fuzzy left h-ideals in hemirings with t-norms, International Journal of Mathematics and Mathematical Sciences., **19**, 3127–3144 (2005).

Novel Digital Signature Scheme with Multiple Private Keys on Non-commutative Division Semirings

G. S. G. N. Anjaneyulu and B. Davvaz

Abstract In this article, we propose a novel signature scheme connecting two private keys and two public keys generated on general non-commutative division semiring. The key notion of our technique engrosses three core steps. In the first step, we assemble polynomials on additive structure of non-commutative division semiring and execute them as underlying base work infrastructure. In the second step, we generate first set of private and public key pair using polynomial symmetrical decomposition problem. In the third step, we generate second set of private and public key pair using discrete logarithm. We use factorization theorem to generate the private key in discrete logarithm problem. By making so, we can execute a new signature scheme on multiplicative algebraic structure of the semiring using multiple private keys. The security of the designed signature scheme is depending on the intractability or hardness of the polynomial symmetrical decomposition problem and discrete logarithmic problem over the designed non-commutative division semiring. Hacking or tracking private keys should cross two mathematical hard problems. Hence, this signature scheme is much stronger than existing protocols in security point of view.

Keywords Digital signature · Factorization · Discrete logarithm problem · Symmetrical decomposition problem · Non-commutative and division semiring

2010 Mathematics Subject Classification: 16Y60, 14G50

G. S. G. N. Anjaneyulu (✉)
Department of Mathematics, VIT - Vellore Institute of Technology, Vellore, Tamil Nadu, India
e-mail: anjaneyulu.gsgn@vit.ac.in

B. Davvaz
Department of Mathematics, Yazd University, Yazd, Iran
e-mail: davvaz@yazd.ac.ir

V. Madhu et al. (eds.), *Advances in Algebra and Analysis*, Trends in Mathematics,
https://doi.org/10.1007/978-3-030-01120-8_8

1 Introduction

At present digital signatures are basically the most important and very widely used cryptographic primitive derived by public key technology, and they are building modules of several modern distributed computer and network security applications, like electronic contract authentication, authorized email and protective web browsing, etc. But mostly existing signature protocols lie in the intractability of issues closely connected to the number theory than group theory.

1.1 Background of Public Key Environment and Protocols Based on Commutative Rings

At present, there is no query that the Internet is affecting each and every aspect of human lives; the best and most significant turning points are occurring in private and public sector organizations that are transforming their conventional operating devices to Internet service-based models, known as eBusiness, eCommerce and eGovernment. Public key infrastructure (PKI) is now basically one of the most important methodologies in the arsenal of security measures that can be brought to withstand against the aforementioned increasing risks, attacks and threats. The design of reliable and secured public key infrastructure presents a compendium challenging issues that have fascinated more researchers in computer science, electrical engineering and mathematics alike for the last few decades and are necessary to continue to do so.

In their seminal path breaking paper "New directions in Cryptography" [3] Diffie and Hellman invited public key infrastructure and in particular, the concept of digital signature for auhentication. The trapdoor one-way functions play the crucial role in the concept of PKC and digital signature schemes. At present, the most successful signature schemes depend on the difficulty of certain issues in particular high cardinality of finite commutative rings. For instance, the difficulty of breaking integer factorization problem (IFP) defined over $\mathbb{Z}_n$ (where n is the product of two large primes) establishes the ground of the basic RSA signature scheme [4], the other variants of RSA and elliptic curve cryptosystem of RSA like KMOV [5]. Another known case is that the ElGamal signature scheme[6] connected with the difficulty of solving the discrete logarithm problem (DLP) designed over a finite field $\mathbb{Z}_p$ (where p is a very large prime), of course also a commutative ring [7, 8].

The theoretical developments and foundations for the above signature schemes related to the intractability of problems very closely connected to the number theory than group theory.

As mentioned in [9], in order to enlighten the cryptography, there have been numerous efforts to develop alternative PKC based on different types of problems. Historically, some efforts were made for a cryptographic primitive construction using more complex algebraic systems instead of regular and traditional finite order cyclic groups or finite fields during the previous decade. The originator in this track was [10], where a proposition to use non-commutative semigroups and groups in general key agreement protocol is depicted. Some practical key agreement protocols are using a methodology with application of the semigroup action level could be identified in [10]. Some concrete design of commutative sub-semigroup is proposed there.

In best of our knowledge, the first signature scheme proposed in an infinite non-commutative group was presented in [11]. This innovation is based on the necessary gap existing of conjugacy decision problem (CDP) with conjugator search problem (CSP) over non-commutative group [12]. In, [13], Z. Cao et al. proposed first time a new DH-like key exchange protocol and ElGamal-like cryptosystems with the polynomials over non-commutative rings.

1.2 Outline of the Paper

The rest of the paper is demonstrated as follows. In Sect. 2, the necessary cryptographic assumptions over non-commutative groups are given. In Sect. 3, first we are giving again polynomial over an arbitrary non-commutative ring and present necessary assumptions over non-commutative division semiring, for smooth understanding of the paper. These things can be seen in [1, 2] in detail. In Sect. 4, we depict and analyse new digital signature scheme on non-commutative division semiring and assumptions. In Sect. 5, we examine the confirmation theorem and security issues of the proposed signature scheme.

2 Assumptions of Cryptography on Non-commutative Groups

2.1 Two Well-Known Assumptions on Cryptography

In this section, we present the necessary intractable problems related to our proposed signature. In any non-commutative group G, two members x, y are conjugate, symbolically $x \sim y$, if $y = z^{-1}xz$ for some $z \in G$. Here, z or z^{-1} is known as conjugator. In a non-commutative group G, the following two cryptographic problems which are related to conjugacy are defined.

- *Conjugator Search Problem (CSP):* Given $(x, y) \in G \times G$, find $z \in G$ such that $y = z^{-1}xz$.
- *Decomposition Problem (DP):* Given $(x, y) \in G \times G$ and $S \subseteq G$, find $z_1, z_2 \in S$ such that $y = z_1xz_2$.

At present, in a general non-commutative group G, both of the above problems CSP and DP are intractable.

2.2 Symmetrical Decomposition and Computational Diffie–Hellman Assumptions over Non-commutative Groups

Enlightened by the above problems, Cao [13] explained the following cryptographic problems over a non-commutative group G.

- *Symmetrical Decomposition Problem (SDP):* Given $(x, y) \in G \times G$ and $m, n \in \mathbb{Z}$, the set of integers, find $z \in G$ such that $y = z^m x z^n$.
- *Generalized symmetrical decomposition problems (GSDP):* Given $(x, y) \in G \times G$, $S \subseteq G$ and $m, n \in \mathbb{Z}$, find $z \in S$ such that $y = z^m x z^n$.
- *Computational Diffie–Hellman (CDH) problem over non-commutative group G:* Compute $x^{z_1 z_2}$ or $x^{z_2 z_1}$ for given x, x^{z_1} and x^{z_2}, where $x \in G$, $z_1, z_2 \in S$, for $S \subseteq G$.

At present, we have no logic to solve this kind of CDH problem without extracting z_1 or z_2 from x and x^{z_1} (or x^{z_2}). Then, the CDH assumption over G says that CDH problem over G is not tractable.

3 Building Levels for Proposed Digital Signature Scheme

3.1 Integral Coefficient Ring Polynomials

Let R be a ring with $(R, +, 0)$ and $(R, \cdot, 1)$ as its additive abelian group and multiple non-abelian semigroup, respectively. Then the positive integral coefficient ring polynomials are defined as follows. Let $f(x) = a_0 + a_1 x + \ldots + a_n x^n$ be a given positive integral coefficient polynomial. We generate this polynomial by using an element r in R and finally obtain $f(r) = \sum_{i=0}^{n} a_i r^i = a_0 + a_1 r + \ldots + a_n r^n$. which is also a member in R (details can be seen in Section 3.4).

Further, if we define r as a variable in R, then $f(r)$ can be treated as polynomial about r. The set of all this kind of polynomials, taking over all $f(x)\mathbb{Z}_{>0}[x]$, can be regarded as the extension of $Z_{>0}$ with r, denoted by $\mathbb{Z}_{>0}[r]$. We say that it is the set of 1-ary positive integral coefficient R-polynomials.

3.2 Semiring

A Semiring R is a non-empty set, in which the operations of addition and multiplication have been assigned such that the following conditions are true.

1. $(R, +)$ is a commutative monoid with 0 as identity element.
2. $(R, \cdot)$ is a monoid with 1 as identity element.
3. Multiplication operation distributes over addition from either side.
4. $0 \cdot r = r \cdot 0$, for every $r \in R$.

3.3 Division Semiring

An element r of a semiring R is called a *unit* if and only if there exists an element r' of R satisfying $r \cdot r' = 1 = r' \cdot r$.

The element r' is called as the *inverse* of r in R. If such an inverse r' exists for a unit r, that must be unique. We will normally represent the inverse of r by r^{-1}. It is straightforward to note that, if r, r' are units of R, then $r \cdot (r')^{-1} = (r')^{-1} \cdot r^{-1}$. In particular, $(r^{-1})^{-1} = r$.

We will represent the set of all units of R, by $U(R)$. This set is always non-empty, since it consists 1 and is not all of R, since it does not contain 0.

It is evident that $U(R)$ is a submonoid of $(R, \cdot)$, which is in fact also a group. If $U(R) = R \setminus \{0\}$, then R is known as a *division semiring*.

3.4 Polynomials on Division Semiring

Let $(R, +, \cdot)$ be a non-commutative division semiring. Let us consider positive integer coefficient polynomials with semiring structure as follows. At first, the operation of multiplication over R is already on hand. For $k \in \mathbb{Z}_{>0}$ and $r \in R$, then $(k)r = \underbrace{r + \ldots + r}_{k\ times}$. For $k = 0$, it is very clear to define $(k)r = 0$.

Property 1 For all $a, b, m, n \in \mathbb{Z}$ and $r \in R$, we have

$$(a)r^m \cdot (b)r^n = (ab) \cdot r^{m+n} = (b)r^n \cdot (a)r^m,$$

Remark 1 Note that in general, $(a)r \cdot (b)s \neq (b)s \cdot (a)r$ when $r \neq s$, since the reason multiplication in R is non-commutative.

Now, positive integral coefficient semiring polynomials are defined as follows. Suppose that $f(x) = a_0 + a_1x + \ldots + a_nx^n \in \mathbb{Z}_{>0}[x]$ and $h(x) = b_0 + b_1x + \ldots + b_mx^m \in \mathbb{Z}_{>0}[x]$, where $n \geq m$, are given positive integral coefficient polynomials. We generate this polynomial by using an element r in R, and finally, we obtain

$f(r) = a_0+a_1r+\ldots+a_nr^n \in R$. Similarly, we have $h(r) = b_0+b_1r+\ldots+b_mr^m \in R$. Then, we have the following.

Proposition 1 *We have* $f(r) \cdot h(r) = h(r) \cdot f(r)$, *for* $f(r), h(r) \in R$.

Remark 2 If r and s are two distinct variables in R, then $f(r) \cdot h(s) \neq h(s) \cdot f(r)$, in generic way.

3.5 Further Assumptions of Cryptography on Non-commutative Division Semirings

Suppose that $(R, +, \cdot)$ be a non-commutative division semiring. For any $a \in R$, a set is defined as follows: $P_a \subseteq R$ by $P_a = \{f(a) \mid f(x) \in \mathbb{Z}_{>0}[x]\}$. Then, let us agree the new variants of GSD and CDH problems over (R,?.) with respect to its subset P_a and call them as polynomial symmetrical decomposition (PSD) problem and polynomial Diffie–Hellman (PDH) problem, respectively:

- *Polynomial symmetrical decomposition(PSD) problem over non-commutative division semiring R:* Given $(a, x, y) \in R^3$ and $m, n \in \mathbb{Z}$, find $z \in P_a$ such that $y = z^m x z^n$.
- *Polynomial Diffie–Hellman (PDH) problem over non-commutative division semiring R:* Compute $x^{z_1z_2}$ (or $x^{z_2z_1}$) for given x, x^{z_1} and x^{z_2}, where $x \in R$ and $z_1, z_2 \in P_a$.

Accordingly, the PSD (PDH) cryptographic assumption states that PSD (PDH) problem over $(R, \cdot)$ is intractable, i.e. there does not exist probabilistic polynomial time algorithm which can track PSD (PDH) problem over $(R, \cdot)$.

4 Proposed Signature Scheme

4.1 Signature Scheme on Non-commutative Division Semiring

This digital signature scheme includes the following main steps:

- *Initial setup:* Let $(S, +, \cdot)$ be the non-commutative division semiring and is the essential work fundamental infrastructure in which PSD and conjugacy problem are not tractable on the non-commutative group $(S, \cdot)$. Choose two small integers $m, n \in \mathbb{Z}$. Let $H : S \rightarrow S$ be a cryptographic hash function that maps S to the message space S. Choose $0 \neq m, n \in \mathbb{Z}$. Then, the public parameters of the signature would be the tuple $< S, m, n, S, H >$.

  Remark 3 In this case, we must choose message space is also in S.

- *Key generation:* Alice requires to sign and send a message M to Bob for verification. First Alice chooses two random elements $p, q \in S$ and a polynomial $f(x) \in \mathbb{Z}_{>0}[x]$ randomly such that $0 \neq f(p) \in S$ and then she keeps $f(p)$ as her private key, calculates $g = f(p)^m q f(p)^n$ and announces this as her public key.

  Let k be the product of two very large secure primes a, b. Its security is based on the difficulty of factoring k, such that $1 < e < \phi(k) = (a-1)(b-1)$ and $gcd(e, \phi(k)) = 1$. Since $(a-1)(b-1)$ is even, it follows that e is always odd. So, we can compute second private key d with $1 < e < \phi(k) = (a-1)(b-1)$ and $de \equiv 1(modphi(k))$. Then, we calculate second public key by discrete logarithm $y = g^d$. So, that the private and public key pairs are $(f(p), d)$ and (g, y, e).
- *Signature generation:* Alice performs the following simultaneously by taking a message M from message space.

  Alice chooses randomly another polynomial $h(x) \in \mathbb{Z}_{>0}[x]$ such that $h(p) \in S$. Then, she defines $h(p)$ as salt and computes $u = h(p)^{-m} q h(p)^{-n}$ and $r = f(p)^m \cdot H(M + du) \cdot f(p)^n, \alpha = f(p)h(p), s = f(p)^{-n}[(H(M+du))^{-1} \cdot q] \cdot f(p)^n v = \alpha^m \cdot u \cdot \alpha^n$.

  Then, (v, r, s) is the signature of Alice on the message M and sends it to Bob for acceptance, and it needs verification.
- *Verification:* On receiving the signature (v, r, s) from Alice, Bob will perform the following. For this, he generates $z = r \cdot s$ and $w = y^e$.

  Bob accepts Alice's signature if and only if $g^{-1}v = wz^{-1}$; otherwise, he rejects the signature.

5 Confirmation Theorem

5.1 Completeness

Let (p, q, g, y, e) be the public parameters for $p, q, g, y \in S$. Given a signature (v, r, s), if Alice agrees signature verification algorithm, then Bob always receives (v, r, s) as a valid signature.

In verification, the parameters are v, r, s, z and w. Then, $g^{-1}v = wz^{-1} \rightarrow v \cdot z = g \cdot w$. Now $LHS = v \cdot z = v \cdot r \cdot s = (\alpha^m \cdot u \cdot \alpha^n) \cdot r \cdot s$ on simplification, we obtain $vz = [f(p)^m q f(p)^n] \cdot [f(p)^m q f(p)^n] = [f(p)^m q f(p)^n] \cdot [f(p)^m q f(p)^n]^{ed} = g.(g^d)^e = gw$, which is RHS, as the reason $de \equiv 1(mod\phi(k))$.

5.2 Security Analysis

Assume that an active eavesdropper "Eve" can retrieve, delete, forge and retransmit any message, Alice sends to Bob. Any forgered data d, we symbolize it by d_f.

We investigate the security of the signature scheme for three main attacks and that are forgering data on valid signature and signature repudiation by valid data, existential forgering.

- *Data forgering*. Suppose Eve swaps the original message M, with forgered one M_f. Then, Bob receives the signature $(u, s, \alpha, \beta, v_1)$. Using forgered data Mf or $H(Mf)$, verifying the equation $u^{-1} \cdot v_1 = s^{-1} \cdot v_2$ is impossible, because Mf or $H(Mf)$ is completely involved in the signature generation, but not in the verification algorithm.

  Hence, $u^{-1} \cdot v_1 = s^{-1} \cdot v_2$ is true only for the original message. Data forgery without extracting signature is not possible.

  Another attempt is to find Mf, for valid $H(M)$. But this is impossible, because we know that hash function H is cryptographically secure. So, the invalid data can't be verified with a valid signature.
- *Signature repudiation.* Assume Alice wants to refuse recognition of his signature on some valid data. Then, it agrees that valid signature $(u, s, \alpha, \beta, v_1)$ can be forged by Eve and she will sign the message M, with the forgered signature $(u_f, s_f, \alpha_f, \beta_f, v_{1f})$ instead. The verification procedure is as follows:

  $V_2 = \alpha_f \cdot y^{-1} \cdot \beta_f = [h(p)^m \cdot rf(p)^n]_f [f(p)^{-n} \cdot q^{-1} \cdot f(p)^{-m}][f(p)^m H(M) \cdot h(p)^n]_f$. Since $[f(p)^n]_f \cdot [f(p)^n] \neq 1, [f(p)^{-m}] \cdot [f(p)^m]_f \neq 1$, where 1 is the identity element in the multiplicative structure of the defined division semiring. Consequently, $[u^{-1} \cdot v_1]_f \neq [s^{-1} \cdot v_2]_f$. So, this signature scheme assures that the non-repudiation property.
- *Existential forgery.* Suppose Eve is trying to sign a forgered message M_f. Then, she must forge the private key by replacing with some $[f(p)]_f$. Immediately, she faces a difficult with the public key, as we believe that PSD is not tractable on non-commutative division semiring. Also, note that all the structures in this signature scheme are constructed on non-commutative division semiring and based on PSD. Exact identification these structures are almost intractable as long as PSD is so hard on this underlying work structure. Consequently construction new valid signatures, without proper knowledge of private key are impossible. So, Eve is not able to compute forgered signatures.

5.3 Soundness

The key notion is that selecting a polynomial $f(x)$ randomly, with semiring assignment and for any $p \in S$, such that $0 \neq f(p) \in S$. A cheating prover P^* has no way to identify the polynomial $f(x) \in \mathbb{Z}_{>0}[x]$ such that $0 \neq f(p) \in S$, even if he has huge computational power. Let n be the number of members of S, P^* best strategy is to guess the value of p, and there are n choices for p. Hence, even with potential computing power, the cheating prover P^* with a negligible probability to extract the exact private key $f(p) \in S$, so as to provide a valid response for an invalid signature. Hence, this signature scheme is clearly sound.

6 Conclusion

In this paper, we depicted a new signature scheme on general non-commutative division semiring. The key notion behind our signature lies that we take polynomials over the given non-commutative algebraic system as the essential work structure for constructing signature scheme. The security and strength of the proposed scheme depend on the intractability of polynomial symmetrical decomposition problem over the given non-commutative division semirings.

References

1. Anjaneyulu, G.S.G.N., Vasudeva Reddy, P., Reddy, U.M., Secured digital signature scheme using polynomials over non-commutative division semirings. International Journal of Computer Science and Network Security. **8(8)**, 278–284 (2008).
2. Anjaneyulu, G.S.G.N., Venkateswarlu, B., Reddy, U.M., Diffie-Hellman-Like key agreement protocol using polynomials over non-commutative division semirings, International Journal of Computer Information Systems, **5(1)**, 37–41 (2012).
3. Diffie, W., Hellman, M.E., *New directions in cryptography*, IEEE Transaction on Information Theory, 22 (1976) 644–654.
4. Rivest, R.L., Shamir, A., and Adleman, L., *A method for obtaining digital signatures and public key cryptosystems*, Communications of the ACM, 27 (1978) 120–126.
5. Komaya,K., Maurer,V., Okamoto, T., Vanstone,S. *New PKC based on elliptic curves over the ring* $\mathbb{Z}_n$, LNCS 516, PP. 252–266, Springer-verlag 1992.
6. Elgamal,T., *A public key cryptosystem and a signature scheme based on discrete logarithms*, IEEE Transactions on Information Theory, 31 (1985)469–472.
7. Maglivers,S.S., Stinson, D.R., Van Trungn, T. *New approaches to designing public key cryptosystems using one-way functions and trapdoors in finite groups*,, Journal of Cryptology, 15 (2002) 285–297.
8. Shor, P. *Polynomial-time algorithms for prime factorization and discrete logarithms on a quantum computer*, SIAM J. Computing, 5 (1997) 31484–1509.
9. Lee, E. *Braid groups in cryptography*, IEICE Trans. Fundamentals, E87-A (5) (2004) 986–992.
10. Sidelnikov,V., Cherepnev, M., Yaschenko, V.,*Systems of open distribution of keys on the basis of non-commutation semigroups*, Russian Acad. Sci. Dok L. Math., 48(2) (1993) 566–567.
11. Ko K.H.,Choi,D.H.,Cho,M.S., Lee J.W *New signature scheme using conjugacy problem*, Cryptology e print Archive: Report 2002/168, 2002.
12. Ko,K.H.,Lee,J.S.,Cheon J.H.,Han,Kang, J.S., Park C., *New public-key cryptosystem using Braid Groups*, Advances in cryptology, Proc. CRYPTO 2000. LNCS 1880, PP. 166–183, Springer-verlag, 2000.
13. Cao, Z. Dong, X., Wang, L., *New public key cryptosystems using polynomials over non-commutative rings*, Cryptography e-print archive, http://eprint.iacr.org/2007/

Cozero Divisor Graph of a Commutative Rough Semiring

B. Praba, A. Manimaran, V. M. Chandrasekaran, and B. Davvaz

Abstract In this paper, we define the ideal generated by an element in the commutative rough semiring (T, Δ, ∇). The characterization of this ideal along with its properties are also studied. The cozero divisor graph of a commutative rough semiring is defined using this ideal. These concepts are illustrated through examples.

Keywords Semiring · Ideal · Principal ideal · Cozero divisor

1 Introduction

Fundamentals of semigroups were discussed by Howie [4] in his classical book in 2003. In 1982 Pawlak [10] defined the concept of rough set as pair of sets called lower and upper approximation. The authors Bonikowaski [2] and Iwinski [5] in 1994, Kondo [6] in 2006, Kuroki [7] in 1997, Chinram [15] in 2009, and Liu [16] in 2011 described some structures of algebra on rough sets. Zadeh [17] initiated the idea of fuzzy sets in his paper.

Hong et al. [3] dealt with some resultants over commutative idempotent semirings in 2017. Praba et al. [11–14] described the set of all rough sets $T = \{RS(X) \mid X \subseteq U\}$ as a rough semiring on the given information system $I = (U, A)$ by defining the two new operations Praba Δ and Praba ∇ on T. Also, we established zero divisor graph structure of the rough semiring (T, Δ, ∇). In 2017, Manimaran

B. Praba
SSN College of Engineering, Chennai, Tamil Nadu, India
e-mail: prabab@ssn.edu.in

A. Manimaran (✉) · V. M. Chandrasekaran
School of Advanced Sciences, VIT, Vellore, India
e-mail: marans2011@gmail.com; vmcsn@yahoo.com

B. Davvaz
Department of Mathematics, Yazd University, Yazd, Iran
e-mail: davvaz@yazd.ac.ir

V. Madhu et al. (eds.), *Advances in Algebra and Analysis*, Trends in Mathematics,
https://doi.org/10.1007/978-3-030-01120-8_9

et al.[8, 9] discussed about characterization of rough semirings and explored the rough homomorphism between them. In 2012, Afkhami and Khashyarmanesh [1] introduced cozero divisor graph structure in a commutative ring. In this view, we describe the concept of cozero divisor graph of the rough semiring (T, Δ, ∇) more elaborately.

This paper is organized as follows. We give some basic concepts of rough sets and algebraic structures in Sect. 2. In Sect. 3, we define the ideal generated by an element in the commutative rough semiring (T, Δ, ∇) and cozero divisor graph of (T, Δ, ∇), and it is illustrated through examples. In Sect. 4, we give the conclusion.

2 Preliminaries

In this section we give some basic definitions in rough sets and algebraic structures.

2.1 Rough Sets

An information system is a pair $I = (U, A)$ where U is a nonempty finite set of objects, called universal set, and A is a nonempty set of fuzzy attributes defined by $\mu_a : U \rightarrow [0, 1]$, $a \in A$, is a fuzzy set. *Indiscernibility* is a core concept of rough set theory, and it is defined as an equivalence between objects. Formally any set $P \subseteq A$, there is an associated equivalence relation called $P - Indiscernibility$ relation defined as follows:

$$IND(P) = \{(x, y) \in U^2 \mid \forall a \in P, \mu_a(x) = \mu_a(y)\}.$$

The partition induced by $IND(P)$ consists of equivalence classes defined by

$$[x]_p = \{y \in U \mid (x, y) \in IND(P)\}.$$

Definition 1 (Rough Set) For any arbitrary subset X of U, $RS(X) = (\underline{P}(X), \overline{P}(X))$ is said to be a rough set of X where $\underline{P}(X) = \{x \in U \mid [x]_p \subseteq X\}$ is the lower approximation space, and $\overline{P}(X) = \{x \in U \mid [x]_p \cap X \neq \phi\}$ is the upper approximation space.

Example 1 ([11])

Let $X = \{x_1, x_3, x_5, x_6\}$ and $P = A$. Then the equivalence classes induced by $IND(P)$ are given below.

$$X_1 = [x_1]_p = \{x_1, x_3\} \quad (1)$$

$$X_2 = [x_2]_p = \{x_2, x_4, x_6\} \quad (2)$$

$$X_3 = [x_5]_p = \{x_5\} \quad (3)$$

Hence $\underline{P}(X) = \{x_1, x_3, x_5\}$ and $\overline{P}(X) = \{x_1, x_2, x_3, x_4, x_5, x_6\}$. Therefore, we have $RS(X) = (\{x_1, x_3, x_5\}, \{x_1, x_2, x_3, x_4, x_5, x_6\})$.

Definition 2 ([11]) The number of equivalence classes (Induced by IND(P)) contained in X is called as the Ind. weight of X where $X \subseteq U$. It is denoted by $IW(X)$.

Example 2 ([11]) Let $U = \{x_1, x_2, \cdots, x_6\}$ as in Table 1. The equivalence classes induced by $IND(P)$ are

$$[x_1]_p = \{x_1, x_3\}$$

$$[x_2]_p = \{x_2, x_4, x_6\}$$

$$[x_5]_p = \{x_5\}$$

Let $X = \{x_1, x_4, x_5\} \subseteq U$ then by definition, Ind. weight of $X = IW(X) = 1$ (since there is only one equivalence class $[x_5]_p = \{x_5\}$ present in X).

Definition 3 ([11]) Let $X, Y \subseteq U$. The Praba Δ is defined as $X\Delta Y = X \cup Y$, if $IW(X \cup Y) = IW(X) + IW(Y) - IW(X \cap Y)$.

If $IW(X \cup Y) > IW(X) + IW(Y) - IW(X \cap Y)$, then identify the new equivalence class formed by $X \cup Y$. Then delete the elements of that class belonging to Y. Call the new set as Y. Presently, get $X\Delta Y$. Repeat this process until $IW(X \cup Y) = IW(X) + IW(Y) - IW(X \cap Y)$.

Example 3 ([11]) Let $U = \{x_1, x_2, \cdots, x_6\}$ as in Table 1.

Let $X = \{x_2, x_4, x_5\}$, $Y = \{x_1, x_6\} \subseteq U$ then by definition,

$$IW(X) = 1;\ IW(Y) = 0;\ IW(X \cup Y) = 2;\ IW(X \cap Y) = 0$$

Here,

$$IW(X \cup Y) > IW(X) + IW(Y) - IW(X \cap Y).$$

The new equivalence class formed in $X \cup Y$ is $[x_2]_p$. As $x_6 \in Y$ and x_6 is an element of $[x_2]_p$, by deleting x_6 from Y, we have Y as $\{x_1\}$. Now for $X = \{x_2, x_5, x_6\}$ and

Table 1 Information system

A/U	a_1	a_2	a_3	a_4
x_1	0	0.1	0.3	0.2
x_2	1	0.6	0.7	0.3
x_3	0	0.1	0.3	0.2
x_4	1	0.6	0.7	0.3
x_5	0.8	0.5	0.2	0.4
x_6	1	0.6	0.7	0.3

$Y = \{x_1\}$, we observe that $IW(X \cup Y) = IW(X) + IW(Y) - IW(X \cap Y)$. Therefore, $X \Delta Y = X \cup Y = \{x_1, x_2, x_4, x_5\}$.

Definition 4 ([11]) An element $x \in U$ is called as a pivot element, if $[x]_p \not\subseteq X \cap Y$, but $[x]_p \cap X \neq \phi$ and $[x]_p \cap Y \neq \phi$ where $X, Y \subseteq U$.

Definition 5 ([11]) The set of pivot elements of X and Y is called the pivot set of X and Y and it is denoted by $P_{X \cap Y}$ where $X, Y \subseteq U$.

Definition 6 ([11]) Praba ∇ of X *and* Y is denoted by $X \nabla Y$, and it is defined as

$$X \nabla Y = \{x \mid [x]_p \subseteq X \cap Y\} \cup P_{X \cap Y} \text{ where } X, Y \subseteq U.$$

Note that each pivot element in $P_{X \cap Y}$ is the representative of that particular class.

Example 4 ([11]) Let $U = \{x_1, x_2, \cdots, x_6\}$ as in Table 1 and let $X = \{x_1, x_2, x_4, x_5\}$ and $Y = \{x_3, x_5, x_6\} \subseteq U$ then $X \cap Y = \{x_5\}$. Here, $[x_1]_p \not\subseteq X \cap Y$, but $[x_1]_p \cap X \neq \phi$ and $[x_1]_p \cap Y \neq \phi$. Therefore x_1 is a pivot element

Similarly x_2 is a pivot element. Also pivot set $P_{X \cap Y} = \{x_1, x_2\}$. Therefore $X \cap Y = \{x_1, x_2, x_5\}$. Similarly $Y \nabla X = \{x_3, x_5, x_6\}$. Therefore, we have $X \nabla Y \neq Y \nabla X$ and $RS(X \nabla Y) = ([x_5]_p, [x_1]_p \cup [x_2]_p \cup [x_5]_p)$ and $RS(Y \nabla X) = ([x_5]_p, [x_1]_p \cup [x_2]_p \cup [x_5]_p)$.

Thus, $RS(X \nabla Y) = RS(Y \nabla X)$.

Definition 7 (Binary Operation as Δ [12]) The binary operation $\Delta : T \times T \to T$ is defined as $\Delta(RS(X), RS(Y)) = RS(X \Delta Y)$ where T is the set of all rough sets.

Theorem 1 ([12]) *Let $I = (U, A)$ be an information system where U be the universal (finite) set and A be the set of attributes and T be the set of all rough sets then (T, Δ) is a commutative monoid of idempotents.*

Theorem 2 ([12]) *(T, Δ) is a regular rough monoid of idempotents.*

Definition 8 (Binary Operation as ∇ [8]) The binary operation $\nabla : T \times T \to T$ is defined as $\nabla(RS(X), RS(Y)) = RS(X \nabla Y)$ where T is the set of all rough sets.

Theorem 3 ([8]) *(T, ∇) is a commutative regular rough ∇ monoid of idempotents.*

Theorem 4 ([13]) *(T, Δ, ∇) is a rough semiring.*

Theorem 5 ([13]) *The pivot rough set is an ideal of the semiring (T, Δ, ∇).*

Theorem 6 ([14]) *If a subset X of U is not dominant, then $RS(X)$ is a zero divisor of the rough semiring (T, Δ, ∇).*

Theorem 7 *[14] Let (T, Δ, ∇) be a semiring. If a subset X of U is dominant then $RS(X)$ is not a zero divisor in T.*

2.2 *Algebraic Structures*

Definition 9 (Semiring) A nonempty set S together with the binary operations "+" and "." satisfies the following conditions:

(i) $(S, +)$ and $(S, .)$ are semigroups.
(ii) $p(q+r) = pq + pr$ and $(p+q)r = pr + qr$ for any $p, q, r \in S$.

Definition 10 (Ideal of a Semiring) A nonempty set $I \subseteq S$ is said to be an ideal of a semiring S if it satisfies

(i) $p + q \in I$ for any $p, q \in I$ and
(ii) $ap \in I$ *and* $pa \in I$ for any $p \in I$ and $a \in S$

In the following section, we discuss the ideal generated by an element and cozero divisor graph of a commutative rough semiring (T, Δ, ∇).

3 Ideal and Cozero Divisor Graph of a Commutative Rough Semiring

In this section, we consider the set of all rough sets $T = \{RS(X)|X \subseteq U\}$ and let $E = \{X_1, X_2, \ldots X_n\}$ be the set of equivalence classes induced by $IND(P)$.

Theorem 8 *$RS(X)\nabla T$ is an ideal in the rough semiring (T, Δ, ∇).*

Proof For any subset X of U, $RS(X)\nabla T = \{RS(X)\nabla RS(Y) \mid RS(Y) \in T\}$

(i) For $RS(X)\nabla RS(Y_1)$ and $RS(X)\nabla RS(Y_2) \in RS(X)\nabla T$,

$$\begin{aligned}(RS(X)\nabla RS(Y_1))\Delta(RS(X)\nabla RS(Y_2)) &= RS(X\nabla Y_1)\Delta RS(X\nabla Y_2)\\ &= RS(X\nabla(Y_1\Delta Y_2))\\ &= RS(X)\nabla RS(Y_1\Delta Y_2) \in RS(X)\nabla T.\end{aligned}$$

(ii) Let $RS(Y_2) \in T$ and for $RS(X)\nabla RS(Y_1) \in RS(X)\nabla T$,

$$\begin{aligned}RS(X)\nabla RS(Y_1)\nabla RS(Y_2) &= RS(X\nabla(Y_1\nabla Y_2))\\ &= RS(X)\nabla RS(Y_1\nabla Y_2) \in RS(X)\nabla T.\end{aligned}$$

Therefore $RS(X)\nabla T$ is an ideal in the rough semiring T. We call this ideal as a principal rough ideal generated by $RS(X)$.

In the following theorem, we give the characterization of the elements of this ideal $RS(X)\nabla T$. Let E_X be the set of equivalence classes contained in X, P_X be the set of pivot elements of X and $Z_X = \{x \in U \mid [x]_p \cap X \neq \phi\}$. That is, Z_X contains the pivot element of each equivalence class having nonempty intersection with X.

Theorem 9 (Characterization Theorem) *For any arbitrary subset X of U, the principal rough ideal generated by $RS(X)$ in T is given by* $RS(X)\nabla T = \{RS(Y) \mid Y \in (\mathscr{P}(E_X) \cup \mathscr{P}(Z_X))\}$.

Proof For $RS(Y) \in RHS$ and $Y = Z_1 \cup Z_2$ where $Z_1 \in \mathscr{P}(E_X)$ and $Z_2 \in \mathscr{P}(Z_X)$ then $RS(Y) = RS(Z_1 \cup Z_2)$ and $RS(X\nabla(Z_1 \cup Z_2)) = RS(X)\nabla RS(Z_1 \cup Z_2) = RS(Z_1 \cup Z_2) = RS(Y) \in LHS$. On the other hand, let $RS(X\nabla Y) \in LHS$, where $RS(Y) \in T$. If $X \cap Y = \phi$ then $RS(X\nabla Y) = RS(\phi) \in RHS$. If $X \cap Y$ contains one or more equivalence classes then $RS(X\nabla Y) \in RHS$ as $X\nabla Y \in \mathscr{P}(E_X)$. If $X \cap Y$ contains one or more pivot elements then $RS(X\nabla Y) \in RHS$ as $X\nabla Y \in \mathscr{P}(Z_X)$. If $X \cap Y$ contains one or more equivalence classes and one or more pivot elements then $RS(X\nabla Y) \in RHS$ as $X\nabla Y \in \mathscr{P}(E_X) \cup \mathscr{P}(Z_X)$. This proves the theorem.

3.1 Properties

1. If x, y belongs to the same equivalence class then $RS(x)\nabla T = RS(y)\nabla T$
2. If they do not belong to the same equivalence class, then $RS(x)\nabla T \neq RS(y)\nabla T$
3. $RS(U) \notin RS(X)\nabla T$ for $X \neq U$ and $RS(\phi) \in RS(X)\nabla T$ for $X \neq \phi$
4. If $X \subseteq Y$ then $RS(X)\nabla T \subseteq RS(Y)\nabla T$
5. If X_i and X_j are two equivalence classes in U such that $|X_i|$ *and* $|X_j| > 1$ and if $x_i \in X_i$ and $x_j \in X_j$ then $|RS(x_i)\nabla T| = |RS(x_j)\nabla T|$

3.2 Cozero Divisor Graph of a Commutative Rough Semiring

Definition 11 (Cozero Divisor Graph) Cozero divisor graph of a commutative rough semiring is $\Gamma'(T) = (V, E)$ where V is the set of vertices consisting of the elements $T^* = T \setminus \{RS(\phi), RS(U)\}$ and two elements $RS(X)$ and $RS(Y)$ in T^* are adjacent if and only if $RS(X) \notin RS(Y)\nabla T$ and $RS(Y) \notin RS(X)\nabla T$.

Theorem 10 *An element $RS(Y) \in T$ is adjacent to $RS(X) \in T$ if and only if $Y \in \mathscr{P}(U \setminus \overline{P}(X) \setminus \phi)$ where $\mathscr{P}$ denotes the power set of $U \setminus \overline{P}(X) \setminus \phi$.*

Proof Let $Y \in \mathscr{P}(U \setminus \overline{P}(X) \setminus \phi)$ iff Y does not contain any of the equivalence class in X and also Y does not contain any of the equivalence class having a nonempty intersection with X iff $RS(Y) \notin RS(X)\nabla T$ and $RS(X) \notin RS(Y)\nabla T$. Hence, $RS(X)$ and $RS(Y)$ are adjacent.

Table 2 Information system

A/U	a_1	a_2	a_3	a_4
x_1	0.2	0.3	1	0
x_2	0.8	0.4	0.1	0.9
x_3	0.2	0.3	1	0
x_4	0.8	0.4	0.1	0.9

3.3 Examples

Example 5 Let $I = (U, A)$ be an information system where $U = \{x_1, x_2, x_3, x_4\}$ and $A = \{a_1, a_2, a_3, a_4\}$ where each $a_i\,(i = 1 \text{ to } 4)$ is a fuzzy set of attributes whose membership values are shown in Table 2. Let $X = \{x_1, x_2, x_3, x_4\} \subseteq U$, then the equivalence classes induced by $IND(P)$ are given below:

$$X_1 = [x_1]_p = \{x_1, x_3\} \tag{4}$$

$$X_2 = [x_2]_p = \{x_2, x_4\} \tag{5}$$

$$T = \{RS(\phi), RS(U), RS(X_1), RS(X_2), RS(\{x_1\}), RS(\{x_2\}), RS(X_1 \cup \{x_1\}), RS(\{x_1\} \cup X_2), RS(\{x_1\} \cup \{x_2\})\}$$

Example 6 Let $X = \{x_1, x_2, x_3\}$ be an arbitrary subset of the finite universal set U. From Example (5), $E_X = \{X_1\}$ and $Z_X = \{x_1, x_2\}$ = pivot element of each equivalence class having nonempty intersection with X. Then,

$$\mathscr{P}(E_X) = \{\phi, X_1\} \text{ and } \mathscr{P}(Z_X) = \{\phi, \{x_1\}, \{x_2\}, \{x_1, x_2\}\}.$$

Now, $\{RS(Y) \mid Y \in \mathscr{P}(E_X) \cup \mathscr{P}(Z_X)\} = \{RS(\phi), RS(\{x_1\}), RS(\{x_2\}), RS(X_1), RS(X_1 \cup \{x_2\})\}$ and $RS(X)\nabla T = \{RS(\phi), RS(\{x_1\}), RS(\{x_2\}), RS(X_1), RS(X_1 \cup \{x_2\})\}$.

Therefore,

$$RS(X)\nabla T = \{RS(Y) \mid Y \in \mathscr{P}(E_X) \cup \mathscr{P}(Z_X)\}.$$

Example 7 From Example (5),

1. The principal rough ideal generated by $RS(X_1)$ is $RS(\phi)$, $RS(\{x_1\})$, and $RS(X_1)$.
2. The principal rough ideal generated by $RS(\{x_1\})$ is $RS(\phi)$ and $RS(\{x_1\})$.
3. The principal rough ideal generated by $RS(U)$ is $RS(\phi)$, $RS(U)$, $RS(X_1)$, $RS(X_2)$, $RS(\{x_1\})$, $RS(\{x_2\})$, $RS(X_1 \cup \{x_2\})$, $RS(\{x_1\} \cup X_2)$, and $RS(\{x_1\} \cup \{x_2\})$.

Example 8 From Eq. (5), we have the vertex set of the cozero divisor graph $\Gamma'(T)$ as $V = \{RS(X_1), RS(X_2), RS(\{x_1\}), RS(\{x_2\}), RS(X_1 \cup \{x_1\}), RS(\{x_1\} \cup X_2), RS(\{x_1\} \cup \{x_2\})\}$, and then the cozero divisor graph of a rough semiring T is given below (Fig. 1).

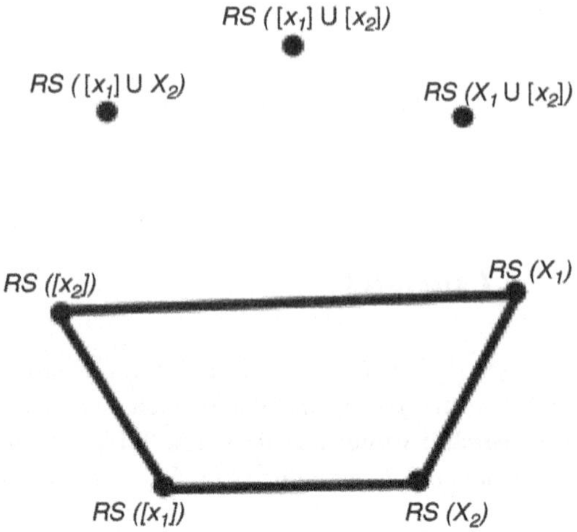

Fig. 1 Cozero divisor graph of a rough semiring T

4 Conclusion

In this paper, we defined the ideal generated by an element in the commutative rough semiring (T, Δ, ∇), and we discussed some characterization of this ideal along with its properties. Using this ideal, we defined the cozero divisor graph of the commutative rough semiring. These derived concepts are illustrated through examples. Future work in this direction is to study the properties of this cozero divisor graph.

References

1. Afkhami, M., Khashyarmanesh, K.: On the Cozero-Divisor Graphs of Commutative Rings and Their Complements. Bull. Malays. Math. Sci. Soc. **(2) 35 (4)**, 935–944 (2012).
2. Bonikowaski, Z.: Algebraic structures of rough sets, Rough sets, fuzzy sets and knowledge discovery. Springer, London, 242-247 (1994).
3. Hong, H., Kim, Y., Scholten, G., Sendra, J. R.: Resultants Over Commutative Idempotent Semirings I : Algebraic Aspect. Journal of Symbolic Computation. **79 (2)**, 285–308 (2017).
4. Howie, J.M.: Fundamentals of Semigroup Theory. Oxford University Press. New York, (2003).
5. Iwinski, T. B.: Algebraic approach to Rough Sets. Bulletin of the Polish Academy of Sciences Mathematics. **35**, 673–683 (1987).
6. Kondo, M.: On The Structure of Generalized Rough sets. Information Sciences. **176**, 586–600 (2006).
7. Kuroki, N.: Rough Ideals in semigroups. Information Sciences. **100**, 139–163 (1997).
8. Manimaran, A., Praba, B., Chandrasekaran, V. M.: Regular Rough ∇ Monoid of idempotents. International Journal of Applied Engineering and Research. **9(16)**, 3469–3479 (2014).
9. Manimaran, A., Praba, B., Chandrasekaran, V. M.: Characterization of rough semiring. Afrika Matematika. **28**, 945–956 (2017).
10. Pawlak,Z.: Rough Sets. International Journal of Computer & Information Sciences. **11**, 341–356 (1982).

11. Praba, B., Mohan, R.: Rough Lattice. International Journal of Fuzzy Mathematics and System. **3(2)**, 135–151 (2013).
12. Praba, B., Chandrasekaran, V. M., Manimaran, A.: Commutative Regular Monoid on Rough Sets. Italian Journal of Pure and Applied Mathematics. **31**, 307–318 (2013).
13. Praba, B., Chandrasekaran, V. M., Manimaran, A.: Semiring on Rough sets. Indian Journal of Science and Technology. **8(3)**, 280–286 (2015).
14. Praba, B., Manimaran, A., Chandrasekaran, V. M.: The Zero Divisor Graph of a Rough Semiring. International Journal of Pure and Applied Mathematics. **98(5)**, 33–37 (2015).
15. Ronnason Chinram.: Rough Prime Ideals and Rough Fuzzy Prime Ideals in Gamma Semigroups. Korean Mathematical Society. **24 (3)**, 341–351 (2009).
16. Yonghong Liu.: Special Lattice of Rough Algebras. Applied Mathematics. **2**, 1522–1524 (2011).
17. Zadeh, L. A.: Fuzzy Sets. Information and Control. **8**, 338–353 (1965).

Gorenstein FI-Flat Complexes and (Pre)envelopes

V. Biju

Abstract In this paper, Gorenstein FI-flat complexes are introduced, and their characteristics are studied over a $\mathscr{GFI}_{\mathscr{F}}$-closed ring. Also this paper proves that every complex of R-modules has a Gorenstein FI-flat complex preenvelope over a $\mathscr{GFI}_{\mathscr{F}}$-closed ring.

1 Introduction and Preliminaries

Homological algebra emerged as one of the interesting areas of study since the early 1800s and it proved to be more applicable in various other fields such as algebraic topology, group theory, commutative ring theory and algebraic geometry, etc.

In basic homological algebra, A left R-module K is called FP-injective (or absolutely pure) [3] if $Ext_1(L, K) = 0$ for every finitely presented left R-module L. In this fashion, FI-injective and FI-flat modules also were introduced by Mao et al. in [8].

Further, Selvaraj et al. introduced Gorenstein FI-injective and Gorenstein FI-flat modules in [10] and identified the covers of "Gorenstein FI-flat modules" in [12] and Tate homology in [14]. Gangyang et al. surfaced some significant results of Gorenstein flat complexes in [11].

This paper is written in three sections. The second section introduces Gorenstein FI-flat complexes and also investigates the characteristics of them. Section 3 concludes the existence of Gorenstein FI-flat complex preenvelopes for every complex of R-modules.

R is to be considered as an associative ring. For category of modules, character complex for any complex M is denoted by M^+. For the introduction of Gorenstein homological algebra and also for all other "undiscussed definitions and notations," I refer the readers to [1, 2, 4, 7, 15]. FI-cotorsion module was defined in [13].

V. Biju (✉)
Department of Mathematics, School of Advanced Sciences, VIT, Vellore, Tamil Nadu, India
e-mail: bijuwillwin@gmail.com; biju.v@vit.ac.in

V. Madhu et al. (eds.), *Advances in Algebra and Analysis*, Trends in Mathematics,
https://doi.org/10.1007/978-3-030-01120-8_10

$\mathcal{GFI}_{\mathcal{F}}$ -closed ring is discussed in [12]. Covers, precovers, envelopes, and preenvelopes were analyzed by Enochs in [5]. Gorenstein FI-flat and Gorenstein FI-injective modules were introduced by Selvaraj et al. in [10] as follows:

Definition 1 A left R-module M is known as Gorenstein FI-flat if there exists an FI-injective right R-module A, such that $A \otimes -$ preserves the exactness of the sequence $\cdots \longrightarrow F_1 \longrightarrow F_0 \longrightarrow F^0 \longrightarrow F^1 \longrightarrow \cdots$. of FI-flat left R-modules with $M \cong Ker(F^0 \longrightarrow F^1)$.

Definition 2 A right R-module N is called as Gorenstein FI-injective if there is an FI-injective right R-module A such that $Hom_R(A, -)$ preserves the exactness of the sequence $\cdots \longrightarrow E_1 \longrightarrow E_0 \longrightarrow E^0 \longrightarrow E^1 \longrightarrow \cdots$. of FI-injective right R-modules with $N \cong Ker(E^0 \longrightarrow E^1)$.

2 Gorenstein FI-Flat Complexes

This section introduces the Gorenstein FI-flat complexes and Gorenstein FI-injective complexes as follows:

Definition 3 Let $\mathcal{GFI}_{\mathcal{F}}$ be the class in which every element is a Gorenstein FI-flat module. A complex $\mathcal{X}$ is said to be a Gorenstein FI-flat complex if $\mathcal{X}$, $Z(\mathcal{X})$, $B(\mathcal{X})$, and $H(\mathcal{X})$ are all in $\mathcal{C}(\mathcal{GFI}_{\mathcal{F}})$. Where $\mathcal{C}(\mathcal{GFI}_{\mathcal{F}})$ denotes the class of complexes such that every element is in $\mathcal{GFI}_{\mathcal{F}}$.

Definition 4 Let $\mathcal{GFI}_{\mathcal{I}}$ be the class of all Gorenstein FI-injective modules. Any complex $\mathcal{Y}$ is called a Gorenstein FI-injective complex if $\mathcal{Y}$, $Z(\mathcal{Y})$, $B(\mathcal{Y})$, and $H(\mathcal{Y})$ are all in $\mathcal{C}(\mathcal{GFI}_{\mathcal{I}})$. Where $\mathcal{C}(\mathcal{GFI}_{\mathcal{I}})$ denotes the class of complexes with each component in $\mathcal{GFI}_{\mathcal{I}}$.

Definition 5 Let A be a right R-module. Then A is known an Gorenstein FI-cotorsion module if $Ext^1(F, A) = 0$ for any F which is a Gorenstein FI-flat right R-module.

Lemma 1 *Let R be a "$\mathcal{GFI}_{\mathcal{F}}$ "closed ring" and $\mathcal{A}$ be a complex of right R-modules. Then $\mathcal{A}$ is Gorenstein FI-injective if and only if it's character complex A^+ is Gorenstein FI-flat.*

Lemma 2 *A complex $\mathcal{X}$ is Gorenstein FI-flat in $\mathcal{C}(R\text{-}Mod)$ if and only if $-\otimes \mathcal{X}$ is exact for any short Gorenstein FI-flat exact sequence of complexes of right R-modules.*

Proof Suppose $\mathcal{X}$ be a Gorenstein FI-flat complex and $0 \longrightarrow F_1 \longrightarrow F_2 \longrightarrow F_3 \longrightarrow 0$ be a short Gorenstein FI-flat exact sequence of complexes of right R-modules. Then $\mathcal{X} = \underline{lim} P_i$ with P_i are Gorenstein projective complexes by [9, Theorem 7.2]. Hence, by natural isomorphism, we get that $- \otimes \mathcal{X}$ is exact for any short Gorenstein FI-flat exact sequence of complexes of right R-modules. Conversely suppose $- \otimes \mathcal{X}$ be exact for any short Gorenstein FI-flat exact sequence. By Lemma 1, we are supposed to prove that $X^+ = Hom(\mathscr{X}, Q/Z)$ is Gorenstein FI-injective in $\mathscr{C}(Mod - R)$. For any complex A of right R-modules, we let $0 \longrightarrow Y \longrightarrow P \longrightarrow A \longrightarrow 0$ be a short Gorenstein FI-flat exact sequence in $\mathscr{C}(Mod - R)$ with P Gorenstein projective. Then we will get the following commutative diagram

$$\begin{array}{ccccccc} 0 & \longrightarrow & Hom(P, X^+) & \longrightarrow & Hom(Y, X^+) & \longrightarrow & 0 \\ & & \downarrow & & \downarrow & & \\ 0 & \longrightarrow & P \otimes X^+ & \longrightarrow & Y \otimes X^+ & \longrightarrow & 0 \end{array}$$

in which the downward arrows are isomorphisms by natural isomorphisms of functors. Thus, the morphism $Hom(P, X^+) \longrightarrow Hom(Y, X^+)$ is epic, and so $Hom(P, X^+) \longrightarrow Hom(Y, X^+) \longrightarrow 0$ is exact. On the other hand, we get the sequence $Hom(P, X^+) \longrightarrow Hom(Y, X^+) \longrightarrow Ext^1(A, X^+) \longrightarrow Ext^1(P, X^+)$ is exact, where $Ext^1(P, X^+) = 0$. This will lead to $Ext^1(A, X^+) = 0$, and so X^+ is Gorenstein FI-injective in $\mathscr{C}(Mod - R)$.

Proposition 1 *Every pure subcomplex of a Gorenstein FI-flat complex is Gorenstein FI-flat.*

Proof Let $Y \leq X$ be a pure subcomplex of Gorenstein FI-flat complex X. Given a short Gorenstein FI-flat exact sequence $0 \longrightarrow F_1 \longrightarrow F_2 \longrightarrow F_3 \longrightarrow 0$ in $\mathscr{C}(Mod - R)$, it will lead to the commutative diagram given below

$$\begin{array}{ccccccccc} & & 0 & & 0 & & 0 & & \\ & & \downarrow & & \downarrow & & \downarrow & & \\ 0 & \longrightarrow & F_1 \otimes Y & \longrightarrow & F_2 \otimes Y & \longrightarrow & F_3 \otimes Y & \longrightarrow & 0 \\ & & \downarrow & & \downarrow & & \downarrow & & \\ 0 & \longrightarrow & F_1 \otimes X & \longrightarrow & F_2 \otimes X & \longrightarrow & F_3 \otimes Y & \longrightarrow & 0 \end{array}$$

where the bottom row is exact by Lemma 2. Here all the columns are exact since Y is pure in X. Then we get $F_1 \otimes Y \longrightarrow F_2 \otimes Y$ is a monomorphism; therefore Y is Gorenstein FI-flat by Lemma 2.

Proposition 2 *Let X be a complex, then the statements given below are equivalent.*

(1) X is Gorenstein FI-flat.
(2) Every short Gorenstein FI-flat exact sequence $0 \longrightarrow Y \longrightarrow P \longrightarrow X \longrightarrow 0$ is pure.
(3) There exists a pure exact sequence "$0 \longrightarrow Y \longrightarrow P \longrightarrow X \longrightarrow 0$" such that P is Gorenstein Projective.

Proof (1) $\Rightarrow$ (2). Let $0 \longrightarrow Y \longrightarrow P \longrightarrow X \longrightarrow 0$ be a short Gorenstein FI-flat exact sequence, and let $\mathscr{C}$ be a complex of right R-modules. If $Q \longrightarrow \mathscr{C}$ is a Gorenstein projective precover of $\mathscr{C}$, then we have a Gorenstein FI-flat exact sequence of complexes $0 \longrightarrow L \longrightarrow Q \longrightarrow \mathscr{C} \longrightarrow 0$ by Enochs [9]. Now we get a commutative diagram as follows:

$$\begin{array}{ccccccccc}
 & & 0 & & 0 & & 0 & & \\
 & & \downarrow & & \downarrow & & \downarrow & & \\
0 & \longrightarrow & L\otimes Y & \longrightarrow & L\otimes P & \longrightarrow & L\otimes X & \longrightarrow & 0 \\
 & & \downarrow & & \downarrow & & \downarrow & & \\
0 & \longrightarrow & Q\otimes Y & \longrightarrow & Q\otimes P & \longrightarrow & Q\otimes X & \longrightarrow & 0 \\
 & & \downarrow & & \downarrow & & \downarrow & & \\
0 & \longrightarrow & \mathscr{C}\otimes Y & \longrightarrow & \mathscr{C}\otimes P & \longrightarrow & \mathscr{C}\otimes X & \longrightarrow & 0 \\
 & & \downarrow & & \downarrow & & \downarrow & & \\
 & & 0 & & 0 & & 0 & &
\end{array}$$

As all Gorenstein projective complexes are Gorenstein FI-flat, we find that the right-hand column and the center row in the above commutative diagram are exact by Lemma 2. Thus, we get that $0 \longrightarrow \mathscr{C} \otimes Y \longrightarrow \mathscr{C} \otimes P \longrightarrow \mathscr{C} \otimes X \longrightarrow 0$ is exact by the snake lemma. Hence the Gorenstein FI-flat exact sequence "$0 \longrightarrow Y \longrightarrow P \longrightarrow X \longrightarrow 0$" is pure.

(2) $\Rightarrow$ (3) follows from [9].

(3) $\Rightarrow$ (1). Let $0 \longrightarrow Y \longrightarrow P \longrightarrow X \longrightarrow 0$ be a pure exact sequence such that P is Gorenstein projective, and let "$0 \longrightarrow A \longrightarrow B \longrightarrow \mathscr{C} \longrightarrow 0$" be a Gorenstein FI-flat exact sequence in $\mathscr{C}(Mod-R)$. Then we identify a commutative diagram as follows:

$$\begin{array}{ccccccccc}
 & & 0 & & 0 & & 0 & & \\
 & & \downarrow & & \downarrow & & \downarrow & & \\
0 & \longrightarrow & A\otimes Y & \longrightarrow & A\otimes P & \longrightarrow & A\otimes X & \longrightarrow & 0 \\
 & & \downarrow & & \downarrow & & \downarrow & & \\
0 & \longrightarrow & B\otimes Y & \longrightarrow & B\otimes P & \longrightarrow & B\otimes X & \longrightarrow & 0 \\
 & & \downarrow & & \downarrow & & \downarrow & & \\
0 & \longrightarrow & \mathscr{C}\otimes Y & \longrightarrow & \mathscr{C}\otimes P & \longrightarrow & \mathscr{C}\otimes X & \longrightarrow & 0 \\
 & & \downarrow & & \downarrow & & \downarrow & & \\
 & & 0 & & 0 & & 0 & &
\end{array}$$

Since all the rows and the center column in the above commutative diagram are exact by hypothesis, we get by the snake lemma that the right-hand column is exact. Hence, X is concluded as Gorenstein FI-flat by Lemma 2.

Corollary 1 *"If $0 \longrightarrow A \longrightarrow B \longrightarrow \mathscr{C} \longrightarrow 0$ be an exact sequence with $\mathscr{C}$ is Gorenstein FI-flat, then A is Gorenstein FI-flat if and only if B is Gorenstein FI-flat."*

Proof Assume that $0 \longrightarrow F_1 \longrightarrow F_2 \longrightarrow F_3 \longrightarrow 0$ is a Gorenstein FI-flat exact sequence in $\mathscr{C}(R\text{-}Mod)$. Then we will get a commutative diagram as given below:

$$\begin{array}{ccccccccc}
 & & 0 & & 0 & & 0 & & \\
 & & \downarrow & & \downarrow & & \downarrow & & \\
0 & \longrightarrow & F_1 \otimes A & \longrightarrow & F_1 \otimes B & \longrightarrow & F_1 \otimes \mathscr{C} & \longrightarrow & 0 \\
 & & \downarrow & & \downarrow & & \downarrow & & \\
0 & \longrightarrow & F_2 \otimes A & \longrightarrow & F_2 \otimes B & \longrightarrow & F_2 \otimes \mathscr{C} & \longrightarrow & 0 \\
 & & \downarrow & & \downarrow & & \downarrow & & \\
0 & \longrightarrow & F_3 \otimes A & \longrightarrow & F_3 \otimes B & \longrightarrow & F_3 \otimes \mathscr{C} & \longrightarrow & 0 \\
 & & \downarrow & & \downarrow & & \downarrow & & \\
 & & 0 & & 0 & & 0 & &
\end{array}$$

Here every row is exact by Proposition 2. Also, the right extreme column is exact as $\mathscr{C}$ is Gorenstein FI-flat. This will lead to the conclusion that the middle column is exact if and only if the left extreme column is exact. Thus, A is Gorenstein FI-flat if and only if B is Gorenstein FI-flat by Lemma 2.

Proposition 3 *Let R be a "$\mathscr{GFI}_{\mathscr{F}}$-closed ring." Then a complex A in $\mathscr{C}(R\text{-}Mod)$ is such that A and $A/B(A)$ are in $\mathscr{C}(\mathscr{GFI}_{\mathscr{F}})$ if and only if A^+ is Gorenstein FI-injective in $\mathscr{C}(R - Mod)$.*

Proof Assume that A and $A/B(A)$ are in $\mathscr{C}(\mathscr{GFI}_{\mathscr{F}})$, then all right R-modules $Hom_Z(A_{-n}, Q/Z)$ and $Hom_Z(A_{-n}/B_{-n}(A, Q/Z)$ are Gorenstein FI-injective, but

$$Hom_Z(A_{-n}/B_{-n}(A), Q/Z) = Z_n(A^+)$$

by natural homomorphism and clearly $Hom_Z(A_{-n}, Q/Z) = (A^+)_n$. Now using the exact sequences "$0 \longrightarrow Z_n(A^+) \longrightarrow (A^+)_n \longrightarrow B_{-1}(A^+) \longrightarrow 0$ and $0 \longrightarrow B_n(A^+) \longrightarrow Z_n(A^+) \longrightarrow H_n(A^+) \longrightarrow 0$," we get that all right R-modules $B_n(A^+)$ and $H_n(G^+)$ are Gorenstein FI-injective, and so A^+ is "Gorenstein FI-injective" in $\mathscr{C}(R\text{-}Mod)$. Conversely, suppose A^+ is "Gorenstein FI-injective" in

$\mathscr{C}(R\text{-}Mod)$. Then we get that each $(A^+)_n = Hom_Z(A_{-n}, Q/Z)$, and $Z_n(A^+)$, which is isomorphic to $Hom_Z(A_{-n}/B_{-n}(A), Q/Z)$ by natural homomorphism, is Gorenstein FI-injective, and so A_{-n} and $A_{-n}/B_{-n}(A)$ are Gorenstein FI-flat. This proves that A and A/B(A) are in $\mathscr{C}(\mathscr{G}\mathscr{F}\mathscr{I}_{\mathscr{F}})$.

3 Existence of Preenvelopes

Lemma 3 *"If R is a left $\mathscr{G}\mathscr{F}\mathscr{I}_{\mathscr{F}}$-closed ring, then the class of Gorenstein FI-flat complexes is closed under direct limits."*

Proof This is analogous to the proof of Lemma 4.6 [12].

Lemma 4 *Let $\mathscr{K}$ be a Kaplansky class [6, Theorem 2.9]. If "$\mathscr{K}$ contains the projective modules and it is closed under extensions and direct limits", then $(\mathscr{K}, \mathscr{K}^{\perp})$ is a perfect cotorsion pair in R-Mod.*

Lemma 5 *The class of Gorenstein FI-flat complexes over any ring is a Kaplansky class.*

Proof Since the class $\mathscr{G}\mathscr{F}\mathscr{I}_{\mathscr{F}}$ is a Kaplansky class [12], so does the class $\mathscr{C}(\mathscr{G}\mathscr{F}\mathscr{I}_{\mathscr{F}})$.

Corollary 2 *Let R be a $\mathscr{G}\mathscr{F}\mathscr{I}_{\mathscr{F}}$-closed ring, then $(\mathscr{C}(\mathscr{G}\mathscr{F}\mathscr{I}_{\mathscr{F}}), \mathscr{C}(\mathscr{G}\mathscr{F}\mathscr{I}_{\mathscr{F}})^{\perp})$ is perfect cotorsion pair.*

Proof By Lemmas 3, 4, and 5 we find that $(\mathscr{C}(\mathscr{G}\mathscr{F}\mathscr{I}_{\mathscr{F}}), \mathscr{C}(\mathscr{G}\mathscr{F}\mathscr{I}_{\mathscr{F}})^{\perp})$ is a perfect cotorsion pair. Since "the class of Gorenstein FI-flat modules is projectively resolving by [12, Theorem 3.10]", we conclude that $(\mathscr{C}(\mathscr{G}\mathscr{F}\mathscr{I}_{\mathscr{F}}), \mathscr{C}(\mathscr{G}\mathscr{F}\mathscr{I}_{\mathscr{F}})^{\perp})$ is hereditary.

Corollary 3 *If R is $\mathscr{G}\mathscr{F}\mathscr{I}_{\mathscr{F}}$-closed ring, then every complex has a Gorenstein FI-flat complex preenvelope.*

Proof by Corollary 2 the cotorsion pair $(\mathscr{C}(\mathscr{G}\mathscr{F}\mathscr{I}_{\mathscr{F}}), \mathscr{C}(\mathscr{G}\mathscr{F}\mathscr{I}_{\mathscr{F}})^{\perp})$ is hereditary perfect complete cotorsion pair, and the class of all Gorenstein FI-flat complexes $\mathscr{C}(\mathscr{G}\mathscr{F}\mathscr{I}_{\mathscr{F}})$ is closed under direct limits. So we conclude that every complex has a Gorenstein FI-flat complex preenvelope.

References

1. Auslander. M.: Anneaux de Gorenstein, et torsion en algebre commutative. Seminaire d'Algebre Commutative dirige par Pierre Samuel. Secretariat mathematique. Paris (1967)
2. Auslander. M and Bridger.M.: Stable Module Theory. Memoirs. Amer. Math. Soc. Vol. 94, Providence, RI: Amer. Math. Soc., (1969).
3. Meggiben. C.:Absolutely Pure modules, Proc. Amer. Math. Soc., **26**, 561–566 (1970).
4. Rotman J. J.: An Introduction to Homological Algebra. Academic Press, 1979.

5. Enochs E.: Injective and flat covers, envelopes and resolvents. Israel J. of Math. **39** 189–209 (1981).
6. Enochs E. and Lopez J.A.-Ramos: Kapalansky classes. Rend. Semin. Mat. Univ.Padova. **107**, 67–79 (2002).
7. Holm H.: Gorenstein homological dimensions. J. Pure Appl. Algebra. **189**, 167–193 (2004).
8. Mao L. and Ding N.: FI-injective and FI-flat modules. J. Algebra. **209**, 367–385 (2007).
9. Enochs E.: Cartan-Eilenberg, complexes and resolutions. J.Algebra, **342**,16–39 (2011).
10. Selvaraj C., Biju V. and Udhayakumar R.: Stability of Gorenstein F I-flat mod-ules. Far East J. of Math. **95**, (2), 159–168 (2014).
11. Gangyang and Li Liang : Carten-Eilenberg Gorenstein Flat complexes. Math. Scand. **114**, 5–25 (2014).
12. Selvaraj C., Biju V. and Udhayakumar R.: Gorenstein FI-flat (pre)covers. Gulf J. of Math. **3**, 46–58 (2015).
13. Biju V. and Udhayakumar R.: FI-flat Resolutions and Dimensions. Global Journal of Pure and Applied Mathematics. **12**, 808–811 (2016).
14. Selvaraj C., Biju V. and Udhayakumar R.: Gorenstein FI-flat Dimension and Tate Homology. Vietnam. J. Math. **44**, 679–695 (2016).
15. Vasudevan B., Udhayakumar R. and Selvaraj C.: Gorenstein FI-flat dimension and Relative Homology. Afrika Matematika. **28**, 1143–1156 (2017).

Bounds of Extreme Energy of an Intuitionistic Fuzzy Directed Graph

B. Praba, G. Deepa, V. M. Chandrasekaran, Krishnamoorthy Venkatesan, and K. Rajakumar

Abstract We are considering the website network http://www.pantechsolutions.net/ of the web navigation of the customers. This website network can be representing as an intuitionistic fuzzy directed graph by means of considering the navigation of the customers. In this intuitionistic fuzzy directed graph, the links are considering as vertices and the path between the links are considering as edges. The weightage of each edge are considering as number of visitors getting the link from one link to another link (membership value), number of visitors not getting the link, i.e. under traffic from one link to another link (non-membership value) and drop off case (intuitionistic fuzzy index). For this graph we are determining the maximum, minimum energies and its upper, lower bounds.

Keywords Energy of a graph · Energy of a fuzzy graph · Energy of an intuitionistic fuzzy graph

---

B. Praba
SSN College of Engineering, Chennai, Tamil Nadu, India
e-mail: prabab@ssn.edu.in

G. Deepa (✉) · V. M. Chandrasekaran
School of Advanced Sciences, VIT University, Vellore, Tamil Nadu, India
e-mail: deepa.g@vit.ac.in; vmcsn@vit.ac.in

K. Venkatesan
College of Natural Sciences, Arba Minch University, Arba Minch, Ethiopia
e-mail: krishnamoorthy.venkatesan@amu.edu.et

K. Rajakumar
SCOPE, VIT University, Vellore, Tamil Nadu, India
e-mail: rajakumar.krishnan@vit.ac.in

V. Madhu et al. (eds.), *Advances in Algebra and Analysis*, Trends in Mathematics,
https://doi.org/10.1007/978-3-030-01120-8_11

1 Introduction

The motivation about the study of graph energy originates from chemistry. Gutman [7] introduced the concept of "graph energy" as the sum of the absolute of the eigenvalues of the adjacency matrix of the graph. Gutman and Polansky [9] proposed the chemical applications on energy of a graph. Koolen et al. [10] discussed the maximal energy of a graph. Gutman [8] originated the mathematical properties on energy of a graph. Brankov et al. [3] described the equal energy of a graph. Yeh and Bang [15] introduced various connectedness concepts in fuzzy graphs. McAllister [11] extended a generalization of intersection graphs to fuzzy intersection graphs. Mordeson [12] introduced the concept of fuzzy line graphs and established its basic properties. The first definition of intuitionistic fuzzy relations and intuitionistic fuzzy graphs was proposed by Atanassov [2]. Shannon and Atanassov [14] built up a new generalization of the intuitionistic fuzzy graphs. Chountas et al. [5] explored the intuitionistic fuzzy version of the tree. Chandrashekar and Smitha [4] discussed the maximum degree energy of a graph. The energy of fuzzy graph and its bounds are discussed by Anjali and Mathew [1]. The energy of an intuitionistic fuzzy graph and its bounds are discussed by Praba et al. [13].

2 Preliminaries

In this section maximum, minimum energies and their bounds are discussed for real roots of an intuitionistic fuzzy directed graph.

Definition 1 ([6]) Let $G = (V, E, \mu, \gamma)$ be an intuitionistic fuzzy graph. For every vertex i, define $\alpha_j = \max_i \mu_{ij}$ and $\sigma_j = \min_i \gamma_{ij}$.

Definition 2 ([6]) Let $G = (V, E, \mu, \gamma)$ be an intuitionistic fuzzy graph. The Max-Min intuitionistic fuzzy matrix of an intuitionistic fuzzy graph is defined as $M(G) = \left[(r_{ij}, s_{ij})\right]$. We denote $R = [r_{ij}]$ is the Max degree intuitionistic fuzzy matrix and $S = [s_{ij}]$ is the Min degree intuitionistic fuzzy matrix of an intuitionistic fuzzy graph $G = (V, E, \mu, \gamma)$ where $R = \left\{max\left(\alpha_i, \alpha_j\right), if\, \mu_{ij} \neq 0\, and\; 0,\; otherwise\right\}$ and $S = \left\{min\left(\sigma_i, \sigma_j\right), if\, \gamma_{ij} \neq 0\right.$ $\left. and\; 0,\; otherwise\right\}$.

Definition 3 ([6]) Let G be an intuitionistic fuzzy graph. Two vertices v_i and v_j of G are said to be mutually adjacent if there is an edge from v_i to v_j and there is an edge from v_j to v_i.

Definition 4 ([6]) Let G be an intuitionistic fuzzy graph. Three vertices v_i, v_j and v_k of G are said to be cyclic if there is an edge from v_i to v_j , v_j to v_k and v_k to v_i.

Theorem 1 ([6]) *If $\theta_1, \theta_2, \ldots, \theta_n$ are real or complex eigenvalues of R and if $\lambda_1, \lambda_2, \ldots, \lambda_n$ are real or complex eigenvalues of S, then*

$$(i) \sum_{i=1}^{n} \theta_i^2 = 2 \sum_{1 \le i < j \le n} (r_{ij})^2 \text{ and } (ii) \sum_{i=1}^{n} \lambda_i^2 = 2 \sum_{1 \le i < j \le n} (s_{ij})^2.$$

Definition 5 ([6]) The spectrum of $M(G)$ is defined as (X, Y) where X is the set of eigenvalues of R and Y is the set of eigenvalues of S. It is denoted by $spec(M(G))$.

Definition 6 ([6]) If $\theta_1, \theta_2, \ldots, \theta_n$ are the eigenvalues of R and if $\lambda_1, \lambda_2, \ldots, \lambda_n$ are the eigenvalues of S, then the extreme energy of the intuitionistic fuzzy graph $G = (V, E, \mu, \gamma)$ is defined by

$$E(M(G)) = \left(\sum_{i=1}^{n} |\theta_i|, \sum_{i=1}^{n} |\lambda_i| \right).$$

Theorem 2 ([6]) *Let G be an intuitionistic fuzzy directed graph (without loops) with n vertices, and if $\theta_1, \theta_2, \ldots, \theta_n$ are the real eigenvalues of R and $\lambda_1, \lambda_2, \ldots, \lambda_n$ are the real eigenvalues of S, then*

$$(i) \sqrt{2 \sum_{1 \le i < j \le n} \left(r_{ij}\right)^2 + n\,(n-1)\,|R|^{\frac{2}{n}}} \le E\,(R) \le \sqrt{2n \sum_{i=1}^{n} \left(r_{ij}\right)^2}$$

$$(ii) \sqrt{2 \sum_{1 \le i < j \le n} \left(s_{ij}\right)^2 + n\,(n-1)\,|S|^{\frac{2}{n}}} \le E\,(S) \le \sqrt{2n \sum_{i=1}^{n} \left(s_{ij}\right)^2}$$

where $|R|$ is the determinant of R and $|S|$ is the determinant of S.

Corollary 1 *If $\theta_1, \theta_2, \ldots, \theta_n$ are the real eigenvalues of R and $\lambda_1, \lambda_2, \ldots, \lambda_n$ are the real eigenvalues of S and if $E(R) \ge E(S)$, then*

$$2n \sum_{i=1}^{n} \left(r_{ij}\right)^2 \ge 2 \sum_{1 \le i < j \le n} \left(s_{ij}\right)^2 + n\,(n-1)\,|S|^{\frac{2}{n}}.$$

Corollary 2 *If $\theta_1, \theta_2, \ldots, \theta_n$ are the real eigenvalues of R and $\lambda_1, \lambda_2, \ldots, \lambda_n$ are the real eigenvalues of S and if $E(S) \ge E(R)$, then*

$$2n \sum_{i=1}^{n} \left(s_{ij}\right)^2 \ge 2 \sum_{1 \le i < j \le n} \left(r_{ij}\right)^2 + n\,(n-1)\,|R|^{\frac{2}{n}}.$$

3 Bounds of Extreme Energy for Complex Roots

In this section we described the bounds of extreme energy for the complex roots of an intuitionistic fuzzy directed graph.

Theorem 3 *Let G be an intuitionistic fuzzy directed graph (without loops) with n vertices, and if* $\theta_1, \theta_2, \ldots, \theta_n$ *are complex eigenvalues of R and* $\lambda_1, \lambda_2, \ldots, \lambda_n$ *are complex eigenvalues of S, then*

$$(i)\ \sqrt{\sum_{i=1}^{n} |\theta_i|^2 + n(n-1)|R|^{\frac{2}{n}}} \leq E(R) \leq \sqrt{n\left[\left(\sum_{i=1}^{n} |\theta_i|\right)^2 - 2\sum_{1\leq i<j\leq n} |\theta_i\theta_j|\right]}$$

$$(ii)\ \sqrt{\sum_{i=1}^{n} |\lambda_i|^2 + n(n-1)|S|^{\frac{2}{n}}} \leq E(S) \leq \sqrt{n\left[\left(\sum_{i=1}^{n} |\lambda_i|\right)^2 - 2\sum_{1\leq i<j\leq n} |\lambda_i\lambda_j|\right]}$$

where $|R|$ *is the determinant of R and* $|S|$ *is the determinant of S.*

Proof

(i) Upper bound: Apply Cauchy–Schwarz inequality to the n numbers $1, 1, \ldots, 1$ and $|\theta_1|, |\theta_2|, \ldots, |\theta_n|$, we have

$$\sum_{i=1}^{n} |\theta_i| \leq \sqrt{n}\sqrt{\sum_{i=1}^{n} |\theta_i|^2}. \tag{1}$$

We know that

$$\left(\sum_{i=1}^{n} |\theta_i|\right)^2 = \sum_{i=1}^{n} |\theta_i|^2 + 2\sum_{1\leq i<j\leq n} |\theta_i\theta_j|$$

$$\Rightarrow \sum_{i=1}^{n} |\theta_i|^2 = \left(\sum_{i=1}^{n} |\theta_i|\right)^2 - 2\sum_{1\leq i<j\leq n} |\theta_i\theta_j|. \tag{2}$$

Substitute Eq. (2) in Eq. (1), we get

$$\sum_{i=1}^{n} |\theta_i| \leq \sqrt{n}\sqrt{\left(\sum_{i=1}^{n} |\theta_i|\right)^2 - 2\sum_{1\leq i<j\leq n} |\theta_i\theta_j|}$$

$$\Rightarrow E(R) \leq \sqrt{n\left[\left(\sum_{i=1}^{n} |\theta_i|\right)^2 - 2\sum_{1\leq i<j\leq n} |\theta_i\theta_j|\right]}. \tag{3}$$

(ii) Lower bound: Note that

$$(E(R))^2 = \left(\sum_{i=1}^{n} |\theta_i|\right)^2$$

$$= \sum_{i=1}^{n} |\theta_i|^2 + 2 \sum_{1 \le i < j \le n} |\theta_i \theta_j|$$

$$= \sum_{i=1}^{n} |\theta_i|^2 + \frac{2n(n-1)}{2} AM \left\{|\theta_i \theta_j|\right\}.$$

Since $AM\left\{|\theta_i\theta_j|\right\} \geq GM\left\{|\theta_i\theta_j|\right\}, 1 \leq i < j \leq n$, we get

$$E(R) \geq \sqrt{\sum_{i=1}^{n} |\theta_i|^2 + n(n-1)\, GM\left\{|\theta_i\theta_j|\right\}}.$$

$$\Rightarrow E(R) \geq \sqrt{\sum_{i=1}^{n} |\theta_i|^2 + n(n-1)\, |R|^{\frac{2}{n}}}. \tag{4}$$

From the above Eqs. (3) and (4), we have

$$\sqrt{\sum_{i=1}^{n} |\theta_i|^2 + n(n-1)|R|^{\frac{2}{n}}} \leq E(R) \leq \sqrt{n\left[\left(\sum_{i=1}^{n} |\theta_i|\right)^2 - 2\sum_{1 \le i < j \le n} |\theta_i\theta_j|\right]}.$$

Similarly, we can prove (ii).

Corollary 3 *If $\theta_1, \theta_2, \ldots, \theta_n$ are the complex eigenvalues of R and $\lambda_1, \lambda_2, \ldots, \lambda_n$ are the complex eigenvalues of S and if $E(R) \geq E(S)$, then*

$$n\left[\left(\sum_{i=1}^{n} |\theta_i|\right)^2 - 2\sum_{1 \le i < j \le n} |\theta_i\theta_j|\right] \geq \sum_{i=1}^{n} |\lambda_i|^2 + n(n-1)|S|^{\frac{2}{n}}.$$

Corollary 4 *If $\theta_1, \theta_2, \ldots, \theta_n$ are the complex eigenvalues of R and $\lambda_1, \lambda_2, \ldots, \lambda_n$ are the complex eigenvalues of S and if $E(S) \geq E(R)$, then*

$$n\left[\left(\sum_{i=1}^{n} |\lambda_i|\right)^2 - 2\sum_{1 \le i < j \le n} |\lambda_i\lambda_j|\right] \geq \sum_{i=1}^{n} |\theta_i|^2 + n(n-1)|R|^{\frac{2}{n}}.$$

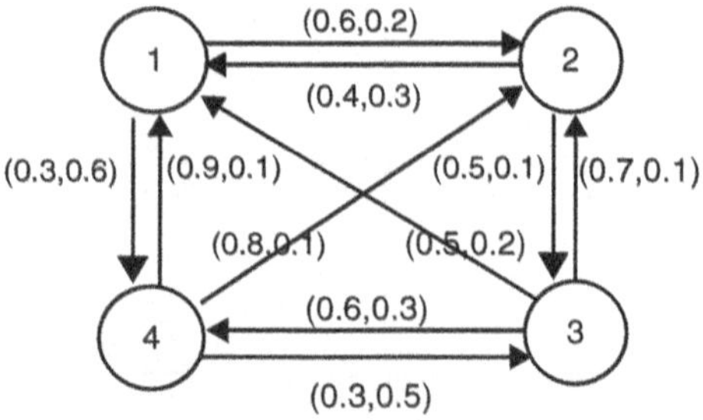

Fig. 1 Intuitionistic fuzzy graph

The above-mentioned concepts are illustrated in the following example.

Example 1 For the Fig. 1, eigenvalues of R and S are given by

$$Spec\,(R) = \{1.9622, -0.9071, -0.4551, -0.6000\}$$

$$Spec\,(S) = \{0.2414, -0.1000, -0.0414, -0.1000\}$$

From Theorem 2, we have (i) $3.6886 \leq 3.9243 \leq 4.5779$ and (ii) $0.4472 \leq 0.4828 \leq 0.5653$. By Corollary 1, we have $4.5779 \geq 0.4472$ (Fig. 1).

4 Numerical Examples

In the real-life problems, eigenvalues are not real always; it may have complex eigenvalues also. So, these concepts are determined by considering the website http://www.pantechsolutions.net/. This website has been represented as an intuitionistic fuzzy graph with four different time periods and for each of these periods maximum, minimum energies and their bounds are calculated.

Example 2 In the above-mentioned website, the four links 1.microcontroller-boards, 2./log-in html, 3./ and 4. project kits are considered for the period July 16 to August 15 in 2013 (Period I). For this period, the eigenvalues of R and S are given by

$$Spec\,(R) = \{0.4372, -0.2000, -0.1372, -0.1000\}$$
$$Spec\,(S) = \{0.4184, -0.1892 + 0.1553i, -0.1892 - 0.1553i, -0.0399\}$$

From Theorem 3, we have (i) $0.8219 \leq 0.8745 \leq 1.0202$ and (ii) $0.8222 \leq 0.9479 \leq 1.0890$. By Corollary 4, we have $1.0890 \geq 0.8219$ (Fig. 2).

Example 3 For the period August 16 to September 15 in 2013 (Period II), the eigenvalues of R and S are given by

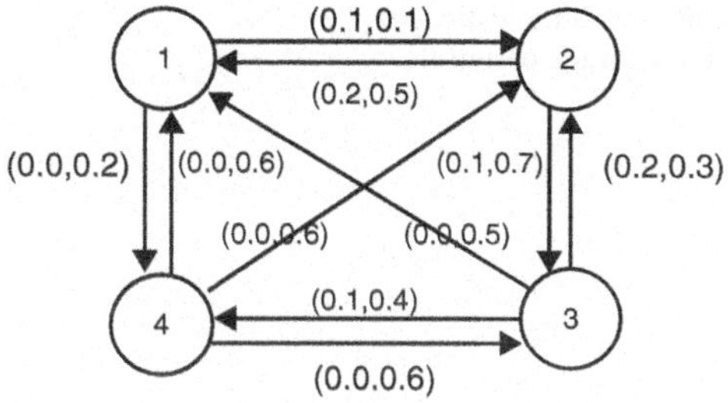

Fig. 2 Intuitionistic fuzzy graph for the period I

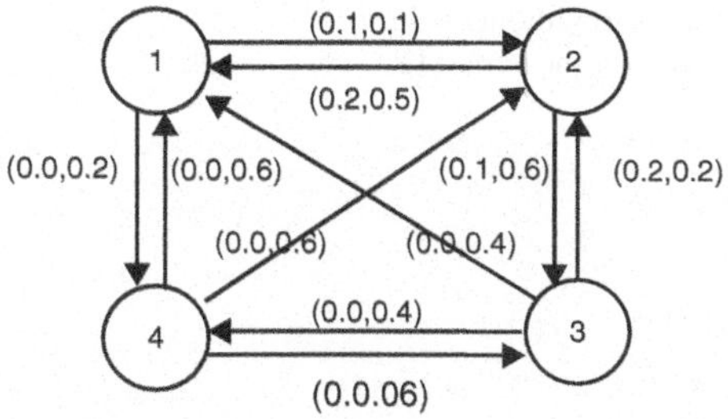

Fig. 3 Intuitionistic fuzzy graph for the period II

$$Spec\,(R) = \{0.4372, -0.2000, -0.1372, -0.1000\}$$

$$Spec\,(S) = \{0.4051, -0.1826 + 0.1273i, -0.1826 - 0.1273i, -0.0398\}$$

From Theorem 3, we have (i) $0.8219 \leq 0.8745 \leq 1.0202$ and (ii) $0.7773 \leq 0.8902 \leq 1.0298$. By Corollary 4, we have $1.0298 \geq 0.8219$ (Fig. 3).

Example 4 For the period September 16 to October 15 in 2013 (Period III), the eigenvalues of R and S are given by

$$Spec\,(R) = \{0.3714, -0.2231, -0.0483, -0.1000\}$$

$$Spec\,(S) = \{0.4184, -0.1892 + 0.1553i - 0.1892 - 0.1553i, -0.0399\}$$

From Theorem 3, we have (i) $0.6633 \leq 0.7427 \leq 0.8939$ and (ii) $0.8222 \leq 0.9479 \leq 1.0890$. By Corollary 4, we have $1.0890 \geq 0.6633$ (Fig. 4).

Example 5 For the period October 16 to November 15 in 2013 (Period IV), the eigenvalues of R and S are given by

$$Spec\,(R) = \{0.3714, -0.2231, -0.0483, -0.1000\}$$

$$Spec\,(S) = \{0.4051, -0.1826 + 0.1273i - 0.1826 - 0.1273i, -0.0398\}$$

From Theorem 3, we have (i) $0.6633 \leq 0.7427 \leq 0.8939$ and (ii) $0.7773 \leq 0.8902 \leq 1.0298$. By Corollary 4, we have $1.0298 \geq 0.6633$ (Fig. 5).

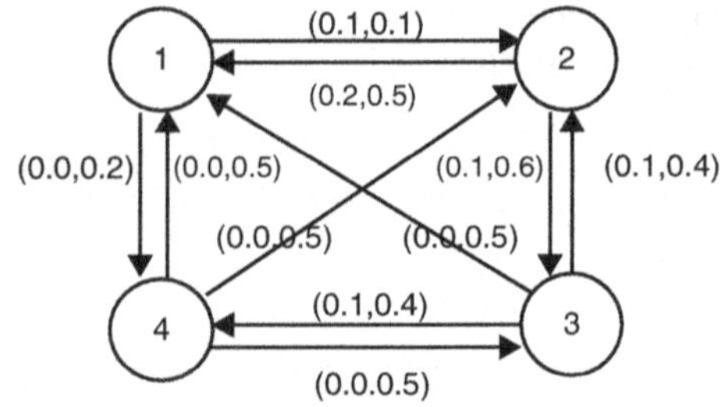

Fig. 4 Intuitionistic fuzzy graph for the period III

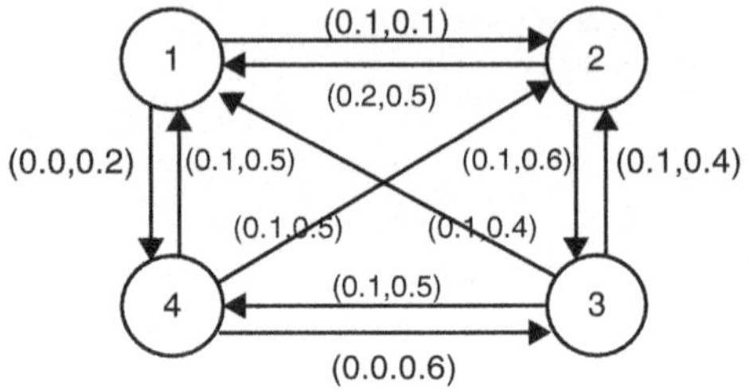

Fig. 5 Intuitionistic fuzzy graph for the period IV

5 Conclusion

Maximum and minimum energies are determined and its bounds are derived. Defined concepts are illustrated through real-time examples.

Acknowledgements I wish to thank my co-authors for their kind help and support in my research work throughout this paper. My heartfelt thanks to my institution for providing me infrastructural facilities and excellent resources to carry out my research in VIT University, Vellore.

References

1. Anjali, N., Mathew, S.: Energy of a fuzzy graph. AFMI. **6**, 455–465 [2013]
2. Atanassov, K.: Intuitionistic fuzzy sets: theory and applications. Heidelberg, New York, Physica-Verlag [1999]
3. Brankov, V., Stevanovic, D., Gutman, I.: Equienergetic chemical trees. J. Seb. Chem. Soc. **69**, 549–553 [2004]
4. Chandrashekar, A., Smitha, M.: On maximum degree energy of a graph. Int. J. Contemp. Math. Sciences. **4(8)**, 385–396 [2009]
5. Chountas, P., Alzebdi, M.S., Shannon, A., Atanassov, K.: On intuitionistic fuzzy trees. 13^{th}ICIFS, Sofia, NIFS. **15(2)**, 30–32 [2009]
6. Deepa, G., Praba, B., Chandrasekaran, V.M.: Max-Min intuitionistic fuzzy matrix of an intuitionistic fuzzy graph. IJPAM **98(3)**, 375–387 [2015]
7. Gutman, I.: The energy of a graph. Ber. Math-Statist. Sekt. Forschungsz. Graz. **103**, 1–22 [1978]
8. Gutman, I.: The energy of graph: old and new results. Betten, A., Kohnert, A., Laue, R., Wassermann A. [Eds.], Algebraic Combinatorics and Applications. Springer-Verlag, Berlin, 196–211 [2001]

9. Gutman, I., Polansky, O.E.: Mathematical concepts in organic chemistry. Springer-Verlag, Berlin [1986]
10. Koolen, J.H., Moulton, V., Gutman, I.: Improving the McClelland inequality for total π-electron energy. Chemical physics Letters. **320**, 213–216 [2000]
11. McAllister, M.L.N.: Fuzzy introduction graphs. Comput. Math. Applic. **15(10)**, 871–886 [1988]
12. Mordeson, J.N.: Fuzzy line graphs. Pattern Recognition Letters. **14(5)**, 381–384, [1993]
13. Praba, B., Chandrasekaran, V.M., Deepa, G.: Energy of an intuitionistic fuzzy graph. Italian journal of pure and applied mathematics. **32**, 431–444 [2014]
14. Shannon, A., Atanassov, K.: On a generalization of intuitionistic fuzzy graphs. NIFS. **12(1)**, 24–29 [2006]
15. Yeh, R.T., Bang, S.Y.: Fuzzy relations, fuzzy graphs and their applications to clustering analysis, cognitive and decision processes in fuzzy sets, Zadeh, L.A., Fu, K.S., Shimara, M.Eds, Academic Press, New York, 125–149, [1975]

Part II
Analysis

On Ultra Separation Axioms via $\alpha\omega$-Open Sets

M. Parimala, Cenap Ozel, and R. Udhayakumar

Abstract The concept of separation of two $\alpha\omega$-*open* sets is done via *ultra-$\alpha\omega$-separation*. Also this separation is used to define $\alpha\omega$-*kernel* and $\alpha\omega$-*closure*. In addition to this, weak separation axioms like $\alpha\omega$-T_0 spaces and $\alpha\omega$-T_1 spaces are introduced, and its properties are studied.

1 Introduction

Parimala et al. [1] introduced the concept of $\alpha\omega$-*closed* sets and studied some of its properties. In this paper, we define a set B is *ultra-$\alpha\omega$-separated* from C if there exists an $\alpha\omega$-*open* set K containing B such that $K \cap C = \phi$. From this separation, we define $\alpha\omega$-*closure* and the $\alpha\omega$-*kernel*, $\alpha\omega$-*derived* set and $\alpha\omega$-*shell* of a set $B \subseteq X$.

Throughout this paper, topological space is denoted by spaces. Let B be a subset of a space X. The closure and the interior of B are denoted by $cl(B)$ and $int(B)$, respectively.

M. Parimala
Department of Mathematics, Bannari Amman Institute of Technology, Sathyamangalam, TN, India
e-mail: rishwanthpari@gmail.com; parimalam@bitsathy.ac.in

C. Ozel
Department of Mathematics, King Abdulaziz University, Jeddah, Saudi Arabia
e-mail: cenap.ozel@gmail.com

R. Udhayakumar (✉)
Department of Mathematics, School of Advanced Sciences, Vellore Institute of Technology, Vellore, TN, India
e-mail: udhayaram_v@yahoo.co.in; udhayakumar.r@vit.ac.in

V. Madhu et al. (eds.), *Advances in Algebra and Analysis*, Trends in Mathematics,
https://doi.org/10.1007/978-3-030-01120-8_12

2 Preliminaries

We collect the following definitions, which are useful in the sequel.

Definition 1 A subset B of a space (X, τ) is called a:

1. a $\omega(=\widehat{g})$-*closed* set [2, 3] if $cl(B) \subseteq U$ whenever $B \subseteq G$ and G is *semiopen* in (X, τ).
2. a $\alpha\omega$-*closed* set [3] if $\omega cl(B) \subseteq G$ whenever $B \subseteq G$ and G is α-*open* in (X, τ).

3 $\alpha\omega$-Kernel and $\alpha\omega$-Closure

Definition 2 $\alpha\omega$-*kernel* of B is defined as the intersection of all subsets which are $\alpha\omega$-*open* in (X, τ) and containing B.

$$\text{i.e. } \alpha\omega - ker(B) = \cap\{K \in \alpha\omega O(X) : B \subseteq K\}.$$

Definition 3 $\alpha\omega$-*kernel* of $a \in X$ is defined by $\alpha\omega$-$ker(\{a\}) = \cap\{K \in \alpha\omega O(X) : a \in K\}$.

Definition 4 Let $a \in X$, X is a topological space. A subset N_a of X is called an $\alpha\omega$-neighbourhood (briefly, $\alpha\omega$-*nbd*) of X if there exists an $\alpha\omega$-*open* set K such that $a \in G \subseteq N_a$.

Theorem 1 *Let X be a topological space. Then for any nonempty subset B of X, $\alpha\omega$-$ker(B) = \{a \in X : \alpha\omega cl(\{a\}) \cap B \neq \phi\}$.*

Proof Let $X \in \alpha\omega$-$ker(B)$. Suppose that $\alpha\omega cl(\{a\}) \cap B = \phi$. Then $B \subseteq X - \alpha\omega cl(\{a\})$ and $X - \alpha\omega cl(\{a\})$ is $\alpha\omega$-*open* set containing B but not a, which is a contradiction.

Conversely, let us assume that $a \notin \alpha\omega$-$ker(B)$ and $\alpha\omega cl(\{a\}) \cap B \neq \phi$. Then there exist a $\alpha\omega$-*open* set D containing B but not a and $b \in \alpha\omega cl(\{a\}) \cap B$. Hence we get a $\alpha\omega$-*nbd* of b, say D with $a \notin D$, which is a contradiction. Hence $a \in \alpha\omega$-$ker(B)$. □

Theorem 2 *For any subset B of X, $X_1 \cap cl(A) \subseteq \alpha\omega$-$ker(B)$.*

Proof Let $a \in X_1 \cap cl(B)$ and suppose that $a \notin \alpha\omega$-$ker(B)$. Then there is a $\alpha\omega$-*open* set K containing B such that $a \notin K$. If $F = X - K$, then F is $\alpha\omega$-*closed*. Since $cl(\{a\}) \subseteq cl(B)$, we have $int(cl(int(\{a\}))) \subseteq B \cup int(cl(int(B)))$ and since $a \in X_1$, we have $a \notin X_1$ and so $int(cl(int(\{a\}))) = \phi$. Therefore, there exist some point $b \in B \cap int(cl(int(\{a\})))$ and hence $b \in F \cap B$, which is a contradiction. □

Definition 5 Let X be a topological space. A set B is said to be *ultra*-$\alpha\omega$-*separated* from C if there exists an $\alpha\omega$-*open* set K such that $B \subseteq K$ and $K \cap C = \phi$ or $B \cap \alpha\omega cl(C) = \phi$.

By the Definition 5 and the Theorem 1, the following statements are true for $a, b \in X$ of a topological space:

(i) $\alpha\omega\text{-}cl(\{a\}) = \{b : b\ is\ not\ \ ultra\text{-}\alpha\omega\text{-}separated\ from\ a\}$
(ii) $\alpha\omega\text{-}ker(\{a\}) = \{b : b\ is\ not\ \ ultra\text{-}\alpha\omega\text{-}separated\ from\ b\}$.

Definition 6 For any point a of a topological space X,

(i) the $\alpha\omega$-*derived* set (briefly, $\alpha\omega\text{-}d(\{a\})$) of a defined as the set $\alpha\omega\text{-}d(\{a\}) = \alpha\omega\text{-}cl(\{a\}) - \{a\} = \{b : b \neq a\ and\ b\ is\ not\ \text{-}ultra\text{-}\alpha\omega\text{-}separated\ from\ a\}$,
(ii) the $\alpha\omega$-shell (briefly, $\alpha\omega\text{-}shl(\{a\})$) of a singleton set $\{a\}$ is defined to be the set $\alpha\omega\text{-}shl(\{a\}) = \alpha\omega\text{-}ker(\{a\}) - \{a\} = \{b : b \neq a\ and\ a\ is\ not\ \text{-}ultra\text{-}\alpha\omega\text{-}separated\ from\ b\}$.

Definition 7 Let X be a topological space. Then we define:

(i) $\alpha\omega\text{-}D = \{a : a \in X\ and\ \alpha\omega\text{-}d(\{a\}) = \phi\}$,
(ii) $\alpha\omega\text{-}S = \{a : a \in X\ and\ \alpha\omega\text{-}shl(\{a\}) = \phi\}$
(iii) $\alpha\omega\text{-}\langle a\rangle = \alpha\omega\text{-}cl(\{a\}) \cap \alpha\omega\text{-}ker(\{a\})$.

Theorem 3 *Let a, b be any two points in X. Then the following conditions are true:*

(i) $b \in \alpha\omega\text{-}ker(\{a\})\ if\ and\ only\ if\ a \in \alpha\omega\text{-}cl(\{b\})$,
(ii) $b \in \alpha\omega\text{-}shl(\{a\})\ if\ and\ only\ if\ a \in \alpha\omega\text{-}d(\{b\})$,
(iii) $b \in \alpha\omega\text{-}cl(\{a\})\ implies\ \alpha\omega\text{-}cl(\{b\}) \subseteq \alpha\omega\text{-}cl(\{a\})$
(iv) $b \in \alpha\omega\text{-}ker(\{a\})\ implies\ \alpha\omega\text{-}ker(\{b\}) \subseteq \alpha\omega\text{-}ker(\{a\})$.

Proof The proof of (i) and (ii) are obvious.

(iii) Let $c \in \alpha\omega\text{-}cl(\{b\})$. Then c is not $ultra\text{-}\alpha\omega\text{-}separated$ from b. So that there exists an $\alpha\omega$-open set $K \supset c$ such that $K \cap \{b\} \neq \phi$. Hence $b \in K$ and by assumption $K \cap \{a\} \neq \phi$. Hence c is not $ultra\text{-}\alpha\omega\text{-}separated$ from the point a. So $c \in \alpha\omega\text{-}cl(\{a\})$. Therefore $\alpha\omega\text{-}cl(\{b\}) \subseteq \alpha\omega\text{-}cl(\{a\})$.
(iv) Let $c \in \alpha\omega\text{-}\ ker(\{b\})$. Then b is not $ultra\text{-}\alpha\omega\text{-}separated$ from c. So $b \in \alpha\omega\text{-}cl(\{c\})$. Hence $\alpha\omega\text{-}cl(\{b\}) \subseteq \alpha\omega\text{-}cl(\{c\})$. By assumption $b \in \alpha\omega\text{-}ker(\{a\})$ and then $a \in \alpha\omega\text{-}cl(\{b\})$. So $\alpha\omega\text{-}cl(\{a\}) \subseteq \alpha\omega\text{-}cl(\{b\})$. Ultimately $\alpha\omega\text{-}cl(\{a\}) \subseteq \alpha\omega\text{-}cl(\{c\})$. Hence $a \in \alpha\omega\text{-}cl(\{c\})$, that is $c \in \alpha\omega\text{-}ker(\{a\})$. Therefore $\alpha\omega\text{-}ker(\{b\}) \subseteq \alpha\omega\text{-}ker(\{a\})$.

□

Let us recall that a subset B of X is called a degenerate set if B is either a null set or a singleton set.

Theorem 4 *Let $a, b \in X$. Then,*

(i) for every $a \in X$, $\alpha\omega\text{-}shl(\{a\})$ is degenerate set iff for every $a, b \in X$, $a \neq b$, $\alpha\omega\text{-}d(\{a\}) \cap \alpha\omega\text{-}d(\{b\}) = \phi$,
(ii) for every $a \in X$, $\alpha\omega\text{-}d(\{a\})$ is degenerate set iff for every $a, b \in X$, $a \neq b$, $\alpha\omega\text{-}shl(\{a\}) \cap \alpha\omega\text{-}shl(\{b\}) = \phi$.

Proof (i) Let $\alpha\omega\text{-}d(\{a\}) \cap \alpha\omega\text{-}d(\{b\}) \neq \phi$. Then, there $\exists$ a $c \in X$ such that $c \in \alpha\omega\text{-}d(\{a\})$ and $c \in \alpha\omega\text{-}d(\{b\})$. Then $c \neq b \neq a$ and $c \in \alpha\omega\text{-}cl(\{a\})$ and $c \in \alpha\omega\text{-}cl(\{b\})$, that is $a, b \in \alpha\omega\text{-}ker(\{c\})$. Hence $\alpha\omega\text{-}ker(\{c\})$ and so $\alpha\omega\text{-}shl(\{c\})$ is not a degenerate set.

Conversely, let $a, b \in \alpha\omega\text{-}shl(\{c\})$. Then we get $a \neq c$, $a \in \alpha\omega\text{-}ker(\{c\})$ and $b \neq c$ and $b \in \alpha\omega\text{-}ker(\{c\})$, and hence c is an element of both $\alpha\omega\text{-}cl(\{a\})$ and $\alpha\omega\text{-}cl(\{b\})$, which is a contradiction.

The proof of (ii) is simple and obvious. □

Proposition 1 *If* $b \in \alpha\omega\text{-}\langle a\rangle$, *then* $\alpha\omega\text{-}\langle a\rangle = \alpha\omega\text{-}\langle b\rangle$.

Proof If $b \in \alpha\omega\text{-}\langle a\rangle$, then $b \in \alpha\omega\text{-}cl(\{a\}) \cap \alpha\omega\text{-}ker(\{a\})$. Hence $b \in \alpha\omega\text{-}cl(\{a\})$ and $b \in \alpha\omega\text{-}ker(\{a\})$, and so we have $\alpha\omega\text{-}cl(\{b\}) \subseteq \alpha\omega\text{-}cl(\{a\})$ and $\alpha\omega\text{-}ker(\{b\}) \subseteq \alpha\omega\text{-}ker(\{a\})$. Then $\alpha\omega\text{-}cl(\{b\}) \cap \alpha\omega\text{-}ker(\{b\}) \subseteq \alpha\omega\text{-}cl(\{a\}) \cap \alpha\omega\text{-}ker(\{a\})$. Hence $\alpha\omega\text{-}\langle b\rangle \subseteq \alpha\omega\text{-}\langle a\rangle$. The fact that $b \in \alpha\omega\text{-}cl(\{a\})$ implies $a \in \alpha\omega\text{-}ker(\{b\})$ and $b \in \alpha\omega\text{-}ker(\{a\})$ implies $a \in \alpha\omega\text{-}cl(\{b\})$. Then we have that $\alpha\omega\text{-}\langle a\rangle \subseteq \alpha\omega\text{-}\langle b\rangle$. So $\alpha\omega\text{-}\langle a\rangle = \alpha\omega\text{-}\langle b\rangle$. □

Corollary 1 *For every* $a, b \in X$, *either* $\alpha\omega\text{-}\langle a\rangle \cap \alpha\omega\text{-}\langle b\rangle = \phi$ *or* $\alpha\omega\text{-}\langle a\rangle = \alpha\omega\text{-}\langle b\rangle$.

Proof If $\alpha\omega\text{-}\langle a\rangle \cap \alpha\omega\text{-}\langle b\rangle \neq \phi$, then there exists $c \in X$ such that $c \in \alpha\omega\text{-}\langle a\rangle$ and $c \in \alpha\omega\text{-}\langle b\rangle$ and by Proposition 1, $\alpha\omega\text{-}\langle c\rangle = \alpha\omega\text{-}\langle a\rangle = \alpha\omega\text{-}\langle b\rangle$. Hence the result. □

Theorem 5 *Let* a, b *be any two points in* X, *then the following two statements are equivalent:*

(i) $\alpha\omega\text{-}ker(\{a\}) \neq \alpha\omega\text{-}ker(\{b\})$
(ii) $\alpha\omega\text{-}cl(\{a\}) \neq \alpha\omega\text{-}cl(\{b\})$.

Proof $(i) \Longrightarrow (ii)$ Let us assume $\alpha\omega\text{-}ker(\{a\}) \neq \alpha\omega\text{-}ker(\{b\})$. Then, there exists a $c \in \alpha\omega\text{-}ker(\{a\})$ but $c \notin \alpha\omega\text{-}ker(\{b\})$. As $c \in \alpha\omega\text{-}ker(\{a\})$, $a \in \alpha\omega\text{-}cl(\{c\})$ and $\alpha\omega\text{-}cl(\{a\}) \subseteq \alpha\omega\text{-}cl(\{c\})$. Also we have taken $c \notin \alpha\omega\text{-}ker(\{b\})$, by Theorem 1, $\alpha\omega\text{-}cl(\{c\}) \cap \{b\} = \phi$, so $\alpha\omega\text{-}cl(\{a\}) \cap \{b\} = \phi$ and so b is *ultra-$\alpha\omega$-separated* from a, and hence we get that $b \notin \alpha\omega\text{-}cl(\{a\})$. Hence $\alpha\omega\text{-}cl(\{b\}) \neq \alpha\omega\text{-}cl(\{a\})$.

$(ii) \Longrightarrow (i)$ Suppose $\alpha\omega\text{-}cl(\{a\}) \neq \alpha\omega\text{-}cl(\{b\})$. Then there exists a point $c \in \alpha\omega\text{-}cl(\{a\})$ but $c \notin \alpha\omega\text{-}cl(\{b\})$. So, we get an *$\alpha\omega$-open* set containing c and a but not b. That is $b \notin \alpha\omega\text{-}ker(\{a\})$. Hence $\alpha\omega\text{-}ker(\{b\}) \neq \alpha\omega\text{-}ker(\{a\})$. □

4 $\alpha\omega$-T_0 and $\alpha\omega$-T_1

Definition 8 A space X is said to be $\alpha\omega$-T_0 space if a, b be any two points of X, $a \neq b$, then there exists $K \in \alpha\omega O(X)$ such that K contains only a or b but not both.

Definition 9 A space X is said to be $\alpha\omega$-T_1 space if a, b be any two points of X, $a \neq b$, then there exists $K, T \in \alpha\omega O(X)$ such that $a \in K$ and $b \in T$ but $b \notin K$ and $a \notin T$.

Remark 1 Every $\alpha\omega$-T_1 space is $\alpha\omega$-T_0.

Theorem 6 *A topological space X is $\alpha\omega$-T_0 if and only if one of the following conditions holds:*

(i) For any $a, b \in X$, $a \neq b$, either a is ultra-$\alpha\omega$-separated from b or b is ultra-$\alpha\omega$-separated from a,
(ii) $b \in \alpha\omega - cl(\{a\})$ implies $a \notin \alpha\omega$-$cl(\{b\})$
(iii) For every $a, b \in X$ if $a \neq b$, then $\alpha\omega$-$cl(\{a\}) \neq \alpha\omega$-$cl(\{b\})$.

Proof

(i) The proof is obvious from the definitions of $\alpha\omega$-T_0 space and *ultra-$\alpha\omega$-separation.*
(ii) By (ii), $b \in \alpha\omega$-$cl(\{a\})$, and so b is not *ultra-$\alpha\omega$-separated* from a sice X is $\alpha\omega$-T_0, a should be *ultra-$\alpha\omega$-separated* from b, that is $a \notin \alpha\omega$-$cl(\{b\})$.
(iii) If X is $\alpha\omega$-T_0, then for all $a, b \in X$ and $a \neq b$, $\alpha\omega$-$cl(\{a\}) \neq \alpha\omega$-$cl(\{b\})$ as clear by (ii). Now, let $\alpha\omega$-$cl(a) \neq \alpha\omega$-$cl(\{b\})$. Then there exists $c \in X$, such that $c \in \alpha\omega$-$cl(\{a\})$ and $c \notin \alpha\omega$-$cl(\{b\})$. If a is not *ultra-$\alpha\omega$-separated* from b, then $a \in \alpha\omega$-$cl(\{b\})$. So $\alpha\omega$-$cl(\{a\}) \subseteq \alpha\omega$-$cl(\{b\})$. Then $c \in \alpha\omega$-$cl(\{b\})$, which is contradicts.

□

Lemma 1 *A topological space is $\alpha\omega$-T_0 if and only if $\alpha\omega$-$d(\{a\}) \cap \alpha\omega$-$shl(\{a\}) = \phi$.*

Proof Let X be $\alpha\omega$-T_0. Suppose we have $\alpha\omega$-$d(\{a\}) \cap \alpha\omega$-$shl(\{a\}) \neq \phi$. Let $c \in \alpha\omega$-$d(\{a\})$ and $c \in \alpha\omega$-$shl(\{a\})$. Then $c \neq a$ and $c \in \alpha\omega$-$cl(\{a\})$ and $c \in \alpha\omega$-$ker(\{a\})$. Then c is not *ultra-$\alpha\omega$-separated* from a and also a is not *ultra-$\alpha\omega$-separated* from c, which is contradicts.

Conversely, let $\alpha\omega$-$d(\{a\}) \cap \alpha\omega$-$shl(\{a\}) = \phi$. Then there exists $c \neq a$, $c \in \alpha\omega$-$cl(\{a\})$ and $c \notin \alpha\omega$-$ker(\{a\})$. Hence if we have c, which is not *ultra-$\alpha\omega$-separated* from a, then a is *ultra-$\alpha\omega$-separated* from c. Hence, X is $\alpha\omega$-T_0. □

Proposition 2 *A space X is $\alpha\omega$-T_1 iff $\{a\}$ is $\alpha\omega$-closed in $X \forall a \in X$.*

Proof If $\{a\}$ is *$\alpha\omega$-closed* in X, for $a \neq b$, $X - \{a\}$, $X - \{b\}$ are *$\alpha\omega$-open* sets such that $b \in X - \{a\}$ and $a \in X - \{b\}$. Therefore, X is $\alpha\omega$-T_1.

Conversely, if X is $\alpha\omega$-T_1 and if $b \in X - \{a\}$ then $a \neq b$. Therefore, there exists an *$\alpha\omega$-open* sets U_a, V_b in X such that $a \in U_a$ but $b \notin U_a$ and $b \in V_b$ but $a \notin V_b$. Let K be the union of all such V_b. Then K is an *$\alpha\omega$-open* set and $K \subseteq X - \{a\} \subseteq X$; therefore, $a - \{a\}$ is a *$\alpha\omega$-open* set in X. □

Corollary 2 *A topological space X is $\alpha\omega$-T_1 if and only if the following conditions holds:*

(i) For arbitrary $a, b \in X$, $a \neq b$, a is $ultra$-$\alpha\omega$-$separated$ from b,
(ii) For every $a \in X$, $\alpha\omega$-$cl(\{a\}) = \{a\}$,
(iii) For every $a \in X$, $\alpha\omega$-$d(\{a\}) = \phi$ or $\alpha\omega$-$D = X$,
(iv) For every $a \in X$, $\alpha\omega$-$ker(\{a\}) = \{a\}$,
(v) For every $a \in X$, $\alpha\omega$-$shl(\{a\}) = \phi$ or $\alpha\omega$-$S = X$,
(vi) For every $a, b \in X$, $a \neq b$, $\alpha\omega$-$cl(\{a\}) \cap \alpha\omega$-$cl(\{b\}) = \phi$
(vii) For arbitrary $a, b \in X$, $a \neq b$, we have $\alpha\omega$-$ker(\{a\}) \cap \alpha\omega$-$ker(\{b\}) = \phi$.

Proof (i) Proof is clear from the definition of $\alpha\omega$-T_1.
(ii) Proof is clear.
(iii), (iv) and (v) are clear.
(vi) Since X is $\alpha\omega$-T_1, $\alpha\omega$-$cl(\{a\}) = \{a\}$ and $\alpha\omega$-$cl(\{b\}) = \{b\}$. So, when $a \neq b$, $\alpha\omega$-$cl(\{a\}) \cap \alpha\omega$-$cl(\{b\}) = \phi$.
(vii) Obvious from the condition (vi). □

References

1. Parimala, M. Udhayakumar, R. Jeevitha, R. Biju, V. : On $\alpha\omega$-closed sets in topological spaces. Int. J. Pure Appl. Math. **115 (5)**, 1049–1056 (2017).
2. Sundaram, P. Shrik John, M. : On ω-closed sets in topology. Acta Ciencia Indica. **4**, 389–392 (2000).
3. Veera kumar, M. K. R. S.: $\widehat{g}$-locally closed sets and $\widehat{G}LC$-functions. Indian J. Math. **43(2)**, 231–247 (2001).

Common Fixed Point Theorems in 2-Metric Spaces Using Composition of Mappings via A-Contractions

J. Suresh Goud, P. Rama Bhadra Murthy, Ch. Achi Reddy, and K. Madhusudhan Reddy

Abstract The paper contains two common fixed point theorems using composition of self-mappings via the notion of A-contractions. The first theorem deals with the common fixed points of three self-maps via A-contractions where the contractive condition depends on the composition of self-maps. Then we proved a common fixed point theorem for an arbitrary set of self-maps. Examples are presented to show the significance of our results.

Keywords A-contractions · Common fixed points · Arbitrary set of self-maps

MSC Subject Classification: 47H10, 54H25

1 Introduction

Akram et al. [1] proposed the new class of contraction mappings in metric spaces named as A-contractions which were extended by the authors in [2, 3]. They showed that these contractions include Bianchini's contractions, Kannan's contractions,

J. Suresh Goud
Department of Mathematics, Institute of Aeronautical Engineering, Hyderabad, Telangana, India
e-mail: j.sureshgoud@iare.ac.in

P. Rama Bhadra Murthy
Department of Mathematics, Osmania University, Hyderabad, Telangana, India
e-mail: badri2502@gmail.com

Ch. Achi Reddy
Department of Mathematics, MLR Institute of Technology, Hyderabad, Telangana, India
e-mail: achireddy.ch@gmail.com

K. Madhusudhan Reddy (✉)
Department of Mathematics, Vellore Institute of Technology, Vellore, Tamil Nadu, India
e-mail: drkmsreddy@yahoo.in; madhu.katta@shct.edu.om

V. Madhu et al. (eds.), *Advances in Algebra and Analysis*, Trends in Mathematics,
https://doi.org/10.1007/978-3-030-01120-8_13

Reich's contractions. The same work is extended to 2-metric space in different directions by the authors in [6, 8] and [4]. For the results in 2-metric space, refer to [5, 7, 9–12] and [13]. The present work contains two common fixed point theorems using the Class of A-contractions. The first theorem deals with the common fixed points of three self-maps via A-contractions where the contractive condition depends on the composition of self-maps while the second theorem deals with the common fixed point theorem for a set of self-maps. Examples are presented to show the significance of our results.

2 Preliminaries

Throughout the paper, let N, N_0 and R_+ denote natural numbers, whole numbers, and nonnegative real numbers, respectively, whereas fT denotes the composition of self-maps f and T.

Definition 1 ([5]) Let X be a non-empty set and $d : X^3 \to \mathrm{R}_+$ be a map satisfying:

1. For each distinct pair $x_1, x_2 \in X$, there exists $x_3 \in X$ such that $d(x_1, x_2, x_3) \neq 0$.
2. $d(x_1, x_2, x_3) \geq 0$, and the equality holds if at least two of the three points x_1, x_2, x_3 are equal.
3. $d(x_1, x_2, x_3) = d(p(x_1, x_2, x_3))$ for all $x_1, x_2, x_3 \in X$ and for all permutations $p(x_1, x_2, x_3)$.
4. $d(x_4, x_2, x_3) \leq d(x_1, x_2, x_4)+d(x_1, x_4, x_3)+d(x_1, x_2, x_3)$ for all $x_1, x_2, x_3, x_4 \in X$ [pyramidal inequality]

Then d is called 2-metric and (X, d) together as 2-metric space.

Definition 2 ([1]) Let A be the set of all functions $\alpha : R_+^3 \to R_+$ satisfying

1. α is continuous on the set R_+^3 of all triplets (w.r.t metric induced on R^3)
2. $x_1 \leq \mathrm{k}x_2$ for some $k \in [0, 1)$ whenever $x_1 \leq \alpha(x_1, x_2, x_2)$or $x_1 \preceq \alpha(x_2, x_1, x_2)$ or $x_1 \leq \alpha(x_2, x_2, x_1)$.

Definition 3 ([8]) A self-map T on a 2-metric space X is said to be an A-contraction, if to each $u \in X$, $d(Tx_1, T_2, u) \leq \alpha\,(d(x_1, x_2, u), d(x_1, Tx_1, u), d(x_2, Tx_2, u))$ holds for all $x_1, x_2 \in X$ and for some $\alpha \in A$.

Definition 4 ([6]) A pair of self-maps f, g defined on a non-empty set X is said to be weakly compatible if $f(gt) = g(ft)$ whenever $ft = gt$ for some $t \in X$.

3 Main Result

Theorem 1 *Let (X, d) be a 2-metric space. Suppose f, g, and T are a set of three self-maps on X such that $f(X) \subseteq g(X)$ with $f(X)$ to be complete subspace of X and*

$$d(fTx, gTy, u) \leq \alpha(d(fy, gx, u), d(fy, gTy, u), d(gx, fTx, u)) \tag{1}$$

for some $\alpha \in A$ *and for all* $x, y, u \in X$. *If the pairs* (f, gT) *and* (g, fT) *are weakly compatible, then* f, g *share a common fixed point. Furthermore, if* T *commutes with* f *or* g *at the common fixed point of* f, g*, then the three self maps has unique common fixed point.*

Proof Take an arbitrary $x_0 \in X$. For all $n \in N_0$, define the sequences $\{x_n\}$ and $\{y_n\}$ as $x_{n+1} = Tx_n$ and $y_{2n} = fx_{2n}$, $y_{2n+1}=gx_{2n+1}$, respectively. Then, using (1)

$$\begin{aligned} d(y_{2n}, y_{2n+1}, u) &= d(fx_{2n}, gx_{2n+1}, u) \\ &= d(fTx_{2n-1}, gTx_{2n}, u) \\ &\leq \alpha(d(y_{2n}, y_{2n-1}, u), d(y_{2n}, y_{2n+1}, u), d(y_{2n-1}, y_{2n}, u)) \end{aligned}$$

Since $\alpha \in A$, for some $k \in [0, 1)$, we get

$$d(y_{2n}, y_{2n+1}, u) \leq kd(y_{2n-1}, y_{2n}, u) \tag{2}$$

Similarly, applying (1) to $d(y_{2n+1}, y_{2n+2}, u)$, we get

$$d(y_{2n+1}, y_{2n+2}, u) \leq kd(y_{2n}, y_{2n+1}, u) \tag{3}$$

From (1) and (2), we have

$$d(y_n, y_{n+1}, u) \leq kd(y_{n-1}, y_n, u) \text{ forall } n \in N \tag{4}$$

The induction process gives, for all $n \in N$,

$$d(y_n, y_{n+1}, u) \leq k^n d(y_0, y_1, u) \tag{5}$$

Now we show that $d(y_n, y_{n+1}, y_{n+2}) = 0$ for all $n \in N_0$. Using (1),

$$\begin{aligned} d(y_{2n}, y_{2n+1}, y_{2n+2}) &= d(y_{2n+1}, y_{2n+2}, y_{2n}) \\ &= d(gx_{2n+1}, fx_{2n+2}, y_{2n}) \\ &= d(fTx_{2n+1}, gTx_{2n}, y_{2n}) \\ &\leq \alpha(d(y_{2n}, y_{2n+1}, y_{2n}), d(y_{2n}, y_{2n+1}, y_{2n}), d(y_{2n+1}, y_{2n+2}, y_{2n})) \\ &= \alpha(0, 0, d(y_{2n+1}, y_{2n+2}, y_{2n})) \end{aligned}$$

Since $\alpha \in A$, for some $k \in [0, 1)$, we get

$$d(y_{2n}, y_{2n+1}, y_{2n+2}) \leq k(0) = 0 \tag{6}$$

correspondingly, we get that

$$d(y_{2n-1}, y_{2n}, y_{2n+1}) \le k(0) = 0 \tag{7}$$

(6) and (7) together can be written as

$$d(y_n, y_{n+1}, y_{n+2}) = 0 \text{ forall } n \in N_0 \tag{8}$$

Using the pyramidal inequality and then (8), we get

$$d(y_n, y_{n+2}, u) \le d(y_n, y_{n+2}, y_{n+1}) + \sum_{r=0}^{1} d(y_{n+r}, y_{n+r+1}, u)$$

$$= \sum_{r=0}^{1} d(y_{n+r}, y_{n+r+1}, u)$$

Continuing, for any positive integer p, we get

$$d(y_n, y_{n+p}, u) \le \sum_{r=0}^{p-1} d(y_{n+r}, y_{n+r+1}, u) \tag{9}$$

Upon expanding (9) and then using (5), we get

$$d(y_n, y_{n+p}, u) \le \frac{k^n}{1-k} d(y_0, y_1, u) \tag{10}$$

Eq. (10) shows that $\{y_n\}$ forms a Cauchy sequence in X. Since $f(X)$ is complete subspace of X, there is a $z \in f(X) \subseteq g(X)$ such that as $n \to \infty$, we have $y_n \to z$. Thus, $y_{2n} = fx_{2n} \to z$ and $y_{2n+1} = gx_{2n+1} \to z$ as $n \to \infty$. Since $z \in f(X) \subseteq g(X)$, we can find $x, y \in X$ such that

$$fx = z \text{ and } gy = z \tag{11}$$

Taking $x = x_{2n+1}, y = x$ in (1), we get

$$d(y_{2n+2}, gTx, u) \le \alpha\,(d(z, y_{2n+2}, u), d(z, gTx, u), d(y_{2n+1}, y_{2n+2}, u))$$

As $n \to \infty$, we have

$$d(z, gTx, u) \le \alpha\,(d(z, z, u), d(z, gTx, u), d(z, z, u))$$

which implies $d(z, gTx, u) \le \alpha\,(0, d(z, gTx, u), 0)$ giving $gTx = z$. Thus we have

$$gTx = fx = z \tag{12}$$

Here x and z are the coincidence point and the point of coincidence of f, gT, respectively. Since (f, gT) is weakly compatible pair, we have $fz = f(gTx) = (gT)fx = (gT)z = gTz$. Thus we have

$$fz = gTz \tag{13}$$

Similarly, with $x = y$ and $y = x_{2n}$ in (1) and because (g, fT) is weakly compatible pair, we have the following:

$$fTy = gy = z \tag{14}$$

and

$$gz = fTz \tag{15}$$

Using (12) to (15) in the following, we have

$$\begin{aligned} d(z, fz, u) &= d(fTy, gTz, u) \\ &\leq \alpha\,(d(fz, gy, u), d(fz, gTz, u), d(gy, fTy, u)) \\ &= \alpha\,(d(fz, z, u), d(fz, fz, u), d(z, z, u)) \\ &= \alpha\,(d(fz, z, u), 0, 0) \end{aligned}$$

Since $\alpha \in A$, we get $fz = z$ or the fixed point of f is z.

Again, using $d(gz, z, u) = d(fTz, gTx, u)$ and with (1), we get that $gz = z$. Thus, the common fixed point of f, g is z. Now suppose that, T commutes with f at z. Then

$$Tz = T(fz) = (Tf)z = (fT)z = gz = z$$

which implies z is the fixed point of T.

Therefore, the common fixed point of f, g, T is z, and hence $fz = gz = Tz = z$. For uniqueness, suppose w is also a common fixed point of the three maps f, g, T. That is $fw = gw = Tw = w$. Then, using (1), we have

$$\begin{aligned} d(z, w, u) &= d(fz, gw, u) \\ &= d(fTz, gTw, u) \\ &\leq \alpha(d(fw, gz, u), d(fw, gTw, u), d(gz, fTz, u)) \\ &= \alpha(d(z, w, u), d(w, w, u), d(z, z, u)) \end{aligned}$$

Since $\alpha \in A$, we have $d(z, w, u) = 0$ for all $a \in X$ proving $z = w$ and hence the uniqueness.

Example 1 Take $X = [0, 1]$. Define $d : X^3 \rightarrow R_+$ as $d(x_1, x_2, x_3) = min\{| x_1 - x_2 |, | x_2 - x_3 |, | x_3 - x_1 |\}$ where $x_1, x_2, x_3 \in X$. Now, (X, d) forms a 2-metric space. Take the three self-maps f, g, T on X as $fx = \frac{x}{4}, gx = \frac{x}{3}$ and $Tx = \frac{x}{2}$. Then it can be seen that the assumptions of Theorem 1 are satisfied. Thus, by Theorem 1, f, g, T has unique common fixed point which is 0.

Example 2 Let $X = \{1, 2, 3, 4\}$. Define the 2-metric space $d : X^3 \rightarrow R_+$ as

$$d(x, y, z) = \begin{cases} 6 & \text{if}\{x, y, z\} = \{1, 2, 3\} \\ 7 & \text{if}\{x, y, z\} = \{1, 2, 4\} \\ 8 & \text{if}\{x, y, z\} = \{1, 3, 4\} \\ 9 & \text{if}\{x, y, z\} = \{2, 3, 4\} \end{cases}$$

Then (X, d) forms a 2-metric space. Now define the self-maps f, g, T on X as follows:

$f_1 = 3, f_2 = 4, f_3 = 1, f_4 = 2$

$g_1 = 4, g_2 = 1, g_3 = 2, g_4 = 3$ and $T(x) = x$ for all $x \in X$.

Taking $x = 1, y = 2$ and $a = 1$ in Theorem 1,we have

$$d(3, 4, 1) \leq \alpha[d((4, 4, 1), d((4, 1, 1), d((4, 3, 1)$$

$$d(3, 4, 1) \leq \alpha[0, 0, d((3, 4, 1)$$

i.e., $8 \leq \alpha(0, 0, 8)$ which is not true since $\alpha \in A$.

Thus the condition (1) is not satisfied by the self-maps. Also f, g, T has no common fixed point in x. This shows the significance of the main Theorem 1.

Theorem 2 *Let $\alpha \in A$ and $\{m_i\}_{i=1}^{\infty}$ be a nonnegative integer sequence. Suppose that $\{T_i\}_{i=1}^{\infty}$ is a set of self-maps on complete 2-metric space (X, d), that satisfies*

$$d(T_i^{m_i} x, T_j^{m_j} y, u) \leq \alpha(d(x, y, u), d(T_i^{m_i} x, x, u), d(T_j^{m_j} y, y, u)) \tag{16}$$

for all $x, y, u \in X$ and $i, j \in N$. Then $\{T_i\}_{i=1}^{\infty}$ share a unique common fixed point in X.

Proof Write $f_i = T_i^{m_i}$ for $i \in N$. Then (1) reduces to

$$d(f_i x, f_j y, u) \leq \alpha(d(x, y, u), d(f_i x, x, u), d(f_j y, y, u)) \tag{17}$$

Take an arbitrary $x_0 \in X$. For each $n \in N_0$, define $\{x_n\}$ as $x_{n+1} = f_{n+1}(x_n)$. Then, to each $u \in X$,

$$\begin{aligned} d(x_n, x_{n+1}, u) &= d(f_n x_{n-1}, f_{n+1} x_n, u) \\ &\leq \alpha\,(d(x_{n-1}, x_n, u), d(f_n x_{n-1}, x_{n-1}, u), d(f_{n+1} x_n, x_n, u)) \\ &= \alpha\,(d(x_{n-1}, x_n, u), d(x_n, x_{n-1}, u), d(x_{n+1}, x_n, u)) \end{aligned}$$

Since $\alpha \in A$, for some $k \in [0, 1)$, we get

$$d(x_n, x_{n+1}, u) \leq k d(x_n, x_{n-1}, u) \tag{18}$$

which is similar form of inequality (4). So, continuing the proof essentially the same steps as in Theorem 1, it can see that $\{x_n\}$ is a Cauchy sequence in X. The completeness of X provide a $z \in X$ such that letting $n \to \infty$, we have $x_n \to z$.

Now, applying (16) to $d(x_{n+1}, f_n z, u)$, we get $d(z, f_n z, u) = 0$ for each $u \in X$ which implies $f_n z = z$, or z is the common fixed point of f_n for all $n \in N$. Using (16), the uniqueness can be proved. Thus we have $z = f_n z = T_n^{m_n} z$.

Now $T_n z = T_n(f_n z) = T_n(T_n^{m_n} z) = T_n^{m_n+1} z = T_n^{m_n}(T_n z) = f_n(T_n z)$
i.e., $T_n z$ is also a fixed point of f_n for all $n \in N$. The uniqueness of common fixed point f_n shows that $T_n z = z$ for all n. Finally, the uniqueness for T_n can also be proved using (16).

Example 3 Let $X = [0, \infty)$. Define the 2-metric $d : X^3 \to [0, \infty)$ as

$$d(x, y, z) = \begin{cases} xy + yz + zx & \text{if } x \neq y \neq z \neq x \\ 0 & \text{otherwise} \end{cases}$$

Then (X, d) forms a complete 2-metric space. Define $T : X \to X$ as follows:
$T_i(x) = \frac{x}{c+1}$, for all $c \in N$. and $x \in X$.

Take $\{n_i\} = \{2\}$.Then all the conditions of Theorem 2 are satisfied. Thus $\{T_i\}_{i=1}^{\infty}$ share a unique common fixed point which is 0.

Example 4 Take $X = \{(0, 0), (1, 0), a_n, b_n\}$ where $a_n = (1 + \frac{1}{n}, 0)$, $b_n = (0, \frac{1}{n})$ for all $n \in N$. Define the 2-metric $d : X^3 \to [0, \infty)$ as

$$d(x, y, z) = \begin{cases} \frac{1}{8} & \text{if } \{x, y, z\} = \{a_n, b_n, a\} \text{ or } \{a_n, b_n, b\} \text{ for some} \\ & n \in N \text{ or } \{a_n, b_n, a_n\} \text{ or } \{a_n, b_n, b_m\} \text{ for some} \\ & m, n \in N \text{ with } m \neq n \\ \Delta & \text{otherwise} \end{cases}$$

where Δ = area of the triangle xyz. Now, (X, d) forms a complete 2-metric space. Define $T : X \to X$ as follows:

$T_i(0,0) = (1,0)$, $T_i(1,0) = (0,0)$, $T_i\left(1+\frac{1}{n},0\right) = \left(0,\frac{1}{n+i}\right)$, $T_i\left(0,\frac{1}{n}\right) = \left(1+\frac{1}{n+i},0\right)$ for all $i \in N$.

Take the sequence $\{m_i\} = \{1\}$. Taking $x = a = (0,0)$, $y = a_3 = \left(\frac{4}{3},0\right)$ with T_2 and T_1 maps in (16), we get

$$d(T_2(0,0), T_1(\frac{4}{3},0),(0,0)) \leq \alpha[d((0,0),(\frac{4}{3},0),(0,0)),$$
$$d((0,0),(1,0),(0,0)), d(\left(\frac{4}{3},0\right),(0,\frac{1}{4}),(0,0)]$$

i.e., $\frac{1}{8} \leq \alpha(0,0,\frac{1}{8})$ which is not true since $\alpha \in A$.

Thus the condition (16) is not satisfied by $\{T_i\}$ with $m_i = 1$ for all $i \in N$. Therefore, $\{T_i\}_{i=1}^{\infty}$ has no common fixed point.

References

1. Akram,M., Siddiqui,A.A. :A fixed point theorem for $A - contractions$ on a class of generalized metric spaces. Korean J. Math. Sciences.**10**, 1–5 (2003).
2. Akram,M.,Zafar,A.A., Siddiqui,A.A. :A general class of contractions: $A - contractions$. Novi Sad J. Math. **38**, 25–33 (2008).
3. Akram,M.,Zafar,A.A., Siddiqui,A.A. :Common fixed point theorems for self maps of a generalized metric space satisfying A-contraction type condition. Int. Journal of Math. Analysis. **5**, 757–763 (2011).
4. Vinod Bharadwaj,K., Vishal Gupta, Raman Deep.:Some Fixed Point Results for $A - contractions$ in 2-metric spaces and their applications.Miskole Mathematical Notes, **16**, 679–694, (2015).
5. Gahler,V.S. :2-metrische Raume und ihre topologisehe Struktur. Math. Nachr. **26**, 115–118 (1962).
6. Vishal Gupta, Ramandeep Kaur.:Some Common Foxed Point Theorems For a Class of $A - contractions$ on 2-metric space.Int. J. Pure and Appl. Math. **78**, 909–916 (2012).
7. Naidu,S.V.R., Rajendra Prasad,J.:Fixed point theorems in 2-metric spaces. Indian J. pure appl. Math. **17**, 974–993 (1986).
8. Mantu Saha, Debashis Dey.:Fixed Point Theorems For a class of $A - contractions$ on a 2-metric space. Novi Sad J. Math. **40**, 3–8 (2010).
9. Kiyoshi Iseki.:Fixed point theorems in 2-metric spaces.Math.scm Notes. **3**, (1975).
10. Khan,M. S.:On Fixed point theorems in 2-metric space.publ.Inst.math.(Beograd)(NS). **41**, 107–112 (1980).
11. Lal,S. N.,Singh,A. K.:An analogue of Banach's contraction principle for 2-metric spaces.Buletin of the Australian mathematical society. **8**, 137–143 (1978).
12. Rhoades,B. E.:contraction type mappings on a 2-metric space. math. Nachr. **91**, 151–155 (1979).
13. Singh,S. L.,Tiwari,B. M. L., Gupta,V. K. :common Fixed points of commuting mappings in 2-metric spaces an application. math. Nachr. **95**, 293–297 (1980).

Coefficient Bounds for a Subclass of m-Fold Symmetric λ-Pseudo Bi-starlike Functions

Jay M. Jahangiri, G. Murugusundaramoorthy, K. Vijaya, and K. Uma

Abstract In this paper we consider a class of λ-pseudo bi-starlike functions defined by subordination and determine the upper bounds for the first two coefficients of m-fold symmetric functions in this class. We also determine upper bounds for the Fekete–Szegö coefficients of such m-fold symmetric functions. Our findings for certain cases improve some of the previously published results.

1 Introduction and Definitions

Denote by $\mathscr{A}$ the class of functions of the form

$$f(z) = z + \sum_{n=2}^{\infty} a_n z^n, \qquad z \in \mathbb{U}, \tag{1}$$

which are *analytic* in the open unit disk $\mathbb{U} = \{z \in \mathbb{C} : |z| < 1\}$, and also denote by $\mathscr{S}$ the class of all functions in $\mathscr{A}$ which are univalent in $\mathbb{U}$ (e.g., see Duren [3]). It is well known that every function $f \in \mathscr{S}$ has an inverse f^{-1}, defined by

$$f^{-1}(f(z)) = z \quad (z \in \mathbb{U}) \tag{2}$$

and

$$f(f^{-1}(w)) = w \quad \left(|w| < r_0(f); r_0(f) \geq \frac{1}{4}\right). \tag{3}$$

J. M. Jahangiri
Mathematical Sciences, Kent State University, Burton, OH, USA
e-mail: jjahangi@kent.edu

G. Murugusundaramoorthy (✉) · K. Vijaya · K. Uma
Department of Mathematics, School of Advanced Sciences, VIT, Vellore, India
e-mail: gmsmoorthy@yahoo.com; gms@vit.ac.in; kvijaya@vit.ac.in; kuma@vit.ac.in

V. Madhu et al. (eds.), *Advances in Algebra and Analysis*, Trends in Mathematics,
https://doi.org/10.1007/978-3-030-01120-8_14

The inverse function $f^{-1}(w)$ has an analytic continuation to $\mathbb{U}$ with the expansion

$$f^{-1}(w) = w - a_2w^2 + (2a_2^2 - a_3)w^3 - (5a_2^3 - 5a_2a_3 + a_4)w^4 + \cdots. \tag{4}$$

Let σ denote the class of functions $f \in \mathscr{A}$ and of the form (1) which is said to be *bi-univalent* in $\mathbb{U}$ if both $f(z)$ and $f^{-1}(z)$ are univalent in $\mathbb{U}$. Some examples of bi-univalent functions are $z/(1-z)$, $log\sqrt{(1+z)/(1-z)}$, and $-\log(1-z)$. The study of *bi-univalent* functions due mainly to the pioneering work of Srivastava et al.[13] is based on this paper; many researchers (see [5, 6, 8, 13–17]) investigated several interesting subclasses of the class σ and found estimates for the first two coefficients of such functions. No estimates for the general coefficient $|a_n|$; $n > 3$ were investigated up until the publication of the article [9] in 2013. Using Faber polynomial expansions of bi-univalent functions, Jahangiri and Hamidi [9] obtained estimates for the general coefficients $|a_n|$; $n > 3$ of the functions f of the form (1) subject to a given gap series as well as provided bounds for early coefficients of such functions.

For each $f \in \mathscr{S}$ of the form (1), the m-fold symmetric function

$$h(z) = \sqrt[m]{f(z^m)} = z + \sum_{k=1}^{\infty} a_{mk+1} z^{mk+1}, \qquad z \in \mathbb{U}, \tag{5}$$

is also univalent in $\mathbb{U}$ (see Duren [3]). We let $\mathscr{S}_m$ denote the class of m-fold symmetric univalent functions of the form (5). In fact, the functions in the class $\mathscr{S}$ are onefold symmetric. In [15] Srivastava et al. defined m-fold symmetric bi-univalent function σ_m analogues to the concept of *m-fold symmetric univalent functions*. Furthermore, for $f(z) \in \mathscr{S}_m$ of the form (5), Srivastava et al. [15] gave the series expansion for $f^{-1} = g$ as follows:

$$\begin{aligned} g(w) &= w - a_{m+1}w^{m+1} + \left[(m+1)a_{m+1}^2 - a_{2m+1}\right]w^{2m+1} \\ &- \left[\frac{1}{2}(m+1)(3m+2)a_{m+1}^3 - (3m+2)a_{m+1}a_{2m+1} + a_{3m+1}\right]w^{3m+1} + \cdots. \end{aligned} \tag{6}$$

where $f^{-1} = g$. Recently Hamidi and Jahangiri [7] (also see [10]) took a new approach to show that the initial coefficients of classes of m-fold symmetric bi-starlike functions are unpredictable as well as provide an estimate for the general coefficients of such functions subject to a given gap series condition. We note that finding an extremal function for any subclass of m-fold symmetric bi-univalent functions is open. For recent study on m-fold symmetric *bi-univalent* functions σ_m, one can refer to the recent papers [2, 4, 7, 9, 10, 15–17].

Let $\mathscr{P}$ be the class of analytic functions of the form

$$p(z) = 1 + p_1z + p_2z^2 + \cdots$$

and $\Re(p(z)) > 0$ in $\mathbb{U}$. In view of the work of Pommerenke[12], the m-*fold* symmetric function $\mathscr{P}$ in the class p is of the form

$$p(z) = 1 + c_m z^m + c_{2m} z^{2m} + c_{3m} z^{3m} + \cdots .$$

For our present investigation,we assumed that $\phi(z) \in \mathscr{P}$ such that

$$\phi(0) = 1 \quad and \quad \phi'(0) > 0$$

and $\phi(\mathbb{U})$ is symmetric with respect to the real axis and has a series expansion

$$\phi(z) = 1 + B_1 z + B_2 z^2 + B_3 z^3 + \cdots , \quad (B_1 > 0). \tag{7}$$

Let $u(z)$ and $v(z)$ be two analytic functions in $\mathbb{U}$ with $u(0) = v(0) = 0$ and $max|u(z)|, |v(z)| < 1$. We suppose also that

$$u(z) = p_m z^m + p_{2m} z^{2m} + p_{3m} z^{3m} + \cdots , \tag{8}$$

$$v(z) = q_m w^m + q_{2m} w^{2m} + q_{3m} w^{3m} + \cdots . \tag{9}$$

We observe that

$$|b_m| \leq 1, \quad |p_{2m}| \leq 1 - |p_m|, \qquad |q_m| \leq 1 \quad and \quad |q_{2m}| \leq 1 - |q_m|. \tag{10}$$

By simple computations, we have

$$\phi(u(z)) = 1 + B_1 p_m z^m + (B_1 p_{2m} + B_2 p_m^2) z^{2m} + \cdots + \tag{11}$$

$$\phi(v(z)) = 1 + B_1 q_m w^m + (B_1 q_{2m} + B_2 q_m^2) w^{2m} + \cdots + \tag{12}$$

Remark 1 For the case when $m = 1$(onefold symmetry)and suitable special choices of $\phi(z)$ we have

$$\phi(z) = \left(\frac{1+z}{1-z}\right)^{\alpha} = 1 + 2\alpha z + 2\alpha^2 z^2 + \frac{4\alpha^2 + 2\alpha}{3} z^3 + \cdots \quad (0 < \alpha \leq 1), \tag{13}$$

we have $B_1 = 2\alpha$, $B_2 = 2\alpha^2$ and $B_3 = \frac{4\alpha^2+2\alpha}{3}$. On the other hand, if we take

$$\phi(z) = \frac{1 + (1-2\beta)z}{1-z} = 1 + 2(1-\beta)z + 2(1-\beta)z^2 + \cdots \quad (0 \leq \beta < 1), \tag{14}$$

then $B_1 = B_2 = B_3 = 2(1-\beta)$.

The class $\mathscr{L}_\lambda(\alpha)$ of λ-pseudo-starlike functions of order $\alpha (0 \leq \alpha < 1)$ were introduced and investigated by Babalola[1] whose geometric conditions satisfy

$\Re\left(\frac{z(f'(z))^\lambda}{f(z)}\right) > \alpha$. In [1] it was shown that all pseudo-starlike functions are Bazilevič functions of type $(1 - 1/\lambda)$ and of order $\alpha^{1/\lambda}$ and univalent in open unit disk $\mathbb{U}$. Babalola [1] remarked that though for $\lambda > 1$, these classes of λ- starlike functions clone the analytic representation of starlike functions, it is not yet known the possibility of any inclusion relations between them.

In this paper we define the following new subclass λ-pseudo bi-starlike functions of σ_m based on subordination and determine the coefficient bounds of $|a_{m+1}|$ and $|a_{2m+1}|$ for functions both f and f^{-1} which are $m-$fold symmetric analytic functions. Further we determine the Fekete–Szegö result for the function class, and the special cases are stated as corollaries which are new and have not been studied so far. Further we note that the special cases of our results improve the results of Joshi et al.[11] for $|a_3|$ by taking onefold symmetry as a special case.

Definition 1 A function $f \in \sigma_m$ given by (5) is said to be in the class $\mathscr{L}^\lambda_{\sigma_m}(\phi)$, the class of λ-bi-Pseudo-starlike functions based on subordination if the following conditions holds

$$\frac{z(f'(z))^\lambda}{f(z)} \prec \phi(z) \quad and \quad \frac{w(g'(w))^\lambda}{g(w)} \prec \phi(w) \tag{15}$$

where $z, w \in \mathbb{U}$, $\lambda \geq 1$, and g is given by (6).

Remark 2 For $\lambda = 1$ a function $f \in \sigma_m$ is in the class $\mathscr{L}^1_{\sigma_m}(\phi) \equiv \mathscr{S}^*_{\sigma_m}(\phi)$ if

$$\frac{zf'(z)}{f(z)} \prec \phi(z) \quad \text{and} \quad \frac{wg'(w)}{g(w)} \prec \phi(w) \tag{16}$$

where the function $z, w \in \mathbb{U}$ and g is given by (6).

Remark 3 For $\lambda = 2$ a function $f \in \sigma_m$ is in the class $\mathscr{L}^2_{\sigma_m} \equiv \mathscr{G}_{\sigma_m}(\phi)$ if

$$f'(z)\frac{zf'(z)}{f(z)} \prec \phi(z) \quad \text{and} \quad g'(w)\frac{wg'(w)}{g(w)} \prec \phi(w) \tag{17}$$

where $z, w \in \mathbb{U}$ and g is given by (6).

Remark 4 A function $f \in \sigma_m$, is in the class $\mathscr{L}^\lambda_{\sigma_m}(\alpha)$ if

$$\left|arg(\frac{z(f'(z))^\lambda}{f(z)})\right| < \frac{\alpha\pi}{2}, \quad and \quad \left|arg\left(\frac{w(g'(w))^\lambda}{g(w)}\right)\right| < \frac{\alpha\pi}{2}$$

where $z, w \in \mathbb{U}$; $\lambda \geq 1$; $0 < \alpha \leqq 1$ the function $g(w)$ is given by (6).

Remark 5 A function $f(z)$ given by (5) and $f \in \sigma_m$is said to be in the class $\mathscr{L}^\lambda_{\sigma_m}(\beta)$ if the following conditions are satisfied:

$$\Re\left(\frac{z(f'(z))^{\lambda}}{f(z)}\right) > \beta \text{ and } \Re\left(\frac{w(g'(w))^{\lambda}}{g(w)}\right) > \beta, \tag{18}$$

where $z, w \in \mathbb{U}$; $0 \leqq \beta < 1$; $\lambda \geq 1$ and the function g is defined by (6).

In the following section, we obtain coefficient bounds of $|a_{m+1}|$ and $|a_{2m+1}|$ for functions in these new classes, in which both f and f^{-1} are m-fold symmetric analytic functions.

2 Coefficient Bounds for the Class $\mathscr{L}_{\sigma_m}^{\lambda}(\phi)$

In this section we state and prove the following theorem.

Theorem 1 *Let $f(z)$ given by (5) be in the class $\mathscr{L}_{\sigma_m}^{\lambda}(\phi)$, $(\lambda \geqq 1)$. Then*

$$|a_{m+1}| \leqq \frac{B_1\sqrt{2B_1}}{\sqrt{\begin{array}{l}2B_1[(m+1)\lambda-1]^2+|B_1^2\{\lambda(\lambda-1)(m+1)^2+[\lambda(2m-1)-1](m+1)+2\}\\ \quad -2B_2[(m+1)\lambda-1]^2|\end{array}}}. \tag{19}$$

and

$$|a_{2m+1}| \leqq \begin{cases} \left[\dfrac{(m+1)}{2}-\dfrac{[(m+1)\lambda-1]^2}{B_1[(2m+1)\lambda-1]}\right] \\ \times\dfrac{2B_1^3}{\begin{array}{l}2B_1[(m+1)\lambda-1]^2+|B_1^2\{\lambda(\lambda-1)(m+1)^2\\ +[\lambda(2m-1)-1](m+1)+2\}-2B_2[(m+1)\lambda-1]^2|\end{array}} \\ +\dfrac{B_1}{[(2m+1)\lambda-1]}. & |B_1| \geq \dfrac{2[(m+1)\lambda-1]^2}{(m+1)[(2m+1)\lambda-1]} \\ \dfrac{B_1}{[(2m+1)\lambda-1]} & |B_1| \leq 1\dfrac{2[(m+1)\lambda-1]^2}{(m+1)[(2m+1)\lambda-1]}s \end{cases} \tag{20}$$

Proof Let f and $g = f^{-1}$ be in $\mathscr{L}_{\sigma_m}^{\lambda}(\phi)$, $(\lambda \geqq 1)$ .Then there are analytic functions u, v and $u, v : \mathbb{U} \to \mathbb{U}$, with $u(0) = 0$ and $v(0) = 0$ satisfying the following conditions:

$$\frac{z(f'(z))^{\lambda}}{f(z)} = \phi(u(z)) \tag{21}$$

and

$$\frac{w(f'(w))^{\lambda}}{f(w)} = \phi(v(w)) \tag{22}$$

where $\phi(u(z)) \in \mathscr{P}$ and $\phi(v(w)) \in \mathscr{P}$ have the series representation given by (11) and (12) respectively. We also have

$$\frac{z[f'(z)]^{\lambda}}{f(z)} = 1 + ((m+1)\lambda - 1)a_{m+1}z^{m}$$
$$+[((2m+1)\lambda-1)a_{2m+1}+\frac{1}{2}\left(\lambda(\lambda-1)(m+1)^2 - 2\lambda(m+1)+2\right)a_{m+1}^2]z^{2m}\ldots \tag{23}$$

$$\frac{w[g'(w)]^{\lambda}}{g(w)} = 1 - ((m+1)\lambda - 1)a_{m+1}w^{m} + [(1-(2m+1)\lambda)a_{2m+1}+\frac{1}{2}$$
$$\left(\lambda(\lambda-1)(m+1)^2 + 2(2m\lambda-1)(m+1)+2\right)a_{m+1}^2]w^{2m}\ldots. \tag{24}$$

Now, equating the coefficients in (21) and (22), we get

$$((m+1)\lambda - 1)a_{m+1} = B_1 p_m, \tag{25}$$

$$((2m+1)\lambda - 1)a_{2m+1} + \frac{1}{2}\left(\lambda(\lambda-1)(m+1)^2 - 2\lambda(m+1)+2\right)a_{m+1}^2$$
$$= B_1 p_{2m} + B_2 p_m^2, \tag{26}$$

$$-((m+1)\lambda - 1)a_{m+1} = B_1 q_m \tag{27}$$

and

$$(1-(2m+1)\lambda)a_{2m+1} + \frac{1}{2}\left(\lambda(\lambda-1)(m+1)^2 + 2(2m\lambda-1)(m+1)+2\right)a_{m+1}^2$$
$$= B_1 q_{2m} + B_2 q_m^2. \tag{28}$$

From (25) and (27), we get

$$p_m = -q_m \tag{29}$$

and

$$2[(m+1)\lambda-1]^2a_{m+1}^2=B_1^2(p_m^2+q_m^2). \tag{30}$$

Equivalently

$$a_{m+1}^2=\frac{B_1^2(p_m^2+q_m^2)}{2[(m+1)\lambda-1]^2}. \tag{31}$$

Also from (26), (28), and (30), a simple computation shows that

$$\begin{aligned}&\{\lambda(\lambda-1)(m+1)^2+[\lambda(2m-1)-1](m+1)+2\}a_{m+1}^2\\&=B_1(p_{2m}+q_{2m})+B_2(p_m^2+q_m^2)\\&=B_1(p_{2m}+q_{2m})+\frac{2B_2[(m+1)\lambda-1]^2}{B_1^2}a_{m+1}^2.\end{aligned}$$

$$\begin{aligned}&\{B_1^2\{\lambda(\lambda-1)(m+1)^2+[\lambda(2m-1)-1](m+1)+2\}\\&\quad-2B_2[(m+1)\lambda-1]^2\}a_{m+1}^2=B_1(p_{2m}+q_{2m})\end{aligned} \tag{32}$$

Substituting for the coefficients p_{2m} and q_{2m}, from (10), we have

$$\begin{aligned}&|B_1^2\{\lambda(\lambda-1)(m+1)^2+[\lambda(2m-1)-1](m+1)+2\}\\&\quad-2B_2[(m+1)\lambda-1]^2||a_{m+1}^2|=2B_1^3(1-|p_m|^2).\end{aligned}$$

$$|a_{m+1}|^2=\frac{2B_1^3}{2B_1[(m+1)\lambda-1]^2+|B_1^2\{\lambda(\lambda-1)(m+1)^2+[\lambda(2m-1)-1](m+1)+2\}-2B_2[(m+1)\lambda-1]^2|} \tag{33}$$

Therefore, we get

$$|a_{m+1}|\leqq\frac{B_1\sqrt{2B_1}}{\sqrt{2B_1[(m+1)\lambda-1]^2+|B_1^2\{\lambda(\lambda-1)(m+1)^2+[\lambda(2m-1)-1](m+1)+2\}-2B_2[(m+1)\lambda-1]^2|}}.$$

This gives the desired estimate on $|a_{m+1}|$ as asserted in (19). Next, in order to find the bound on $|a_{2m+1}|$, by subtracting (28) from (26), we get

$$2[(2m+1)\lambda - 1]a_{2m+1} = (m+1)[(2m+1)\lambda - 1]a_{m+1}^2 + B_1(p_{2m} - q_{2m}). \tag{34}$$

By a simple computation and using (29) – (34), we get

$$|a_{2m+1}| \leqq \frac{(m+1)}{2}|a_{m+1}|^2 + \frac{B_1|p_{2m} - q_{2m}|}{2[(2m+1)\lambda - 1]} \tag{35}$$

$$\leqq \frac{(m+1)}{2}|a_{m+1}|^2 + \frac{B_1(|p_{2m}| + |q_{2m}|)}{2[(2m+1)\lambda - 1]}$$

$$\leqq \frac{(m+1)}{2}|a_{m+1}|^2 + + \frac{B_1(1 - |p_m|^2)}{[(2m+1)\lambda - 1]}. \tag{36}$$

Again substituting for the coefficients p_m, p_{2m}, q_m, and q_{2m},from (10), we get

$$|a_{2m+1}| \leqq \left[\frac{(m+1)}{2} - \frac{[(m+1)\lambda - 1]^2}{B_1[(2m+1)\lambda - 1]}\right]|a_{m+1}|^2 + \frac{B_1}{[(2m+1)\lambda - 1]}.$$

$$|a_{2m+1}| \leqq \left[\frac{(m+1)}{2} - \frac{[(m+1)\lambda - 1]^2}{B_1[(2m+1)\lambda - 1]}\right]$$
$$\times \frac{2B_1^3}{2B_1[(m+1)\lambda - 1]^2 + |B_1^2\{\lambda(\lambda-1)(m+1)^2 + [\lambda(2m-1) - 1](m+1) + 2\} - 2B_2[(m+1)\lambda - 1]^2|}$$
$$+ \frac{B_1}{[(2m+1)\lambda - 1]}.$$

This completes the proof of Theorem 1.

3 Fekete–Szegö Inequalities for the Function Class $\mathscr{L}_{\sigma_m}^{\lambda}(\phi)$

Making use of the values of a_{m+1}^2 and a_{2m+1} and motivated by the recent work of Zaprawa [18], we prove the following Fekete–Szegö result for the function class $\mathscr{L}_{\sigma_m}^{\lambda}(\phi)$.

Lemma 1 ([18]) *Let $k \in \mathbb{R}$ and $z_1, z_2 \in \mathbb{C}$. If $|z_1| < R$ and $|z_2| < R$ then*

$$|(k+1)z_1 + (k-1)z_2| \leq \begin{cases} 2|k|R, & |k| \geq 1 \\ 2R & |k| \leq 1. \end{cases} \tag{37}$$

Lemma 2 ([18]) *Let $k, l \in \mathbb{R}$ and $z_1, z_2 \in \mathbb{C}$. If $|z_1| < R$ and $|z_2| < R$ then*

$$|(k+l)z_1 + (k-l)z_2| \leq \begin{cases} 2|k|R, & |k| \geq |l| \\ 2|l|R & |k| \leq |l|. \end{cases} \tag{38}$$

Theorem 2 *Let the function $f(z)$ given by (5) be in the class $\mathscr{L}_{\sigma_m}^{\lambda}(\phi)$ and $\eta \in \mathbb{R}$. Then*

$$|a_{2m+1} - \eta a_{m+1}^2| \leq \begin{cases} 2B_1|h(\eta)|, & |h(\eta)| > \frac{1}{2[(2m+1)\lambda-1]}, \\ \frac{B_1}{2[(2m+1)\lambda-1]}, & 0 \leq |h(\eta)| \leq \frac{1}{2[(2m+1)\lambda-1]}. \end{cases} \tag{39}$$

where

$$h(\eta) = \frac{B_1^2(m+1-2\eta)}{2B_1^2\{\lambda(\lambda-1)(m+1)^2 + [\lambda(2m-1)-1](m+1)+2\} - 4B_2[(m+1)\lambda-1]^2}.$$

Proof From (34), we get

$$a_{2m+1} = \frac{(m+1)}{2}a_{m+1}^2 + \frac{B_1(p_{2m} - q_{2m})}{2[(2m+1)\lambda - 1]}. \tag{40}$$

Substituting for a_{m+1}^2 given by (32) and by simple calculation, we get

$$a_{2m+1} - \eta a_{m+1}^2 = B_1\left[\left(h(\eta) + \frac{1}{2[(2m+1)\lambda-1]}\right)p_{2m} + \left(h(\eta) - \frac{1}{2[(2m+1)\lambda-1]}\right)q_{2m}\right],$$

where

$$h(\eta) = \frac{B_1^2(m+1-2\eta)}{2B_1^2\{\lambda(\lambda-1)(m+1)^2 + [\lambda(2m-1)-1](m+1)+2\} - 4B_2[(m+1)\lambda-1]^2}.$$

Since all B_j are real and $B_1 > 0$, we have

$$|a_{2m+1} - \eta a_{m+1}^2| \leq \begin{cases} 2B_1|h(\eta)|, & |h(\eta)| > \frac{1}{2[(2m+1)\lambda-1]}, \\ \frac{B_1}{2[(2m+1)\lambda-1]}, & 0 \leq |h(\eta)| \leq \frac{1}{2[(2m+1)\lambda-1]}. \end{cases}$$

which completes the proof.

Concluding Remarks By specializing the parameter λ and various choices of ϕ for functions in each of these four $m - fold$ symmetric bi-univalent function classes stated in Remarks 2–5, we can easily derive the estimates of the Taylor-Maclaurin coefficients $|a_{m+1}|$, $|a_{2m+1}|$ and Fekete–Szegö functional problems for functions belonging to these new subclasses in Remarks 2 to 5. Further by taking $m = 1$ onefold symmetry, the results presented in this paper have been shown to considerably improve the earlier results of Joshi [11]

References

1. Babalola, K. O.: On λ−pseudo starlike functions, *Journal of Classical Analysis,* **3(2)** 137–147(2013)
2. Bulut, S.: Coefficient estimates for general subclasses of m-fold symmetric analytic bi-univalent functions,Turk J Math, **40** 1386–1397(2016)
3. Duren,P. L.: *Univalent functions,* Springer-Verlag, New York, Berlin, Hiedelberg and Tokyo, 1983
4. Eker,S.S.: Coefficient bounds for subclasses of m-fold symmetric bi-univalent functions, Turk. J Math., **40** 641–646 (2016)
5. Frasin, B. A., Aouf,M. K.: New subclasses of bi-univalent functions, Appl. Math. Lett. **24** 1569–1573 (2011)
6. Goyal, S. P., Goswami,P.: Estimate for initial Maclaurin coefficients of bi-univalent functions for a class defined by fractional derivatives, J. Egyptian Math. Soc., **20** 179–182(2012).
7. Hamidi,S.G., Jahangiri,J.M Un-predictability of the coefficients of m-fold symmetric bi-starlike functions.Internat. J Math.,**25** 1450064 (2014) (8pages).
8. Hayami, T., Owa,S.: Coefficient bounds for bi-univalent functions, Panamer. Math. J., **22** (4) 15–26(2012).
9. Jahangiri, J. M., Hamidi,S. G: Coefficient estimates for certain classes of bi-univalent functionsInternat. J. Math., and Math. Sci., (2013), Article ID 190560.
10. Jahangiri, J. M., Hamidi,S. G.: Advances on the coefficient bounds for m-fold symmetric bi-close-to-convex functions, Tbilisi Math. J., **9** (2) 75–82 (2016)
11. Joshi, S., Joshi,S., Pawar,H.: *On some subclasses of bi-univalent functions associated with pseudo-starlike function*, J. Egyptian Math. Soc. **24** 522–525 (2016).
12. Pommerenke,Ch.: On the coefficients of close-to-convex functions, Michigan.Math.J. **9** 259–269(1962).
13. Srivastava,H. M., Mishra A. K., Gochhayat,P.: Certain subclasses of analytic and bi-univalent functions, Appl. Math. Lett. **23**(10) 1188–1192 (2010).
14. Srivastava,H. M., Bulut,S., Cagler, M., Yagmur,N.:Coefficient estimates for a general subclass of analytic and bi-univalent functions, Filomat. **27** 831–842 (2013).
15. Srivastava, H.M., Sivasubramanian,S., Sivakumar,R.:Initial coefficient bounds for a subclass of m-fold symmetric bi-univalent functions. Tbilisi Math. J.,**7**1–10 (2014)
16. Srivastava,H. M., Gaboury, S., Ghanim,F.: Coefficient estimates for some subclasses of m−fold symmetric bi-univalent functions, Acta Univ. Apulensis Math. Inform. **41** 153–164(2015)
17. Tang,H.,Srivastava,H.M.,Sivasubramanian, S.,Gurusamy,P.: The Fekete–Szegö functional problems for some subclasses of m−fold symmetric bi-univalent functions Journal of Mathematical Inequalities **10**,(4)1063–1092 (2016)
18. Zaprawa,P.: On the Fekete–Szegö problem for classes of bi-univalent functions,Bull. Belg. Math. Soc. Simon Stevin **21**(1) 1–192 (2014).

Laplacian and Effective Resistance Metric in Sierpinski Gasket Fractal

P. Uthayakumar and G. Jayalalitha

Abstract Laplacian operator for functions on fractal field plays a vital role in the study of partial differential equations of nonlinear in fractals. In this paper self-similar fractal Sierpinski gasket is considered with regular harmonic structures, and energy renormalization factor and scaling constant are obtained. Effective resistance presents a metric with which the properties of the fractal and the transmission can be discussed. Hausdorff dimension of Sierpinski gasket fractal is obtained by scaling constant. Spectral dimension of Sierpinski gasket fractal is calculated by using Laplacian and effective resistance metric. Finally the dimensions of the Sierpinski gasket are interpreted.

Keywords Energy renormalization factor · Sierpinski gasket fractal · Effective resistance metric · Hausdorff dimension · Spectral dimension

1 Introduction

The word "fractal" was coined by Benoit Mandelbrot in the year 1975. Formal mathematical definition of fractal states that a fractal is a set for which Hausdorff-Besicovitch dimension of an object strictly exceeds its topological dimension [1]. In general, a fractal is defined as a rough or fragmented geometric object that can be subdivided into parts, each of which is reduced- size copy of the whole [2]. Fractals like the von Koch curve and Sierpinski gasket are weakly described by their topological dimension [3]. Fractal objects are normally self-similar and independent of the scale [5]. Sierpinski discovered a set called the Sierpinski gasket (SG) at the

P. Uthayakumar (✉)
Department of Mathematics, PSNA College of Engineering and Technology, Dindigul, Tamil Nadu, India
e-mail: uthaya20@gmail.com; uthaya20@psnacet.edu.in

G. Jayalalitha
Department of Mathematics, Vels University, Pallavaram, Chennai, Tamil Nadu, India
e-mail: ragaji94@yahoo.com; g.jayalalithamaths.sbs@velsuniv.ac.in

V. Madhu et al. (eds.), *Advances in Algebra and Analysis*, Trends in Mathematics,
https://doi.org/10.1007/978-3-030-01120-8_15

beginning of this century [10]. Initially the physicists have started the analysis on fractal in the field of this ordered materials. The heat and water transfer in disorder materials such as polymers, fractured rocks, etc. are modeled into fractals. Related to this work, the mathematicians have developed the analysis on fractals in materials with irregular or fractal structures [2]. Laplacian on a fractal plays a vital role in analyzing materials with fractal structure. Laplacian operators converge to a refined operator under a proper scaling with dense domain, called the Laplacian on the Sierpinski gasket [6]. Kigami, J. has provided a general result that any Laplacian is regular corresponding to the effective resistance metric [7]. In this research paper, Laplacian and effective resistance metric on a finite Sierpinski fractal are discussed. The main idea for applying the effective resistance is so as to each Dirichlet form (Laplacian) over a finite fractal can be connected with an electrical network consists of resistors [4, 6]. He has proved that the effective resistance is shown metric with which the analytic properties of the fractal are discussed. Also he shows that the similarity dimension (S) of the fractal provides a useful intrinsic notion of dimension. This is the Hausdorff fractal dimension of the fractal object with respect to the effective resistance metric. The similarity dimension (S) is defined as the unique solution with respect to the resistance of the ith component. Kigami, J. has proved that for the post-critically finite self-similar fractal like Sierpinski gasket, the similarity dimension can be expressed in terms of the spectral dimension [6]. In Sect. 2, the construction of the Sierpinski gasket by the iterated function system is discussed. In Sect. 3, energy renormalization constant by Laplacian and then the scaling factor are obtained as the reciprocal of the energy renormalization constant and also the scaling constant is directly obtained. In Sect. 4, effective resistance metric and Laplacian on Sierpinski gasket are applied; thus Hausdorff fractal dimension and spectral fractal dimension are obtained.

2 Construction of Sierpinski Gasket

The construction of the Sierpinski gasket fractal starts with a filled-in equilateral triangle with sides of unit length, which is called as G_0. It is subdivided into four smaller triangles with side length $\frac{1}{2}$, by joining the midpoints of the sides, which are also equilateral triangles. As first iteration process the G_1 is obtained by removing the middle triangle which is rotated by 180° compared to other triangles. The boundary of that equilateral triangle is not removed. In each of the remaining three equilateral triangles, we remove the equilateral triangles formed by the midpoints of the three sides and so on. The set G_n contains 3^n triangles with side length 2^{-n}. Continuing this process, we get the Sierpinski gasket as the limiting case of the sequences $G_0, G_1, G_2, \ldots$ which is given in Fig. 1, and the Sierpinski Gasket is given as $G = \bigcap_{n=0}^{\infty} G_n$ [9].

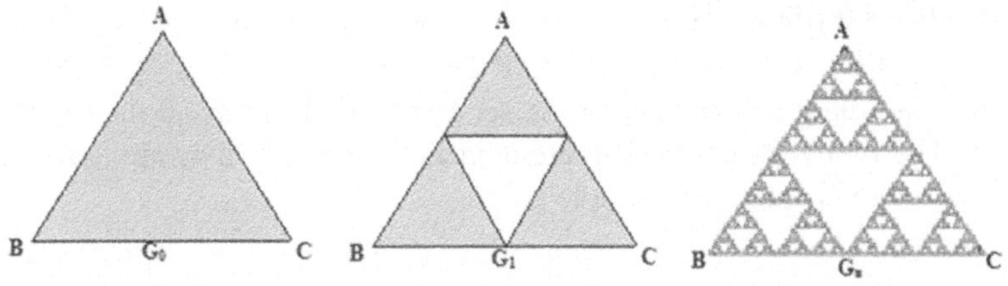

Fig. 1 The Sierpinski gasket

The iterated function system defined on the Sierpinski gasket is given by

$$F_i(x) = \frac{1}{2}(x - p_i) + p_i \quad \text{with fixed point} \quad p_1, p_2, p_3.$$

This yields the following iterated function system. The IFS for the Sierpinski gasket is

$$f_1(x) = Ax$$

$$f_2(x) = Ax + \begin{bmatrix} \frac{1}{2} \\ 0 \end{bmatrix}$$

$$f_3(x) = Ax + \begin{bmatrix} \frac{1}{4} \\ \frac{\sqrt{3}}{4} \end{bmatrix} \quad \text{where } A = \begin{bmatrix} \frac{1}{2} & 0 \\ 0 & \frac{1}{2} \end{bmatrix}$$

Next we apply the IFS on G_1, then we get G_2, and we apply the IFS process endlessly up to infinite times. The resultant graph is called a Sierpinski gasket, and it is the attractor for this IFS process and it is shown in Fig. 1.

3 Laplacian in Sierpinski Gasket

3.1 Laplacian in Finite Graphs [8]

The symmetry operator $H : l(S) \rightarrow l(S)$ which is linear and is known as a Laplacian operator on the set S and $l(S)$ is used for denoting the set of real-valued functions on the set S; then

$$l_0(V_n) = \{f \in l(V_n) : f(p) = 0 \text{ for } p \in V_0\} \tag{1}$$

For two sets S_1 and S_2, it is defined that

$$L(S_1, S_2) = \{H : l(S_1) \rightarrow l(S_2) \text{ and } H \text{ is linear}\}$$

In particular, $L(S)$ means $L(S, S)$.

3.2 *Vertex Degree* [3]

Let G be a simple graph consisting the set $V(G)$ of vertices and the set of $E(G)$ of edges. The two vertices a and b are adjacent vertices if they are associated by exactly one edge $e(a, b)$ in the simple graph.

3.3 *Standard and Normalized Laplacian*

Consider a function $f \in l(V(G))$, the graph (standard) Laplacian of the function f at any vertex x with Laplacian scaling constant $c(a, b)$, is defined by

$$\Delta f(a) = \sum_{e(a,b)\in E(G)} c(a, b)[f(a) - f(b)] \tag{2}$$

The notation $\hat{\Delta}$ denotes discrete Laplacian in a graph which is defined by

$$\hat{\Delta} f(a) = \frac{1}{\deg(a)} \sum_{e(a,b)\in E(G)} [f(a) - f(b)] \tag{3}$$

The matrix H is called the Laplacian matrix, which is symmetric and corresponds to Δ which is defined as

$$H_{i,j} = \begin{cases} 1 & \text{if } i \neq j \\ -\deg(a_i) & \text{if } i = j \text{ and } e(a_i, a_j) \in E(G) \\ 0 & \text{if otherwise} \end{cases} \tag{4}$$

for $x_i, x_j \in V(G)$.
The set (H_n, r) with weight r is the generalized standard Laplacian in the graph G_n, the nth iteration of the fractal graph. The matrix H_n is decomposed into

$$H_n = \begin{pmatrix} T_n & J_n^T \\ J_n & X_n \end{pmatrix} \tag{5}$$

Here $T_n \in L(G_0)$, $J_n \in L(G_0, G_n)$, and $X_n \in L(G_n)$, where G_0 and G_n are the initial and the nth iteration of the fractal graph, respectively. In particular $T = T_1$, $J = J_1$, and $X = X_1$.

Lemma 1 *The renormalization equation which relates the difference operator H_0 on G_0 with the difference operator H_1 on G_1 is defined by* [2]

$$\lambda H_0 = T - J^T X^{-1} J \ .$$

where λ is the renormalization constant.

Proof Let D be the Laplacian matrix on G_0. Then from Eq. (4), we get

$$H_0 = \begin{pmatrix} -2 & 1 & 1 \\ 1 & -2 & 1 \\ 1 & 1 & -2 \end{pmatrix}$$

The matrix corresponding to the standard Laplacian on G_1 is

$$H_1 = \begin{pmatrix} -2 & 0 & 0 & 0 & 1 & 1 \\ 0 & -2 & 0 & 1 & 0 & 1 \\ 0 & 0 & -2 & 1 & 1 & 0 \\ 0 & 1 & 1 & -4 & 1 & 1 \\ 1 & 0 & 1 & 1 & -4 & 1 \\ 1 & 1 & 0 & 1 & 1 & -4 \end{pmatrix}$$

Now from Eq. (5), by considering

$$H_1 = \begin{pmatrix} T & J^T \\ J & X \end{pmatrix}$$

$$T = \begin{pmatrix} -2 & 0 & 0 \\ 0 & -2 & 0 \\ 0 & 0 & -2 \end{pmatrix}, \quad J = \begin{pmatrix} 0 & 1 & 1 \\ 1 & 0 & 1 \\ 1 & 1 & 0 \end{pmatrix}, \quad X = \begin{pmatrix} -4 & 1 & 1 \\ 1 & -4 & 1 \\ 1 & 1 & -4 \end{pmatrix}$$

Applying the above matrices in the Lemma 1, we have

$$\begin{aligned} \lambda \begin{pmatrix} -2 & 1 & 1 \\ 1 & -2 & 1 \\ 1 & 1 & -2 \end{pmatrix} &= \begin{pmatrix} -2 & 0 & 0 \\ 0 & -2 & 0 \\ 0 & 0 & -2 \end{pmatrix} - \begin{pmatrix} 0 & 1 & 1 \\ 1 & 0 & 1 \\ 1 & 1 & 0 \end{pmatrix} \begin{pmatrix} -4 & 1 & 1 \\ 1 & -4 & 1 \\ 1 & 1 & -4 \end{pmatrix} \begin{pmatrix} 0 & 1 & 1 \\ 1 & 0 & 1 \\ 1 & 1 & 0 \end{pmatrix} \\ &= \begin{pmatrix} -2 & 0 & 0 \\ 0 & -2 & 0 \\ 0 & 0 & -2 \end{pmatrix} - \begin{pmatrix} -0.8 & -0.6 & -0.6 \\ -0.6 & -0.8 & -0.6 \\ -0.6 & -0.6 & -0.8 \end{pmatrix} \\ &= \begin{pmatrix} -1.2 & 0.6 & 0.6 \\ 0.6 & -1.2 & 0.6 \\ 0.6 & 0.6 & -1.2 \end{pmatrix} \end{aligned}$$

Solving above matrices, we obtain $\lambda = \frac{3}{5}$, which is the renormalization constant of the Sierpinski gasket. Then H_0 together with the scaling $r = (1, 1, 1)$ is a harmonic structure. And the scaling constant is $c = \frac{1}{\lambda} = \frac{5}{3}$.

Lemma 2 *$[N(1), N(2), N(3)]$ is the gasket formed by the resistances (r_1, r_2, r_3) and corresponds with a self-similar energy with $N = N(1) + N(2) + N(3)$, where c is a constant such that* [10]

$$cr_1 = r_1 + \frac{N(2)N(3)}{N}(r_2 + r_3)$$
$$cr_2 = r_1 + \frac{N(1)N(3)}{N}(r_2 + r_3)$$
$$cr_3 = r_1 + \frac{N(1)N(2)}{N}(r_2 + r_3) \quad (6)$$

Proof For the Sierpinski gasket which is (1,1,1) gasket, and $r_1 = r_2 = r_3$ hence from Eq. (6), we obtain the scaling constant $c = \frac{5}{3}$.

4 Effective Resistance Metric and Laplacian in Sierpinski Gasket

The term effective resistance comes from electrical network analysis. The effective resistance $R(a, b)$ is defined as the resistance relating any two points a and b of the network after restricting that network to just those two points. The effective resistance relating any two vertices of a network or circuit is defined as the ratio of voltage across the nodes to the current flow injected into them. Consider the regular harmonic structure (D, r) on the Sierpinski gasket G and (ϵ, F). The intrinsic metric R between two points $a, b \in G$, called the effective resistance metric, is defined in terms of the Dirichlet form as

$$R(a, b) = [\min\{E(u, u) : u \in F, u(a) = 1, u(b) = 0\}]^{-1}$$

On the other hand, the effective resistance $R(a, b)$ is symbolized as the minimum value such that

$$|u(a) - u(b)|^2 \leq cE(u, u)$$

where the constant c represents the scaling factor of the Laplacian form. Using energy functions the effective resistance metric is defined by

$$R_{eff}(a, b) = \max\left\{\frac{(u(a) - u(b))^2}{\xi(u)};\ u \in Dom\xi \text{ and } \xi(u) > 0\right\} \quad (7)$$

where $\xi(u)$ is the energy of u, and it is defined by

$$\xi(u) = \lim_{m\to\infty} \xi(u)$$

Similarly for any two vertices u and v, it is defined as

$$\lim_{m\to\infty} \xi_m(u,u) = \xi(u,v)$$

The Hausdorff dimension D_H of the Sierpinski gasket fractal is the same as the similarity dimension (S), and then the similarity dimension (S) is described by the unique solution of the equation

$$\alpha = \left\{ s : \sum_{i=1}^{N} r_i^s = 1 \right\}$$

where r_i is the resistance of the ith component. And the sum of the components is denoted by N.

The resistance scaling constant (renormalization factor) is used in the computation of the Hausdorff fractal dimension which corresponds to the effective resistance metric. The Hausdorff fractal dimension of Sierpinski gasket corresponding to the effective resistance metric has been defined as

$$D_H = \frac{\log_e N}{\log_e c} \tag{8}$$

This Hausdorff fractal dimension is depending on the construction of Sierpinski gasket which means that it depends on the inner connection between the parts of Sierpinski gasket. For the existence of the limit, the spectral dimension of the Dirichlet form or Laplacian is defined by

$$D_S = \lim_{x\to\infty} \frac{2\log_e \lambda(x)}{\log_e x} \tag{9}$$

where the eigenvalue $\lambda(x)$ is the Laplacian counting function.
For standard Laplacian or Dirichlet form on Sierpinski gasket, spectral dimension can be calculated by using Hausdorff dimension value of the self-similar fractal:

$$D_S = \frac{2D_H}{D_H + 1} \tag{10}$$

Spectral dimension by the resistance scaling constant (renormalization factor) value is

$$D_S = \frac{2\log_e N}{\log_e(Nc)} \tag{11}$$

By Eq. (8), Hausdorff fractal dimension of Sierpinski Gasket which corresponds to the effective resistance metric is

$$D_H = \frac{\log_e 3}{\log_e(\frac{5}{3})} = \frac{\log_e 3}{\log_e 5 - \log_e 3} = 2.1507$$

This value of Hausdorff fractal dimension which corresponds to the effective resistance metric is different from Hausdorff fractal dimension of Sierpinski gasket which corresponds to the Euclidean metric. By Eq. (10), the spectral dimension of Sierpinski gasket using Hausdorff fractal dimension (D_H) of F which corresponds to the effective resistance metric R is

$$D_s(v) = \frac{2D_H}{D_H + 1} = \frac{2(2.1507)}{2.1507} = 1.3651$$

Now using the scaling constant of Sierpinski Gasket, $c = \frac{5}{3}$, in Eq. (11), the spectral dimension of Sierpinski gasket fractal is

$$D_S = \frac{2\log_e N}{\log_e(Nc)} = \frac{2\log_e 3}{\log_e 5} = 1.3651$$

The spectral fractal dimensions of Sierpinski gasket fractal obtained in both the cases are equal.

5 Conclusion

Here energy renormalization constant is calculated as the scaling factor by using Laplacian matrix method. Resistance scaling factor is obtained as the reciprocal of energy renormalization constant. Finally Hausdorff fractal dimension and spectral fractal dimension are obtained by effective resistance metric and Laplacian in Sierpinski gasket.

References

1. Barlow, M.T., Bass, R.F.: On resistance of the sierpinski carpet. Proc. Roy. Soc. Londan A. **431**,354–360 (1992)
2. Boyle, B., Cekala, K., Ferrone, D., Rifkin, N., Teplyaev, A.: Electrical resistance of N-gasket fractal networks. Pacific Journal of Mathematics, **233**,1 (2007).
3. Ericson, Josh., Pietro Poggi-Corradini., Hainan Zhang.: Effective resistance on graphs and the epidemic quasimetric. arXiv preprint arXiv:1210.1460,(2012).
4. Jorgensen, Palle ET., Erin Peter James Pearse.: A hilbert space approach to effective resistance metric. Complex Analysis and Operator Theory. 4.4, 975–1013(2010)
5. Kenneth Falconer.: Fractal geometry: Mathematical Foundations and Applications. second edition. John Willy & Sons, 2003.
6. Kigami, J.: Effective resistances for harmonic structures on P.C.F. self similar sets. Mathematical Proceedings of the Cambridge philosophical Society. **115**, 291–303(1994)

7. Kigami, Jun.: Harmonic analysis for resistance forms. Journal of Functional Analysis **204** (2),399–444 (2003)
8. Strichartz, R.: Differential equations on fractals. Princeton University Press, 2006.
9. Uthayakumar, P., Jayalalitha, G.: Spectral decimation functions and forbidden eigen values in the graph of level sierpinski triangles. South Asian Journal of Engineering and Technology. **2** 14,47–57 (2016)
10. Yamaguti, M., Hata, M., Kigami, J.: Mathematics of fractals. American Mathematical society, **167**, (1997).

Some Properties of Certain Class of Uniformly Convex Functions Defined by Bessel Functions

V. Srinivas, P. Thirupathi Reddy, and H. Niranjan

Abstract The aim of the present paper is to investigate some characterization for generalized Bessel functions of first kind that is to be subclass of analytic functions. Furthermore, we studied coefficient estimates, radius of starlikeness, convexity, close-to-convexity, and convex linear combinations for the class $UB(\gamma, k, c)$. Finally we proved integral means inequalities for the class.

1 Introduction

Let A be the class of functions f normalized by

$$f(z) = z + \sum_{n=2}^{\infty} a_n z^n \tag{1}$$

which are analytic in the open unit disk $U = \{z : z \in C \text{ and } |z| < 1\}$. We denote by T the subclass of A consisting of functions of the form

$$f(z) = z - \sum_{n=2}^{\infty} a_n z^n, \quad (a_n \geq 0). \tag{2}$$

V. Srinivas
Department of Mathematics, S.R.R. Government Arts & Science College, Karimnagar, Telangana, India
e-mail: vsrinivas079@gmail.com

P. Thirupathi Reddy
Department of Mathematics, Kakatiya Univeristy, Warangal, Telangana, India
e-mail: reddypt2@gmail.com

H. Niranjan (✉)
Department of Mathematics, School of Advanced Sciences, VIT University, Vellore, Tamil Nadu, India
e-mail: niranjan.hari@vit.ac.in

V. Madhu et al. (eds.), *Advances in Algebra and Analysis*, Trends in Mathematics,
https://doi.org/10.1007/978-3-030-01120-8_16

This subclass was introduced and extensively studied by Silverman [4]. Let $T^*(\alpha)$ and $C(\alpha)$ denote the subclasses of T consisting of starlike and convex functions of order α, $(0 \leq \alpha < 1)$, respectively.

In [1], Kanas and Wisniowska introduced the classes $UCV(\alpha, \beta)$ which consists of uniform β-convex functions of order α and $SP(\alpha, \beta)$ which consists of parabolic β-starlike functions of order α, $-1 < \alpha \leq 1, \beta \geq 0$, which generalizes the class UCV and SP, respectively.

The function $f \in A$ belongs to $UCV(\alpha, \beta)$ if it satisfies the condition

$$Re\left\{1 + \frac{zf''(z)}{f'(z)} - \alpha\right\} > \beta\left|\frac{zf''(z)}{f'(z)}\right|, \quad z \in U. \tag{3}$$

The function $f \in A$ belongs to $SP(\alpha, \beta)$ if it satisfies the condition

$$Re\left\{\frac{zf'(z)}{f(z)} - \alpha\right\} > \beta\left|\frac{zf'(z)}{f(z)} - 1\right|, \quad z \in U. \tag{4}$$

Indeed, it follows from (3) and (4) that $f \in UCV(\alpha, \beta)$ if and only if $zf'(z) \in SP(\alpha, \beta)$. The generalized Bessel function of the first kind $\omega = \omega_{p,b,c}$ is defined as the particular solution of the second-order linear homogeneous differential equation.

$$z^2\omega''(z) + bz\omega'(z) + [cz^2 - p^2 + (1-b)p]\omega(z) = 0, \quad p, b, c \in C \tag{5}$$

which is natural generalization of Bessel's equation. This function has the representation

$$\omega(z) = \omega_{p,b,c}(z) = \sum_{n=0}^{\infty} \frac{(-1)^n c^n}{n!\Gamma\left(p + n + \frac{b+1}{2}\right)}\left(\frac{z}{2}\right)^{2n+p}, \quad z \in C \tag{6}$$

where $p, b, c, z \in C$ and $c \neq 0$.

The differential equation (5) permits the study of Bessel, modified Bessel, and spherical Bessel functions all together. Solutions of (5) are referred as generalized Bessel functions of order p. The particular solution given by (6) is called the generalized Bessel function of the first kind of order p. Although the series defined above is convergent everywhere, the function $\omega_{p,b,c}$ is generally not univalent in the open unit disk $U = \{z \in C : |z| < 1\}$. It is worth mentioning that, in particular, when $b = c = 1$.

We obtain the Bessel function of the first kind $\omega_{p,1,1} = J_p$ and $b = 1, c = -1$ the function $\omega_{p,1,-1}$ becomes the modified Bessel function of the first kind I_p. Now consider the function $u_{p,b,c} : C \to C$ defined by the transformation

$$u_{p,b,c}(z) = 2^p\Gamma\left(p + \frac{b+1}{2}\right)z^{1-\frac{p}{2}}\omega_{p,b,c}(\sqrt{z}) \tag{7}$$

Using the well-known Pochhammer symbol (or the shifted factorial) is defined in terms of the Euler Γ function by

$$(a)_n = \frac{\Gamma(a+n)}{\Gamma(a)} = \begin{cases} 1, & \text{if } n = 0; \\ a(a+1)\cdots(a+n-1), & \text{if } n \in N; \end{cases} \quad (a)_0 = 1.$$

We obtain for the function $u_{p,b,c}(z)$ the following representation:

$$u_{p,b,c}(z) = z + \sum_{n=1}^{\infty} \frac{\left(\frac{-c}{4}\right)^n}{n!\left(p+\frac{b+1}{2}\right)_n} z^{n+1} \tag{8}$$

where $\left(p+\frac{b+1}{2}\right) \neq 0, -1, -2, \cdots$. For convenience, we write $u_{p,b,c}(z) = u_{\kappa,c}(z)$.

We have the given below operator $B_\kappa^c : A \to A$ defined by the Hadamard product:

$$\begin{aligned} B_\kappa^c f(z) &= u_{\kappa,c}(z) * f(z) = z + \sum_{n=1}^{\infty} \frac{(-c)^n a_{n+1}}{4^n (\kappa)_n n!} a_n z^{n+1} \\ &= z + \sum_{n=2}^{\infty} \frac{(-c)^{n-1}}{4^{n-1}(\kappa)_{n-1}(n-1)!} a_n z^n = z + \sum_{n=2}^{\infty} D(c,\kappa,n) a_n z^n \end{aligned} \tag{9}$$

$$\text{where } D(c,\kappa,n) = \frac{(-c)^{n-1}}{4^{n-1}(\kappa)_{n-1}(n-1)!}, \quad \kappa = \left(p+\frac{b+1}{2}\right) \neq 0, -1, -2, \cdots. \tag{10}$$

We can easily see that from (9) $z\left[B_{\kappa+1}^c f(z)\right]' = \kappa B_\kappa^c f(z) - (\kappa-1) B_{\kappa+1}^c f(z)$.

The function $B_\kappa^c f(z)$ in (9) is an elementary transformation of the generalized hypergeometric function, so that $B_\kappa^c f(z) = z_0F_1(\kappa; \frac{-c}{4}z) * f(z)$ and $u_{\kappa,c}\left(\frac{-c}{4}z\right) = z_0F_1(\kappa; z)$. For $f \in A$ is given by (1) and $g \in A$ is given by $g(z) = z + \sum_{n=1}^{\infty} b_{n+1} z^{n+1}$, the Hadamard product or convolution of $f(z)$ and $g(z)$ is defined by $(f*g)(z) = (g*f)(z) = z + \sum_{n=1}^{\infty} a_{n+1} b_{n+1} z^{n+1}$, $z \in E$. In this paper, using the operator $B_\kappa^c f(z)$, we define the following new subclass motivated by Ramachandran et al. [3].

Definition 1 Let $c > 1, 0 \leq \gamma < 1, k \geq 0$ and $z \in E$, $f(z) \in UB(\gamma, k, c)$, where f is in the form (1). Then

$$Re\left\{\frac{z\left(B_\kappa^c f(z)\right)'}{B_\kappa^c f(z)} - \gamma\right\} > k\left|\frac{z\left(B_\kappa^c f(z)\right)'}{B_\kappa^c f(z)} - 1\right|. \tag{11}$$

2 Coefficient Estimates

In this section, we obtain the coefficient bounds of function $f(z)$.

Theorem 1 *If $f(z) \in UB(\gamma, k, c)$, where f is in the form (1), then*

$$\sum_{n=2}^{\infty}[n(1+k)-(\gamma+k)]D(c,\kappa,n)|a_n| \leq 1-\gamma, \tag{12}$$

where $0 \leq \gamma < 1$, $k \geq 0$ and $D(c,\kappa,n)$ is given by (10)

Proof It is enough to show that

$$k\left|\frac{z\left(B_\kappa^c f(z)\right)'}{B_\kappa^c f(z)}-1\right| - Re\left\{\frac{z\left(B_\kappa^c f(z)\right)'}{B_\kappa^c f(z)}-1\right\} \leq 1-\gamma.$$

$$\begin{aligned}\text{We have } k\left|\frac{z(B_\kappa^c f(z))'}{B_\kappa^c f(z)}-1\right| - Re\left\{\frac{z(B_\kappa^c f(z))'}{B_\kappa^c f(z)}-1\right\} \\ \leq \quad (1+k)\left|\frac{z(B_\kappa^c f(z))'}{B_\kappa^c f(z)}-1\right| \\ \leq \quad \frac{(1+k)\sum_{n=2}^{\infty}(n-1)D(c,\kappa,n)|a_n||z|^{n-1}}{1-\sum_{n=2}^{\infty}D(c,\kappa,n)|a_n||z|^{n-1}} \\ \leq \quad \frac{(1+k)\sum_{n=2}^{\infty}(n-1)D(c,\kappa,n)|a_n|}{1-\sum_{n=2}^{\infty}D(c,\kappa,n)|a_n|}.\end{aligned}$$

The last expression is bounded above by $(1-\gamma)$ if

$$\sum_{n=2}^{\infty}[n(1+k)-(\gamma+k)]D(c,\kappa,n)|a_n| \leq (1-\gamma)$$

and the proof is complete.

Theorem 2 *Let $0 \leq \gamma < 1$, $k \geq 0$. Then $f \in UB(\gamma, k, c)$, where f is in the form (2), if and only if*

$$\sum_{n=2}^{\infty}[n(1+k)-(\gamma+k)]D(c,\kappa,n)|a_n| \leq (1-\gamma), \tag{13}$$

where $D(c,\kappa,n)$ is given by (10).

Proof In view of the above theorem, it is enough to prove the necessity. If $f \in UB(\gamma, k, c)$ and z is real, then

$$\frac{1-\sum\limits_{n=2}^{\infty} nD(c,\kappa,n)a_n z^{n-1}}{1-\sum\limits_{n=2}^{\infty} D(c,\kappa,n)a_n z^{n-1}} - \gamma > k \left| \frac{\sum\limits_{n=2}^{\infty}(n-1)D(c,\kappa,n)a_n z^{n-1}}{1-\sum\limits_{n=2}^{\infty} D(c,\kappa,n)a_n z^{n-1}} \right| .$$

Along the real axis, $z \to 1$, we get the desired inequality

$$\sum_{n=2}^{\infty}[n(1+k)-(\gamma+k)]D(c,\kappa,n)|a_n| \le (1-\gamma),$$

where $0 \le \gamma < 1,\ k \ge 0$, and $D(c,\kappa,n)$ are given by (10).

Corollary 1 *If $f \in UB(\gamma, k, c)$ then*

$$|a_n| \le \frac{(1-\gamma)}{[n(1+k)-(\gamma+k)]D(c,\kappa,n)}, \tag{14}$$

where $0 \le \gamma < 1,\ k \ge 0$ and $D(c,\kappa,n)$ are given by (10). Equality holds for the function

$$f(z) = z - \frac{(1-\gamma)}{[n(1+k)-(\gamma+k)]D(c,\kappa,n)} z^n. \tag{15}$$

3 Convex Linear Combinations

In this section, we prove that the class $UB(\gamma, k, c)$ is a convex set. And also we prove that if $f \in UB(\gamma, k, c)$ then $f(z)$ is close-to-convex of order δ and starlike of order $\delta,\ 0 \le \delta < 1$.

Theorem 3 *Let $f_1(z) = z$ and*

$$f_n(z) = z - \frac{(1-\gamma)}{[n(1+k)-(\gamma+k)]D(c,\kappa,n)} z^n,\ n \ge 2. \tag{16}$$

Then $f(z) \in UB(\gamma, k, c)$ if and only if it can be in the form

$$f(z) = \sum_{n=1}^{\infty} \omega_n f_n(z),\ \ \omega_n \ge 0,\ \sum_{n=1}^{\infty} \omega_n = 1. \tag{17}$$

Proof Suppose that $f(z)$ can be written as in (17). Then

$$f(z) = z - \sum_{n=2}^{\infty} \omega_n \frac{(1-\gamma)}{[n(1+k)-(\gamma+k)]D(c,\kappa,n)} z^n.$$

Now $\sum_{n=2}^{\infty} \omega_n \frac{(1-\gamma)[n(1+k)-(\gamma+k)]D(c,\kappa,n)}{(1-\gamma)[n(1+k)-(\gamma+k)]D(c,\kappa,n)} = \sum_{n=2}^{\infty} \omega_n = (1-\omega_1) \leq 1.$

Thus $f(z) \in UB(\gamma, k, c)$.

Conversely suppose that $f(z) \in UB(\gamma, k, c)$. Then by using (14), setting,

$$\omega_n = \frac{[n(1+k)-(\gamma+k)]D(c,\kappa,n)}{1-\gamma} a_n, \ n \geq 2 \ \text{and} \ \omega_1 = 1 - \sum_{n=2}^{\infty} \omega_n.$$

Then we have $f(z) = \sum_{n=1}^{\infty} \omega_n f_n(z)$. Hence the theorem.

Theorem 4 *The class $UB(\gamma, k, c)$ is a convex set.*

Proof Let the function

$$f_j(z) = z - \sum_{n=2}^{\infty} a_{n,j} z^n, \quad a_{n,j} \geq 0, \ j = 1, 2. \tag{18}$$

be in the class $UB(\gamma, k, c)$. It is enough to show that the function $h(z)$ defined by $h(z) = \xi f_1(z) + (1-\xi) f_2(z), 0 \leq \xi < 1$ is in the class $UB(\gamma, k, c)$.

Since $h(z) = z - \sum_{n=2}^{\infty} [\xi a_{n,1} + (1-\xi) a_{n,2}] z^n$, with the help of Theorem 2, and by an easy computation, we get

$$\sum_{n=2}^{\infty} [n(1+k)-(\gamma+k)] \xi D(c,\kappa,n) a_{n,1} + \sum_{n=2}^{\infty} [n(1+k)-(\gamma+k)](1-\xi)$$
$$\times D(c,\kappa,n) a_{n,2} \leq \xi(1-\gamma) + (1-\xi)(1-\gamma) \leq (1-\gamma)$$

which implies that $h \in UB(\gamma, k, c)$. Hence, the $UB(\gamma, k, c)$ is convex.

Theorem 5 *If $f \in UB(\gamma, k, c)$, where $f(z)$ is in the form (2), then it is close-to-convex of order δ, $(0 \leq \delta < 1)$ in the disk $|z| < r_1$, where*

$$r_1 = \inf_{n \geq 2} \left[\frac{(1-\delta) \sum_{n=2}^{\infty} [n(1+k)-(\gamma+k)] D(c,\kappa,n)}{n(1-\gamma)} \right]^{\frac{1}{n-1}}, \quad n \geq 2. \tag{19}$$

The result is sharp with the extremal function $f(z)$ by (15).

Proof Given $f \in T$ and f is close-to-convex of order δ, we have

$$|f'(z) - 1| < (1 - \delta). \tag{20}$$

For the left-hand side of (20), we have $|f'(z) - 1| < \sum_{n=2}^{\infty} na_n|z|^{n-1}$.
The right-hand side of the above inequality is less than $(1 - \delta)$. Then

$$\sum_{n=2}^{\infty} \frac{n}{1-\delta} a_n|z|^{n-1} \leq 1.$$

We have $f(z) \in UB(\gamma, k, c)$ if and only if $\sum_{n=2}^{\infty} \frac{[n(1+k)-(\gamma+k)]D(c,\kappa,n)}{(1-\gamma)} a_n \leq 1$.
Thus, (20) will be is true if $\frac{n}{1-\delta} z|^{n-1} \leq \frac{[n(1+k)-(\gamma+k)]D(c,\kappa,n)}{(1-\gamma)}$
or equivalently $|z| \leq \left[\frac{(1-\delta)[n(1+k)-(\gamma+k)]D(c,\kappa,n)}{n(1-\gamma)}\right]^{\frac{1}{n-1}}$
and hence the proof of the theorem.

Theorem 6 *If $f \in UB(\gamma, k, c)$, then $f(z)$ is starlike of order δ, $(0 \leq \delta < 1)$ in the disk $|z| < r_2$, where*

$$r_2 = \inf_{n\geq 2} \left[\frac{(1-\delta) \sum_{n=2}^{\infty} [n(1+k) - (\gamma+k)]D(c, \kappa, n)}{(n-\delta)(1-\gamma)} \right]^{\frac{1}{n-1}}, \quad n \geq 2. \tag{21}$$

The result is sharp with the extremal function given by (15).

Proof Given $f \in T$ and f is starlike of order δ, we have

$$\left| \frac{zf'(z)}{f(z)} - 1 \right| < (1 - \delta). \tag{22}$$

For the left-hand side of (22), we have $\left|\frac{zf'(z)}{f(z)} - 1\right| \leq \frac{\sum_{n=2}^{\infty}(n-1)a_n|z|^{n-1}}{1-\sum_{n=2}^{\infty} a_n|z|^{n-1}}$.
The right-hand side of the above inequality is less than $(1 - \delta)$ if

$$\sum_{n=2}^{\infty} \frac{(n-\delta)}{(1-\delta)} a_n|z|^{n-1} < 1.$$

We have $f(z) \in UB(\gamma, k, c)$ if and only if $\sum_{n=2}^{\infty} \frac{[n(1+k)-(\gamma+k)]D(c,\kappa,n)}{(1-\gamma)} a_n \leq 1$,

(22) is true if $\frac{(n-\delta)}{(1-\delta)}|z|^{n-1} \leq \frac{[n(1+k)-(\gamma+k)]D(c,\kappa,n)}{(1-\gamma)}$
or equivalently $|z|^{n-1} \leq \frac{(1-\delta)[n(1+k)-(\gamma+k)]D(c,\kappa,n)}{(n-\delta)(1-\gamma)}$.
which yield starlikeness of the family.

4 Integral Means Inequalities

In [5], Silverman found that the function $f_2(z) = z - \frac{z^2}{2}$ is often extremal over the family T. He applied this function to resolve his integral means inequality conjectured [5] and settled in [6], that

$$\int_0^{2\Pi} \left|f(re^{i\phi})\right|^\eta d\phi \leq \int_0^{2\Pi} \left|f_2(re^{i\phi})^\eta\right| d\phi,$$

for all $f \in T, \eta > 0$ and $0 < r < 1$. In [6], he also proved his conjecture for the subclasses $T^*(\alpha)$ and $C(\alpha)$ of T.

Now we prove Silverman's conjecture for the class of functions $UB(\gamma, k, c)$. We need the concept of subordination between analytic functions and a subordination theorem of Littlewood [2]. Two functions f and g are analytic in E; the function f is said to be subordinate to g in E if there exists a function ω analytic in E with $\omega(0) = 0, |\omega(z)| < 1, (z \in E)$ such that $f(z) = g(\omega(z)), (z \in E)$. We denote this subordination by $f(z) \prec g(z)$.

Lemma 1 ([2]) *If the functions f and g are analytic in E with $f(z) \prec g(z)$, then for $\eta > 0$ and $z = re^{i\phi}, 0 < r < 1$, $\int_0^{2\Pi} |g(re^{i\phi})|^\eta d\phi \leq \int_0^{2\Pi} |f(re^{i\phi})|^\eta d\phi$.*

Now we discuss the integral means inequalities for functions $f \in UB(\gamma, k, c)$ and $\int_0^{2\Pi} |g(re^{i\phi})|^\eta d\phi \leq \int_0^{2\Pi} |f(re^{i\phi})|^\eta d\phi$.

Theorem 7 *Let $f(z) \in UB(\gamma, k, c), 0 \leq \gamma < 1, k \geq 0$ and $f_2(z)$ be defined by*

$$f_2(z) = z - \frac{1-\gamma}{\phi_2(\gamma, k)} z^2. \tag{23}$$

Then $\int_0^{2\Pi} |f(z)|^\eta d\phi \leq \int_0^{2\Pi} |f_2(z)|^\eta d\phi$, where $z = re^{i\phi}, 0 < r < 1$.

Proof For $f(z) = z - \sum\limits_{n=2}^{\infty} a_n z^n$, (23) is equivalent to

$$\int_0^{2\pi} \left| 1 - \sum_{n=2}^{\infty} a_n z^{n-1} \right|^{\eta} d\phi \leq \int_0^{2\pi} \left| 1 - \frac{1-\gamma}{\phi_2(\gamma, k)} z \right|^{\eta} d\phi.$$

By Lemma 1, it is enough to prove that $1 - \sum\limits_{n=2}^{\infty} a_n z^{n-1} \prec 1 - \frac{1-\gamma}{\phi_2(\gamma,k)} z$.

Assuming $1 - \sum\limits_{n=2}^{\infty} a_n z^{n-} \prec 1 - \frac{1-\gamma}{\phi_2(\gamma,k)} \omega(z)$ and using (13), we obtain

$$\omega(z) = \left| \sum_{n=2}^{\infty} \frac{\phi_2(\gamma, k)}{1-\gamma} a_n z^{n-1} \right| \leq |z| \sum_{n=2}^{\infty} \frac{\phi_2(\gamma, k)}{1-\gamma} a_n \leq |z|,$$

where $\phi_n(\gamma, k) = [n(1+k) - (\gamma + k)] D(c, \kappa, n)$.

Hence, the proof is completed.

References

1. Kanas, S., Wisniowska,A.: Conic regions and K-uniforn convexity. Comput. Appl. Math. **105**, 327–336 (1999)
2. Littlewood, J. E.: On inequalities in the theory of functions. Proc. London Math. Soc. **23(2)**, 481–519 (1925)
3. Ramachandran, Ch., Dhanalakshmi, K., Lakshminarayanan Vanitha.: Certain aspects of univalent function with negative coefficients defined by Bessel function. Brazilian archives of biology and technology. **59**, 1–14 (2016)
4. Silverman, H.: Univalent functions with negative coefficients. Proc. Amer. Math. Soc. **51**, 109–116 (1975)
5. Silverman, H.: A survey with open problems on univalent functions whose coefficient are negative. Rocky Mountain J. Math. **21(3)**, 1099–1125 (1991)
6. Silverman, H.: Integral means for univalent functions with negative coefficient, Houston J. Math., **23(1)**, 169–174 (1997)

A New Subclass of Uniformly Convex Functions Defined by Linear Operator

A. Narasimha Murthy, P. Thirupathi Reddy, and H. Niranjan

Abstract In this paper, we define a new subclass of uniformly convex functions with negative coefficients and obtain coefficient estimates, extreme points, closure and inclusion theorems, and the radii of starlikeness and convexity for the new subclass. Furthermore, results on partial sums are discussed.

1 Introduction

Let A be the class of functions f normalized by

$$f(z) = z + \sum_{n=2}^{\infty} a_n z^n \tag{1}$$

which are analytic in the open unit disk $U = \{z \mid z \in C \text{ and } |z| < 1\}$. We denote by T the subclass of A consisting of functions of the form

$$f(z) = z - \sum_{n=2}^{\infty} a_n z^n, \; a_n \geq 0 \tag{2}$$

A. Narasimha Murthy
Department of Mathematics, Government A. V. V. College, Warangal, Telangana, India
e-mail: sparsha.adluru@gmail.com

P. Thirupathi Reddy
Department of Mathematics, Kakatiya Univeristy, Warangal, Telangana, India
e-mail: reddypt2@gmail.com

H. Niranjan (✉)
Department of Mathematics, School of Advanced Sciences, VIT University, Vellore, Tamil Nadu, India
e-mail: niranjan.hari@vit.ac.in

V. Madhu et al. (eds.), *Advances in Algebra and Analysis*, Trends in Mathematics,
https://doi.org/10.1007/978-3-030-01120-8_17

Following Goodman [2, 3], Ronning [4, 5] introduced and studied the following subclasses: This subclass was introduced and extensively studied by Silverman [8] and Schild and Silverman [7].

1. A function $f \in A$ is said to be in the class $S_p(\alpha, \beta)$ uniformly $\beta-$starlike functions if it satisfies the condition:

$$Re\left\{\frac{zf'(z)}{f(z)} - \alpha\right\} > \beta\left|\frac{zf'(z)}{f(z)} - 1\right|, \quad z \in U \text{ and } -1 < \alpha \leq 1, \beta \geq 0. \tag{3}$$

2. A function $f \in A$ is said to be in the class $UCV(\alpha, \beta)$ uniformly $\beta-$starlike functions if it satisfies the condition:

$$Re\left\{\frac{1 + zf''(z)}{f'(z)} - \alpha\right\} > \beta\left|\frac{zf''(z)}{f'(z)}\right|, \quad z \in U \text{ and } -1 < \alpha \leq 1, \beta \geq 0. \tag{4}$$

Indeed, it follows from (3) and (4) that

$$f \in UCV(\alpha, \beta) \text{ if and only if } zf' \in S_p(\alpha, \beta). \tag{5}$$

For functions $f \in A$ given by (1) and $g(z) \in A$ given by $g(z) = z + \sum_{n=2}^{\infty} b_n z^n$, we define the Hadamard product (or convolution) of f and g by

$$(f * g)(z) = z + \sum_{n=2}^{\infty} a_n b_n z^n. \tag{6}$$

Let $\phi(a, c; z)$ be the incomplete beta function defined by

$$\phi(a, c; z) = z + \sum_{n=2}^{\infty} \frac{(a)_{n-1}}{(c)_{n-1}} z^n, c \neq 0, -1, -2 \cdots, \tag{7}$$

where $(\lambda)_n$ is the Pochhammer symbol defined in terms of the gamma functions, by

$$(\lambda)_n = \frac{\Gamma(\lambda + n)}{\Gamma(\lambda)} = \begin{cases} 1, & \text{if } n = 0; \\ \lambda(\lambda + 1) \cdots (\lambda + n - 1), & \text{if } n \in N; \end{cases} \tag{8}$$

Further, for $f \in A$

$$L(a, c)f(z) = \phi(a, c; z) * f(z) = z + \sum_{n=2}^{\infty} \frac{(a)_{n-1}}{(c)_{n-1}} a_n z^n \tag{9}$$

where $L(a, c)$ is called Carlson–Shaffer operator [1] and the operator $*$ stands for the Hadamard product (or convolution product) of two power series which are given by (6). We notice that $L(a, a)f(z) = f(z)$, $L(2, 1)f(z) = zf'(z)$.

Now, we define a generalized Carlson–Shaffer operator $L(a, c; \gamma)$ by

$$L(a, c; \gamma) f(z) = \phi(a, c; z) * D_\gamma f(z) \tag{10}$$

For a function $f \in A$ where $D_\gamma f(z) = (1 - \gamma) f(z) + \gamma z f'(z),\ (n \geq 0,\ z \in E.)$. So, we have

$$L(a, c; \gamma) f(z) = z - \sum_{n=2}^{\infty} [1 + (n-1)\gamma] \frac{(a)_{n-1}}{(c)_{n-1}} a_n z^n. \tag{11}$$

It is easy to observe that for $\gamma = 0$, we get the Carlson–Shaffer linear operator [1]. For $-1 \leq \alpha < 1$ and $\beta \geq 0$, we let $S(\alpha, \beta, \gamma)$ be the subclass of functions of the form (1), satisfying the analytic criterion.

$$Re\left\{\frac{z\,(L(a, c; \gamma) f(z))'}{L(a, c; \gamma) f(z)} - \alpha\right\} > \beta \left|\frac{z\,(L(a, c; \gamma) f(z))'}{L(a, c; \gamma) f(z)} - 1\right|$$

where $L(a, c; \gamma) f(z)$, we also let (11) and $TS(\alpha, \beta, \gamma) = S(\alpha, \beta, \gamma) \cap T$.

By suitably specializing the values of (a) and (c), the class $S(\alpha, \beta, \gamma)$ can reduce to the class studied earlier by Ronning [5, 6]. Also choosing (a) and (c), the class coincides with the class studied in [11] and [12], respectively.

2 Main Results

Theorem 1 *A function $f(z)$ of the form (1) is in $S(\alpha, \beta, \gamma)$ if*

$$\sum_{n=2}^{\infty} [n(1+\beta) - (\alpha + \beta)][1 + (n-1)\gamma] \frac{(a)_{n-1}}{(c)_{n-1}} |a_n| \leq (1 - \alpha), \quad -1 \leq \alpha < 1, \beta \geq 0, \gamma \geq 0. \tag{12}$$

Proof It suffices to show that

$$\beta \left|\frac{z\,(L(a, c; \gamma) f(z))'}{L(a, c; \gamma) f(z)} - 1\right| - Re\left\{\frac{z\,(L(a, c; \gamma) f(z))'}{L(a, c; \gamma) f(z)} - 1\right\} \leq (1 - \alpha).$$

We have $\beta \left|\frac{z\,(L(a, c; \gamma) f(z))'}{L(a, c; \gamma) f(z)} - 1\right| - Re\left\{\frac{z\,(L(a, c; \gamma) f(z))'}{L(a, c; \gamma) f(z)} - 1\right\}$

$$\leq (1+\beta) \left|\frac{z\,(L(a, c; \gamma) f(z))'}{L(a, c; \gamma) f(z)} - 1\right|$$

$$\leq \frac{(1+\beta) \sum_{n=2}^{\infty} (n-1)[1 + (n-1)\gamma] \frac{(a)_{n-1}}{(c)_{n-1}} |a_n|}{1 - \sum_{n=2}^{\infty} [1 + (n-1)\gamma] \frac{(a)_{n-1}}{(c)_{n-1}} |a_n|}.$$

This last expression is bounded above by $(1 - \alpha)$ if

$$\sum_{n=2}^{\infty}[n(1+\beta)-(\alpha+\beta)][1+(n-1)\gamma]\frac{(a)_{n-1}}{(c)_{n-1}}|a_n| \le (1-\alpha)$$

and this completes the proof of the theorem.

Theorem 2 *A necessary and sufficient condition for $f(z)$ of the form (2) to be in the class $TS(\alpha, \beta, \gamma)$, $-1 \le \alpha < 1, \beta \ge 0, \gamma \ge 0$ is that*

$$\sum_{n=2}^{\infty}[n(1+\beta)-(\alpha+\beta)][1+(n-1)\gamma]\frac{(a)_{n-1}}{(c)_{n-1}}a_n \le (1-\alpha), \text{ and the result is sharp.} \tag{13}$$

Proof In view of Theorem 1, we need only to prove the necessity. If $f(z) \in TS(\alpha, \beta, \gamma)$ and z is real, then

$$\frac{1-\sum_{n=2}^{\infty} n[1+(n-1)\gamma]\frac{(a)_{n-1}}{(c)_{n-1}}a_n z^{n-1}}{1-\sum_{n=2}^{\infty}[1+(n-1)\gamma]\frac{(a)_{n-1}}{(c)_{n-1}}a_n z^{n-1}} - \alpha$$

$$\ge \beta\left|\frac{\sum_{n=2}^{\infty}(n-1)[1+(n-1)\gamma]\frac{(a)_{n-1}}{(c)_{n-1}}a_n z^{n-1}}{1-\sum_{n=2}^{\infty}[1+(n-1)\gamma]\frac{(a)_{n-1}}{(c)_{n-1}}a_n z^{n-1}}\right|.$$

Letting $z \to 1$ along the real axis, we obtain the desired inequality

$$\sum_{n=2}^{\infty}[n(1+\beta)-(\alpha+\beta)][1+(n-1)\gamma]\frac{(a)_{n-1}}{(c)_{n-1}}a_n \le (1-\alpha).$$

Corollary 1 *If $f(z) \in TS(\alpha, \beta, \gamma)$, then*

$$a_n \le \frac{(1-\alpha)}{[n(1+\beta)-(\alpha+\beta)][1+(n-1)\gamma]\frac{(a)_{n-1}}{(c)_{n-1}}}, \text{ for } n \ge 2.$$

The result is sharp for the function

$$f(z) = z - \frac{(1-\alpha)}{[n(1+\beta)-(\alpha+\beta)][1+(n-1)\gamma]\frac{(a)_{n-1}}{(c)_{n-1}}}z^n, \quad n \ge 2. \tag{14}$$

If $\gamma = 0$, we get the result the following result of [4]

Corollary 2 *If* $f(z) \in TS(\alpha, \beta, \gamma)$*, then* $a_n \leq \frac{(1-\alpha)}{[n(1+\beta)-(\alpha+\beta)]\frac{(a)_{n-1}}{(c)_{n-1}}}$*, for* $n \geq 2$*.*

The result is sharp for the function

$$f(z) = z - \frac{(1-\alpha)}{[n(1+\beta)-(\alpha+\beta)]\frac{(a)_{n-1}}{(c)_{n-1}}} z^n, \; n \geq 2. \tag{15}$$

Theorem 3 *Let* $f(z)$ *defined by (2) and* $g(z)$ *defined* $g(z) = z - \sum_{n=2}^{\infty} b_n z^n$ *be in the class* $TS(\alpha, \beta, \gamma)$*. The then function* $h(z)$ *defined by* $h(z) = (1-\lambda)f(z) + \lambda g(z) = z - \sum_{n=2}^{\infty} q_n z^n$*, where* $q_n = (1-\lambda)a_n + \lambda b_n$*,* $0 \leq \lambda < 1$*, is also in the class* $TS(\alpha, \beta, \gamma)$*.*

The following theorem is similar to the proof of the theorem on extreme points given in Silverman [9].

Theorem 4 *Let* $f_1(z) = z$ *and* $f_1(z) = z - \frac{(1-\alpha)(c)_{n-1}}{[n(1+\beta)-(\alpha+\beta)][1+(n-1)\gamma](a)_{n-1}}$ *for* $n = 2, 3, \cdots$*. Then* $f(z) \in TS(\alpha, \beta, \gamma)$ *if and only if* $f(z)$ *can be expressed in the form* $f(z) = \sum_{n=1}^{\infty} \lambda_n f_n(z)$*, where* $\lambda \geq 0$ *and* $\sum_{n=1}^{\infty} \lambda_n = 1$*.*

We prove the following theorem by defining $f_j(z)$ $(j = 1, 2 \cdots m)$ of the form

$$f_j(z) = z - \sum_{n=2}^{\infty} a_{n,j} z^n \text{ for } a_{n,j} \geq 0, z \in E \tag{16}$$

Theorem 5 *Let the function* $f_j(z)$ $(j = 1, 2 \cdots m)$ *defined by (16) be in the class* $TS(\alpha_j, \beta, \gamma)$*, respectively. Then the function* $h(z)$ *defined by* $h(z) = z - \frac{1}{m}\sum_{n=2}^{\infty}\left(\sum_{j=1}^{m} a_{n,j}\right) z^n$ *is in the class* $TS(\alpha, \beta, \gamma)$*, where* $\alpha = \min_{1 \leq j \leq m} \{\alpha_j\}$*, where* $-1 \leq \alpha_j < 1$*.*

Proof Since $f_j(z) \in TS(\alpha_j, \beta, \gamma)$, $(j = 1, 2 \cdots m)$ by applying Theorem 2 to (16), we observe that

$$\sum_{n=2}^{\infty}[n(1+\beta)-(\alpha+\beta)][1+(n-1)\gamma]\frac{(a)_{n-1}}{(c)_{n-1}}\left(\frac{1}{m}\sum_{j=1}^{m} a_{n,j}\right)$$

$$= \frac{1}{m}\sum_{j=1}^{m}\left(\sum_{n=2}^{\infty}[n(1+\beta)-(\alpha+\beta)][1+(n-1)\gamma]\right)\frac{(a)_{n-1}}{(c)_{n-1}} a_{n,j}$$

$$\leq \frac{1}{m}\sum_{j=1}^{m}(1-\alpha) \leq (1-\alpha).$$

which in view of Theorem 2, again implies that $h(z) \in TS(\alpha, \beta, \gamma)$.
Hence the theorem follows.

Theorem 6 *Let the function $f(z)$ defined by (2) be in the class $TS(\alpha, \beta, \gamma)$. Then $f(z)$ close to convex of order δ $(0 \leq \delta < 1)$ in $|z| < r_1$, where*

$$r_1 = \inf_{n \geq 2} \left[\frac{(1-\delta)[n(1+\beta)-(\alpha+\beta)][1+(n-1)\gamma]\frac{(a)_{n-1}}{(c)_{n-1}}}{n(1-\alpha)} \right]^{\frac{1}{n-1}}. \tag{17}$$

The result is sharp, with the extremal function $f(z)$ given by (14).

Proof We must show that

$$|f'(z)-1| \leq 1-\delta, \text{ for } |z| < r_1, \text{ where , } r_1 \text{ is given by (17).} \tag{18}$$

Indeed we have $|f'(z)-1| \leq \sum\limits_{n=2}^{\infty} na_n|z|^{n-1}$. Thus

$$|f'(z)-1| \leq 1-\delta \text{ if } \sum_{n=2}^{\infty} \left(\frac{n}{1-\delta}\right) a_n|z|^{n-1} \leq 1. \tag{19}$$

Using the fact, $f \in TS(\alpha, \beta, \gamma)$ if and only if $\sum\limits_{n=2}^{\infty} \frac{[n(1+\beta)-(\alpha+\beta)][1+(n-1)\gamma]\frac{(a)_{n-1}}{(c)_{n-1}}}{1-\alpha}$

$a_n \leq 1$.

we can say (19) is true if $\left(\frac{n}{1-\delta}\right)|z|^{n-1} \leq \frac{[n(1+\beta)-(\alpha+\beta)][1+(n-1)\gamma]\frac{(a)_{n-1}}{(c)_{n-1}}}{1-\alpha}$.

That is, if $|z| \leq \left[\frac{(1-\delta)[n(1+\beta)-(\alpha+\beta)][1+(n-1)\gamma]\frac{(a)_{n-1}}{(c)_{n-1}}}{n(1-\alpha)}\right]^{\frac{1}{n-1}}$, for $n \geq 2$.

This completes the proof of the theorem.

The following proof of the theorem is similar to above Theorem 6. So we omit the proof.

Theorem 7 *Let the function $f(z)$ defined by (2) be in the class $TS(\alpha, \beta, \gamma)$. Then $f(z)$ is starlike of order δ $(0 \leq \delta < 1)$ in $|z| < r_2$, where*

$$r_2 = \inf_{n \geq 2} \left[\frac{(1-\delta)[n(1+\beta)-(\alpha+\beta)][1+(n-1)\gamma]\frac{(a)_{n-1}}{(c)_{n-1}}}{(n-\delta)(1-\alpha)} \right]^{\frac{1}{n-1}}.$$

The result is sharp, with the extremal function $f(z)$ given by (14).

Using the fact that $f(z)$ is convex if and only if $zf'(z)$ is starlike, we get the following corollary:

Corollary 3 *Let the function $f(z)$ defined by (2) be in the class $TS(\alpha, \beta, \gamma)$. Then $f(z)$ is convex of order $\delta(0 \le \delta < 1)$ in $|z| < r_3$, where*

$$r_3 = \inf_{n\ge2}\left[\frac{(1-\delta)[n(1-\beta)-(\alpha+\beta)][1+(n-1)\gamma]\frac{(a)_{n-1}}{(c)_{n-1}}}{n(n-\delta)(1-\alpha)}\right]^{\frac{1}{n-1}}.$$

The result is sharp with external function given by (14).

3 Partial Sums

Motivated by Silverman [9] and Silvia [10] on partial sums of analytic functions, we consider in this section partial sums of functions in this class $TS(\alpha, \beta, \gamma)$ and obtain sharp lower bounds for the ratios of the real part of $f(z)$ to $f_k(z)$ and $f'(z)$ to $f_k'(z)$.

Theorem 8 *Let $f(z) \in TS(\alpha, \beta, \gamma)$ be given by (1) and define the partial sums $f_1(z)$ and $f_k(z)$ by*

$$f_1(z) = z \text{ and } f_k(z) = z + \sum_{n=2}^{\infty} a_n z^n, \quad (k \in N/I). \tag{20}$$

$$\text{Suppose that } \sum_{n=2}^{\infty} d_n|a_n| \le 1,$$

$$\text{where } d_n = \left(\frac{n(\alpha+\beta)-(\alpha+\beta)}{1-\alpha}\right)[1+(n-1)\gamma]\frac{(a)_{n-1}}{(c)_{n-1}}. \tag{21}$$

$$\text{Then } f \in TS(\alpha, \beta, \gamma). \text{ Further more, } Re\left[\frac{f(z)}{f_k(z)}\right] > 1 - \frac{1}{d_{k+1}},\ z \in E,\ k \in N \tag{22}$$

$$\text{and } Re\left[\frac{f_k(z)}{f(z)}\right] > \frac{d_{k+1}}{1+d_{k+1}}. \tag{23}$$

Proof For the coefficients d_n given by (21), it is not difficult to verify that

$$d_{n+1} > d_n > 1. \tag{24}$$

$$\text{Therefore we have } \sum_{n=2}^{k} |a_n| + d_{k+1}\sum_{n=k+1}^{\infty} |a_n| \le \sum_{n=2}^{\infty} d_n|a_n| \le 1 \tag{25}$$

by using the hypothesis (21). By setting

$$g_1(z) = d_{k+1}\left[\frac{f(z)}{f_k(z)} - \left(1 - \frac{1}{d_{k+1}}\right)\right] = 1 + \frac{d_{k+1}\sum_{n=k+1}^{\infty} a_n z^{n-1}}{1+\sum_{n=2}^{\infty} a_n z^{n-1}} \tag{26}$$

and applying (25), we find that

$$\left|\frac{g_2(z)-1}{g_2(z)+1}\right| \leq \frac{d_k+1\sum\limits_{n=k+1}^{\infty}|a_n|}{2-2\sum\limits_{n=k+1}^{\infty}|a_n|-d_{k+1}\sum\limits_{n=2}^{\infty}|a_n|} \leq 1 \tag{27}$$

which readily yields the assertion (22) of Theorem 8. In order to see that

$$f(z)=z+\frac{z^{k+1}}{d_{k+1}} \text{ gives sharp result, we observe that for } z=re^{\frac{i\pi}{k}} \text{ that} \tag{28}$$

$$\frac{f(z)}{f_k(z)}=1+\frac{z^k}{d_{k+1}} \to 1-\frac{1}{d_{k+1}} \text{ as } z\to 1^-.$$

similarly, if we take

$$g_2(z)-(1+d_{k+1})\left(\frac{f_k(z)}{f(z)}-\frac{d_{k+1}}{1+d_{k+1}}\right)=1-\frac{(1+d_{n+1})\sum\limits_{n=k+1}^{\infty}a_nz^{n-1}}{1+\sum\limits_{n=2}^{\infty}a_nz^{n-1}} \tag{29}$$

and making use of (25), we can deduce that $\left|\frac{g_2(z)-1}{g_2(z)+1}\right| \leq \frac{(1+d_{k+1})\sum\limits_{n=k+1}^{\infty}|a_n|}{2-2\sum\limits_{n=2}^{\infty}|a_n|-(1-d_{k+1})\sum\limits_{n=k+1}^{\infty}|a_n|}$

which leads immediately to the assertion (23) of Theorem 8.

The bound in (23) is sharp for each $k \in N$ with the external function $f(z)$ given by (28). The proof of the Theorem 8 is thus complete.

Theorem 9 *If $f(z)$ of the form (1) satisfies the condition (12), then*

$$Re\left[\frac{f'(z)}{f_k'(z)}\right] \geq 1-\frac{k+1}{d_{k+1}} \tag{30}$$

Proof By setting

$$g(z)=d_{k+1}\left[\frac{f'(z)}{f_k'(z)}\right]-\left(1-\frac{k+1}{d_{k+1}}\right)=\frac{1+\frac{d_{k+1}}{k+1}\sum\limits_{n=k+1}^{\infty}na_nz^{n-1}+\sum\limits_{n=2}^{\infty}na_nz^{n-1}}{1+\sum\limits_{n=2}^{\infty}na_nz^{n-1}}$$

$$=\frac{1+\frac{d_{k+1}}{k+1}\sum\limits_{n=k+1}^{\infty}na_nz^{n-1}}{1+\sum\limits_{n=2}^{\infty}na_nz^{n-1}}$$

$$\left|\frac{g(z)-1}{g(z)+1}\right| \leq \frac{\frac{d_{k+1}}{k+1}\sum_{n=k+1}^{\infty} n|a_n|}{2-2\sum_{n=2}^{\infty} n|a_n|-\frac{d_{k+1}}{k+1}\sum_{n=k+1}^{\infty} n|a_n|}. \tag{31}$$

Now $\left|\frac{g(z)-1}{g(z)+1}\right| \leq 1$ if $\sum_{n=2}^{k} n|a_n| + \frac{d_{k+1}}{k+1}\sum_{n=k+1}^{\infty} n|a_n| \leq 1.$ (32)

Since the left hand side of (32)is bounded above by

$$\sum_{n=2}^{\infty} d_n|a_n| \text{ if } \sum_{n=2}^{\infty}(d_n - n)|a_n| + \sum_{n=k+1}^{\infty} d_n - \frac{d_{k+1}}{k+1} n|a_n| \geq 0. \tag{33}$$

and the proof is complete.

The result is sharp for the extremal function $f(z) = z + \frac{z^{k+1}}{d_{k+1}}$

Theorem 10 *If $f(z)$ of the form (1) satisfies the condition (12), then* $Re\left[\frac{f_k'(z)}{f'(z)}\right] \geq \frac{d_{k+1}}{k+1+d_{k+1}}$.

References

1. Carlson, B. C., Shaffer, S. B.: Starlike and prestarlike hypergeometric functions. SIAM J. Math. Anal. **15**, 737–745 (2002)
2. Goodman, A. W.: On uniformly convex functions. Ann. Polon. Math. **56**, 87–92 (1991)
3. Goodman, A. W.: On uniformly starlike functions. J. Math. Anal. & Appl. **155**, 364–370 (1991)
4. Murugusundaramoorthy, G., Magesh, N.: Linear operators associated with a subclass of uniformly convex functions. Internat. J. Pure Appl. Math. Sci. **4**, 113–125 (2006)
5. Ronning, F.: Uniformly convex functions and a corresponding class of starlike functions. Proc. Am. Math. Soc. **1(18)**, 189–196 (1993)
6. Ronning, F.: Integral representations for bounded starlike functions. Annal. Polon. Math. **60** , 289–297 (1995)
7. Schild, A., Silverman, H.: Convolution of univalent functions with negative co-efficient. Ann. Univ. Marie Curie-Sklodowska Sect. A. **29**, 99–107 (1975)
8. Silverman, H.: Univalent functions with negative coefficients. Proc. Amer. Math. Soc. **51**, 109–116 (1975)
9. Silverman, H.: Partial sums of starlike and convex functions. J. Math. Anal. & Appl. **209**, 221–227 (1997)
10. Silvia, E. M.: Partial sums of convex functions of order α. Houston J. Math. **11**, 517–522 (1985)
11. Subramanian, K. G., Murugusundaramoorthy, G., Balasubrahmanyam, P., Silverman, H.: Sublcasses of uniformly convex and uniformly starlike functions. Math Japonica. **42**, 517–522 (1995)
12. Subramanian, K. G., Sudharsan, T.V., Balasubrahmanyam, P., Silverman, H.: Classes of uniformly starlike functions. Publ. Math. Debrecen. **53**, 309–315 (1998)

Coefficient Bounds of Bi-univalent Functions Using Faber Polynomial

T. Janani and S. Yalcin

Abstract In this research article, we study a bi-univalent subclass Σ related with Faber polynomial and investigate the coefficient estimate $|a_n|$ for functions in the considered subclass with a gap series condition. Also, we obtain the initial two coefficient estimates $|a_2|$, $|a_3|$ and find the Fekete–Szegö functional $|a_3 - a_2^2|$ for the considered subclass. New results which are further examined are also pointed out in this article.

1 Introduction

Let $\mathscr{A}$ denote the class of analytic functions $f(z)$ which is of the form

$$f(z) = z + \sum_{n=2}^{\infty} a_n z^n \tag{1}$$

in open unit disk, $\mathbb{U}$ [$z \in \mathbb{C}$ with $|z| < 1$] (see [8]). Also, let the class of univalent and normalized analytic function in $\mathbb{U}$ be denoted by $\mathscr{S}$. The normalization conditions are $f(0) = 0$ and $f'(0) = 1$.

We know very well that the inverse function f^{-1} exists for every function in $\mathscr{S}$, which are of the form

$$f^{-1}(w) = g(w) = w - a_2 w^2 + (2a_2^2 - a_3)w^3 - (5a_2^3 - 5a_2 a_3 + a_4)w^4 + \cdots, \tag{2}$$

T. Janani (✉)
School of Computer Science and Engineering, Vellore Institute of Technology, Vellore, India
e-mail: janani.t@vit.ac.in

S. Yalcin
Department of Mathematics, Faculty of Arts and Science, Uludag University, Bursa, Turkey
e-mail: syalcin@uludag.edu.tr

V. Madhu et al. (eds.), *Advances in Algebra and Analysis*, Trends in Mathematics,
https://doi.org/10.1007/978-3-030-01120-8_18

where $|w| < r_0(f)$ and $r_0(f) \geq \frac{1}{4}$. Let the bi-univalent function class be denoted by Σ, where both the functions and its inverses in $\mathbb{U}$ are univalent. The initial estimates $|a_2|$ and $|a_3|$ were obtained [5, 20] for bi-starlike $\mathcal{S}^*_\Sigma(\alpha)$ and bi-convex $\mathcal{K}_\Sigma(\alpha)$ subclasses of order α.

Recently, Srivastava et al. [18] analyzed and studied the bi-univalent function class Σ and estimated the first two intial coefficients $|a_2|$ and $|a_3|$ in (1). Though, obtaining coefficients $|a_n|$ $(n = 4, 5, 6\ldots)$ is yet an open problem (see [5, 16, 17, 20]). In order to solve the open problem, many bi-univalent subclasses were studied [6, 13, 15, 18, 19], but the initial two estimates only were obtained. More recently, researchers [3, 10–12] used Faber polynomials [9](also see [4]) to study bi-univalent subclass and obtained the nth coefficient with certain gap series. In [14], new analytic criteria for a subclass of univalent functions were introduced by Janani and Murugusundaramoorthy.

With these motivations, in this work, we consider a bi-univalent subclass and provide bounds for generalized coefficient $|a_n|$ by involving Faber polynomials with a certain gap series. Also, we estimate initial coefficients $|a_2|$, $|a_3|$ and find the Fekete–Szeogö functional $|a_3 - a_2^2|$. The bounds provided in this paper are better estimates than the results provided in [5, 20].

2 Bi-univalent Subclass $\mathcal{G}_\Sigma(\lambda, \alpha)$

Definition 1 A bi-univalent function f of the form (1) is in class $\mathcal{G}_\Sigma(\lambda, \alpha)$, satisfying the below analytic criteria

$$\Re\left(\frac{zf'(z) + z^2 f''(z)}{(1-\lambda)z + \lambda z f'(z)}\right) > \alpha \tag{3}$$

and

$$\Re\left(\frac{wg'(w) + w^2 g''(z)}{(1-\lambda)w + \lambda w g'(w)}\right) > \alpha, \tag{4}$$

where $0 \leq \alpha < 1,\ 0 \leq \lambda \leq 1,\ z, w \in \mathbb{U}$ and g is given by (2).

Example 1 A bi-univalent function f of the form (1) is in class $\mathcal{G}_\Sigma(0, \alpha) \equiv \mathcal{F}_\Sigma(\alpha)$ with $\lambda = 0$, satisfying the below analytic criteria

$$\Re\left(f'(z) + zf''(z)\right) > \alpha$$

and

$$\Re\left(g'(w) + wg''(w)\right) > \alpha$$

where $0 \leq \alpha < 1,\ z, w \in \mathbb{U}$ and g is given by (2).

Example 2 A bi-univalent function f of the form (1) is in class $\mathcal{G}_\Sigma(1,\alpha) \equiv \mathcal{K}_\Sigma(\alpha)$ with $\lambda = 1$, satisfying the below analytic criteria

$$\Re\left(1 + \frac{zf''(z)}{f'(z)}\right) > \alpha$$

and

$$\Re\left(1 + \frac{wg''(w)}{g'(w)}\right) > \alpha$$

where $0 \leq \alpha < 1$, $z, w \in \mathbb{U}$ and g is given by (2).

For functions $f \in \mathcal{K}_\Sigma(\alpha)$ the bounds $|a_2|$ and $|a_3|$ were obtained in [5, 20].

3 Faber Polynomial Expansion for $\mathcal{G}_\Sigma(\lambda, \alpha)$

From literature (see [1, 2] or [21]), we consider the following for our study:

Let $f \in \Sigma$ given by (1) be univalent; it has an inverse $f^{-1} = g$ that has coefficients given by Faber polynomial:

$$g(w) = w + \sum_{n=2}^{\infty} \frac{1}{n} K_{n-1}^{-n}(a_2, \ldots, a_n) w^n, \; w \in \mathbb{U}, \tag{5}$$

where

$$\begin{aligned} K_{n-1}^{-n} &= \frac{(-n)!}{(-2n+1)!\,(n-1)!} a_2^{n-1} \\ &+ \frac{(-n)!}{(2(-n+1))!\,(n-3)!} a_2^{n-3} a_3 \\ &+ \frac{(-n)!}{(-2n+3)!\,(n-4)!} a_2^{n-4} a_4 \\ &+ \frac{(-n)!}{(2(-n+2))!\,(n-5)!} a_2^{n-5} [a_5 + (-n+2) a_3^2] \\ &+ \frac{(-n)!}{(-2n+5)!\,(n-6)!} a_2^{n-6} [a_6 + (-2n+5) a_3 a_4] \\ &+ \sum_{j \geq 7} a_2^{n-j} V_j \end{aligned} \tag{6}$$

and V_j is a jth degree homogeneous polynomial with $7 \leq j \leq n$.

Initial terms of K_{n-1}^{-n} are $-2a_2$, $3(2a_2^2 - a_3)$, $-4(5a_2^3 - 5a_2a_3 + a_4)$ with $n = 1, 2, 3$, respectively.

Considering $D_n^p = D_n^p(a_2, a_3 \ldots)$, the generalized expansion is given as

$$K_n^p = pa_n + \frac{p(p-1)}{2} D_n^2 + \frac{p!}{(p-3)!\,3!} D_n^3 + \ldots + \frac{p!}{(p-n)!\,n!} D_n^n, \quad \text{for any } p \in \mathbb{N}.$$

(For details see [1, 2, 4, 9, 21]).

For functions $f \in \mathcal{G}_\Sigma(\lambda, \alpha)$, we have below form

$$\frac{zf'(z) + z^2 f''(z)}{(1-\lambda)z + \lambda z f'(z)} = 1 + \sum_{n=2}^{\infty} F_{n-1}(a_2, a_3, \ldots, a_n) z^{n-1}, \tag{7}$$

where

$$F_1 = -2(\lambda - 2)a_2 \tag{8}$$

$$F_2 = 4\lambda(\lambda - 2)a_2^2 - 3(\lambda - 3)a_3 \tag{9}$$

$$F_3 = -8\lambda^2(\lambda - 2)a_2^3 + 6\lambda(2\lambda - 5)a_2a_3 - 4(\lambda - 4)a_4 \tag{10}$$

$$\begin{aligned} F_4 = {} & 16\lambda^3(\lambda - 2)a_2^4 - 12\lambda^2(3\lambda - 7)a_2^2a_3 + 9\lambda(\lambda - 3)a_3^2 + 8\lambda(2\lambda - 6)a_2a_4 \\ & -5(\lambda - 5)a_5 \end{aligned} \tag{11}$$

$$\begin{aligned} F_5 = {} & -32\lambda^4(\lambda - 2)a_2^5 + 24\lambda^3(4\lambda - 9)a_2^3a_3 - 18\lambda^2(3\lambda - 8)a_2a_3^2 \\ & -16\lambda^2(3\lambda - 8)a_2^2a_4 \\ & +10\lambda(2\lambda - 7)a_2a_5 + 12\lambda(2\lambda - 7)a_3a_4 - 6(\lambda - 6)a_6 \end{aligned} \tag{12}$$

In general,

$$\begin{aligned} F_{n-1} = {} & \sum_{m_1 + 2m_2 + \ldots + (n-1)m_{n-1} = n-1} \\ & \times (-1)^{(m_1 + m_2 + \ldots m_{n-1})} [2^{m_1} 3^{m_2} \ldots n^{m_{n-1}}] \lambda^{(m_1 + m_2 + \ldots m_{n-1} - 1)} \\ & \times \left\{ \frac{(m_1 + m_2 + \ldots m_{n-1})!}{m_1! m_2! \ldots m_{n-1}!} \lambda - [2P_{m_1} + 3P_{m_2} + \ldots nP_{m_{n-1}}] \right\} \\ & \times a_2^{m_1} a_3^{m_2} \ldots a_n^{m_{n-1}} \end{aligned} \tag{13}$$

is a $(n-1)$th degree Faber polynomial and $F_{n-1} = F_{n-1}(a_2, a_3, \ldots, a_n)$, where

$$P_0 = 0 \text{ and } P_{m_j} = \frac{(m_1 + m_2 + \ldots m_{n-1} - 1)!}{m_1! m_2! \ldots m_{j-1}!(m_j - 1)! m_{j+1}! \ldots m_{n-1}!}, m_j \neq 0, \forall j.$$

Theorem 1 *Let f of the form (1) be bi-univalent function in $\mathcal{G}_\Sigma(\lambda,\alpha)$, then*

$$|a_n| \leq \frac{2(1-\alpha)}{n(n-\lambda)}, \text{ with } a_k = 0, (2 \leq k \leq n-1) \tag{14}$$

where $0 \leq \lambda \leq 1$, $0 \leq \alpha < 1$ and $n \geq 2$.

Proof For every bi-univalent functions of the form (1), when $a_k = 0, (2 \leq k \leq n-1)$, we can write below the expression:

$$\frac{zf'(z) + z^2 f''(z)}{(1-\lambda)z + \lambda z f'(z)} = 1 + \sum_{n=2}^{\infty} n(n-\lambda)a_n z^{n-1}. \tag{15}$$

Its inverse function, g have below the expression:

$$\frac{wg'(w) + w^2 g''(z)}{(1-\lambda)w + \lambda w g'(w)} = 1 + \sum_{n=2}^{\infty} n(n-\lambda)b_n w^{n-1}$$
$$= 1 + \sum_{n=2}^{\infty} n(n-\lambda)\frac{1}{n}K_{n-1}^{-n}(a_2, a_3, \ldots, a_n)w^{n-1}. \tag{16}$$

By definition, as f is in $\mathcal{G}_\Sigma(\lambda,\alpha)$, there exist below functions which have positive real part:

$$p(z) = 1 + \sum_{n=1}^{\infty} p_n z^n,\ \Re\, p(z) > 0 \text{ and } q(w) = 1 + \sum_{n=1}^{\infty} q_n w^n,\ \Re\, q(w) > 0,\ z, w \in \mathbb{U}.$$

Hence, we write

$$\frac{zf'(z) + z^2 f''(z)}{(1-\lambda)z + \lambda z f'(z)} = 1 + (1-\alpha)\sum_{n=1}^{\infty} \mathrm{K}_n^1(p_1, p_2, \ldots, p_n)z^n \tag{17}$$

and

$$\frac{wg'(w) + w^2 g''(z)}{(1-\lambda)w + \lambda w g'(w)} = 1 + (1-\alpha)\sum_{n=1}^{\infty} \mathrm{K}_n^1(q_1, q_2, \ldots, q_n)w^n. \tag{18}$$

From (15) and (17), we get

$$n(n-\lambda)a_n = (1-\alpha)\mathrm{K}_{n-1}^1(p_1, p_2, \ldots, p_{n-1}). \tag{19}$$

and from (16) and (18), we get

$$-n(n-\lambda)\frac{1}{n}\mathrm{K}_{n-1}^{-n}(a_2, a_3, \ldots, a_n) = (1-\alpha)\mathrm{K}_{n-1}^{1}(q_1, \ldots, q_{n-1}). \tag{20}$$

We note here that $b_n = -a_n$, for $a_k = 0$, $(2 \leq k \leq n-1)$. Thus, we get

$$n(n-\lambda)a_n = (1-\alpha)p_{n-1}, \tag{21}$$

$$-n(n-\lambda)a_n = (1-\alpha)q_{n-1}. \tag{22}$$

From (21) and (22), we get

$$a_n = \frac{(1-\alpha)(p_{n-1} - q_{n-1})}{2n(n-\lambda)}.$$

By Caratheodory Lemma [7],

$$|p_n| \leq 2 \text{ and } |q_n| \leq 2 \quad \text{for each} \quad n = 1, 2, 3 \ldots \tag{23}$$

Hence, we have

$$|a_n| \leq \frac{2(1-\alpha)}{n(n-\lambda)}, (n \geq 2). \tag{24}$$

For $\lambda = 0$ and 1, we state below the corollaries:

Corollary 1 *Let f be function in $\mathscr{F}_\Sigma(\alpha)$ is as given in (1), then*

$$|a_n| \leq \frac{2(1-\alpha)}{n^2}, \ \textit{with } a_k = 0, (2 \leq k \leq n-1) \tag{25}$$

where $0 \leq \alpha < 1$ and $n \geq 2$.

Corollary 2 *Let f be function in $\mathscr{K}_\Sigma(\alpha)$ is as given in (1), then*

$$|a_n| \leq \frac{2(1-\alpha)}{n(n-1)}, \ \textit{with } a_k = 0, (2 \leq k \leq n-1) \tag{26}$$

where $0 \leq \alpha < 1$ and $n \geq 2$.

Theorem 2 *Let f of the form (1) be bi-univalent function in $\mathscr{G}_\Sigma(\lambda, \alpha)$,, then*

$$(i)\ |a_2| \leq \begin{cases} \sqrt{\frac{2(1-\alpha)}{3(3-\lambda)}}, & 0 \leq \alpha < \frac{1+5\lambda-2\lambda^2}{3(3-\lambda)}, \\ \frac{1-\alpha}{2-\lambda}, & \frac{1+5\lambda-2\lambda^2}{3(3-\lambda)} \leq \alpha < 1 \end{cases}$$

$$(ii)\ |a_3| \leq \frac{2(1-\alpha)}{3(3-\lambda)}$$

$$(iii)\ |a_3 - 2a_2^2| \leq \frac{2(1-\alpha)}{3(3-\lambda)}$$

where $0 \leq \lambda \leq 1$ and $0 \leq \alpha < 1$.

Proof For functions f of the form (1), from (19) and (20), when $n = 2$ and 3, we get,

$$2(2-\lambda)a_2 = (1-\alpha)p_1, \tag{27}$$

$$3(3-\lambda)a_3 = (1-\alpha)p_2, \tag{28}$$

$$-2(2-\lambda)a_2 = (1-\alpha)q_1, \tag{29}$$

$$3(3-\lambda)(2a_2^2 - a_3) = (1-\alpha)q_2. \tag{30}$$

From Eqs. (27) and (29), we get

$$4(2-\lambda)a_2 = (1-\alpha)(p_1 - q_1). \tag{31}$$

By taking absolute values and by Eq. (23), we get

$$|a_2| \leq \frac{1-\alpha}{2-\lambda}.$$

From Eqs. (28) and (30), we have

$$6(3-\lambda)a_2^2 = (1-\alpha)(p_2 + q_2). \tag{32}$$

Applying Caratheodory Lemma [7], we get

$$|a_2| \leq \sqrt{\frac{2(1-\alpha)}{3(3-\lambda)}}.$$

The $|a_2|$ estimate bound obtained in Theorem 2 (i) follows, if $\frac{1+5\lambda-2\lambda^2}{3(3-\lambda)} \leq \alpha < 1$, then

$$\frac{1-\alpha}{2-\lambda} \leq \sqrt{\frac{2(1-\alpha)}{3(3-\lambda)}}.$$

From Eq. (28), we get

$$a_3 = \frac{(1-\alpha)p_2}{3(3-\lambda)}. \tag{33}$$

Again, by taking absolute values and by Eq. (23), we get

$$|a_3| \leq \frac{2(1-\alpha)}{3(3-\lambda)}.$$

From Eq. (30), we get

$$2a_2^2 - a_3 = \frac{(1-\alpha)q_2}{3(3-\lambda)}. \tag{34}$$

Using Caratheodory Lemma [7], we obtain

$$|a_3 - 2a_2^2| \leq \frac{2(1-\alpha)}{3(3-\lambda)}.$$

Corollary 3 *For* $\lambda = 0$ and $0 \leq \alpha < 1$, *let* f *of the form (1) be function in* $\mathscr{F}_\Sigma(\alpha)$, *then*

$$(i) \qquad |a_2| \leq \begin{cases} \sqrt{\frac{2(1-\alpha)}{9}}, & 0 \leq \alpha < \frac{1}{9}, \\ \frac{1-\alpha}{2}, & \frac{1}{9} \leq \alpha < 1. \end{cases}$$

$$(ii) \qquad |a_3| \leq \frac{2(1-\alpha)}{9}.$$

$$(iii) \qquad |a_3 - 2a_2^2| \leq \frac{2(1-\alpha)}{9}.$$

Corollary 4 *For* $\lambda = 1$ and $0 \leq \alpha < 1$, *let* f *of the form (1) be function in* $\mathscr{K}_\Sigma(\alpha)$, *then*

$$(i) \qquad |a_2| \leq \begin{cases} \sqrt{\frac{1-\alpha}{3}}, & 0 \leq \alpha < \frac{2}{3}, \\ 1-\alpha, & \frac{2}{3} \leq \alpha < 1. \end{cases}$$

$$(ii) \qquad |a_3| \leq \frac{1-\alpha}{3}.$$

$$(iii) \qquad |a_3 - 2a_2^2| \leq \frac{1-\alpha}{3}.$$

Remark 1 The estimates $|a_2|$ and $|a_3|$ stated in the Corollary 4 improved the estimates obtained in [5, 20].

References

1. Airault, H., Bouali, A.: Differential calculus on the Faber polynomials. Bulletin des Sciences Mathematiques. **130**, no 3, 179–222 (2006)
2. Airault, H., Ren, J.: An algebra of differential operators and generating functions on the set of univalent functions. Bulletin des Sciences Mathematiques. **126**, no 5, 343–367 (2002)
3. Altinkaya, S., Yalcin, S.: Faber polynomial coefficient bounds for a subclass of bi-univalent functions. Comptes Rendus Mathematique. **353**, no 12, 1075–1080 (2015)
4. Bouali, A.: Faber polynomials, Cayley-Hamilton equation and Newton symmetric functions. Bulletin des Sciences Mathematiques. **130**, no 1, 49–70 (2006)
5. Brannan, D.A., Taha, T.S.: On some classes of bi-univalent functions. Studia Universitatis Babes-Bolyai Mathematica. **31**, no 2, 70–77 (1986)
6. Çağlar, M., Orhan, H., Yağmur, N.: Coefficient bounds for new subclasses of bi-univalent functions. Filomat. **27**, 1165–1171 (2013)
7. Duren, P.L.: Coefficients of meromorphic schlicht functions. Proceedings of the American Mathematical Society. **28**, 169–172 (1971)
8. Duren, P.L.: Univalent functions. Springer Science and Business Media. **259**, (2001)
9. Faber, G.: About polynomial evolutions. Mathematische Annalen. **57**, no 3, 389–408 (1903)
10. Hamidi, S.G., Halim, S.A., Jahangiri, J.M.: Coefficient estimates for a class of meromorphic bi-univalent functions. Comptes Rendus Mathematique. **351**, no 9, 349–352 (2013)
11. Hamidi, S.G., Halim, S.A., Jahangiri, J.M.: Faber Polynomial Coefficient Estimates for Meromorphic Bi-Starlike Functions. International Journal of Mathematics and Mathematical Sciences. **2013**, 1–4 (2013)
12. Hamidi, S.G., Janani, T., Murugusundaramoorthy, G., Jahangiri, J.M.: Coefficient estimates for certain classes of meromorphic bi-univalent functions. Comptes Rendus Mathematique. **352**, no 4, 277–282 (2014)
13. Hussain, S., Khan, S., Zaighum, M.A., Darus, M., Shareef, Z.: Coefficients bounds for certain subclass of biunivalent functions associated with Ruscheweyh-Differential operator. Journal of Complex Analysis. article no 2826514 (2017).
14. Janani, T., Murugusundaramoorthy, G.: Inclusion results on subclasses of Starlike and Convex functions associated with Struve functions. Italian Journal of Pure and Applied Mathematics. **32**, 467–476 (2014)
15. Khan, S., Khan, N., Hussain, S., Ahmad, Q.Z., Zaighum, M.A.: Some subclasses of bi-univalent functions associated with srivastava-attiya operator. Bulletin of Mathematical Analysis and Applications. **9**, no 2, 37–44 (2017)
16. Lewin, M.: On a coefficient problem for bi-univalent functions. Proceedings of the American Mathematical Society. **18**, 63–68 (1967)
17. Netanyahu, E.: The minimal distance of the image boundary from the origin and the second coefficient of a univalent function in $|z| < 1$. Archive for Rational Mechanics and Analysis. **32**, 100–112 (1969)
18. Srivastava, H.M., Mishra, A.K., Gochhayat, P.: Certain subclasses of analytic and bi-univalent functions. Applied Mathematics Letters. **23**, 1188–1192 (2010)
19. Srivastava, H.M., Bulut, S., Çağlar, M., Yağmur, N.: Coefficient estimates for a general subclass of analytic and bi-univalent functions. Filomat. **27**, 831–842 (2013)
20. Taha, T.S.: Topics in Univalent Function Theory. Ph.D. Thesis. University of London (1981)
21. Todorov, P.G.: On the Faber polynomials of the univalent functions of class. Journal of Mathematical Analysis and Applications. **162**, no 1, 268–276 (1991)

Convexity of Polynomials Using Positivity of Trigonometric Sums

Priyanka Sangal and A. Swaminathan

Abstract Positivity of trigonometric polynomials is of interest for more than a century because of its applications. In this work, we use positivity of trigonometric sine and cosine sums to find the convexity of a polynomial $f(z) = \sum_{k=1}^{n} a_k z^k$. Further, we also investigate the radius of convexity r such that $f(\mathbb{D}_\rho)$ is convex where $\mathbb{D}_\rho = \{z; |z| \leq \rho, 0 < \rho < 1\}$.

1 Preliminaries

Let $\mathscr{A}$ be the class of analytic function on the unit disc $\mathbb{D} := \{z : |z| < 1\}$. Let $\mathscr{S}$ be the class of functions which are one to one in the unit disc $\mathbb{D}$ having the normalization $f(0) = 0 = f'(0) - 1$. A function $f \in \mathscr{S}$ has the following power series representation:

$$f(z) = z + \sum_{k=2}^{\infty} a_k z^k, \quad z \in \mathbb{D}.$$

Among several subclasses of $\mathscr{S}$ having nice geometric properties, we are interested in the class of convex functions. A domain $\Omega \in \mathbb{C}$ is called convex if for every pair of points in Ω, the line segment joining these two points also lie in Ω. If a function maps $\mathbb{D}$ onto a convex domain, then it is said to be a convex function. Analytically, a function $f(z) \in \mathscr{S}$ is convex in $\mathbb{D}$ if it satisfies

$$\text{Re}\left(1 + \frac{zf''(z)}{f'(z)}\right) > 0, \quad z \in \mathbb{D}.$$

P. Sangal (✉) · A. Swaminathan
Indian Institute of Technology Roorkee, Roorkee, India
e-mail: sangal.priyanka@gmail.com; priyadma@iitr.ac.in; mathswami@gmail.com; swamifma@iitr.ac.in

V. Madhu et al. (eds.), *Advances in Algebra and Analysis*, Trends in Mathematics,
https://doi.org/10.1007/978-3-030-01120-8_19

The set of all convex univalent functions is denoted by $\mathscr{C}$. The radius of convexity of function $f \in \mathscr{S}$ is the number r which is the largest radius such that $f(z)$ is convex in $|z| < r$. It is known [2, Theorem 2.13] that the radius of convexity for class $\mathscr{S}$ is $2 - \sqrt{3} = 0.267\ldots$. In this manuscript, we are interested in finding the convexity of polynomials of the form $p(z) = z + a_2z^2 + \cdots + a_nz^n$ in the unit disc $\mathbb{D}$. For fulfilling this objective, the positivity of trigonometric polynomials will be used as an important tool. The relation between positivity of trigonometric sums and univalent function theory can be seen in [3]. In the direction of positivity of trigonometric sums, the following result of Koumandos [5] is of much importance.

Theorem 1 (Koumandos [5]) *Let $\{a_k\}$ be a sequence of positive real numbers such that*

$$a_0 \geq a_1 \geq a_2 \geq \cdots \geq a_n > 0, \tag{1}$$

$$\text{and} \quad a_{2k} \leq \left(1 - \frac{\beta}{k}\right) a_{2k-1}, \tag{2}$$

then for all positive integers n and $0 < \theta < \pi$,

$$\sum_{k=0}^{n} a_k \cos k\theta > 0 \quad for \quad \beta_0 \leq \beta < 1.$$

Moreover, for the sine sums for $0 < \theta < \pi$,

$$\sum_{k=1}^{2n+1} a_k \sin k\theta > 0 \quad for \quad \beta_0 \leq \beta < 1,$$

$$\sum_{k=1}^{2n} a_k \sin k\theta > 0 \quad for \quad 1/2 \leq \beta < 1,$$

where β_0 is the unique solution in $(0, 1)$ of the equation $\displaystyle\int_0^{3\pi/2} \frac{\cos t}{t^{\beta}} dt = 0.$

Similar type of result has also been obtained by Brown et al. [1]. In comparison with the results obtained by Brown et al. and Theorem 1, the latter one is better, particularly in the context of sine sums. For $\beta = 1/2$, Theorem 1 reduces to the following result by Vietoris [6].

Theorem 2 (Vietoris [6]) *Let $\{a_k\}$ be any sequence of nonnegative real numbers satisfying condition (1) and*

$$a_{2k} \leq \left(1 - \frac{1}{2k}\right) a_{2k-1}, \quad k \geq 1. \tag{3}$$

Then for all positive integers n and $0 < \theta < \pi$, *we have*

$$\sum_{k=0}^{n} a_k \cos k\theta > 0 \quad and \quad \sum_{k=1}^{n} a_k \sin k\theta > 0.$$

For odd sine sums, Theorem 1 is stronger, whereas for even sine sums, Theorem 2 is best possible. We will use these two results as a tool in finding the convexity of polynomials.

2 Main Results

In this section, we apply Theorem 1 to obtain the conditions on the coefficients such that the odd degree polynomial is convex.

Theorem 3 *Let* $\beta_0 = 0.308443\ldots$ *denote the Littlewood-Salem-Izumi number that is the solution of* $\int_0^{3\pi/2} \frac{\cos t}{t^\beta} dt = 0$. *Assume that n is odd and the coefficients of the polynomial*

$$p_n(z) = z + b_2 z^2 + \cdots + b_n z^n$$

satisfy

$$1 = b_1 \geq 2b_2 \geq 3b_3 \geq \cdots \geq nb_n > 0. \tag{4}$$

Let us denote

$$\rho_1 = \min \left\{ \left(\frac{k+1}{k+2} \right)^2 \frac{b_{k+1}}{b_{k+2}}; \quad k \in \{0, 1, 2 \cdots, n-2\} \right\}$$

and

$$\rho_2 = \min \left\{ \left(1 - \frac{\alpha_0}{k}\right) \left(\frac{2k}{2k+1} \right)^2 \frac{b_{2k}}{b_{2k+1}}; \quad k \in \{1, 2, \cdots, [n/2]\} \right\}.$$

Then $p_n(z)$ *is convex in* $|z| < \rho$ *where* $\rho = \min\{\rho_1, \rho_2\}$.

Proof Assume n is odd and $p_n(z) = z + b_2 z^2 + \cdots + b_n z^n = \sum_{k=1}^{n} b_k z^k$ where $b_1 = 1$. Then $p_n'(z) = \sum_{k=1}^{n} k b_k z^{k-1}$ and $p_n''(z) = \sum_{k=1}^{n} k(k-1) b_k z^{k-2}$. Let the coefficients of $p_n(z)$ satisfy the condition

$$1 = b_1 \geq 2b_2 \geq 3b_3 \geq \cdots \geq nb_n > 0.$$

Then Kakeya Enestöm theorem [4] yields that $p'_n(z)$ does not vanish in $\mathbb{D}$. Thus $1 + \dfrac{zp''_n(z)}{p'_n(z)}$ is well-defined analytic function in $\mathbb{D}$. We have to find radius $\rho > 0$ such that

$$\mathrm{Re}\left\{1 + \frac{zp''_n(z)}{p'_n(z)}\right\} > 0 \quad \text{for} \quad |z| < \rho.$$

Now,

$$\begin{aligned}
\mathrm{Re}\left(1 + \frac{zp''_n(z)}{p'_n(z)}\right) &= \mathrm{Re}\left(1 + \frac{\sum_{k=1}^{n} k(k-1)b_k z^{k-1}}{\sum_{k=1}^{n} kb_k z^{k-1}}\right) \\
&= \mathrm{Re}\left(\frac{\sum_{k=1}^{n} k^2 b_k z^{k-1}}{\sum_{k=1}^{n} kb_k z^{k-1}}\right) \\
&= \mathrm{Re}\left(\frac{\sum_{k=0}^{n-1} (k+1)^2 b_{k+1} z^k}{\sum_{k=0}^{n-1} (k+1) b_{k+1} z^k}\right)
\end{aligned}$$

Putting $z = \rho e^{i\theta}$, for $0 < \theta < 2\pi$, we obtain

$$= \frac{(\sum_{k=0}^{n-1} c_k(\rho)\cos k\theta)\cdot(\sum_{k=0}^{n-1} d_k(\rho)\cos k\theta) + (\sum_{k=0}^{n-1} c_k(\rho)\sin k\theta)\cdot(\sum_{k=0}^{n-1} d_k(\rho)\sin k\theta)}{(\sum_{k=0}^{n-1} d_k(\rho)\cos k\theta)^2 + (\sum_{k=0}^{n-1} d_k(\rho)\sin k\theta)^2}$$

where $c_k(\rho) = (k+1)^2 b_{k+1}\rho^k$ and $d_k(\rho) = (k+1)b_{k+1}\rho^k$. To prove our theorem, it is enough to show that all the sums inside the bracket are positive. Since the coefficients of $p_n(z)$ are real, so $p_n(z) = \overline{p_n(\bar{z})}$, i.e., $p_n(z)$ is symmetric with respect to real axis. So using Schwarz reflection principle, it is sufficient to prove the result for $0 < \theta < \pi$. So the sequence $\{c_k(\rho)\}$ and $\{d_k(\rho)\}$ satisfy the conditions of (1) and (2) if

$$\begin{aligned}
c_{k+1}(\rho) \leq c_k(\rho) &\Longrightarrow r^{k+1}(k+2)^2 b_{k+2} \leq (k+1)^2 b_{k+1} r^k \\
&\Longrightarrow r \leq \left(\frac{k+1}{k+2}\right)^2 \frac{b_{k+1}}{b_{k+2}} \quad \text{for} \quad k \in \{0, 1, \cdots, n-2\}.
\end{aligned}$$

Let

$$\rho_1 := \min\left\{\left(\frac{k+1}{k+2}\right)^2 \frac{b_{k+1}}{b_{k+2}}; \quad k \in \{0, 1, 2 \cdots, n-2\}\right\}.$$

Further,

$$c_{2k}(\rho) \le \left(1 - \frac{\beta_0}{k}\right) c_{2k-1}(\rho) \Longrightarrow r^{2k}(2k+1)^2 b_{2k+1} \le \left(1 - \frac{\beta_0}{k}\right)(2k)^2 b_{2k} r^{2k+1}$$

$$\Longrightarrow r \le \left(1 - \frac{\beta_0}{k}\right)\left(\frac{2k}{2k+1}\right)^2 \frac{b_{2k}}{b_{2k+1}}.$$

Let

$$\rho_2 = \min\left\{\left(1 - \frac{\beta_0}{k}\right)\left(\frac{2k}{2k+1}\right)^2 \frac{b_{2k}}{b_{2k+1}}; k \in \{1, 2, \cdots, [n/2]\}\right\}.$$

Define $\rho = \min\{\rho_1, \rho_2\}$. Then, for such ρ, the trigonometric sums are positive in $|z| < \rho$. Hence $p_n(z)$ is convex in $|z| < \rho$. □

Corollary 1 *For odd* n, $p_n(z) = z + qz^2 + q^2z^3 + \cdots + q^{n-1}z^n$ *is convex in* $|z| < \frac{1}{4q}$ *where* $0 < q \le 1$.

Proof For odd n, if we choose $b_k = q^{k-1}$ where $0 < q \le 1$, then b_k satisfy the assumption of Theorem 3; we have

$$\rho_1 = \min\left\{\left(\frac{k+1}{k+2}\right)^2 \frac{b_{k+1}}{b_{k+2}}; k \in \{0, 1, 2 \cdots, n-2\}\right\}$$

$$= \min\left\{\frac{1}{4q}, \frac{4}{9q}, \cdots, \left(\frac{n-1}{n}\right)^2 \cdot \frac{1}{q}\right\} = \frac{1}{4q}$$

and

$$\rho_2 = \min\left\{\left(1 - \frac{\beta_0}{k}\right)\left(\frac{2k}{2k+1}\right)^2 \frac{1}{q}; k \in \{1, 2, \cdots, [n/2]\}\right\}$$

$$= \left\{(1-\beta_0)\frac{4}{9q}, \cdots, \frac{(n-2\beta_0)}{nq}.\frac{n^2}{(n+1)^2}\right\} = \frac{(1-\beta_0)4}{9q}$$

Then $p_n(z)$ is convex in $|z| < \rho = \min\{\rho_1, \rho_2\} = \frac{1}{4q}$.

By choosing particular values of q, Corollary 1 leads us to the following interesting examples:

Example 1 Let $n > 1$ be any odd positive integer. Then $p_n(z) = z + \frac{z^2}{2} + \frac{z^3}{4} + \cdots + \frac{z^n}{2^{n-1}}$ is convex in $|z| < 1/2 = 0.5$.

Example 2 Let $n > 1$ be any odd positive integer. Then $p_n(z) = z + \frac{z^2}{3} + \frac{z^3}{9} + \cdots + \frac{z^n}{3^{n-1}}$ is convex in $|z| < 3/4 = 0.75$.

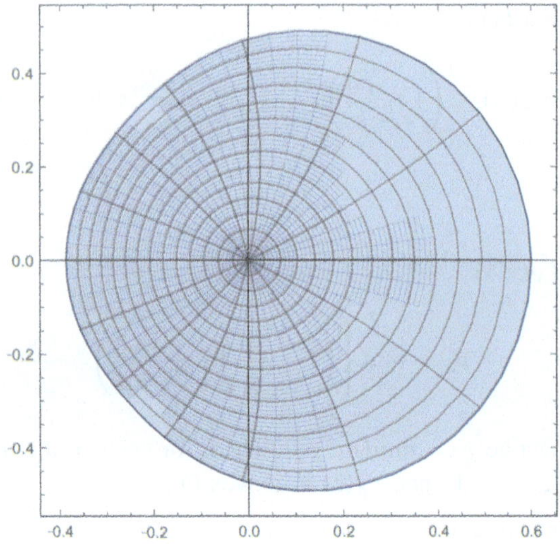

Fig. 1 $p_3(z) = z + \dfrac{z^2}{2} + \dfrac{z^3}{3}$ in $|z| < 0.4610373\ldots$

Example 3 Let $n > 1$ be any odd positive integer. Then $p_n(z) = z + \frac{z^2}{4} + \frac{z^3}{16} + \cdots + \frac{z^n}{4^{n-1}}$ is convex in $|z| < 1$.

Corollary 2 *Let* $b_k = \frac{1}{k}$ *and* $n > 1$ *be any odd positive integer and* $p_n(z) = z + \frac{z^2}{2} + \frac{z^3}{3} + \cdots + \frac{z^n}{n}$. *Then* $p_n(z)$ *is convex in* $|z| < 0.4610373\ldots$.

We know that as $n \to \infty$, $p_n(z) \to -\log(1-z)$ and the family of convex functions is normal family. Hence $-\log(1-z)$ is convex function in $|z| < 0.4610373\ldots$. Figure 1 shows the graph of $p_3(z) = z + \dfrac{z^2}{2} + \dfrac{z^3}{3}$ in $|z| < 0.4610373\ldots$ which is clearly a convex domain.

The next theorem is for convexity of polynomials of even degree.

Theorem 4 *Let n be even and the coefficients of the polynomial* $q_n(z) = z + b_2 z^2 + \cdots + b_n z^n$ *satisfy* (4). *Let us denote*

$$\rho_3 = \min\left\{\frac{2k(2k-1)}{(2k+1)^2}\frac{b_{2k}}{b_{2k+1}}; k \in \{1, 2, \cdots, [n/2]\}\right\}.$$

and ρ_1 *as defined in Theorem 3. Then* $q_n(z)$ *is convex in* $|z| < \rho$ *where* $\rho = \min\{\rho_1, \rho_3\}$.

Proof Applying Theorem 2 and using the same procedure as in Theorem 3, we obtain the required result. We omit the details of the proof. □

Corollary 3 *For even n,* $q_n(z) = z + qz^2 + q^2z^3 + \cdots + q^{n-1}z^n$ *is convex in* $|z| < \frac{2}{9q}$ *where* $0 < q \leq 1$.

Proof For even n, if we choose $b_k = q^{k-1}$ where $0 < q \leq 1$ in Theorem 4, then b_k satisfy the assumption of Theorem 4; we have $\rho_1 = \frac{1}{4q}$ and

$$\rho_3 = \min\left\{\frac{2k(2k-1)}{(2k+1)^2}\frac{1}{q}; k \in \{1, 2, \cdots, [n/2]\}\right\} = \left\{\frac{2}{9q}, \cdots, \frac{n(n-1)}{(n+1)^2 q}\right\} = \frac{2}{9q}$$

Then $q_n(z)$ is convex in $|z| < \rho = \min\{\rho_1, \rho_3\} = \frac{2}{9q}$.

By choosing particular values of q, Corollary 3 leads us to the following interesting examples:

Example 4 Let $n > 1$ be any even positive integer. Then $q_n(z) = z + \frac{z^2}{2} + \frac{z^3}{4} + \cdots + \frac{z^n}{2^{n-1}}$ is convex in $|z| < 4/9$.

Example 5 Let $n > 1$ be any even positive integer. Then $q_n(z) = z + \frac{z^2}{3} + \frac{z^3}{9} + \cdots + \frac{z^n}{3^{n-1}}$ is convex in $|z| < 4/3$.

Example 6 Let $n > 1$ be any even positive integer. Then $q_n(z) = z + \frac{z^2}{4} + \frac{z^3}{16} + \cdots + \frac{z^n}{4^{n-1}}$ is convex in $|z| < 16/9$.

Corollary 4 *If we choose $b_k = \frac{1}{k}$, let n be even positive integer and $q_n(z) = z + \frac{z^2}{2} + \frac{z^3}{3} + \cdots + \frac{z^n}{n}$. Then $q_n(z)$ is convex in $|z| < 1/3$.*

Using an argument similar to the earlier case, we can see that $-\log(1-z)$ is convex function in $|z| < 1/3$. Figure 2 shows the graph of $q_4(z) = z + \dfrac{z^2}{2} + \dfrac{z^3}{3} + \dfrac{z^4}{4}$ in $|z| < 1/3$.

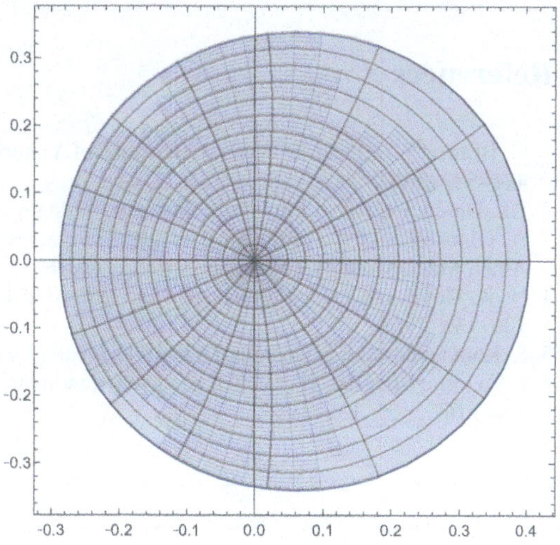

Fig. 2 $q_4(z) = z + \dfrac{z^2}{2} + \dfrac{z^3}{3} + \dfrac{z^4}{4}$ in $|z| < 1/3$

Fig. 3 $q_4(z)$ in $|z| < 4/5$

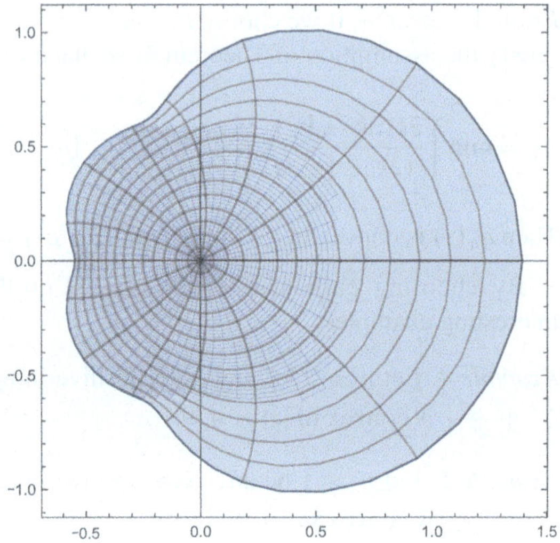

The result obtained in Theorem 4 is not sharp. For example, if we choose $\rho = 4/5$ in Corollary 4, then it is no longer convex, as shown in Fig. 3. Hence we conclude with the following open problem:

Problem 1 To find the sharp value of ρ for which $p_n(z) = z + b_2z^2 + \cdots + b_nz^n$, $n > 1$ is convex in $|z| < \rho < 1$.

Acknowledgements The first author is thankful to the Council of Scientific and Industrial Research, India (grant code: 09/143(0827)/2013-EMR-1) for financial support to carry out the above research work.

References

1. G. Brown, F. Dai and K. Wang, Extensions of Vietoris's inequalities. I, Ramanujan J. **14**,no. 3, 471–507 (2007)
2. P.L. Duren, Univalent Functions, Springer–Verlag, Berlin, (1983)
3. A. Gluchoff and F. Hartmann, Univalent polynomials and non-negative trigonometric sums, Amer. Math. Monthly **105**, no. 6, 508–522 (1998)
4. N. K. Govil and Q. I. Rahman, On the Eneström-Kakeya theorem, Tôhoku Math. J. (2) **20**, 126–136 (1968)
5. S. Koumandos, An extension of Vietoris's inequalities, Ramanujan J. **14**, no. 1, 1–38 (2007)
6. L.Vietoris, Über das Vorzeichen gewisser trignometrishcher Summen, Sitzungsber, Oest. Akad. Wiss. 167 , 125–135 (1958)

Local Countable Iterated Function Systems

A. Gowrisankar and D. Easwaramoorthy

Abstract This paper presents the extended notion of a local iterated function system (local IFS) to the general case of local countable iterated function system (local CIFS). Further, this paper establishes the approximation process of attractor of the local CIFS in terms of attractors of local IFS and discusses the relation between the attractors of CIFS and local CIFS.

Keywords Fractals · Contraction · Iterated function system

MSC Classification codes: 26E50, 28A80, 47H10

1 Introduction

Mandelbrot has addressed the geometrical structures and properties of irregular objects and coined as fractal, which plays a vital role in the nonlinear analysis. Fractal is defined by fragmented geometric structure that can be divided into parts where each part is a reflection of the whole [2, 6]. Hutchinson constructed a non-empty compact invariant set which is a unique fixed point of a given set of contraction mapping in a complete metric space. This unique fixed point is, in generally, called a deterministic fractal or attractor of the iterated function system (IFS) [1, 2, 6].

The construction of fractal by IFS has been extended to more general spaces and various contraction mappings [4, 5, 7, 10]. In fact, there is a large literature that discussed the IFS in which the noticeable work done by N.A. Secelean is that he has implemented the constructing process of the deterministic fractals through

A. Gowrisankar · D. Easwaramoorthy (✉)
Department of Mathematics, School of Advanced Sciences, Vellore Institute of Technology, Vellore, Tamil Nadu, India
e-mail: gowrisankargri@gmail.com; easandk@gmail.com; easwar@vit.ac.in

V. Madhu et al. (eds.), *Advances in Algebra and Analysis*, Trends in Mathematics,
https://doi.org/10.1007/978-3-030-01120-8_20

countable iterated function system (CIFS) [9]. Further, Barnsley presented the extended notion of an iterated function system to the more general case of local iterated function system in which the iterated functions are defined locally and used in the computer graphics especially in the arena of image compression [3].

This paper explores the existence of attractor of local countable iterated function system. The definition and consequences of the Local IFS are discussed in Sect. 2. In Sect. 3, local attractor of the local CIFS is defined and proved that it is a subset of attractor of CIFS. In Sect. 4, the local attractor of the local CIFS is expressed as the limit of a convergence sequence of attractors of the local IFS. Finally, the concluding remarks are given in Sect. 5.

2 Local Iterated Function Systems

Let $X(\neq \Phi)$ be a complete metric space with respect to the metric d, and $\mathscr{K}(X)$ denotes the associated hyperspace of non-empty compact subsets of X endowed with the Hausdorff metric H_d defined by

$$H_d(A, B) = \max\{\sup\inf d(a, b), \sup\inf d(b, a) : a \in A, b \in B\}.$$

For $k \in \mathbb{N}$, let $\mathbb{N}_n := \{1, 2, \ldots, n\}$. If $\{X_i : i \in \mathbb{N}_n\}$ is a n number of non-empty subsets of X and for each X_i there exists a continuous mapping f_i on X_i to X, then the system $\{X; (X_i, f_i) : i \in \mathbb{N}\}$ is called a local iterated function system (local IFS). Whereas an iterated function system (IFS) is a complete metric space X together with a finite set of contraction mappings, denoted by $\{X; f_k : k \in \mathbb{N}_n\}$, with contraction factors $c_k, k \in \mathbb{N}_n$. If $X_i = X$, then local IFS becomes (global) IFS.

The operator $\mathscr{F}_{loc,n}$ on $\mathscr{K}(X)$ is defined by

$$\mathscr{F}_{loc,n}(B) = \overline{\bigcup_{i \in \mathbb{N}_n} f_i(B \cap X_i)},$$

where $f_i(S \cap X_i) = \{f_i(x) : x \in S \cap X_i\}$.

The set-valued map $\mathscr{F} : \mathscr{K}(X) \to \mathscr{K}(X)$ defined by

$$\mathscr{F}(B) = \bigcup_{k \in \mathbb{N}_n} f_k(B)$$

is contraction on $\mathscr{K}(X)$ with contraction factor $c = \max\{c_k : k \in \mathbb{N}_n\}$, and hence it has a unique fixed point, say A, in $\mathscr{K}(X)$. The fixed point A is termed as a deterministic fractal generated by the IFS $\{X; f_k : k \in \mathbb{N}_n\}$. Further, for any $B \in \mathscr{K}(X)$, $\lim_{k\to\infty} \mathscr{F}^{ok}(B) = A$, the limit being taken with respect to the Hausdorff metric H_d, where $\mathscr{F}^{ok}$ denotes the k-fold composition $\mathscr{F} \circ \mathscr{F} \circ \cdots \circ \mathscr{F}$ (k-times).

By the notion of IFS theory, Secelean [9] presented the construction of deterministic fractals through countable iterated function system as follows. An IFS is extended to countable numbers of contraction mappings as $\{X; f_i : i \in \mathbb{N}\}$ and called as countable iterated function system (CIFS). Define the self-mapping $\mathscr{W}$ on $\mathscr{K}(X)$ by

$$\mathscr{W}(B) = \overline{\bigcup_{i \in \mathbb{N}} f_i(B)},$$

for all $B \in \mathscr{K}(X)$, where the bar means the closure of the corresponding set. The self-map $\mathscr{W}$ has an unique fixed point $\mathscr{A}$. Moreover, $\lim_{k\to\infty} \mathscr{W}^{ok}(B) = \mathscr{A}$ for any $B \in \mathscr{K}(X)$. The fixed point $\mathscr{A}$ is a union of countable copies of itself. It is often convenient to call the space $\mathscr{K}(X)$ as the space of fractals in X. The attractor of CIFS is approximated by the attractor of IFS as follows.

Theorem 1 ([8]) *If $B \in \mathscr{K}(X)$, then*

$$\mathscr{W}(B) = \lim_{n\to\infty} \bigcup_{i \in \mathbb{N}_n} f_i(B).$$

In particular, if $\mathscr{A}$ is the attractor of CIFS $\{X; f_i : i \in \mathbb{N}\}$, then

$$\mathscr{A} = \mathscr{W}(\mathscr{A}) = \lim_{n\to\infty} \lim_{k\to\infty} \mathscr{F}^{ok}(A_n),$$

where A_n is the attractor of IFS $\{X; f_i : i \in \mathbb{N}_n\}$.

Theorem 2 ([8]) *Let X be a complete metric space with Hausdorff metric. Let $(E_n)_{n=1}^{\infty}$ be a sequence of compact subsets of X such that $E_n \subset E_{n+1}$ and $E = \bigcup_{n=1}^{\infty} E_n$. Then $\bar{E} = \lim_{n\to\infty} E_n$*

3 Local Countable Iterated Function System

Suppose $\{X_i : i \in \mathbb{N}\}$ is a sequence of non-empty subsets of X. Further assume that for each X_i there exists a continuous mapping $f_i : X_i \longrightarrow X, i \in \mathbb{N}$. Then $\{X; (X_i, f_i) : i \in \mathbb{N}\}$ is called a local countable iterated function system (local CIFS). If $X_i = X$, then local CIFS becomes global CIFS. A local CIFS is called contractive or hyperbolic if each f_i is contraction on their respective domains.

Let $\mathscr{P}(X)$ be the power set of X, i.e., $\mathscr{P}(X) = \{S : S \subset X\}$. Define $\mathscr{W}_{loc} : \mathscr{P}(X) \longrightarrow \mathscr{P}(X)$ by

$$\mathscr{W}_{loc}(B) = \overline{\bigcup_{i \in \mathbb{N}} f_i(B \cap X_i)},$$

where $f_i(S \cap X_i) = \{f_i(x) : x \in S \cap X_i\}$.

Every local CIFS has at least one local attractor (fixed point of $\mathscr{W}_{loc}$), namely, $A = \emptyset$, empty set. Largest local attractor, union of all distinct local attractor, is called the local attractor of local CIFS, $\{X; (X_i, f_i) : i \in \mathbb{N}_n\}$. If X is compact, $X_i, i \in \mathbb{N}$, is closed (compact) in X, and local CIFS $\{X; (X_i, f_i) : i \in \mathbb{N}_n\}$ is contractive, then the local attractor is computed as follows:

Let $K_0 = X$ and set

$$K_n = \mathscr{W}_{loc}(K_{n-1}) = \overline{\bigcup_{i \in \mathbb{N}} f_i(K_{i-1} \cap X_i)}, n \in \mathbb{N}.$$

Then $\{K_n : n \in \mathbb{N}\}$ is a decreasing nested sequence of compact sets. If each K_n is non-empty, then by the Cantor intersection theorem, we get

$$K = \bigcap_{n \in \mathbb{N}} K_n \neq \emptyset$$

$$K = \lim_{n \to \infty} K_n,$$

where the above limit taken with respect to the Hausdorff metric H_d on $\mathscr{K}(X)$.

$$\begin{aligned} K = \lim_{n \to \infty} K_n &= \lim_{n \to \infty} \overline{\bigcup_{i \in \mathbb{N}} f_i(K_{n-1} \cap X_i)} \\ &= \overline{\bigcup_{i \in \mathbb{N}} f_i(K \cap X_i)} \\ &= \mathscr{W}_{loc}(K). \end{aligned}$$

Thus, $K = A_{loc}$. It is noted that $f_i(X_i) \subset X_i, i \in \mathbb{N}$ is the condition, which guaranteed that each K_n is non-empty.

Theorem 3 *Let X be a compact metric space and let $X_i, i \in \mathbb{N}$, be closed subset of X. If $\mathscr{A}$ is attractor of CIFS and $\mathscr{A}^*$ is local attractor of Local CIFS, then $\mathscr{A}^*$ is a subset of $\mathscr{A}$.*

Proof Consider the sequence $\{K_n : n \in \mathbb{N}\}$ such that $K_0 = X$ and $K_n = \mathscr{W}_{loc}(K_{n-1}) = \overline{\bigcup_{i \in \mathbb{N}} f_i(K_{i-1} \cap X_i)}, n \in \mathbb{N}$. The unique attractor $\mathscr{A}$ is obtained as the limit of the sequence $\{K_n : n \in \mathbb{N}\}$. Let $\{X; f_i : i \in \mathbb{N}_n\}$ be the contractive CIFS associated with the set-valued map $\mathscr{W}$ on $\mathscr{K}(X)$ defined by $\mathscr{W}(B) = \overline{\bigcup_{i \in \mathbb{N}} f_i(B)}$. Then, the unique attractor $\mathscr{A}$ of the CIFS is obtained as the limit of the sequence $\{A_n : n \in \mathbb{N}\}$ such that $A_0 = X$ and $A_n = \mathscr{W}(A_{n-1}), n \in \mathbb{N}$. Assume $K_{n-1} \subseteq A_{n-1}, n \in \mathbb{N}$, we have

$$\mathscr{A}^* = \lim_{n \to \infty} K_n = \lim_{n \to \infty} \overline{\bigcup_{i \in \mathbb{N}} f_i(K_{n-1} \cap X_i)}$$

$$\subseteq \lim_{n\to\infty} \overline{\bigcup_{i\in\mathbb{N}} f_i(K_{n-1})}$$

$$\subseteq \lim_{n\to\infty} \overline{\bigcup_{i\in\mathbb{N}} f_i(A_{n-1})} = \lim_{n\to\infty} A_n = \mathscr{A}.$$

4 Approximation of Local CIFS Attractor by the Family of Local IFS Attractors

In this section, it is proved that the local attractor of the local CIFS is expressed as the limit of a convergence sequence of attractors of the Local IFS.

Theorem 4 *Let X be a compact metric space. Let $\{X; (X_i, f_i) : i \in \mathbb{N}\}$ be a local CIFS and $\{X; (X_i, f_i) : i \in \mathbb{N}_n\}$ be a local IFS. Suppose* $\lim_{n\to\infty} E_n = E \neq \emptyset$, *where each n, $E_n \subseteq X$. Then the local attractor $\mathscr{A}^*$ of CIFS is approximated by the local attractors $\mathscr{A}$'s of local IFSs.*

$$\lim_{n\to\infty} \lim_{k\to\infty} \mathscr{F}^k_{loc,n}(E_n) = \mathscr{A}^*.$$

Proof Let $\mathscr{F}_{loc,n}(B) = \bigcup_{i\in\mathbb{N}_n} f_i(B \cap X_i)$, for $n \in \mathbb{N}$. Then it is enough to prove that

$$\lim_{n\to\infty} \mathscr{F}^k_{loc,n}(E_n) = \mathscr{W}^k_{loc}(E),$$

where the limit is taken with respect to Hausdorff metric h. As $\{X; (X_i, f_i) : i \in \mathbb{N}_n\}$ is a local IFS, for each X_i there exists a contraction mapping $f_i : X_i \to X$ with contraction factors c_i, $i \in \mathbb{N}_n$. Denote $f_{i_1\ldots i_k} = f_{i_1} \circ \cdots \circ f_{i_k}$, for each $k \geq 1$, $i_1, i_2, \ldots, i_k$ are positive integers. Clearly $f_{i_1\ldots i_k}$ is a contraction mapping with contraction factor $c_{i_1} c_{i_2} \cdots c_{i_k}$.

$$H_d\left(\mathscr{F}^k_{loc,n}(E_n), \mathscr{W}^k_{loc}(E)\right) \leq H_d\left(\mathscr{F}^k_{loc,n}(E_n), \mathscr{F}^k_{loc,n}(E)\right) + H_d\left(\mathscr{F}^k_{loc,n}(E), \mathscr{W}^k_{loc}(E)\right). \quad (1)$$

Now,

$$H_d\left(\mathscr{F}^k_{loc,n}(E_n), \mathscr{F}^k_{loc,n}(E)\right)$$

$$= H_d\left(\bigcup_{i_1,\ldots,i_k\in\mathbb{N}_n} f_{i_1\ldots i_k}(E_n \cap X_n), \bigcup_{i_1,\ldots,i_k\in\mathbb{N}_n} f_{i_1\ldots i_k}(E \cap X_n)\right)$$

$$\leq \sup_{i_1,\dots,i_k \in \mathbb{N}_n} H_d(f_{i_1\dots i_k}(E_n \cap X_n), f_{i_1\dots i_k}(E \cap X_n))$$
$$\leq c_{i_1} \cdots c_{i_k} H_d(E_n \cap X_n, E \cap X_n)$$
$$\leq H_d(E_n, E).$$

Since $\lim_{n\to\infty} E_n = E$, so $H_d(E_n, E) \to 0$ as $n \to \infty$.
By definition,

$$\mathscr{W}_{loc}^k(B) = \overline{\bigcup_{i_1,\dots,i_k \in \mathbb{N}} f_{i_1\dots i_k}(B)}.$$

Since each f_i's are continuous, then

$$\mathscr{W}_{loc}^{k+1}(E) = \mathscr{W}_{loc}\left(\overline{\bigcup_{i_1,\dots,i_k \in \mathbb{N}} f_{i_1\dots i_k}(E)}\right)$$
$$= \overline{\bigcup_{i=1}^{\infty} f_i\left(\overline{\bigcup_{i_1,\dots,i_k \in \mathbb{N}} f_{i_1\dots i_k}(E \cap X_i)}\right)}$$
$$\subset \overline{\overline{\bigcup_{i=1}^{\infty} f_i\left(\bigcup_{i_1,\dots,i_k \in \mathbb{N}} f_{i_1\dots i_k}(E \cap X_i)\right)}}$$
$$= \overline{\bigcup_{i=1}^{\infty} f_i\left(\bigcup_{i_1,\dots,i_k \in \mathbb{N}} f_{i_1\dots i_k}(E \cap X_i)\right)} = \mathscr{W}_{loc}^{k+1}(E).$$

The sequence of sets $\left(\bigcup_{i_1,\dots,i_k \in \mathbb{N}_n} f_{i_1\dots i_k}(E \cap X_n)\right)_{n \in \mathbb{N}}$ is increasing, and by Theorem 2, we have

$$\lim_{n\to\infty} \mathscr{F}_{loc,n}^k(E) = \lim_{n\to\infty} \bigcup_{i_1,\dots,i_k \in \mathbb{N}_n} f_{i_1\dots i_k}(E \cap X_n)$$
$$= \overline{\bigcup_{n=1}^{\infty} \bigcup_{i_1,\dots,i_k \in \mathbb{N}_n} f_{i_1\dots i_k}(E \cap X_n)}$$
$$= \overline{\bigcup_{i_1,\dots,i_k \in \mathbb{N}} f_{i_1\dots i_k}(E)} = \mathscr{W}_{loc}^k(E).$$

Equation (1) becomes $H_d\left(\mathscr{F}_{loc,n}^k(E_n), \mathscr{W}_{loc}^k(E)\right) = 0$ as $n, k \to 0$. Thus, we conclude,

$$\lim_{n\to\infty}\lim_{k\to\infty}\mathscr{F}_{loc,n}^{k}(E_n)=\lim_{k\to\infty}\mathscr{W}_{loc}^{k}(E)=\mathscr{A}^{*}.$$

This completes the proof.

5 Conclusion

In the paper, we have defined the local attractor of the local CIFS and proved that it is a subset of attractor of CIFS. Further, the local attractor of the local CIFS has been approximated as the limit of a convergence sequence of attractors of the local IFS. This will help us to develop more interesting results on fractals by IFS in more general spaces.

References

1. Barnsley, M.F., Hurd, L.P.: Fractal Image Compression. AK Peters, Ltd., Wellesley, Massachusetts (1993).
2. Barnsley, M.F.: Fractals Everywhere. 3rd Edition, Dover Publications (2012).
3. Barnsley, M.F., Hegland, M., Massopust, P.: Numerics and fractals. Bulletin of the Institute of Mathematics. **9(3)**, 389–430 (2014).
4. Easwaramoorthy, D., Uthayakumar, R.: Analysis on Fractals in Fuzzy Metric Spaces, Fractals. **19(3)**, 379–386 (2011).
5. Gowrisankar, A., Uthayakumar, R.: Fractional calculus on fractal interpolation function for a sequence of data with countable iterated function system. Mediterranean Journal of Mathematics. **13(6)**, 3887–3906 (2016).
6. Hutchinson, J.E.: Fractals and self similarity. Indiana University Mathematics Journal. **30**, 713–747 (1981).
7. Secelean, N.A.: Countable iterated function systems. Far East Journal of Dynamical Systems. **3(2)**, 149–167 (2001).
8. Secelean, N.A.: Approximation of the attractor of a countable iterated function system. General Mathematics. **17(3)**, 221–231 (2009).
9. Secelean, N.A.: The Existence of the Attractor of Countable Iterated Function Systems. Mediterranean Journal of Mathematics. **9** 61–79 (2012).
10. Uthayakumar, R., Gowrisankar, A.: Fractals in product fuzzy metric space. Fractals, Wavelets and Their Applications. Springer Proceedings in Mathematics & Statistics. **92**, 157–164 (2014).

Conclusion

In the paper, we have defined the local attractor of the local CGS and proved that it is a subset of attractor of CIFS. Further, the local attractor of the local CGS has been approximated as the limit of a convergence sequence of attractors of the local CIFS. This will help us to develop more interesting results of fractals by CGS in metric spaces.

References

[illegible]

On Intuitionistic Fuzzy $\mathscr{C}$-Ends

T. Yogalakshmi and Oscar Castillo

Abstract Basic concepts related to disconnectedness in an intuitionistic fuzzy $\mathscr{C}$-ends are constructed. The conceptual ideas related to the intuitionistic fuzzy $\mathscr{C}$-centred system is introduced, and properties related to it are studied. Several preservation properties and characterizations concerning extremally disconnectedness in intuitionistic fuzzy $\mathscr{C}$-ends are discussed. Moreover, Tietze extension theorem is established with respect to the intuitionistic fuzzy $\mathscr{C}$-ends.

1 Introduction

Zadeh [16] designed the fuzzy sets. The role of fuzziness has played a vital part in various mathematical fields such as engineering, economics, and medicine. Fuzzy sets have several applications in the information [11] and control [12] systems. Oscar Castillo and his team work with the high performance computers [9] and movement of a wheeled mobile robot [10] in fuzzy systems. Chang [3] introduced the topological structures [6] of set theory dealing with uncertainties. Atanassov [1] published his article based on the ideas of intuitionistic fuzzy set, and many of his works appeared in the literature [1, 2]. Later, several properties of the intuitionistic fuzzy topological spaces were studied by the author Coker [4, 5]. Iliadis and Fomin [8] introduced the centred systems which deserved serious attention in medical field. Connectedness in topological spaces using fuzziness was established by Fatteh and Bassan [7]. Further, Yogalakshmi et al. [14] studied the concepts of $\mathscr{C}$-open sets and discussed various properties of the disconnectedness [15] in it.

In this paper, the basic notions of intuitionistic fuzzy $\mathscr{C}$-centred systems is introduced, and properties related to it are studied. Several properties of the

T. Yogalakshmi (✉)
SAS, Vellore Institute of Technology, Vellore, Tamil Nadu, India
e-mail: yogalakshmithangaraj@vit.ac.in

O. Castillo
Tijuana Institute Technology, Tijuana, Mexico

V. Madhu et al. (eds.), *Advances in Algebra and Analysis*, Trends in Mathematics,
https://doi.org/10.1007/978-3-030-01120-8_21

extremally disconnectedness [13] in intuitionistic fuzzy $\mathscr{C}$-ends are discussed. Moreover, Tietze extension theorem with respect to the intuitionistic fuzzy $\mathscr{C}$-ends is established.

2 Preliminaries

Definition 1 ([16]) A fuzzy set, $\mu : X \to [0, 1]$, is a mapping from a non-empty set X into $[0, 1]$. $\mu' = 1 - \mu$ is called as the complement of μ .

Definition 2 ([1, 2]) Let the fuzzy sets λ_B and μ_B be the degrees of membership (namely, $\lambda_B(x)$) and non-membership (namely, $\mu_B(x)$), respectively, to the non-empty set B such that $0 \leq \mu_B(x) + \lambda_B(x) \leq 1$, for all $x \in X$. An intuitionistic fuzzy set (inshort. IFS) B is of the form $B = \{\langle x, \lambda_B(x), \mu_B(x)\rangle : x \in X\}$. The symbol $B = \langle X, \lambda, \mu\rangle$ for the IFS $\{\langle x, \lambda_B(x), \mu_B(x)\rangle : x \in X\}$ shall be used for the sake of simplicity. The complement of IFS B is defined as $\bar{A} = \langle X, \mu, \lambda\rangle$.

Definition 3 ([2]) Define an intuitionistic fuzzy point (inshort. IFP) $x_{\langle X,\alpha,\beta\rangle}$ of a non-empty set X as

$$x_{\langle X,\alpha,\beta\rangle}(y) = \begin{cases} \langle x, \alpha(x), \beta(x)\rangle, & \text{if } x = y; \\ \langle x, 0, 1\rangle, & \text{if } x \neq y. \end{cases}$$

Then, x, $\alpha \in I = [0, 1]$ and $\beta \in I = [0, 1]$ is said to be support, value and non-value of $x_{\langle X,\alpha,\beta\rangle}$, respectively.

Definition 4 ([1]) If $P = \langle X, \lambda, \mu\rangle$ and $D = \langle X, \delta, \gamma\rangle$ are the intuitionistic fuzzy sets, then

(1) $P \subseteq D \Leftrightarrow \lambda \leq \delta$ and $\mu \geq \gamma$.
(2) $P \supseteq D \Leftrightarrow \lambda \geq \delta$ and $\mu \leq \gamma$.
(3) $P \cap D \Leftrightarrow \lambda \wedge \delta$ and $\mu \vee \gamma$.
(4) $A \cup D \Leftrightarrow \lambda \vee \delta$ and $\mu \wedge \gamma$.
(5) $0_\sim = \langle X, 0, 1\rangle$; $1_\sim = \langle X, 1, 0\rangle$

Definition 5 ([4]) An intuitionistic fuzzy topology (inshort. IFT) is a collection τ of IFSs in a non-empty set X having the axioms:

(i) $0_\sim, 1_\sim \in \tau$.
(ii) $A_1 \cap A_2 \in \tau$, for any $A_1, A_2 \in \tau$.
(iii) $\cup_i A_i \in \tau$, for any $A_i \in \tau$.

Now, (X, τ) is said to be an intuitionistic fuzzy topological space (inshort. IFTS), and any member of τ is called as an intuitionistic fuzzy open set (inshort. IFOS) of X. The complement of IFOS is an intuitionistic fuzzy closed set (inshort. IFCS).

Definition 6 ([5]) $IFint(A) = \cup\{G : G$ is an IFOS in X and $G \subseteq A\}$ and $IFcl(A) = \cap\{K : K$ is an IFCS in X and $K \supseteq A\}$ are defined as the interior and closure of an IFS, $A = \langle X, \lambda, \mu\rangle$, respectively.

Definition 7 ([4]) Let $f : X \to Y$ be any function. The pre-image of $B = \langle Y, \delta, \gamma \rangle$ is defined as $f^{-1}(B) = \langle X, f^{-1}(\delta), f^{-1}(\gamma) \rangle$, and the image of $A = \langle X, \lambda, \mu \rangle$ is defined as $f(A) = \langle Y, f(\lambda), f(\mu) \rangle$
where

$$f(\lambda)(y) = \begin{cases} sup_{x \in f^{-1}(y)} \lambda(x), & \text{if } f^{-1}(y) \neq \phi; \\ 0, & \text{otherwise} \end{cases}$$

and

$$f(\mu)(y) = \begin{cases} inf_{x \in f^{-1}(y)} \mu(x), & \text{if } f^{-1}(y) \neq \phi; \\ 1, & \text{otherwise} \end{cases}$$

Definition 8 ([4]) If the inverse image of every intuitionistic fuzzy open set in (Y, σ) is an intuitionistic fuzzy open set in (X, τ), then $f : (X, \tau) \to (Y, \sigma)$ is said to be an intuitionistic fuzzy continuous function.

3 On Intuitionistic Fuzzy $\mathscr{C}$-Ends $\mathfrak{S}_{\mathscr{C}}$

Throughout this article X or (X, τ) represents the intuitionistic fuzzy topological space and I represents [0,1].

Definition 9 If $IFint(A) = IFint(IFcl(IFint(A)))$, then the intuitionistic fuzzy set A is said to be an intuitionistic fuzzy α^*-open set (inshort. IFα^*OS).

Definition 10 Let (X, τ) be an IFTS and P be an intuitionistic fuzzy set. If G is an IFOS and A is an IFα^*OS with $P = G \cap A$, then P is said to be an intuitionistic fuzzy $\mathscr{C}$-open set (inshort. IFcOS). The complement of it is to be called as an intuitionistic fuzzy $\mathscr{C}$-closed set (inshort. IFcCS).

Definition 11 The intuitionistic fuzzy $\mathscr{C}$-interior and intuitionistic fuzzy $\mathscr{C}$-closure of an IFS P are, respectively, defined as $IFint_c(P) = \cup\{N : N \text{ is an IFcOS in } X \text{ and } N \subseteq P\}$ and $IFcl_c(P) = \cap\{L : L \text{ is an IFcCS in } X \text{ and } L \supset P\}$.

Definition 12 Let J be an indexed set. An intuitionistic fuzzy $\mathscr{C}$-centred system is a system $\mathfrak{S}_{\mathscr{C}} = \{A_i\}_{i \in J}$ of intuitionistic fuzzy $\mathscr{C}$-open sets in an intuitionistic fuzzy Hausdorff space **R** such that $\cap_{i=1}^{n} A_i \neq 0_\sim$. The system $\mathfrak{S}_{\mathscr{C}}$ is called as an intuitionistic fuzzy $\mathscr{C}$-end if it is maximal.

Proposition 1 *Let $\mathfrak{S}_{\mathscr{C}}$ be the intuitionistic fuzzy $\mathscr{C}$-end. Then,*

(1) If $A_i \in \mathfrak{S}_{\mathscr{C}}$, for i=1,2,... n, then $\cap_{i=1}^{n} A_i \in \mathfrak{S}_{\mathscr{C}}$.
(2) If $0_\sim \neq A \in \mathfrak{S}_{\mathscr{C}}$ and P is an intuitionistic fuzzy $\mathscr{C}$-open set such that $A \subseteq P$, then $P \in \mathfrak{S}_{\mathscr{C}}$.

(3) *If* $\mathfrak{S}_{\mathscr{C}}$ *is the intuitionistic fuzzy* $\mathscr{C}$*-end and* P *is an intuitionistic fuzzy* $\mathscr{C}$*-open set, then* $P \notin \mathfrak{S}_{\mathscr{C}}$ *iff there is an IFS* $D \in \mathfrak{S}_{\mathscr{C}}$ *with* $P \cap D = 0_{\sim}$.
(4) *If* $P \cup Q \in \mathfrak{S}_{\mathscr{C}}$ *and* P, Q *are the IFcOSs such that* $P \cap Q = 0_{\sim}$*, then either* $P \in \mathfrak{S}_{\mathscr{C}}$ *or* $Q \in \mathfrak{S}_{\mathscr{C}}$.
(5) *If* $IFcl_c(A) = 1_{\sim}$*, then* $A \in \mathfrak{S}_{\mathscr{C}}$*, for any intuitionistic fuzzy* $\mathscr{C}$*-end* $\mathfrak{S}_{\mathscr{C}}$.

Definition 13 Let $\mathscr{E}_{\mathscr{C}}(\mathbf{R})$ be the collection of all intuitionistic fuzzy $\mathscr{C}$-ends belonging to $\mathbf{R}$. If $\mathfrak{S}_{\mathscr{C}}(A)$ is the set of all intuitionistic fuzzy $\mathscr{C}$-ends which includes IFcOS A of $\mathbf{R}$ as a member of it, then the collection of intuitionistic fuzzy neighbourhoods of each intuitionistic fuzzy $\mathscr{C}$-end contained in $\mathfrak{S}_{\mathscr{C}}(A)$ forms an intuitionistic fuzzy topology § in $\mathscr{E}_{\mathscr{C}}(\mathbf{R})$. Thus, the pair $(\mathscr{E}_{\mathscr{C}}(\mathbf{R}), \S)$ or $\mathscr{E}_{\mathscr{C}}(\mathbf{R})$ is called as an intuitionistic fuzzy $\mathscr{C}$-centred space.

Note 1 For each intuitionistic fuzzy $\mathscr{C}$-open set B of $\mathbf{R}$, there corresponds an intuitionistic fuzzy neighbourhood $\mathfrak{S}_{\mathscr{C}}(B)$ in $\mathscr{E}_{\mathscr{C}}(\mathbf{R})$. That is, $\mathfrak{S}_{\mathscr{C}}(B)$ is an intuitionistic fuzzy open subset of $\mathscr{E}_{\mathscr{C}}(\mathbf{R})$, and its complement is said to be an intuitionistic fuzzy closed subset of $\mathscr{E}_{\mathscr{C}}(\mathbf{R})$, denoted by $\mathscr{E}_{\mathscr{C}}(\mathbf{R}) - \mathfrak{S}_{\mathscr{C}}(B)$.

Definition 14 Let $\mathbf{R}$ be an intuitionistic fuzzy Hausdorff space and $\mathscr{E}_{\mathscr{C}}(\mathbf{R})$ be an intuitionistic fuzzy $\mathscr{C}$-centred space. If G is any intuitionistic fuzzy $\mathscr{C}$-open subset of $\mathbf{R}$, then intuitionistic fuzzy interior of $\mathscr{E}_{\mathscr{C}}(\mathbf{R})$ and intuitionistic fuzzy closure of $\mathscr{E}_{\mathscr{C}}(\mathbf{R})$ are defined as $Int_{\mathscr{E}_{\mathscr{C}}(\mathbf{R})}(\mathfrak{S}_{\mathscr{C}}(G)) = \cup\{\mathfrak{S}_{\mathscr{C}}(P) : \mathfrak{S}_{\mathscr{C}}(P)$ is an intuitionistic fuzzy open subset of $\mathscr{E}_{\mathscr{C}}(\mathbf{R})$ and $G \supseteq P\}$ and $Cl_{\mathscr{E}_{\mathscr{C}}(\mathbf{R})}(\mathfrak{S}_{\mathscr{C}}(G)) = \cap\{\mathfrak{S}_{\mathscr{C}}(Q) : \mathfrak{S}_{\mathscr{C}}(Q)$ is an intuitionistic fuzzy closed subset of $\mathscr{E}_{\mathscr{C}}(\mathbf{R})$ and $G \subseteq Q\}$, respectively.

Proposition 2 *Let* $\mathbf{R}$ *be an intuitionistic fuzzy Hausdorff space and* $\mathscr{E}_{\mathscr{C}}(\mathbf{R})$ *be an intuitionistic fuzzy* $\mathscr{C}$*-centred space. Let* P, Q *be the IFcOSs in* $\mathbf{R}$*. Then,*

(i) $\mathfrak{S}_{\mathscr{C}}(P \cup Q) = \mathfrak{S}_{\mathscr{C}}(P) \cup \mathfrak{S}_{\mathscr{C}}(Q)$.
(ii) $\mathfrak{S}_{\mathscr{C}}(P) = \mathscr{E}_{\mathscr{C}}(\mathbf{R}) - \mathfrak{S}_{\mathscr{C}}(\overline{(IFcl_c(P))})$.
(iii) *If* $P \subseteq Q$*, then* $\mathfrak{S}_{\mathscr{C}}(P) \subseteq \mathfrak{S}_{\mathscr{C}}(Q)$.
(iv) *If* P *and* H *are the IF* $\mathscr{C}$*-open and IF* $\mathscr{C}$*-closed subsets of* $\mathbf{R}$*, then* $Int_{\mathscr{E}_{\mathscr{C}}(\mathbf{R})}(\mathfrak{S}_{\mathscr{C}}(P)) = \mathfrak{S}_{\mathscr{C}}(P)$ *and* $Cl_{\mathscr{E}_{\mathscr{C}}(\mathbf{R})}(\mathfrak{S}_{\mathscr{C}}(H)) = \mathfrak{S}_{\mathscr{C}}(H)$*, respectively.*
(v) $Int_{\mathscr{E}_{\mathscr{C}}(\mathbf{R})}(\mathfrak{S}_{\mathscr{C}}(P)) \subseteq \mathfrak{S}_{\mathscr{C}}(P) \subseteq Cl_{\mathscr{E}_{\mathscr{C}}(\mathbf{R})}(\mathfrak{S}_{\mathscr{C}}(P))$.

Proof (i) Let $\mathfrak{S}_{\mathscr{C}} \in \mathfrak{S}_{\mathscr{C}}(P) \cup \mathfrak{S}_{\mathscr{C}}(Q)$. Therefore $P \in \mathfrak{S}_{\mathscr{C}}$ or $Q \in \mathfrak{S}_{\mathscr{C}}$. Hence by the Proposition 1, we have $P \cup Q \in \mathfrak{S}_{\mathscr{C}}$. That is, $\mathfrak{S}_{\mathscr{C}} \in \mathfrak{S}_{\mathscr{C}}(P \cup Q)$. Thus $\mathfrak{S}_{\mathscr{C}}(P) \cup \mathfrak{S}_{\mathscr{C}}(Q) \subseteq \mathfrak{S}_{\mathscr{C}}(P \cup Q)$. On the other hand, let $\mathfrak{S}_{\mathscr{C}} \in \mathfrak{S}_{\mathscr{C}}(P \cup Q)$. That is, $P \cup Q \in \mathfrak{S}_{\mathscr{C}}$. This implies that $P \in \mathfrak{S}_{\mathscr{C}}$ or $Q \in \mathfrak{S}_{\mathscr{C}}$. Thus, $\mathfrak{S}_{\mathscr{C}} \in \mathfrak{S}_{\mathscr{C}}(P)$ or $\mathfrak{S}_{\mathscr{C}} \in \mathfrak{S}_{\mathscr{C}}(Q)$. This shows that $\mathfrak{S}_{\mathscr{C}}(P) \cup \mathfrak{S}_{\mathscr{C}}(Q) \supseteq \mathfrak{S}_{\mathscr{C}}(P \cup Q)$. Hence, $\mathfrak{S}_{\mathscr{C}}(P \cup Q) = \mathfrak{S}_{\mathscr{C}}(P) \cup \mathfrak{S}_{\mathscr{C}}(Q)$.
Proofs of (ii) to (v) are simple.

Proposition 3 *The intuitionistic fuzzy* $\mathscr{C}$*-centred space,* $\mathscr{E}_{\mathscr{C}}(\mathbf{R})$ *has a base of intuitionistic fuzzy neighbourhoods that are both intuitionistic fuzzy open and intuitionistic fuzzy closed.*

Proof Proof is clear from the Proposition 2.

Proposition 4 *The intuitionistic fuzzy $\mathscr{C}$-centred space, $\mathscr{E}_{\mathscr{C}}(\mathbf{R})$ is a intuitionistic fuzzy extremally disconnected centred space.*

Proof If $A \subseteq B$, it follows that $\mathfrak{S}_{\mathscr{C}}(A) \subseteq \mathfrak{S}_{\mathscr{C}}(B)$ and therefore $\cup_i \mathfrak{S}_{\mathscr{C}}(A)_i \subseteq \mathfrak{S}_{\mathscr{C}}(\cup_i A_i)$. By the Proposition 3., $\mathfrak{S}_{\mathscr{C}}(\cup_i A_i)$ is a intuitionistic fuzzy closed set in $\mathscr{E}_{\mathscr{C}}(\mathbf{R})$ and therefore, $Cl_{\mathscr{E}_{\mathscr{C}}(\mathbf{R})}(\cup_i \mathfrak{S}_{\mathscr{C}}(A)_i) \subseteq \mathfrak{S}_{\mathscr{C}}(\cup_i A_i)$. Let $\mathfrak{S}_{\mathscr{C}}$ be an arbitrary element of $\mathfrak{S}_{\mathscr{C}}(\cup_i A_i)$. Then, $\cup_i A_i \in \mathfrak{S}_{\mathscr{C}}$. Let $A \in \mathfrak{S}_{\mathscr{C}}$. Then, $A \cap (\cup_i A_i) \neq 0_{\sim}$. Hence there exists i such that $A \cap A_i \neq 0$. But $\mathfrak{S}_{\mathscr{C}}(A) \cap \mathfrak{S}_{\mathscr{C}}(A)_i \neq \phi$ and since $A \in \mathfrak{S}_{\mathscr{C}}$ is arbitrary, this means that $\mathfrak{S}_{\mathscr{C}} \in Cl_{\mathscr{E}_{\mathscr{C}}(\mathbf{R})}(\cup_i \mathfrak{S}_{\mathscr{C}}(A)_i)$. Thus it follows that, $\mathfrak{S}_{\mathscr{C}}(\cup_i A_i) = Cl_{\mathscr{E}_{\mathscr{C}}(\mathbf{R})}(\cup_i \mathfrak{S}_{\mathscr{C}}(A)_i)$. Hence the proposition.

Proposition 5 *The following statements are equivalently true, for any intuitionistic fuzzy $\mathscr{C}$-centred space $\mathscr{E}_{\mathscr{C}}(\mathbf{R})$. Then,*

(a) $\mathscr{E}_{\mathscr{C}}(\mathbf{R})$ is an intuitionistic fuzzy extremally disconnected centred space.
(b) If P is an IFcCS in $\mathbf{R}$, then $Int_{\mathscr{E}_{\mathscr{C}}(\mathbf{R})}(\mathfrak{S}_{\mathscr{C}}(P))$ is an IF closed subset of $\mathscr{E}_{\mathscr{C}}(\mathbf{R})$.
(c) If Q is an IFcOS in $\mathbf{R}$, then $Int_{\mathscr{E}_{\mathscr{C}}(\mathbf{R})}(Cl_{\mathscr{E}_{\mathscr{C}}(\mathbf{R})}(\mathfrak{S}_{\mathscr{C}}(Q))) = Cl_{\mathscr{E}_{\mathscr{C}}(\mathbf{R})}(\mathfrak{S}_{\mathscr{C}}(Q))$.
(d) For each pair of IF$\mathscr{C}$-open sets M and N in $\mathbf{R}$ with $Int_{\mathscr{E}_{\mathscr{C}}(\mathbf{R})}(\mathscr{E}_{\mathscr{C}}(\mathbf{R}) - \mathfrak{S}_{\mathscr{C}}(M)) = \mathfrak{S}_{\mathscr{C}}(N)$, we have $\mathscr{E}_{\mathscr{C}}(\mathbf{R}) - Cl_{\mathscr{E}_{\mathscr{C}}(\mathbf{R})}(\mathfrak{S}_{\mathscr{C}}(M)) = Cl_{\mathscr{E}_{\mathscr{C}}(\mathbf{R})}(\mathfrak{S}_{\mathscr{C}}(N))$.

Proposition 6 *An intuitionistic fuzzy $\mathscr{C}$-centred space, $\mathscr{E}_{\mathscr{C}}(\mathbf{R})$ is an intuitionistic fuzzy extremally disconnected centred space iff for each IFcOS, P and IFcCS, Q such that $\mathfrak{S}_{\mathscr{C}}(P) \subseteq \mathfrak{S}_{\mathscr{C}}(Q)$ in $\mathscr{E}_{\mathscr{C}}(\mathbf{R})$, we have $Cl_{\mathscr{E}_{\mathscr{C}}(\mathbf{R})}(\mathfrak{S}_{\mathscr{C}}(P)) \subseteq Int_{\mathscr{E}_{\mathscr{C}}(\mathbf{R})}(\mathfrak{S}_{\mathscr{C}}(Q))$.*

Remark 1 Let $\mathscr{E}_{\mathscr{C}}(\mathbf{R})$ be an intuitionistic fuzzy extremally disconnected centred space. Let $\{\mathfrak{S}_{\mathscr{C}}(P)_i, \mathfrak{S}_{\mathscr{C}}(Q)_j : i, j \in \mathbb{N}\}$ be a family such that each P_i's and Q_j's are the IFcOS and IFcCS in $\mathbf{R}$, respectively. If $\mathfrak{S}_{\mathscr{C}}(P), \mathfrak{S}_{\mathscr{C}}(Q)$ are the intuitionistic fuzzy open and intuitionistic fuzzy closed subsets of $\mathscr{E}_{\mathscr{C}}(\mathbf{R})$ with $\mathfrak{S}_{\mathscr{C}}(P)_i \subseteq \mathfrak{S}_{\mathscr{C}}(P) \subseteq \mathfrak{S}_{\mathscr{C}}(Q)_j$ and $\mathfrak{S}_{\mathscr{C}}(P)_i \subseteq \mathfrak{S}_{\mathscr{C}}(B) \subseteq \mathfrak{S}_{\mathscr{C}}(B)_j$, for $i, j \in \mathbb{N}$, then there is an IF$\mathscr{C}$-clopen set, G such that $Cl_{\mathscr{E}_{\mathscr{C}}(\mathbf{R})}(\mathfrak{S}_{\mathscr{C}}(P)_i) \subseteq \mathfrak{S}_{\mathscr{C}}(G) \subseteq Int_{\mathscr{E}_{\mathscr{C}}(\mathbf{R})}(\mathfrak{S}_{\mathscr{C}}(Q)_j)$, for $i, j \in \mathbb{N}$.

Proposition 7 *Let $\mathscr{E}_{\mathscr{C}}(\mathbf{R})$ be an intuitionistic fuzzy extremally disconnected centred space. Let $\{\mathfrak{S}_{\mathscr{C}}(A)_q\}_{q \in \mathbb{Q}}$ and $\{\mathfrak{S}_{\mathscr{C}}(B)_q\}_{q \in \mathbb{Q}}$ be the monotone increasing collections of IFOSs and IFCSs in $\mathscr{E}_{\mathscr{C}}(\mathbf{R})$, where $\mathbb{Q}$ is the set of all rational numbers. If $\mathfrak{S}_{\mathscr{C}}(A)_{q_1} \subseteq \mathfrak{S}_{\mathscr{C}}(B)_{q_2}$, whenever $q_1 < q_2 (q_1, q_2 \in \mathbb{Q})$, then there is a monotone increasing collection $\{(G)_q\}_{q \in \mathbb{Q}}$ of both IFcOS and IFcCS in $\mathbf{R}$ with $Cl_{\mathscr{E}_{\mathscr{C}}(\mathbf{R})}(\mathfrak{S}_{\mathscr{C}}(A)_{q_1}) \subseteq \mathfrak{S}_{\mathscr{C}}(G)_{q_2}$ and $\mathfrak{S}_{\mathscr{C}}(G)_{q_1} \subseteq Int_{\mathscr{E}_{\mathscr{C}}(\mathbf{R})}(\mathfrak{S}_{\mathscr{C}}(B)_{q_2})$ whenever $q_1 < q_2$.*

Definition 15 Let $\mathscr{E}_{\mathscr{C}}(\mathbf{R})$ be an intuitionistic fuzzy $\mathscr{C}$-centred space. The intuitionistic fuzzy real line $\mathbf{R}^*(\mathbb{I} \times \mathbb{I})$ in intuitionistic fuzzy $\mathscr{C}$-centred system is the set of all monotone decreasing intuitionistic fuzzy $\mathscr{C}$-end $\mathfrak{S}_{\mathscr{C}}(P)$ satisfying $\cup\{(\mathfrak{S}_{\mathscr{C}}(P))(k) : k \in \mathbf{R}\} = 1_{\sim}$ and $\cap\{(\mathfrak{S}_{\mathscr{C}}(P))(k) : k \in \mathbf{R}\} = 0_{\sim}$, after the identification of $\mathfrak{S}_{\mathscr{C}}(P)$, $\mathfrak{S}_{\mathscr{C}}(Q)$ iff $(\mathfrak{S}_{\mathscr{C}}(P))(k-) = (\mathfrak{S}_{\mathscr{C}}(Q))(k-)$ and $(\mathfrak{S}_{\mathscr{C}}(P))(k+) = (\mathfrak{S}_{\mathscr{C}}(Q))(k+)$ for all $k \in \mathbf{R}$, where $(\mathfrak{S}_{\mathscr{C}}(P))(k-) = \cap\{(\mathfrak{S}_{\mathscr{C}}(P))(l) : l < k\}$ and $(\mathfrak{S}_{\mathscr{C}}(P))(k+) = \cup\{(\mathfrak{S}_{\mathscr{C}}(P))(l) : l > k\}$. The natural intuitionistic fuzzy topology on $\mathbf{R}^*(\mathbb{I} \times \mathbb{I})$ is generated from the sub-basis $\{L_k, R_k : k \in \mathbf{R}\}$ where $L_k[\mathfrak{S}_{\mathscr{C}}(P)] = (\mathfrak{S}_{\mathscr{C}}(P))(k-)$ and $R_k[\mathfrak{S}_{\mathscr{C}}(P)] = (\mathfrak{S}_{\mathscr{C}}(P))(k+)$. A partial order on $\mathbf{R}^*(\mathbb{I} \times \mathbb{I})$ is defined by $[\mathfrak{S}_{\mathscr{C}}(P)] \subseteq [\mathfrak{S}_{\mathscr{C}}(Q)]$ iff $(\mathfrak{S}_{\mathscr{C}}(P))(k-) \subseteq (\mathfrak{S}_{\mathscr{C}}(Q))(k-)$ and $(\mathfrak{S}_{\mathscr{C}}(P))(k+) \subseteq (\mathfrak{S}_{\mathscr{C}}(Q))(k+)$, for all $k \in \mathbf{R}$.

Definition 16 A mapping $f : \mathscr{E}_{\mathscr{C}}(\mathbf{R}) \to \mathbf{R}^*(\mathbb{I} \times \mathbb{I})$ is called as an intuitionistic fuzzy $\mathscr{C}$-centred lower (upper) continuous function, if $f^{-1}R_k$ (respy. $f^{-1}L_k$) is an IF open subset (IF closed subset) of $\mathscr{E}_{\mathscr{C}}(\mathbf{R})$, for each $k \in \mathbf{R}$.

Proposition 8 *For all $\mathfrak{S}_{\mathscr{C}} \in \mathscr{E}_{\mathscr{C}}(\mathbf{R})$, define a mapping $f : \mathscr{E}_{\mathscr{C}}(\mathbf{R}) \to \mathbf{R}^*(\mathbb{I} \times \mathbb{I})$ as*

$$f(\mathfrak{S}_{\mathscr{C}})(k) = \begin{cases} 1_{\sim}, & \text{if } k < 0 \\ P, & \text{if } k \in [0, 1] \\ 0_{\sim}, & \text{if } k > 1 \end{cases}$$

Then, f is an intuitionistic fuzzy $\mathscr{C}$-centred lower (resp. upper) continuous function iff P is an IFcOS (resp. IFcCS) in $\mathbf{R}$.

Proposition 9 *The following statements are identically true for any intuitionistic fuzzy $\mathscr{C}$-centred space, $\mathscr{E}_{\mathscr{C}}(\mathbf{R})$. Then,*

(a) $\mathscr{E}_{\mathscr{C}}(\mathbf{R})$ is an intuitionistic fuzzy extremally disconnected centred space.
(b) Let $g, h : \mathscr{E}_{\mathscr{C}}(\mathbf{R}) \to \mathbf{R}^*(\mathbb{I} \times \mathbb{I})$. If g is intuitionistic fuzzy $\mathscr{C}$-centred lower continuous function and h is intuitionistic fuzzy $\mathscr{C}$-centred upper continuous function with $g \subseteq h$, then there is an intuitionistic fuzzy continuous function, $f : \mathscr{E}_{\mathscr{C}}(\mathbf{R}) \to \mathbf{R}^*(\mathbb{I} \times \mathbb{I})$ such that $g \subseteq f \subseteq h$.
(c) If $\mathfrak{S}_{\mathscr{C}}(P), \mathfrak{S}_{\mathscr{C}}(Q)$ are the subsets of $\mathscr{E}_{\mathscr{C}}(\mathbf{R})$ such that $\mathfrak{S}_{\mathscr{C}}(Q) \subseteq \mathfrak{S}_{\mathscr{C}}(P)$, then there is an intuitionistic fuzzy continuous function, f such that $\mathfrak{S}_{\mathscr{C}}(Q) \subseteq \overline{L_1 f} \subseteq R_0 f \subseteq \mathfrak{S}_{\mathscr{C}}(P)$.

Definition 17 Let $\mathfrak{S}_{\mathscr{C}}$ and $\mathfrak{S}_{\mathscr{C}}(A)$ be any intuitionistic fuzzy $\mathscr{C}$-end and any subset of $\mathscr{E}_{\mathscr{C}}(\mathbf{R})$, respectively. Then, the characteristic function of $\mathfrak{S}_{\mathscr{C}}(A)$, $\chi_{\mathfrak{S}_{\mathscr{C}}(A)}$ is defined as

$$\chi_{\mathfrak{S}_{\mathscr{C}}(A)}(\mathfrak{S}_{\mathscr{C}}) = \begin{cases} \mathscr{E}_{\mathscr{C}}(\mathbf{R}), & \text{if } \mathfrak{S}_{\mathscr{C}} \in \mathfrak{S}_{\mathscr{C}}(A) \\ \phi, & \text{if } \mathfrak{S}_{\mathscr{C}} \notin \mathfrak{S}_{\mathscr{C}}(A) \end{cases}$$

for all $\mathfrak{S}_{\mathscr{C}} \in \mathscr{E}_{\mathscr{C}}(\mathbf{R})$.

Proposition 10 *Let $\mathscr{E}_{\mathscr{C}}(\mathbf{R})$ be an IF extremally disconnected centred space. Let $\mathfrak{S}_{\mathscr{C}}(A) \subseteq \mathscr{E}_{\mathscr{C}}(\mathbf{R})$ such that $\chi_{\mathfrak{S}_{\mathscr{C}}(A)}$ is an IFOS in $\mathscr{E}_{\mathscr{C}}(\mathbf{R})$. Let f be an intuitionistic fuzzy continuous function on $\mathfrak{S}_{\mathscr{C}}(A)$. Then, f has an intuitionistic fuzzy continuous extension over $\mathscr{E}_{\mathscr{C}}(\mathbf{R})$.*

Proof Let $g, h : \mathscr{E}_{\mathscr{C}}(\mathbf{R}) \to \mathbf{R}^*(\mathbb{I} \times \mathbb{I})$ be such that $g = f = h$ on $\mathfrak{S}_{\mathscr{C}}(A)$ and $g(\mathfrak{S}_{\mathscr{C}}) = 1_{\sim}$, if $\mathfrak{S}_{\mathscr{C}} \notin \mathfrak{S}_{\mathscr{C}}(A)$ and $h(\mathfrak{S}_{\mathscr{C}}) = 0_{\sim}$, if $\mathfrak{S}_{\mathscr{C}} \notin \mathfrak{S}_{\mathscr{C}}(A)$. We now have,

$$R_k g = \begin{cases} (\mathfrak{S}_{\mathscr{C}}(B)) \cap \chi_{\mathfrak{S}_{\mathscr{C}}(A)}, & \text{if } k > 0 \\ \mathscr{E}_{\mathscr{C}}(\mathbf{R}), & \text{if } k \leq 0 \end{cases}$$

where, $\mathfrak{S}_{\mathscr{C}}(B)$ is an IFOS in $\mathscr{E}_{\mathscr{C}}(\mathbf{R})$ and for all $k \in \mathbf{R}$,

$$L_k h = \begin{cases} (\mathfrak{S}_{\mathscr{C}}(G)) \cap \chi_{\mathfrak{S}_{\mathscr{C}}(A)}, & \text{if } k < 1 \\ \phi, & \text{if } k \geq 1 \end{cases}$$

where, $\mathfrak{S}_{\mathscr{C}}(G)$ is an IFCS in $\mathscr{E}_{\mathscr{C}}(\mathbf{R})$. Thus, g is an intuitionistic fuzzy $\mathscr{C}$-centred lower continuous function, and h is an intuitionistic fuzzy $\mathscr{C}$-centred upper continuous function with $g \subseteq h$. Now, by the Proposition 9, there exists an intuitionistic fuzzy continuous function, $\mathscr{F} : \mathscr{E}_{\mathscr{C}}(\mathbf{R}) \to \mathbf{R}^*(\mathbb{I} \times \mathbb{I})$ such that $g \subseteq \mathscr{F} \subseteq h$. Hence, $\mathscr{F} \equiv f$ on $\mathscr{E}_{\mathscr{C}}(\mathbf{R})$.

4 Conclusion

The method of centred systems has many applications in the fields of medical, technology, etc. In this paper, the notion of an intuitionistic fuzzy $\mathscr{C}$-end was introduced, and several characterizations related to the intuitionistic fuzzy extremally disconnected centred spaces were studied in the Sect. 3. In Sects. 1 and 2, introduction and some basic definitions related to the concepts of intuitionistic fuzzy $\mathscr{C}$-ends were provided.

References

1. Atanassov,K.: Intuitionistic fuzzy sets. Fuzzy Sets and Systems. **20**, 87–96 (1986).
2. Atanassov,K., Gargov,G.: Elements of intuitionistic fuzzy logic. Fuzzy Sets and Systems. **95**, 39–52 (1998).
3. Chang,C.L.: Fuzzy topological spaces. J. Math. Anal. Appl.. **24**, 182–190 (1968).
4. Coker,D.: An introduction to intuitionistic fuzzy topological spaces. Fuzzy Sets and Systems. **88**, 81–89 (1997).
5. Coker,D., Demirci, M.: On intuitionistic fuzzy points. Notes IFS. **1**, 79–84 (1995).
6. Dugundji,J.: Topology, Prentice Hall of India Private Limited, New Delhi (1975).
7. Fatteh,U.V., Bassan,D.S. : Fuzzy connectedness and its stronger forms. J. Math. Anal. Appl. **111**, 449–464 (1985).
8. Iliadis,S., Fomin,S.: The method of centred systems in the theory of topological spaces.UMN. **21**, 47–66 (1966).
9. Montiel Ross,O., Sepulveda Cruz,R., Castillo,O., Alper Basturk: High performance fuzzy systems for real world problems. Advances in fuzzy systems. (2012) https://doi.org/10.1155/2012/316187, 1–2.
10. Montiel Ross,O., Camacho,J., Sepulveda Cruz,R., Castillo,O. : Fuzzy system to control the movement of a wheeled mobile robot. Soft computing for intelligent control and mobile robotics. (2011) https://doi.org/10.1007/978-3-642-15534-5, 445–463.
11. Smets,P.: The degree of belief in a fuzzy event. Information Sciences. **25**, 1–19 (1981).

12. Sugeno,M.: An introductory survey of fuzzy control. Information Sciences. **36**, 59–83 (1985).
13. Uma,M.K., Roja,E., Balasubramanian,G.: A new characterization of fuzzy extremally disconnected spaces. Atti. Sem. Mat. Fis. Univ. Modenae Reggio Emilia. **L III**, 289–297 (2005).
14. Yogalakshmi,T., Roja,E., Uma,M.K.: A view on soft fuzzy C-continuous function. The Journal of Fuzzy Mathematics. **21(2)**, 349–370 (2013).
15. Yogalakshmi,T.: Disconnectedness in soft fuzzy centred systems. International Journal of Pure and Applied Mathematics. **115(9)**, 223–229 (2017).
16. Zadeh,L.A.: Fuzzy sets. Inform and Control. **8**, 338–353 (1965).

Generalized Absolute Riesz Summability of Orthogonal Series

K. Kalaivani and C. Monica

Abstract In this paper, for $1 \le k \le 2$ and a sequence $\gamma := \{\gamma(n)\}_{n=1}^{\infty}$ that is quasi β-power monotone decreasing with $\beta > 1 - \dfrac{1}{k}$, we prove the $|A, \gamma|_k$ summability of an orthogonal series, where A is Riesz matrix. For $\beta > \frac{1}{2}$, we give a necessary and sufficient condition for $|A, \gamma|_k$ summability, where A is Riesz matrix. Our result generalizes the result of Moricz (Acta Sci Math 23:92–95, 1962) for absolute Riesz summability of an orthogonal series.

Notation $\mathbb{N}$ = Natural numbers, $\mathbb{C}$ = Complex numbers, $\mathbb{R}$ = Real numbers, $\mathbb{Z}^+ = \mathbb{N} \cup \{0\}$, $\underline{c} = \{c_n\} \subset \mathbb{C}$. For $n \in \mathbb{Z}^+$, S_n denotes nth partial sum of the series $\sum\limits_{n=0}^{\infty} b_n$.

1 Introduction

Definition 1 Let $A = (a_{n\,m})_{n,m \in \mathbb{Z}^+}$ be a matrix of complex numbers. For $\sum\limits_{k=0}^{\infty} s_k$, we associate a sequence $\{\sigma_n\}_{n=0}^{\infty}$, given by $\sigma_n = \sum\limits_{k=0}^{\infty} a_{n\,k} s_k$, $n \in \mathbb{Z}^+$, provided the series converges for each n. We call σ_n the nth A-mean of the series.

Definition 2 Let $\gamma : [1, \infty) \to [0, \infty)$ be a nondecreasing function, $k \ge 1$. We say that the series $\sum\limits_{n=0}^{\infty} b_n$ is $|A, \gamma|_k$ summable, if the series

$$\sum_{n=1}^{\infty} \gamma(n)^k n^{k-1} |\sigma_n - \sigma_{n-1}|^k \text{ converges,}$$

K. Kalaivani (✉) · C. Monica
Vellore Institute of Technology, Vellore, Tamilnadu, India
e-mail: kalai_ashbom@yahoo.com; kalaivani.kannan@vit.ac.in; monicariasm@yahoo.com

V. Madhu et al. (eds.), *Advances in Algebra and Analysis*, Trends in Mathematics,
https://doi.org/10.1007/978-3-030-01120-8_22

where σ_n is the nth A-mean of $\sum\limits_{n=0}^{\infty} b_n$.

For example,

(i) $|A, \gamma|_k = |E, q, \gamma|_k$ for $q > 0, k \geq 1$ where $a_{n\,m} = \binom{n}{m} q^{n-m}(1+q)^{-n}$ if $0 \leq m \leq n$ and 0 otherwise.

Definition 3 Let $\gamma := \{\gamma(n)\}_{n=1}^{\infty}$ be a positive sequence. For $\beta \in \mathbb{R}$, we call a sequence γ is quasi β-power monotone decreasing if $\exists\ M = M(\beta, \gamma) \geq 1$ such that

$$n^{\beta}\gamma(n) \leq M m^{\beta}\gamma(m) \text{ for any } m \leq n.$$

For $\beta \in \mathbb{R}$, let $\Gamma_\beta = \{\gamma \mid \gamma : [1, \infty) \to [0, \infty)$ be nondecreasing such that $\{\gamma(n)\}_{n=1}^{\infty}$ is quasi β-power monotone decreasing$\}$.

In 1995, Leindler[4] proved that for any orthonormal system $\{\psi_n\}_{n=1}^{\infty}$ and $\underline{c} \in \ell_2(\mathbb{N})$, the condition

$$\sum_{m=0}^{\infty} \gamma(2^m)^k \left\{ \sum_{n=2^m+1}^{2^{m+1}} |c_n|^2 \right\}^{\frac{k}{2}} < \infty$$

is necessary and sufficient for the orthogonal series $\sum\limits_{n=1}^{\infty} c_n \psi_n$ to be $|C, \alpha, \gamma|_k$ summable a.e where $\gamma \in \Gamma_\beta$ with $\beta > -1,\ \alpha > \frac{1}{2}$, and $1 \leq k \leq 2$.

In [4], he also gave necessary and sufficient for the orthogonal series $\sum\limits_{n=1}^{\infty} c_n \psi_n$ to be $|C, \alpha, \gamma|_k$ summable a.e where $\gamma \in \Gamma_\beta$ with $\beta > -1, \alpha \leq \frac{1}{2}$, and $1 \leq k \leq 2$.

In [1], Moricz proved that for any orthonormal system $\{\psi_n\}_{n=1}^{\infty}$ and $\underline{c} \in \ell_2(\mathbb{N})$, the condition

$$\sum_{m=1}^{\infty} \left\{ \sum_{n=\nu_m+1}^{\nu_{m+1}} |c_\nu|^2 \right\}^{\frac{k}{2}} < \infty \tag{1}$$

is sufficient for the orthogonal series $\sum\limits_{n=1}^{\infty} c_n \psi_n$ to be absolute Riesz summable a.e where $\nu = \Lambda(2^m)$, $\Lambda(x)$ is the inverse of the increasing function $\lambda(x)$.

2 Preliminaries

Definition 4 A positive sequence $\gamma := \{\gamma(n)\}_{n=1}^{\infty}$ is said to be quasi geometrically decreasing if $\exists\ \mu \in \mathbb{N}$ and $M = M(\gamma) \geq 1$ such that

$$\gamma(n+\mu) \leq \frac{1}{2}\gamma(n) \text{ and } \gamma(n+1) \leq M\gamma(n) \;\; \forall\, n \in \mathbb{N}.$$

Theorem 1 *Let $\gamma := \{\gamma(n)\}_{n=1}^{\infty}$ be a positive sequence.*

(i) If γ is quasi β-power monotone decreasing with $\beta > 0$, then $\{\gamma(2^n)\}_{n=0}^{\infty}$ is quasi geometrically decreasing.

(ii) The sequence γ is quasi geometrically decreasing iff

$$\sum_{n=m}^{\infty} \gamma(n) \leq M\gamma(m) \text{ for some } M \geq 1, \;\forall\, m \in \mathbb{N}.$$

For proofs of (i) and (ii), we refer to [3] and [2], respectively.

Lemma 1 ([5]) *Let $\underline{c} \in \ell_2(\mathbb{Z}^+)$. Then*

(i) $f(x) = \sum_{n=0}^{\infty} c_n r_n(x)$ is in $L_2[0,1]$ where $r_n(x) = \text{sign}\sin(2^n \pi x)$ for $0 < x < 1$.

(ii) For any $r > 0$, $\exists\, A(r), B(r) > 0$ such that

$$A(r)\|\underline{c}\|_2 \leq \|f\|_r \leq B(r)\|\underline{c}\|_2 \tag{2}$$

holds.

(iii) For any Lebesgue measurable $E \subset (0,1)$, $\exists\, N = N(E) \in \mathbb{N}$ such that for any $n \geq m \geq N$,

$$\int_E \left|\sum_{i=m}^{n} c_i r_i(x)\right|^2 d\lambda(x) \leq 2\lambda(E)\sum_{i=m}^{n} |c_i|^2. \tag{3}$$

3 Main Theorems

Lemma 2 *If $\{\gamma(n)\}_{n=1}^{\infty}$ is quasi β-power monotone decreasing with $\beta > 0$, then for any $k > 0$, $\exists\, M \geq 1$ such that $\sum_{n=m}^{\infty} \gamma(2^n)^k \leq M\gamma(2^m)^k$, $m \in \mathbb{Z}^+$.*

Proof Let $\{\gamma(n)\}_{n=1}^{\infty}$ be quasi β-power monotone decreasing with $\beta > 0$. It is easy to see that for any $k > 0$, $\{\gamma(n)^k\}_{n=1}^{\infty}$ is quasi $k\beta$-power monotone decreasing. Then by Theorem 2(i), $\{\gamma(2^n)^k\}_{n=0}^{\infty}$ is quasi geomentrically decreasing. Hence, by Theorem 2(ii), $\exists\, M \geq 1$ such that $\sum_{n=m}^{\infty} \gamma(2^n)^k \leq M\gamma(2^m)^k$, $m \in \mathbb{Z}^+$.

Riesz Matrix Let $0 < \lambda_1 < \lambda_2 < \lambda_3 < \ldots$ with $\lim_{n\to\infty} = 0$. For $n, k \in \mathbb{Z}^+$, we define

$$a_{n\,k} := \begin{cases} 1 - \dfrac{\lambda_k}{\lambda_{n+1}} & \text{if } 0 \le k \le n, \\ \\ 0 & \text{otherwise.} \end{cases}$$

Proposition 1 *Let* $\{\phi_n\}_{n=0}^{\infty} \subset L_2[0, 1]$ *be an orthonormal system and* $(a_{n\,k})_{n,k\in\mathbb{Z}^+}$ *be Riesz matrix. Then for* $n \in \mathbb{N}$ *and*

(i) $\displaystyle\int_0^1 |\sigma_n(x) - \sigma_{n-1}(x)|^2 dx = \left(\frac{1}{\lambda_n} - \frac{1}{\lambda_{n+1}}\right)^2 \sum_{k=0}^{n} \lambda_k^2 c_k^2$

(ii) $\displaystyle\int_0^1 |\sigma_n(x) - \sigma_{n-1}(x)|^2 dx \le 2^2 \sum_{m=0}^{n} |c_m|^2 \ \forall\, c_0, c_1, \ldots c_n \in \mathbb{C}.$

Proof For $n \in \mathbb{N}$, $\sigma_n(x) = \sum\limits_{k=0}^{n} a_{n\,k} c_k \phi_k(x)$, thus

$$\begin{aligned}
\sigma_n(x) - \sigma_{n-1}(x) &= \sum_{k=0}^{n} \left(1 - \frac{\lambda_k}{\lambda_{n+1}}\right) c_k\phi_k - \sum_{k=0}^{n-1} \left(1 - \frac{\lambda_k}{\lambda_n}\right) c_k\phi_k \\
&= \sum_{k=0}^{n} c_k\phi_k - \sum_{k=0}^{n-1} c_k\phi_k + \sum_{k=0}^{n-1} \left(\frac{\lambda_k}{\lambda_n} - \frac{\lambda_k}{\lambda_{n+1}}\right) c_k\phi_k - \frac{\lambda_n}{\lambda_{n+1}} c_n\phi_n \\
&= \frac{\lambda_n}{\lambda_n} c_n\phi_n + \sum_{k=0}^{n-1} \left(\frac{\lambda_k}{\lambda_n} - \frac{\lambda_k}{\lambda_{n+1}}\right) c_k\phi_k - \frac{\lambda_n}{\lambda_{n+1}} c_n\phi_n \\
&= \sum_{k=0}^{n} \left(\frac{\lambda_k}{\lambda_n} - \frac{\lambda_k}{\lambda_{n+1}}\right) c_k\phi_k.
\end{aligned}$$

(i) By Parseval's identity, we have

$$\begin{aligned}
\int_0^1 |\sigma_n(x) - \sigma_{n-1}(x)|^2 dx &= \sum_{k=0}^{n} \left(\frac{1}{\lambda_n} - \frac{1}{\lambda_{n+1}}\right)^2 \lambda_k^2 |c_k|^2 \\
&= \left(\frac{1}{\lambda_n} - \frac{1}{\lambda_{n+1}}\right)^2 \sum_{k=0}^{n} \lambda_k^2 |c_k|^2
\end{aligned}$$

(ii) Using monotonicity of $\{\lambda_k\}_{k=0}^{\infty}$ in Proposition 1(i), we obtain

$$\int_0^1 |\sigma_n(x) - \sigma_{n-1}(x)|^2 dx \le \sum_{k=0}^{n} \left(\frac{\lambda_k}{\lambda_n} + \frac{\lambda_k}{\lambda_{n+1}} \right)^2 |c_k|^2$$
$$\le \sum_{k=0}^{n} 2^2 |c_k|^2.$$

Theorem 2 *Let* $\{\phi_n\}_{n=0}^{\infty} \subset L_2[0,1]$ *be an orthonormal system and* $A = (a_{n\,m})_{n,m\in\mathbb{Z}^+}$ *be Riesz matrix,* $1 \le k \le 2$ *and* $\gamma \in \Gamma_\beta$ *with* $\beta > 1 - \frac{1}{k}$. *Then every orthogonal series* $\sum_{n=0}^{\infty} c_n\phi_n$, $\underline{c} \in \ell_2(\mathbb{Z}^+)$ *is* $|A,\gamma|_k$ *summable a.e.*

Proof By Proposition 1(ii), for $n \in \mathbb{N}$, we have

$$\|\sigma_n - \sigma_{n-1}\|_2^2 \le 2^2 \|\underline{c}\|_2^2 \ \forall\ \underline{c} \in \ell_2(\mathbb{Z}^+). \tag{4}$$

For $k = 2$, we use (4) to obtain

$$\sum_{n=1}^{\infty} \gamma(n)^2 n^{2-1} \int_0^1 |\sigma_n(x) - \sigma_{n-1}(x)|^2 dx \le \sum_{n=1}^{\infty} \gamma(n)^2 n^{2-1} 2 \|\underline{c}\|_2.$$

For $1 \le k < 2$, using Hölder's inequality with $p = \frac{2}{k}$, we have

$$\sum_{n=1}^{\infty} \gamma(n)^k n^{k-1} \int_0^1 |\sigma_n(x) - \sigma_{n-1}(x)|^k dx \le \sum_{n=1}^{\infty} \gamma(n)^k n^{k-1} \{\|\sigma_n - \sigma_{n-1}\|_2^2\}^{\frac{k}{2}}.$$

Hence for any $1 \le k \le 2$ and by (4), we obtain

$$\sum_{n=1}^{\infty} \gamma(n)^k n^{k-1} \int_0^1 |\sigma_n(x) - \sigma_{n-1}(x)|^k dx \le \sum_{n=1}^{\infty} \gamma(n)^k n^{k-1} \left\{2^2 \|\underline{c}\|_2^2\right\}^{\frac{k}{2}}$$
$$\le (\|\underline{c}\|_2 2)^k \sum_{n=1}^{\infty} \gamma(n)^k n^{k-1}$$
$$\le (\|\underline{c}\|_2 2)^k \sum_{n=1}^{\infty} \gamma(2^n)^k (2^n)^{k-1}$$
$$\le (\|\underline{c}\|_2 2)^k C \gamma(2)^k (2)^{k-1}.$$

In arriving at the step above, we have used the facts that the sequence $\{\gamma(n)\}_{n=1}^{\infty}$ is quasi β-power monotone decreasing with $\beta > 1 - \frac{1}{k}$ and the sequence $\{n^{k-1}\gamma(n)^k\}_{n=1}^{\infty}$ is quasi $k\epsilon-$power monotone decreasing with $\epsilon = \beta - 1 + \frac{1}{k}$.

Then by Lemma 2, $\exists\, C \geq 1$ such that

$$\sum_{n=m}^{\infty}(2^n)^{k-1}\gamma(2^n)^k \leq C(2^m)^{k-1}\gamma(2^m)^k, \ m \in \mathbb{N}.$$

Theorem 3 *Let $\{\phi_n\}_{n=0}^{\infty} \subset L_2[0,1]$ be an orthonormal system, $A = (a_{n\ m})_{n,m\in\mathbb{Z}^+}$ be a Riesz matrix, $1 \leq k \leq 2$ and $\gamma \in \Gamma_\beta$ with $\beta > -\frac{3}{4}$. Then for any $\underline{c} \in \ell_2(\mathbb{Z}^+)$ the condition*

$$\sum_{m=0}^{\infty}\gamma(2^m)^k\left\{\sum_{n=2^m+1}^{2^{m+1}} n\,|c_n|^2\right\}^{\frac{k}{2}} < \infty \tag{5}$$

is sufficient for the orthogonal series $\sum\limits_{n=0}^{\infty} c_n\phi_n$ to be $|A,\gamma|_k$ summable a.e.

Proof For $n \in \mathbb{N}$, $\sigma_n(x) = \sum\limits_{k=0}^{n} a_{n\ k}c_k\phi_k(x)$. Then by Proposition 1(i) such that

$$\int_0^1 |\sigma_n(x) - \sigma_{n-1}(x)|^2 dx = \left(\frac{1}{\lambda_n} - \frac{1}{\lambda_{n+1}}\right)^2 \sum_{k=0}^{n}\lambda_k^2|c_k|^2. \tag{6}$$

For $1 \leq k \leq 2$, let $\Theta = \sum\limits_{n=2}^{\infty}\gamma(n)^k n^{k-1}\int_0^1 |\sigma_n(x) - \sigma_{n-1}(x)|^k dx$. Using Hölder's inequality with $p = \frac{2}{k}$ and by (6), we have

$$\begin{aligned}\Theta &\leq \sum_{n=2}^{\infty}\gamma(n)^k n^{k-1}\left\{\int_0^1 |\sigma_n(x) - \sigma_{n-1}(x)|^2 dx\right\}^{\frac{k}{2}} \\ &\leq \sum_{n=2}^{\infty}\gamma(n)^k n^{k-1}\left\{\left(\frac{1}{\lambda_n} - \frac{1}{\lambda_{n+1}}\right)^2 \sum_{k=0}^{n}\lambda_k^2|c_k|^2\right\}^{\frac{k}{2}}\end{aligned} \tag{7}$$

Thus for $k = 2$, from (7) we obtain

$$\begin{aligned}\Theta &\leq \sum_{r=0}^{\infty}\sum_{n=2^r+1}^{2^{r+1}}\gamma(n)^2 n\left\{\sum_{k=0}^{n}\left(\frac{1}{\lambda_n} - \frac{1}{\lambda_{n+1}}\right)^2\lambda_k^2|c_k|^2\right\} \\ &\leq \sum_{r=0}^{\infty}\gamma(2^{r+1})^2 2^{r+1}\sum_{k=0}^{2^r+1}\sum_{n=2^r+1}^{2^{r+1}}\left(\frac{1}{\lambda_n} - \frac{1}{\lambda_{n+1}}\right)^2\lambda_k^2|c_k|^2\end{aligned}$$

$$+\sum_{r=0}^{\infty}\gamma(2^{r+1})^2 2^{r+1}\sum_{k=2^r+1}^{2^{r+1}}\sum_{n=k}^{2^{r+1}}\left(\frac{1}{\lambda_n}-\frac{1}{\lambda_{n+1}}\right)^2\lambda_k^2|c_k|^2$$

$$\cdot \leq \sum_{r=0}^{\infty}\gamma(2^{r+1})^2 2^{r+1}\sum_{k=0}^{2^r+1}|c_k|^2\lambda_k^2\sum_{n=2^r+1}^{2^{r+1}}\left(\frac{1}{\lambda_n}-\frac{1}{\lambda_{n+1}}\right)^2$$

$$+\sum_{r=0}^{\infty}\gamma(2^{r+1})^2 2^{r+1}\sum_{k=2^r+1}^{2^{r+1}}|c_k|^2\lambda_k^2\sum_{n=k}^{2^{r+1}}\left(\frac{1}{\lambda_n}-\frac{1}{\lambda_{n+1}}\right)^2$$

$$\leq 2\sum_{r=0}^{\infty}\gamma(2^{r+1})^2 2^{r+1}\sum_{k=0}^{2^{r+1}}|c_k|^2.$$

as $\lambda_k^2\sum_{n=k}^{\infty}\left(\frac{1}{\lambda_n}-\frac{1}{\lambda_{n+1}}\right)^2 \leq 2$. For $1 \leq k < 2$, using Hölder's inequality with $p=\frac{2}{k}$ in (7), we obtain

$$\Theta \leq \sum_{r=0}^{\infty}\sum_{n=2^r+1}^{2^{r+1}}\gamma(n)^k n^{k-1}\left\{\int_0^1|\sigma_n(x)-\sigma_{n-1}(x)|^2dx\right\}^{\frac{k}{2}}$$

$$\leq \sum_{r=0}^{\infty}\sum_{n=2^r+1}^{2^{r+1}}\gamma(n)^k n^{k-1}\left\{\left(\frac{1}{\lambda_n}-\frac{1}{\lambda_{n+1}}\right)^2\sum_{m=0}^{n}\lambda_m^2 a_m^2\right\}^{\frac{k}{2}}$$

$$\leq \sum_{r=0}^{\infty}\left(\sum_{n=2^r+1}^{2^{r+1}}\gamma(n)^{kq}n^{kq-q}\right)^{\frac{1}{q}}\left\{\sum_{n=2^r+1}^{2^{r+1}}\sum_{m=0}^{n}\left(\frac{1}{\lambda_n}-\frac{1}{\lambda_{n+1}}\right)^2\lambda_m^2|c_m|^2\right\}^{\frac{k}{2}}$$

$$\leq \sum_{r=0}^{\infty}\left(\sum_{n=2^r+1}^{2^{r+1}}\gamma(2^{r+1})^{kq}(2^{r+1})^{kq-q}\right)^{\frac{1}{q}}\left\{\sum_{n=2^r+1}^{2^{r+1}}\sum_{m=0}^{n}\left(\frac{1}{\lambda_n}-\frac{1}{\lambda_{n+1}}\right)^2\lambda_m^2|c_m|^2\right\}^{\frac{k}{2}}$$

$$\leq \sum_{r=0}^{\infty}\gamma(2^{r+1})^k(2^{r+1})^{k-1}(2^r)^{\frac{1}{q}}\left\{\sum_{n=2^r+1}^{2^{r+1}}\sum_{m=0}^{n}\left(\frac{1}{\lambda_n}-\frac{1}{\lambda_{n+1}}\right)^2\lambda_m^2|c_m|^2\right\}^{\frac{k}{2}}$$

$$\leq \sum_{r=0}^{\infty}\gamma(2^{r+1})^k(2^{r+1})^{k-1}(2^r)^{1-\frac{k}{2}}\left\{\sum_{n=2^r+1}^{2^{r+1}}\sum_{m=0}^{n}\left(\frac{1}{\lambda_n}-\frac{1}{\lambda_{n+1}}\right)^2\lambda_m^2|c_m|^2\right\}^{\frac{k}{2}}$$

Next we change the order of summation, to obtain

$$\Theta \leq 2^{k-1}\sum_{r=0}^{\infty}\gamma(2^{r+1})^{k}(2^{r})^{\frac{k}{2}}\left\{\sum_{m=0}^{2^{r}+1}\sum_{n=2^{r}+1}^{2^{r+1}}+\sum_{m=2^{r}+1}^{2^{r+1}}\sum_{n=m}^{2^{r+1}}\left(\frac{1}{\lambda_n}-\frac{1}{\lambda_{n+1}}\right)^{2}\lambda_m^{2}|c_m|^{2}\right\}^{\frac{k}{2}}$$

$$\leq 2^{k-1}\sum_{r=0}^{\infty}\gamma(2^{r+1})^{k}(2^{r})^{\frac{k}{2}}\left\{\sum_{m=0}^{2^{r}+1}\sum_{n=2^{r}+1}^{2^{r+1}}\left(\frac{1}{\lambda_n}-\frac{1}{\lambda_{n+1}}\right)^{2}\lambda_m^{2}|c_m|^{2}\right\}^{\frac{k}{2}}$$

$$+2^{k-1}\sum_{r=0}^{\infty}\gamma(2^{r+1})^{k}(2^{r})^{\frac{k}{2}}\left\{\sum_{m=2^{r}+1}^{2^{r+1}}\sum_{n=m}^{2^{r+1}}\left(\frac{1}{\lambda_n}-\frac{1}{\lambda_{n+1}}\right)^{2}\lambda_m^{2}|c_m|^{2}\right\}^{\frac{k}{2}}$$

$$\leq 2^{k-1}\sum_{r=0}^{\infty}\gamma(2^{r+1})^{k}(2^{r})^{\frac{k}{2}}\left\{\sum_{m=0}^{2^{r}+1}|c_m|^{2}\lambda_m^{2}\sum_{n=2^{r}+1}^{2^{r+1}}\left(\frac{1}{\lambda_n}-\frac{1}{\lambda_{n+1}}\right)^{2}\right\}^{\frac{k}{2}}$$

$$+2^{k-1}\sum_{r=0}^{\infty}\gamma(2^{r+1})^{k}(2^{r})^{\frac{k}{2}}\left\{\sum_{m=2^{r}+1}^{2^{r+1}}|c_m|^{2}\lambda_m^{2}\sum_{n=m}^{2^{r+1}}\left(\frac{1}{\lambda_n}-\frac{1}{\lambda_{n+1}}\right)^{2}\right\}^{\frac{k}{2}}$$

$$\leq 2^{\frac{3k}{2}-1}\sum_{r=0}^{\infty}\gamma(2^{r+1})^{k}(2^{r})^{\frac{k}{2}}\left\{\sum_{m=0}^{2^{r}+1}|c_m|^{2}\right\}^{\frac{k}{2}}$$

$$+2^{\frac{3k}{2}-1}\sum_{r=0}^{\infty}\gamma(2^{r+1})^{k}(2^{r})^{\frac{k}{2}}\left\{\sum_{m=2^{r}+1}^{2^{r+1}}|c_m|^{2}\right\}^{\frac{k}{2}}$$

as $\lambda_k^2 \sum_{n=k}^{\infty}\left(\frac{1}{\lambda_n}-\frac{1}{\lambda_{n+1}}\right)^2 \leq 2.$

$$\Theta \leq 2^{\frac{3k}{2}}\sum_{r=0}^{\infty}\gamma(2^{r+1})^{k}(2^{r})^{\frac{k}{2}}\left\{\sum_{m=0}^{2^{r}+1}|c_m|^{2}\right\}^{\frac{k}{2}}$$

$$\leq 2^{\frac{3k}{2}}\sum_{r=0}^{\infty}\gamma(2^{r+1})^{k}(2^{r})^{\frac{k}{2}}\left\{|c_1|^{2}+\sum_{s=0}^{r}\sum_{m=2^{s}+1}^{2^{s+1}}|c_m|^{2}\right\}^{\frac{k}{2}}$$

$$\leq 2^{\frac{3k}{2}}\sum_{r=0}^{\infty}\gamma(2^{r+1})^{k}(2^{r})^{\frac{k}{2}}\sum_{s=0}^{r}\left\{\sum_{m=2^{s}+1}^{2^{s+1}}|c_m|^{2}\right\}^{\frac{k}{2}}+2^{\frac{3k}{2}}\sum_{r=0}^{\infty}\gamma(2^{r+1})^{k}(2^{r})^{\frac{k}{2}}|c_1|^{k}$$

$$\leq 2^{\frac{3k}{2}} \sum_{s=0}^{\infty} \sum_{r=s}^{\infty} \gamma(2^{r+1})^k (2^r)^{\frac{k}{2}} \left\{ \sum_{m=2^s+1}^{2^{s+1}} |c_m|^2 \right\}^{\frac{k}{2}} + 2^{\frac{3k}{2}} \sum_{r=0}^{\infty} \gamma(2^{r+1})^k (2^r)^{\frac{k}{2}} |c_1|^k$$

$$\leq 2^{\frac{3k}{2}} \sum_{s=0}^{\infty} \gamma(2^s)^k (2^s)^{\frac{k}{2}} \left\{ \sum_{m=2^s+1}^{2^{s+1}} |c_m|^2 \right\}^{\frac{k}{2}} + 2^{\frac{3k}{2}} |c_1|^k \gamma(2)^k (2)^{\frac{k}{2}}.$$

To arrive at the step above, we have used the facts that the sequence $\{\gamma(n)\}_{n=1}^{\infty}$ is quasi β-power monotone decreasing with $\beta > \frac{1}{2}$, and $\{n^{\frac{1}{2}}\gamma(n)\}_{n=1}^{\infty}$ is quasi $\epsilon-$power monotone decreasing with $\epsilon = \beta - \frac{1}{2}$. Then by Lemma 2, $\exists\ C \geq 1$ such that

$$\sum_{r=m}^{\infty} \gamma(2^r)^k (2^r)^{\frac{k}{2}} \leq C\gamma(2^m)^k (2^m)^{\frac{k}{2}}, \ m \in \mathbb{Z}^+.$$

Thus, $\Theta \leq 2^{\frac{3k}{4}} \sum_{s=0}^{\infty} \gamma(2^s)^k \left\{ \sum_{m=2^s+1}^{2^{s+1}} m\ |c_m|^2 \right\}^{\frac{k}{2}} + 2^k |c_1|^k \gamma(2)^k$; and hence

$$\sum_{n=2}^{\infty} \gamma(n)^k n^{k-1} \int_0^1 |\sigma_n(x) - \sigma_{n-1}(x)|^k dx$$

$$\leq 2^{\frac{3k}{4}} \sum_{s=0}^{\infty} \gamma(2^s)^k \left\{ \sum_{m=2^s+1}^{2^{s+1}} m\ |c_m|^2 \right\}^{\frac{k}{2}} + |c_1|^k \gamma(2)^k.$$

4 Conclusion

In this paper, given $\gamma \in \Gamma_\beta$ and $1 \leq k \leq 2$

1. For $\beta > 1 - \dfrac{1}{k}$, we have proved the generalized absolute Riesz summability of an orthogonal series.
2. For $\beta > -\frac{3}{4}$, we have given a necessary and sufficient condition for the generalized absolute Riesz summability of an orthogonal series.

References

1. Ferenc Moricz.: Über die Rieszsche Summation der Orthogonal reihen. Acta Sci. Math. **23**, 92–95 (1962).
2. Leindler, L.: On the converse of inequalities of Hardy and Littlewood. Acta Sci. Math. **58**, 191–196 (1993).

3. Leindler, L., Németh, J.: On the connection of quasi power-monotone and quasi geomentrical sequences with application to integrability theorems for power series. Acta Math. Hungar. **68**, 7–19 (1995).
4. Leindler, L.: On the newly generalized absolute Cesaro summability of orthogonal series. Acta Math. Hungar. **68**, 295–316 (1995).
5. Zygmund, A.: Trigonometric Series. I, Cambridge. (1959).

Holder's Inequalities for Analytic Functions Defined by Ruscheweyh-Type q-Difference Operator

N. Mustafa, K. Vijaya, K. Thilagavathi, and K. Uma

Abstract In this paper, we introduce a new generalized class of analytic functions based on Ruscheweyh-type q-difference operator. We obtain coefficient estimates, Holder's inequality result, and integral means results for $f \in \mathscr{T}\mathscr{J}_{\mu}^{\eta}(\alpha, \beta, \gamma, A, B)$.

1 Introduction and Definitions

Let $\mathscr{A}$ denote the class of functions of the form

$$f(z) = z + \sum_{n=2}^{\infty} a_n z^n \tag{1}$$

which are analytic and univalent in the open disc $\mathbb{U} = \{z : z \in \mathbb{C},\ |z| < 1\}$. Also denote by $\mathscr{T}$ a subclass of $\mathscr{A}$ consisting of functions of the form

$$f(z) = z - \sum_{n=2}^{\infty} |a_n| z^n, \quad z \in \mathbb{U}, \tag{2}$$

introduced and studied by Silverman [17]. Also denote by $\mathscr{S}\mathscr{T}(\alpha)$ the class of starlike functions of order $\alpha (0 \leq \alpha < 1)$ such that $\Re\left(\frac{zf'(z)}{f(z)}\right) > \alpha$, and $\mathscr{C}\mathscr{V}(\alpha)$ is convex of order α $(0 \leq \alpha < 1)$ satisfying the analytic criteria $\Re\left(1 + \frac{zf''(z)}{f'(z)}\right) > \alpha$.

N. Mustafa
Department of Mathematics, Kafkas University, Kars, Turkey
e-mail: nizamimustafa@gmail.com; nizamimustafa@kafkas.edu.tr

K. Vijaya (✉) · K. Thilagavathi · K. Uma
SAS, Department of Mathematics, VIT, Vellore, India
e-mail: kvijaya@vit.ac.in; kthilagavathi@vit.ac.in; kuma@vit.ac.in

V. Madhu et al. (eds.), *Advances in Algebra and Analysis*, Trends in Mathematics,
https://doi.org/10.1007/978-3-030-01120-8_23

For functions $f \in \mathscr{A}$ given by (1) and $g \in \mathscr{A}$ given by $g(z) = z + \sum_{n=2}^{\infty} b_n z^n$, we define the Hadamard product (or convolution) of f and g by

$$(f * g)(z) = z + \sum_{n=2}^{\infty} a_n b_n z^n, \quad z \in \mathbb{U}. \tag{3}$$

We briefly recall here the notion of *q-operators*, i.e., *q-difference operator*, that play a vital role in the theory of hypergeometric series, quantum physics, and in the operator theory. The application of *q-calculus* was initiated by Jackson [5] (also see [3, 6, 12]). Kanas and Răducanu [6] have used the fractional *q-calculus operators* in investigations of certain classes of functions which are analytic in $\mathbb{U}$.

For $0 < q < 1$ the Jackson's *q-derivative* of a function $f \in \mathscr{A}$ is, by definition, given as follows [5]

$$\mathscr{D}_q f(z) = \begin{cases} \dfrac{f(z) - f(qz)}{(1-q)z} & for\ z \neq 0, \\ f'(0) & for\ z = 0, \end{cases} \tag{4}$$

and $\mathscr{D}_q^2 f(z) = \mathscr{D}_q(\mathscr{D}_q f(z))$. From (4), we have $\mathscr{D}_q f(z) = 1 + \sum_{n=2}^{\infty} [n]_q a_n z^{n-1}$ where $[n]_q = \frac{1-q^n}{1-q}$, is sometimes called *the basic number n*. If $q \to 1^-$, $[n]_q \to n$. For a function $h(z) = z^m$, we obtain $\mathscr{D}_q h(z) = \mathscr{D}_q z^m = \frac{1-q^m}{1-q} z^{m-1} = [m]_q z^{m-1}$, and $\lim_{q \to 1^-} \mathscr{D}_q h(z) = \lim_{q \to 1^-} \left([m]_q z^{m-1}\right) = m z^{m-1} = h'(z)$, where h' is the ordinary derivative. For details on *q-calculus* one can refer [2, 5] and also the reference cited therein. Using the definition of Ruscheweyh *differential operator*[13] by for $f \in \mathscr{A}$, Kannas and Raducanu [6] defined and discussed the Ruscheweyh *q-differential operator* as

$$R_q^m f(z) = z + \sum_{n=2}^{\infty} \frac{\Gamma_q(n+m)}{[n-1]!\Gamma_q(1+m)} a_n z^n \quad (m > -1, z \in \mathbb{U}) \tag{5}$$

where

$$F_{q,m+1}(z) = z + \sum_{n=2}^{\infty} \frac{\Gamma_q(n+m)}{[n-1]!\Gamma_q(1+m)} z^n = z + \sum_{n=2}^{\infty} \frac{[m+1]_{n-1}}{[n-1]!} z^n \tag{6}$$

and $*$ stands for the Hadamard product (or convolution).

From (5) we note that $R_q^0 f(z) = f(z)$; $R_q^1 f(z) = z D_q f(z)$ and $R_q^m f(z) = \frac{z D_q^m (z^{m-1} f(z))}{[m]!}$. Making use of (5) and (6), we have

$$R_q^m f(z) = z + \sum_{n=2}^{\infty} \frac{\Gamma_q(n+m)}{[n-1]!\Gamma_q(1+m)} a_n z^n \quad (z \in \mathbb{U}). \tag{7}$$

Note that as $\lim_{q\to 1^-}$ we get

$$F_{m+1}(z) = \frac{z}{(1-z)^{m+1}} \text{ and } R^m f(z) = f(z) * \frac{z}{(1-z)^{m+1}}.$$

Hence,

$$\mathscr{D}_q(R_q^m f(z)) = 1 + \sum_{n=2}^{\infty} [n]_q \Psi_q(n,m) a_n z^{n-1}. \tag{8}$$

where

$$\Psi_q(n,m) = \frac{\Gamma_q(n+m)}{[n-1]!\Gamma_q(1+m)} \tag{9}$$

Let $t \in \mathbb{R}$ and $n \in \mathbb{N}$. Let the q-generalized Pochhammer symbol be defined as $[t]_n = [t][t+1][t+2]\ldots[t+n-1]$, and for $t > 0$ let the q-*gamma* function be defined by $\Gamma_q(t+1) = [t]\Gamma_q(t)$; $\Gamma_q(1) = 1$.

For fixed $-1 \le A \le B \le 1$ and $0 < B \le 1$,let $\mathscr{S}\mathscr{J}_\mu^\eta(\alpha,\beta,\gamma,A,B)$ denote the subclass of $\mathscr{A}$ consisting of functions $f(z)$ of the form (1) and satisfying the condition

$$\left| \frac{\frac{z(\mathscr{D}_q(R_q^m f(z))}{R_q^m f(z)} - 1}{2\gamma(B-A)\left(\frac{z(\mathscr{D}_q(R_q^m f(z))}{R_q^m f(z)} - \alpha\right) - B\left(\frac{z(\mathscr{D}_q(R_q^m f(z))}{R_q^m f(z)} - 1\right)} \right| < \beta, \quad z \in \mathbb{U} \tag{10}$$

where $R_q^m f(z)$ is given by (7), $0 \le \alpha < 1, 0 < \beta \le 1$,

$$\frac{B}{2(B-A)} < \gamma \le \begin{cases} \frac{B}{2(B-A)\alpha} & \alpha \neq 0, \\ 1 & \alpha = 0. \end{cases}$$

We also let $\mathscr{T}\mathscr{J}_\mu^\eta(\alpha,\beta,\gamma,A,B) = \mathscr{S}\mathscr{J}_\mu^\eta(\alpha,\beta,\gamma,A,B) \cap \mathscr{T}$.

For convenience in the entire paper, we consider $0 \le \alpha < 1, 0 < \beta \le 1$,

$$\frac{B}{2(B-A)} < \gamma \le \begin{cases} \frac{B}{2(B-A)\alpha} & \alpha \neq 0, \\ 1 & \alpha = 0. \end{cases}$$

for fixed $-1 \le A \le B \le 1$ and $0 < B \le 1$, one or otherwise stated. By suitably specializing the values of A, B, α, β, and γ, the class $\mathscr{T}\mathscr{J}_\mu^\eta(\alpha,\beta,\gamma,A,B)$ leads to known subclasses studied in [1, 7] and [11] and various new subclasses.

The main object of this paper is to determine the coefficient bound, thereby discussing Holder's inequality and integral means results for $f \in \mathscr{T}\mathscr{J}_\mu^\eta(\alpha,\beta,\gamma,A,B)$.

2 Characterization Properties

In this section we obtain the characterization property $f(z) \in \mathscr{T}\mathscr{J}_{\mu}^{\eta}(\alpha, \beta, \gamma, A, B)$ and state coefficient bounds.

Theorem 1 *Let the function $f(z)$ defined by (2) be in the class $\mathscr{T}\mathscr{J}_{\mu}^{\eta}(\alpha, \beta, \gamma, A, B)$ if and only if*

$$\sum_{n=2}^{\infty}[2\beta\gamma(B-A)([n]_q-\alpha)+(1-B\beta)([n]_q-1)]\Psi_q(n,m)|a_n| \leq 2\beta\gamma(1-\alpha)(B-A), \tag{11}$$

where $\Psi_q(n, m)$ is given by (9).

Proof The proof of Theorem 1 is much akin to the proof of theorems on coefficient bounds established in [14], so we skip the details in this regard.

Corollary 1 *Let the function f defined by (2) be in the class $\mathscr{T}\mathscr{J}_{\mu}^{\eta}(\alpha, \beta, \gamma, A, B)$, then we have*

$$|a_n| \leq \frac{2\beta\gamma(1-\alpha)(B-A)}{\left[2\beta\gamma(B-A)([n]_q-\alpha)+(1-B\beta)([n]_q-1)\right]\Psi_q(n,m)}, \tag{12}$$

the Eq. (12) is attained for the function

$$f(z) = z - \frac{2\beta\gamma(1-\alpha)(B-A)}{\left[2\beta\gamma(B-A)([n]_q-\alpha)+(1-B\beta)([n]_q-1)\right]\Psi_q(n,m)} z^n \quad (n \geq 2), \tag{13}$$

For the sake of brevity, we let

$$\Phi_n(\alpha, \beta, \gamma, A, B) = 2\beta\gamma(B-A)([n]_q-\alpha)+(1-B\beta)([n]_q-1) \tag{14}$$

and

$$\Phi_2(\alpha, \beta, \gamma, A, B) = 2\beta\gamma(B-A)(1+q-\alpha)+(1-B\beta)q \tag{15}$$

Followed by Nishiwaki et al.[10] and Murugusundaramoorthy et al.[9] in this section, we study some results of Holder-type inequalities for $f \in \mathscr{T}\mathscr{J}_{\mu}^{\eta}(\alpha, \beta, \gamma, A, B)$. Now we recall the generalization of the convolution due to Cho et al.[4] as given below

$$\mathscr{H}_m(z) = z - \sum_{n=2}^{\infty}\left(\prod_{j=1}^{m} a_{n,j}^{p_j}\right)z^n, \quad (p_j > 0,\ j = 1, 2, \ldots m). \tag{16}$$

Further for functions $f_j \in \mathcal{T}\mathcal{J}_{\mu}^{\eta}(\alpha, \beta, \gamma, A, B), (j = 1, 2 \ldots m)$ given by the familiar Holder's inequality assumes the following form

$$\sum_{n=2}^{\infty}\Big(\prod_{j=1}^{m} a_{n,j}\Big) \le \prod_{j=1}^{m}\Big(\sum_{n=2}^{\infty} a_{n,j}^{p_j}\Big)^{\frac{1}{p_j}}, \tag{17}$$

$$\Big(p_j > 1, \;\; j = 1, 2 \ldots m, \;\; \sum_{j=1}^{m}\frac{1}{p_j} \ge 1\Big).$$

Theorem 2 *If* $f_j \in \mathcal{T}\mathcal{J}_{\mu}^{\eta}(\alpha, \beta, \gamma, A, B), -1 \le B < A \le 1, 0 \le \alpha < 1\, 0 < \beta \le 1, (j = 1, 2, \ldots m)$*, then* $\mathcal{H}_m(z) \in \mathcal{T}\mathcal{J}_{\mu}^{\eta}(\alpha, \beta, \gamma, A, B)$ *with*

$$\xi \le 1 - \frac{(2\beta\gamma(B-A))^s \prod_{j=1}^{s}(1-\xi_j)^{p_j}\Big(1-(1-B\beta)2\beta\gamma(B-A)\Big)}{\prod_{j=1}^{m}\Big[2\beta\gamma(B-A)(1+q-\xi_j)+(1-B\beta)q\Big]^{p_j}[\Psi_q(2, m)]^{p_j-1} - [2\beta\gamma(B-A)]^s \prod_{j=1}^{m}(1-\xi_j)^{p_j}},$$

where

$$s = \sum_{j=1}^{m} p_j > 1; \;\; p_j \ge \frac{1}{q_j}(j = 1, 2, 3 \ldots m), \;\; q_j > 1(j = 1, 2 \ldots m);$$

$$\sum_{j=1}^{m} q_j \ge 1, \Psi_q(n, m) = \frac{\Gamma_q(n+m)}{[n-1]!\Gamma_q(1+m)}.$$

Proof Let $f_j \in \mathcal{T}\mathcal{J}_{\mu}^{\eta}(\alpha, \beta, \gamma, A, B), (j = 1, 2 \ldots m)$, then we have

$$\sum_{n=2}^{\infty}\frac{\Big[2\beta\gamma(B-A)([n]_q-\xi_j)+(1-B\beta)([n]_q-1)\Big]\Psi_q(n, m)}{2\beta\gamma(1-\xi_j)(B-A))}a_{n,j} \le 1,$$

which in turn implies that

$$\left(\sum_{n=2}^{\infty}\frac{\Big[2\beta\gamma(B-A)([n]_q-\xi_j)+(1-B\beta)([n]_q-1)\Big]\Psi_q(n, m)}{2\beta\gamma(1-\xi_j)(B-A)}a_{n,j}\right)^{\frac{1}{q_j}} \le 1,$$

$$\left(q_j > 1, \;\; (j = 1, 2, 3 \ldots m), \sum_{j=1}^{m}\frac{1}{q_j} = 1\right).$$

Applying the inequality (17) we arrive at the following inequality

$$\sum_{n=2}^{\infty}\left(\prod_{j=1}^{m}\frac{\left[2\beta\gamma(B-A)(n-\xi_j)+(1-B\beta)([n]_q-1)\right]}{2\beta\gamma(1-\xi_j)(B-A)}\Psi_q(n,m)a_{n,j}\right)^{\frac{1}{q_j}}a_{n,j}^{\frac{1}{q_j}}\leq 1.$$

Thus we determine the largest ξ such that

$$\left(\sum_{n=2}^{\infty}\frac{\left[2\beta\gamma(B-A)([n]_q-\xi_j)+(1-B\beta)([n]_q-1)\right]c_n(\eta,\mu)}{2\beta\gamma(1-\xi_j)(B-A)}\Psi_q(n,m)\right)\prod_{j=1}^{m}a_{n,j}^{p_j}\leq 1.$$

That is,

$$\left(\sum_{n=2}^{\infty}\frac{\left[2\beta\gamma(B-A)([n]_q-\xi_j)+(1-B\beta)([n]_q-1)\right]\Psi_q(n,m)}{2\beta\gamma(1-\xi_j)(B-A)}\Psi_q(n,m)\right)\prod_{j=1}^{m}a_{n,j}^{p_j}\leq$$

$$\sum_{n=2}^{\infty}\left[\left(\prod_{j=1}^{m}\frac{\left[2\beta\gamma(B-A)([n]_q-\xi_j)+(1-B\beta)([n]_q-1)\right]}{2\beta\gamma(1-\xi_j)(B-A)}\Psi_q(n,m)a_{n,j}\right)^{\frac{1}{q_j}}\right]a_{n,j}^{\frac{1}{q_j}}.$$

Since

$$\prod_{j=1}^{m}\left(\frac{\left[2\beta\gamma(B-A)([n]_q-\xi_j)+(1-B\beta)([n]_q-1)\right]}{2\beta\gamma(1-\xi_j)(B-A)}\Psi_q(n,m)\right)^{p_j-\frac{1}{q_j}}a_{n,j}^{p_j-\frac{1}{q_j}}\leq 1,$$

$$\left(p_j-\frac{1}{q_j}\geq 0,\ \ j=1,2,3\ldots m\right).$$

We see that

$$\prod_{j=1}^{m}a_{n,j}^{p_j-\frac{1}{q_j}}\leq\frac{1}{\prod_{j=1}^{m}\left(\frac{\left[2\beta\gamma(B-A)([n]_q-\xi_j)+(1-B\beta)([n]_q-1)\right]}{2\beta\gamma(1-\xi_j)(B-A)}\Psi_q(n,m)\right)^{p_j-\frac{1}{q_j}}}.\tag{18}$$

This last inequality (18) implies that

$$2\beta\gamma(B-A)\prod_{j=1}^{m}(2\beta\gamma(B-A))^{p_j-1}(1-\xi_j)^{p_j-1}$$

$$-\sum_{j=1}^{m}\left[2\beta\gamma([n]_q-\xi_j)(B-A)+(1-B\beta)([n]_q-1)\right]^{p_j}(\Psi_q(n,m))^{p_j-1}(1-\xi)$$

$$\leq\left(-([n]_q-1)(1-B\beta)\prod_{j=1}^{m}(2\beta\gamma(B-A))^{p_j-1}(1-\xi_j)^{p_j}\right)$$

$$+\left(([n]_q-1)2\beta\gamma(B-A)\prod_{j=1}^{m}(2\beta\gamma(B-A))^{p_j-1}(1-\xi_j)^{p_j}\right),$$

where

$$\Upsilon_j=\prod_{j=1}^{m}(2\beta\gamma(B-A))^{p_j}(1-\xi_j)^{p_j}.$$

which implies

$$\left[\Upsilon_j-\sum_{j=1}^{m}\left[2\beta\gamma([n]_q-\xi_j)(B-A)+(1-B\beta)([n]_q-1)\right]^{p_j}(\Psi_q(n,m))^{p_j-1}\right](1-\xi)$$

$$\leq-\left[([n]_q-1)\Upsilon_j+(1-B\beta)([n]_q-1)\prod_{j=1}^{m}(2\beta\gamma(B-A))^{p_j-1}(1-\xi_j)^{p_j}\right].$$

That is,

$$\xi\leq 1-\frac{\left[([n]_q-1)\Upsilon_j+(1-B\beta)([n]_q-1)\prod_{j=1}^{m}(2\beta\gamma(B-A))^{p_j-1}(1-\xi_j)^{p_j}\right]}{\sum_{j=1}^{m}\left[2\beta\gamma([n]_q-\xi_j)(B-A)+(1-B\beta)([n]_q-1)\right]^{p_j}-\Upsilon_j}.$$

Let

$$\Phi(n)\leq 1-\frac{\left[([n]_q-1)\Upsilon_j+(1-B\beta)([n]_q-1)\prod_{j=1}^{m}(2\beta\gamma(B-A))^{p_j-1}(1-\xi_j)^{p_j}\right]}{\sum_{j=1}^{m}\left[2\beta\gamma([n]_q-\xi_j)(B-A)+(1-B\beta)([n]_q-1)\right]^{p_j}-\Upsilon_j},$$

which is an increasing function in n; hence we have

$$\xi \leq \Phi(2)$$
$$= 1 - \frac{(2\beta\gamma(B-A))^s \prod_{j=1}^{s}(1-\xi_j)^{p_j}\Big(1-(1-B\beta)2\beta\gamma(B-A)\Big)}{\prod_{j=1}^{m}\Big[2\beta\gamma(B-A)(1+q-\xi_j)+(1-B\beta)q\Big]^{p_j}[\Psi_q(2,m)]^{p_j-1}-[2\beta\gamma(B-A)]^s\prod_{j=1}^{m}(1-\xi_j)^{p_j}}.$$

Hence the proof.

Now, we obtain integral means inequalities for the functions in the family $\mathcal{T}\mathcal{J}^{\eta}_{\mu}(\alpha,\beta,\gamma,A,B)$ due to Silverman[16].

Lemma 1 (Littlewood[8]) *If the functions f and g are analytic in $\mathbb{U}$ with $g \prec f$, then for $\eta > 0$, and $0 < r < 1$,*

$$\int_0^{2\pi}\left|g(re^{i\theta})\right|^{\eta}d\theta \leq \int_0^{2\pi}\left|f(re^{i\theta})\right|^{\eta}d\theta. \tag{19}$$

In 1975, Silverman[17] found that the function $f_2(z) = z - \frac{z^2}{2}$ is often extremal over the family $\mathcal{T}$ and applied this function to resolve his integral means inequality, conjectured in Silverman[15] and settled in Silverman [16], that

$$\int_0^{2\pi}\left|f(re^{i\theta})\right|^{\eta}d\theta \leq \int_0^{2\pi}\left|f_2(re^{i\theta})\right|^{\eta}d\theta,$$

for all $f \in \mathcal{T}$, $\eta > 0$ and $0 < r < 1$. Silverman[16] also proved his conjecture for the subclasses $\mathcal{T}^*(\gamma)$ and $\mathcal{C}(\gamma)$ of $\mathcal{T}$.

Applying Lemma 1and Theorem 1 , we prove the following result.

Theorem 3 *Suppose $f \in \mathcal{T}\mathcal{J}^{\eta}_{\mu}(\alpha,\beta,\gamma,A,B)$, $\eta > 0$, $0 \leq \lambda < 1$, $0 \leq \gamma < 1$, $\beta \geq 0$ and $f_2(z)$ is defined by*

$$f_2(z) = z - \frac{2\beta\gamma(1-\alpha)(B-A)}{\Phi_2(\alpha,\beta,\gamma,A,B)}z^2,$$

where $\Phi_2(\alpha,\beta,\gamma,A,B)$ is given by (15). Then for $z = re^{i\theta}$, $0 < r < 1$, we have

$$\int_0^{2\pi}|f(z)|^{\eta}d\theta \leq \int_0^{2\pi}|f_2(z)|^{\eta}d\theta. \tag{20}$$

Proof Let f be of the form (2) and from (20) it is equivalent to prove that

$$\int_0^{2\pi} \left|1 - \sum_{n=2}^{\infty} |a_n| z^{n-1}\right|^{\eta} d\theta \leq \int_0^{2\pi} \left|1 - \frac{2\beta\gamma(1-\alpha)(B-A)}{\Phi_2(\alpha,\beta,\gamma,A,B)} z\right|^{\eta} d\theta.$$

By Lemma 1, it suffices to show that

$$1 - \sum_{n=2}^{\infty} |a_n| z^{n-1} \prec 1 - \frac{2\beta\gamma(1-\alpha)(B-A)}{\Phi_2(\alpha,\beta,\gamma,A,B)} z.$$

Setting

$$1 - \sum_{n=2}^{\infty} |a_n| z^{n-1} = 1 - \frac{2\beta\gamma(1-\alpha)(B-A)}{\Phi_2(\alpha,\beta,\gamma,A,B)} w(z) \tag{21}$$

and using (11), we obtain

$$\begin{aligned} |w(z)| &= \left| \sum_{n=2}^{\infty} \frac{\Phi_n(\alpha,\beta,\gamma,A,B)}{2\beta\gamma(1-\alpha)(B-A)} |a_n| z^{n-1} \right| \\ &\leq |z| \sum_{n=2}^{\infty} \frac{\Phi_n(\alpha,\beta,\gamma,A,B)}{2\beta\gamma(1-\alpha)(B-A)} |a_n| \\ &\leq |z|, \end{aligned}$$

where $\Phi_n(\alpha,\beta,\gamma,A,B)$ is given by (14).

This completes the proof by Theorem 3.

References

1. Aghalary, R.,S. Kullkarni.: Some theorems on univalent functions, J. Indian Acad. Math, **24**,1,81–93 (2002)
2. Araci,S., Duran,U., Acikgoz M., Srivastava, H. M.: A certain (p, q)-derivative operator and associated divided differences,J. Inequal. Appl.**301** (2016)
3. Aral, A., Gupta, V. , Agarwal,R. P.: Applications of q-calculus in operator theory, Springer, New York, (2013)
4. Cho, N. E., Kim, T. H., Owa ,S.: Generalizations of hadamard products of functions with negative coefficients, J. Math. Anal. Appl., **199**, 495–501 (1996)
5. Jackson F. H. :On q-functions and a certain difference operator, Transactions of the Royal Society of Edinburgh,**46**, 253–281 (1908)
6. Kanas S. , Răducanu, D.: Some subclass of analytic functions related to conic domains, Math. Slovaca,**64**, 5, 1183–1196 (2014)

7. Khairanar, S. M. , Meena. : Certain family of analytic and univalent functions with normalized conditions, Acta Math. Acade. Paeda. Nyire, **24**, 333–344 (2008)
8. Littlewood, J. E.: On inequalities in theory of functions, Proc. London Math. Soc., **23**, 481–519 (1925)
9. Murugusundaramoorthy, G., Vijaya, K., Deepa ,K.: Holder inequalities for a subclass of univalent functions involving Dzoik Srivastava operator,Global Journal of Mathematical Analysis, **1**(3),74–82 (2013)
10. Nishiwaki, J., Owa, S., Srivastava ,H. M.:Convolution and Holder type inequalities for a certain class of analytic functions, Math. Inequal. Appl, **11**, 717–727 (2008)
11. Owa, S., Nishiwaki,J.: Coefficient estimates for certain classes of analytic functions, J. Inequal. Pure. Appl. Math, **3**(5), Art.72 (2002)
12. . Purohit S. D., Raina,R. K : Fractional q-calculus and certain subclasses of univalent analytic functions, Mathematica **55**,78, 1, 62–74 (2013)
13. Ruscheweyh St.: New criteria for univalent functions, Proc. Amer. Math. Soc. **49**, 109–115 (1975) M.
14. Salah, J., Darus,M.: A subclass of uniformly convex functions associated with a fractional calculus operator involving Caputo's fractional differentiation. Acta Universitatis Apulensis. No. **24**, 295–304 (2010)
15. Silverman, H.: A survey with open problems on univalent functions whose coefficients are negative, Rocky Mt. J. Math., **21**, 1099–1125 (1991)
16. Silverman, H.: Integral means for univalent functions with negative coefficients,Houston J. Math., **23**, 169–174 (1997)
17. Silverman, H.: Univalent functions with negative coefficients, Proc. Amer. Math. Soc., 51 109–116 (1975)

Fuzzy Cut Set-Based Filter for Fixed-Value Impulse Noise Reduction

P. S. Eliahim Jeevaraj, P. Shanmugavadivu, and D. Easwaramoorthy

Abstract This paper explores the efficient filter to reduce the noises in the digital images corrupted highly with the fixed-value impulse noise using fuzzy α-cut sets and median measure. The efficiency of the proposed filter is analyzed and proved that it is a high-performing fixed-value impulse noise filter qualitatively in terms of peak signal-to-noise ratio (PSNR) and mean structural similarity matrix (MSSIM) values. The human visual perception (HVP) of the filtered images is too validated the merit of the proposed method. It is also proved additionally that the proposed filter has less time complexity and assures higher degree of edge and detail preservation.

Keywords Highly corrupted images · Fixed-value impulse noises · Noise reduction · Median filter · Fuzzy α-cut sets

MSC Classification Codes 03E72, 62H35, 68U10

P. S. Eliahim Jeevaraj
Department of Computer Science, Bishop Heber College, Tiruchirappalli, Tamil Nadu, India
e-mail: eliahimps@gmail.com

P. Shanmugavadivu
Department of Computer Science and Applications, The Gandhigram Rural Institute (Deemed to be University), Gandhigram, Tamil Nadu, India
e-mail: psvadivu67@gmail.com

D. Easwaramoorthy (✉)
Department of Mathematics, School of Advanced Sciences, Vellore Institute of Technology, Vellore, Tamil Nadu, India
e-mail: easandk@gmail.com; easwar@vit.ac.in

V. Madhu et al. (eds.), *Advances in Algebra and Analysis*, Trends in Mathematics,
https://doi.org/10.1007/978-3-030-01120-8_24

1 Introduction

The digital images are represented as function that constitutes the matrix with coordinates and intensity values of the image. The digital image processing consists of three levels of processing. The mid-level processing is pivotal that concentrates on the preprocessing after acquisition. The image restoration is a vital preprocessing technique which recovers the corrupted image from the noise and blurring [4, 6].

A noise is an addition of impurity pixels in the image that is irrelevant to their neighborhood pixels. Such a noise affects the quality of the image and misinterprets the image details to the researchers/experts. The noise creeps into image based on the mathematical and statistical model that categorized as thermal noise, Gaussian noise, exponential noise, Poisson noise, Rayleigh noise, and impulse noise. Due to faulty sensors of capturing devices, inconsistent in atmospheric conditions and error-prone in transmission medium/channel, the impulses are distributed throughout image. The impulse noise is commonly found in all kind of images. Impulse noise is categorized into two types, namely, random-value impulse noise and fixed-value impulse noise. Random-value impulse noise corrupts the images with the intensity values which lie between maximum and minimum intensity values of the image. The fixed-value impulse noise is crept into image by adding the maximum and minimum intensity values of the image. The probability distribution of fixed-value impulse noise of $p\%$ in a corrupted image I' is described as:

$$I' = \begin{cases} 0 & \text{with probability} \quad (p/2)\% \\ 255 & \text{with probability} \quad (p/2)\% \\ U_{i,j} & \text{with probability} \quad (1-p)\% \end{cases} \tag{1}$$

where $U_{i,j}$ is the probability of the uncorrupted pixels in I' [7].

Researchers developed many noise filters for denoising purpose over the few decades. The noise filters are developed nonlinearly and adaptively for suppressing the noise effectively. From the literature, the nonlinear filter is superior to linear filters in terms of the efficiency of the restoration rate. The adaptive nonlinear filters find the noisy pixels and those noisy pixels only undergo the treatment of denoising and preserve the edge and details of the image. This paper focuses on design of fuzzy-based adaptive filter for fixed-value impulse noise. Fuzzy systems give the vital role in the development of noise filters. This fuzzy technique provides the platform for efficient and computationally simple filters. The paper uses simple fuzzy and median technique. So, the proposed filter has less complexity and computationally simple.

This paper utilizes the fuzzy α-cut set- and median-based techniques to develop a novel filter for eliminating the fixed-value impulse noise. The existing techniques are explained in Sect. 2, and the algorithmic description of the proposed filter is detailed in Sect. 3. The results and discussions are presented in Sect. 4, and the concluding remarks are in Sect. 5.

2 Existing Techniques

Generally, noise filters are developed by using mathematical, statistical and soft computing concepts and intuitive algorithms.

2.1 *Median Filters*

The standard median (SM) filter works by dividing the noisy image in sub-image using sliding window and computing the median of the sub-image. The computed median is replaced in the central pixel of the window [4]. This filter is the basis of all median-based filters. The center-weighted median filter (CWM) gives the weightage to center pixel while finding median and replaces the center pixel with computed median [13]. Iterative median filter applies median filtering technique iteratively which is more powerful than CWM and SM [3]. Signal-dependent rank order median (SDROM) filter performs efficiently than other filters, focuses on the false signals, and replaces those signals based on their neighboring true information [1]. Adaptive CWM filters suppress the noise that processes the impulse pixel in the window with only remaining pixels unaltered. This adaptive filter works in the image recursively and non-recursively [15]. An improved median filter works with impulse pixel and eliminates the noise with predefined thresholds. The decision based non-local mean filter is used for detecting the corrupted pixels and the correction is carried by the non-local mean of reference image [2, 12].

2.2 *Partial Differential Equation-Based Filters*

The PDE-based filters works in two distinct phases, namely, noise detection and noise correction. In the first phase, the noisy pixels are to be found and construct the flag matrix for the noisy image. By using the flag matrix, the corrupted pixels are selectively processed in the noise correction phase. The numerical solution formulae of PDE are used for noise correction. Five-point standard formula, five-point diagonal formula, and explicit method are used in ATSM, MPATS, and LEAM filters, respectively [9–11].

2.3 *Switching Filters*

The mechanism of switching median filters is done by correction of noisy pixel based on the predefined thresholds. These filters changed the strategy based on pixel and given condition. After that, the found values are compared with predefined threshold value. TSM and PSM are superior to other median-based filters [5, 14, 16].

2.4 Fuzzy-Based Filters

The fuzzy-based filters have good potential than other ordinary linear filters. These filters converted the image into fuzzy values and framed the rules for the processing. The fuzzy inferences betray the noisy pixels, and using normalization, the identified pixel is processed. After completing all the process, the values are converted into real pixel values [8].

3 Proposed Filter

The proposed filter used fuzzy systems for denoising the image from the impulse noise. The proposed filter comprises two phases, namely, noise detection and noise correction. First, the noisy image is transformed into the fuzzy values. The membership function is used for fuzzification of the pixels in the noisy image which is found in Eq. (2). Each pixel of the noisy image is considered as fuzzy members in fuzzy systems.

$$\mu(i,j) = \frac{1}{1 + \left[\dfrac{p(i,j)}{Max}\right]^{a}} \tag{2}$$

where $\mu(i,j)$ represents the membership of fuzzy set, $p(i,j)$ represents noisy image pixels, Max is the maximum gray-level intensity of the image, and a is arbitrary constant which value is any positive real numbers.

In noise detection phase, the α-cut is found in the fuzzy set itself. The α values are calculated from the image pixels values. α_1 value is the first quartile Q_1 of the image intensity range, and also α_2 value is calculated from the last quartile of the image pixel range. The α values are 0.52 and 0.94. The processed fuzzy set divides into three α-cut sets. The values which belong to first and last set are said to be noise. Using the noise-prone values, the flag matrix is constructed as per the fuzzy rules mentioned below.

Fuzzy Rules

1. If the $\mu(i,j)$ belongs to α_1 or α_3, then the fuzzy value is said to be 1
2. If the $\mu(i,j)$ belongs to α_2, the fuzzy value is said to be 0.

In normalization, the impulse pixel is found using the flag matrix of the noisy image. Those impulse pixels replace the median of the nearest uncorrupted pixel of the noisy pixel in both directions, namely, horizontal and vertical directions. This process is carried over the entire image. Finally, the normalized valued is transformed as real intensity values. The process is taken over by the inverse function of the membership function in Eq. (3).

$$r(i,j) = \sqrt[a]{\frac{Max^a}{\mu(i,j)} - Max^a} \tag{3}$$

where $r(i, j)$ represents the restored image pixel.

3.1 Algorithm of the Proposed Filter

The algorithmic description of the proposed median filter is as follows:

Input : Noisy image, I'
Output : Restored image, R

1. Read the corrupted image (I').
2. Noisy image is converted into fuzzy members set (μ) using membership function.
3. α-cut values are calculated. The fuzzy set is divided into three sets based on α-cut values.
4. If $\mu_{i,j}$ is uncorrupted, go to next $\mu_{i,j}$.
5. If $\mu_{i,j}$ is corrupted which is found by $F_X = 1$, then find the nearest uncorrupted neighboring pixel in horizontal and vertical directions, and then compute the median (MED) of the identified uncorrupted pixels.
6. Replace the corrupted pixel by MED.
7. Repeat steps from 4 to 7.
8. Defuzzify the values and construct restored image (R).

4 Results and Discussions

The newly devised filter developed in MATLAB 7.8 was tested on standard images like Lena, Mandril, Saturn, Cameraman, and Peppers and was also tested on some real images too. The size of the test images is 256 × 256. The performance of the proposed filter is quantitatively in terms of peak signal-to-noise ratio (PSNR) and mean structural similarity matrix (MSSIM) values as well as qualitatively in terms of HVP (human visual perception).

The PSNR values of the Lena images for various filters and proposed filter are recorded in Table 1. From Table 1, the proposed filter produces the highest PSNR values than that of the other high-performing filters except the JM filter. The proposed filter is superior to JM filter for the noise densities **10%–30%**. For the rest of the noise densities, the proposed filter gives comparable PSNR values with JM filter.

Table 2 enlisted the PSNR values of the Mandril image and shows that the proposed filter gives the higher PSNR values. The proposed filter produces the highest values for the noise densities 10–20% than JM filter.

Table 1 PSNR values of Lena image

Methods	Noise densities 10%	20%	30%	40%	50%	60%	70%	80%	90%
Corrupted	15.5	12.4	10.7	9.4	8.5	7.7	7.0	6.4	5.9
SM	28.7	26.4	22.6	18.3	15.0	12.2	9.8	8.1	6.5
CWM	29.7	24.1	19.5	15.7	13.0	10.8	8.9	7.6	6.3
PSM	30.7	28.7	26.9	23.7	20.0	15.2	11.1	8.3	6.4
IMF	27.2	26.7	26.1	25.1	23.9	21.2	16.6	12.1	8.0
SDROM	30.3	26.7	22.0	17.6	14.4	11.7	9.4	7.8	6.4
ACWM	30.9	27.2	22.4	18.1	14.8	12.1	9.7	8.1	6.5
RUSSO	31.0	27.6	24.9	22.7	20.3	17.6	14.7	11.7	8.6
ZHANG	32.8	28.2	23.3	18.6	15.3	12.5	10.0	8.3	6.7
SUN	31.0	27.5	23.0	18.4	15.0	12.2	9.8	8.1	6.5
IRF	30.2	27.0	22.5	18.2	14.9	12.2	9.7	8.1	6.5
TSM	30.3	24.4	19.6	15.5	12.7	10.4	8.4	7.1	6.0
ATSM	31.6	28.3	26.3	24.5	23.4	22.6	21.6	20.5	17.9
LEAM	31.7	28.5	26.6	24.6	23.1	21.5	19.6	17.8	14.9
MPATS	32.3	28.9	26.9	24.8	23.5	21.6	19.6	17.9	15.0
JM	37.4	34.7	32.3	30.7	29.2	27.7	26.5	24.7	22.8
Proposed filter	**38.6**	**34.9**	**32.4**	**30.6**	**28.5**	**26.9**	**24.9**	**21.5**	**17.1**

Table 2 PSNR values of Mandril image

Methods	Noise densities 10%	20%	30%	40%	50%	60%	70%	80%	90%
Corrupted	15.7	12.6	10.9	9.6	8.6	7.9	7.2	6.6	6.1
SM	23.8	23.0	20.7	17.7	14.7	12.3	10.9	8.3	6.8
CWM	25.5	22.7	18.9	15.6	12.9	10.9	9.1	7.7	6.6
PSM	27.8	26.4	25.0	22.9	19.5	15.3	11.0	8.9	6.5
IMF	23.0	22.8	22.6	22.3	21.5	20.0	16.0	12.0	8.3
SDROM	26.0	24.0	20.8	17.4	14.3	11.8	9.6	7.9	6.6
ACWM	27.1	24.8	21.4	17.9	14.8	12.3	9.9	8.2	6.8
RUSSO	29.5	26.7	24.3	22.4	20.2	17.7	15.0	11.9	8.9
ZHANG	28.8	26.1	22.1	18.5	15.2	12.7	10.2	8.4	6.9
SUN	26.1	24.4	21.3	18.0	14.8	12.3	9.9	8.2	6.7
IRF	325.8	24.1	21.2	17.8	14.7	12.3	9.9	8.2	6.8
TSM	25.4	22.9	18.9	15.4	12.6	10.5	8.6	7.2	6.2
ATSM	28.4	25.6	23.6	22.7	21.5	20.6	20.0	19.1	17.2
LEAM	29.2	26.6	24.5	23.6	22.3	21.1	19.8	18.6	17.1
MPATS	29.6	27.0	24.8	23.8	22.4	21.1	19.9	18.2	17.1
JM	33.6	30.1	28.3	26.8	25.5	24.2	23.1	21.9	20.5
Proposed filter	**33.6**	**30.3**	**27.9**	**26.4**	**25.0**	**23.5**	**22.4**	**20.1**	**16.9**

Table 3 MSSIM values of Lena image

	Noise densities								
Filters	10%	20%	30%	40%	50%	60%	70%	80%	90%
TSM	0.9548	0.8511	0.6477	0.4024	0.2270	0.1250	0.0680	0.0372	0.0180
ATSM	0.9857	0.9653	0.9379	0.8888	0.8249	0.7125	0.5180	0.4410	0.3221
LEAM	0.9831	0.9612	0.9316	0.8888	0.8298	0.7355	0.6019	0.4357	0.3194
MPATS	0.9792	0.9527	0.9152	0.8643	0.7993	0.6947	0.5505	0.4157	0.2795
Proposed filter	**0.9864**	**0.9702**	**0.9502**	**0.9267**	**0.8935**	**0.8512**	**0.7874**	**0.6910**	**0.4674**

Table 4 Average run-times of the filters (in seconds/25 trials)

	Noise densities								
Filters	10%	20%	30%	40%	50%	60%	70%	80%	90%
TSM	4.61	4.58	4.63	4.63	4.68	4.69	4.71	4.80	4.93
ATSM	0.61	1.02	1.45	1.85	2.28	2.69	3.08	3.12	3.66
LEAM	0.53	1.00	1.53	2.14	2.87	3.70	4.65	5.55	8.02
MPATS	0.53	0.87	1.19	1.52	1.87	2.21	2.55	2.77	3.41
Proposed filter	**0.29**	**0.54**	**0.8**	**1.05**	**1.32**	**1.56**	**1.83**	**2.07**	**2.36**

Table 3 is constructed based on the MSSIM values of the filters. It is evident that the proposed filter possesses the good restoration potential than the comparing filters like ATSM, MPATS, and LEAM filters. The restored images are most identical to the original images for the lower noise densities. For the remaining higher noise densities, moreover the restored image seems to be an original image.

The time complexity of the proposed filter is calculated based on conducting the 25 trails and finding out the average of the run-times. The proposed filter tests in the machine with Intel Core 2 Duo Processor 2.33 GHz. The time complexity values are recorded in Table 4. The proposed filter utilizes less time when compared with the other existing filters. Hence, the proposed filter has the less complexity than other filters. From Fig. 1, the proposed filter is supremum than other filters for all the noise densities.

Figure 2a is described as (a) Lena original image, (b) 30% noise image, (c) restored image of (b), (d) 50% noise image, (e) restored image of (d), (f). 70% noise image, and (g) restored image of (f). Likewise, Fig. 2b is described as (a) Mandril original image, (b) 10% noise image, (c) restored image of (b), (d) 30% noise image, (e) restored image of (d), (f). 50% noise image, and (g) restored image of (f). From Fig. 2a, b, the proposed filter is proved that it efficiently preserves the edges and image details in terms of human visual perception.

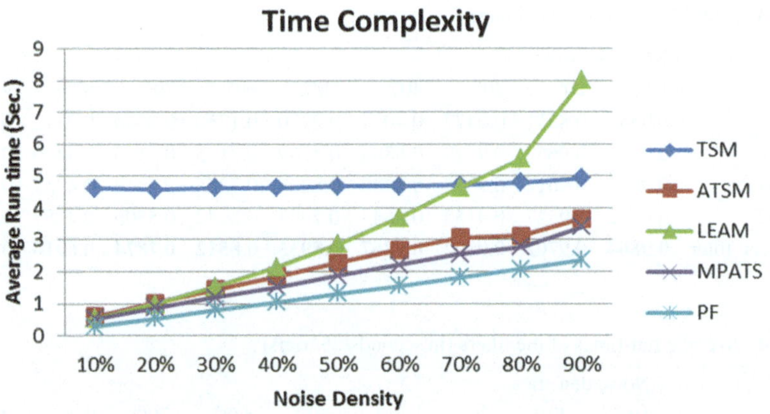

Fig. 1 Time complexity graphs

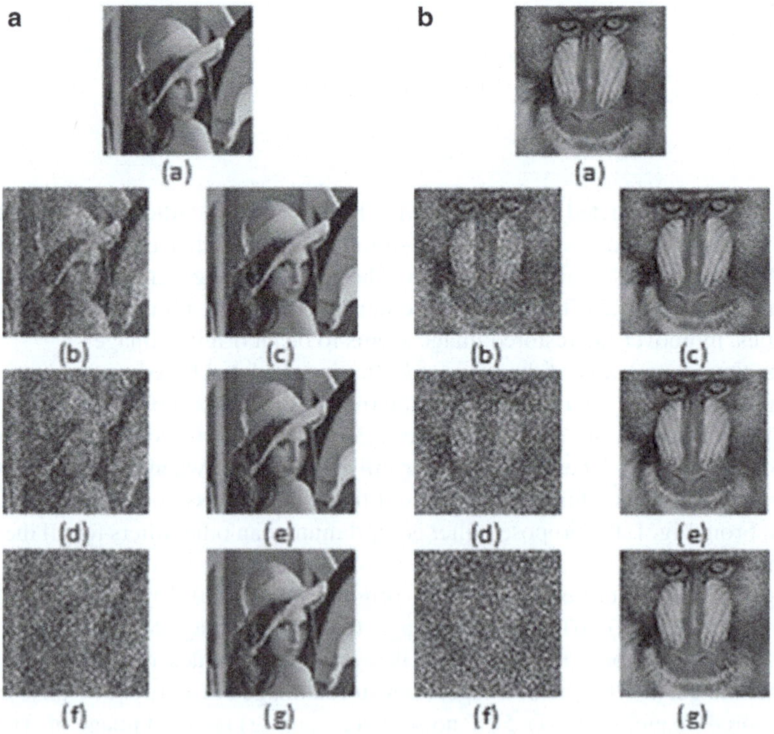

Fig. 2 Original, corrupted at various noise density levels, and restored images. (**a**) Lena image, (**b**) Mandril

5 Conclusion

The proposed filter effectively denoises the images corrupted with fixed-value impulse noise of probabilities 10–90% using fuzzy systems. This filter is found to be more efficient than other high-performing fixed-value impulse noise filters in terms of PSNR and MSSIM values. The human visual perceptions of the filtered images too endorse the merit of the proposed filter. The proposed filter has less time complexity. This newly devised filter assures higher degree of edge and detail preservation, in addition to computational simplicity.

References

1. Abreu, Eduardo, Mitra, Sanjit, K.: A signal-dependent rank ordered mean (SD-ROM) filter - a new approach for removal of impulses from highly corrupted images. Proceedings of the ICASSP. **4**, 2371–2374 (1995).
2. Besdok, E., Emin Yuksel, M.: Impulsive Noise Suppression for Images with Jarque-Bera Test Based Median Filter. International Journal of Electronics and Communications. **59**, 105–110 (2005).
3. Forouzan, Amir, R., Araabi, Babak Nadjar: Iterative median filtering for restoration of images with impulsive noise. Proceedings of ICECS. **1**, 232–235 (2003).
4. Gonzalez, R.C., Woods, R.E.: Digital Image Processing. 3rd Edition, Pearson Prentice Hall (2009).
5. How-Lung Eng, Kai-Kuang Ma: Noise Adaptive Soft-Switching Median Filter. IEEE Transactions on Image Processing. **10(2)**, 242–251 (2001).
6. Milan Sonka, Vaclav Hlavac, Roger Boyle: Digital Image Processing and Computer Vision. Cengage Learning (2008).
7. Mohammed Mansor Roomi. S.: Impulse Noise Detection and Removal. ICGST-GVIP Journal. **7(2)**, 51–56 (2007).
8. Russo, Fabirzio, Ramponi, Giovanni, F.: A Fuzzy Filter for Images Corrupted by Impulse Noise. IEEE Signal Processing Letters. **3(6)**, 168–170, (1996).
9. Shanmugavadivu, P., Eliahim Jeevaraj P.S.: Fixed-value impulse noise suppression for images using PDE based Adaptive Two-Stage Median Filter. Conference Proceedings of ICCCET. 290–295 (2011).
10. Shanmugavadivu, P., Eliahim Jeevaraj P.S.: Modified Partial Differential Equations Based Adaptive Two-Stage Median Filter for Images Corrupted with High Density Fixed-Value Impulse Noise. Conference Proceedings of CCSEIT-11. 376–383 (2011).
11. Shanmugavadivu, P., Eliahim Jeevaraj P.S.: Laplace Equation-Based Adaptive Median Filter for Highly Corrupted Images. International Conference on Computer Communication and Informatics. 47–51 (2012).
12. Somasundaram, K., Shanmugavadivu, P.: Impulsive Noise Detection by Second Order Differential Image and Noise Removal using Nearest Neighbourhood Filter. International Journal of Electronics and Communications. **62(6)**, 472–477 (2007).
13. Sung-Jea Jea Ko, Yong-Hoon Hoon Lee: Center weighted median filters and their applications to image enhancement. IEEE Transactions on Circuits and Systems. **38(9)**, 984–993 (1991).
14. Tao Chen, Kai-Kuang Kuang Ma, Li-Hui H Chen: Tri State Median Filter. IEEE Transactions on Image Processing. **8(12)**, (1999).
15. Tzu-Chao Lin: A New Adaptive Center Weighted Median Filter for Suppressing Impulsive Noise in Images. Information Sciences. **177(4)**, 1073–1087 (2007).
16. Zhou Wang, David Zhang: Progressive Switching Median Filter for the Removal of Impulse Noise from Highly Corrupted Images. IEEE Transactions on Circuits And Systems II: Analog And Digital Signal Processing. **46(1)**, (1999).

On (p, q)-Quantum Calculus Involving Quasi-Subordination

S. Kavitha, Nak Eun Cho, and G. Murugusundaramoorthy

Abstract Let $(p, q) \in (0, 1)$. Let the function f be analytic in $|z| < 1$. Further, let the (p, q) be differential operator defined as $\partial_{p,q} f(z) = \frac{f(pz)-f(qz)}{z(p-q)}, \quad |z| < 1$. In the current investigation, the authors apply the (p, q)-differential operator for few subclasses of univalent functions defined by quasi-subordination. Initial coefficient bounds for the defined new classes are obtained.

1 Introduction

Let $\mathscr{A}$ be the class of all analytic functions f whose Taylor's expansion in the open unit disk $\mathbb{D}$ is of the form

$$f(z) = z + \sum_{n=2}^{\infty} a_n z^n. \tag{1}$$

Further, let $\mathscr{S}$ be the subclass of $\mathscr{A}$ consisting of univalent functions. A function $f \in \mathscr{S}$ is starlike of order α $(0 \leq \alpha < 1)$, if and only if it satisfies the analytic criteria $\Re\left(\frac{zf'(z)}{f(z)}\right) > \alpha \quad (z \in \mathbb{D})$.This function class is denoted by $\mathscr{S}^*(\alpha)$. We also write $\mathscr{S}^*(0) =: \mathscr{S}^*$, where $\mathscr{S}^*$ denotes the class of functions f that are starlike in $\mathbb{D}$ with respect to the origin. A function $f \in \mathscr{S}$ is said to be convex of order α

S. Kavitha
Department of Mathematics, S.D.N.B Vaishnav College for Women, Chennai, India
e-mail: sivasaisastha@gmail.com

N. E. Cho
Department of Applied Mathematics, Pukyong National University, Busan, Republic of Korea
e-mail: necho@pknu.ac.kr

G. Murugusundaramoorthy (✉)
Department of Mathematics, School of Advanced Sciences, VIT, Vellore, India
e-mail: gmsmoorthy@yahoo.com; gms@vit.ac.in

V. Madhu et al. (eds.), *Advances in Algebra and Analysis*, Trends in Mathematics,
https://doi.org/10.1007/978-3-030-01120-8_25

$(0 \leq \alpha < 1)$ if and only if $\Re\left(1 + \frac{zf''(z)}{f'(z)}\right) > \alpha \quad (z \in \mathbb{D})$. This class is denoted by $\mathscr{K}(\alpha)$. Further, $\mathscr{K} = \mathscr{K}(0)$, the well-known standard class of convex functions. It is an established fact that $f \in \mathscr{K}(\alpha) \iff zf' \in \mathscr{S}^*(\alpha)$. Let ϕ be a function analytic with positive real part on $\mathbb{D}$ with $\phi(0) = 1$, $\phi'(0) > 0$ that map $\mathbb{D}$ onto a region starlike with respect to 1 and symmetric with respect to x-axis. Further, a function f is subordinate to the function g, written as $f \prec g$, provided that there is an analytic function $w(z)$ defined on $\mathbb{D}$ with $w(0) = 0$ and $|w(z)| < 1$ such that

$$f(z) = g(w(z)), z \in \mathbb{D}.$$

In particular if g is univalent in $\mathbb{D}$, then $f \prec g$ is equivalent to $f(0) = g(0)$ and $f(\mathbb{D}) \subset g(\mathbb{D})$. In 1995, Ma and Minda gave a technique to find a unified treatment of various subclasses consisting of starlike and convex functions for which $\frac{zf'}{f}$ or $1 + \frac{zf''}{f'}$ is subordinate to a more general superordinate function of the class of convex and starlike functions, respectively. The unified class $\mathscr{S}^*(\phi)$ introduced by Ma and Minda [6] consists of functions $f \in \mathscr{S}$ satisfying $\frac{zf'(z)}{f(z)} \prec \phi(z), \ , \ (z \in \mathbb{D})$. They also investigated the corresponding class $\mathscr{K}(\phi)$ of convex functions $f \in \mathscr{S}$ satisfying $1 + \frac{zf''(z)}{f'(z)} \prec \phi(z), \quad (z \in \mathbb{D})$. Robertson [10] introduced a new concept of quasi-subordination which is defined as follows: An analytic function f is quasi-subordinate to an analytic function g in the open unit disk if there exist two analytic functions φ and w, with $|\varphi(z)| \leq 1$, $w(0) = 0$ and $|w(z)| < 1$ such that $f(z) = \varphi(z)g[w(z)]$ and is denoted by $f(z) \prec_q g(z)$. Observe that when $\varphi(z) = 1$, then $f(z) = g[w(z)]$ so that $f(z) \prec g(z)$ in $\mathbb{D}$. If $w(z) = z$ then $f(z) = \varphi(z)g(z)$ and in this case, we say that $f(z)$ is majorized by $g(z)$ and is written as $f \prec\prec g$ in $\mathbb{D}$. Therefore, it is clear that quasi-subordination is a generalization concept of subordination as well as majorization. Related works on quasi-subordination were done by many, to name a few we refer the interested reader to [5]. The theory of quantum calculus known as q-calculus is equivalent to traditional infinitesimal calculus without the notion of limits. The q-calculus was started by Euler and Jacobi, who found many interesting applications in various areas of mathematics, physics, and engineering sciences. On a recent investigation done by Sahai and Yadav in the theory of special functions by Sahai and Yadav [9], quantum calculus based on two parameters (p, q) was quoted. Indeed generalization of q-calculus is the post-quantum calculus, denoted (p, q)-calculus. The (p, q)-integer was introduced in order to give a generalization or to unify several forms of q-oscillator algebras, well known in the earlier physics literature related to the representation theory of single parameter quantum algebras [4]. Throughout this article, we will use basic notations and definitions of the (p, q)- calculus as follows: Let $p > 0, q > 0$. For any non-negative integer n, the (p, q)-*integer number* n, denoted by $[n]_{p,q}$, is defined as

$$[n]_{p,q} = \frac{p^n - q^n}{p - q}, \qquad [0]_{p,q} = 0. \tag{2}$$

The twin-basic number is a natural generalization of the q-number, that is, $[n]_q = \frac{1-q^n}{1-q}$ $(q \neq 1)$. Similarly, the (p, q)-*differential operator* of a function f, analytic in $|z| < 1$, is defined by

$$\partial_{p,q} f(z) = \frac{f(pz) - f(qz)}{z(p-q)} \quad (p \neq q,\ z \in \mathbb{D} = \{z \in \mathbb{C} : |z| < 1\}). \tag{3}$$

One can easily show that $\partial_{p,q} f(z) \to f'(z)$ as $p \to 1^-$ and $q \to 1^-$. It is clear that q-integers and (p, q)-integers differs, that is, we cannot obtain (p, q)-integers just by replacing q by q/p in the definition of q-integers. However, the definition (2) reduces to quantum calculus for the case $p = 1$. Thus, we can say that (p, q)-calculus can be taken as a generalization of q-calculus. The (p, q)-*factorial* is defined by $[0]_{p,q}! = 1$, $[n]_{p,q}! = \prod_{k=1}^{n} [k]_{p,q}!$ $(n \geq 1)$. Note that for $p \to 1^-$, the (p, q)-factorial reduces to the q-factorial. Also, clearly $\lim_{p \to 1^-} \lim_{q \to 1^-} [n]_{p,q} = n$, and $\lim_{p \to 1^-} \lim_{q \to 1^-} [n]_{p,q}! = n!$. For details on q-calculus and (p, q)-calculus, one can refer to [3, 4] and [7, 9].

The aim of this work is to introduce three new subclasses by using the concept of (p, q) calculus with quasi-subordination. The first two initial coefficient estimates are obtained for those classes.

Definition 1 Let the class $\mathcal{R}^*_{p,q,Q}(\phi)$ consists of functions $f \in \mathcal{S}$ satisfying the quasi-subordination

$$\partial_{p,q}(f(z)) - 1 \prec_Q \phi(z) - 1. \tag{4}$$

Similarly, the class $\mathcal{S}^*_{p,q,Q}(\phi)$ consists of functions $f \in \mathcal{S}$ satisfying the quasi-subordination

$$\frac{z\partial_{p,q}(f(z))}{f(z)} - 1 \prec_Q \phi(z) - 1. \tag{5}$$

and the class $\mathcal{C}_{p,q,Q}(\phi)$ consists of functions $f \in \mathcal{S}$ satisfying the quasi-subordination

$$\frac{\partial_{p,q}\left(z\partial_{p,q}(f(z))\right)}{\partial_{p,q}(f(z))} - 1 \prec_Q \phi(z) - 1. \tag{6}$$

To prove the main results, we need the following lemma:

Lemma 1 ([8]) *Let the function $\omega \in \Omega$ be given by $\omega(z) = \omega_1 z + \omega_2 z^2 + \cdots$ $(z \in \mathbb{D})$. Then for every complex number t,*

$$|\omega_2 - t\omega_1^2| \leq 1 + (|t| - 1)|\omega_1|^2 \leq \max\{1, |t|\}.$$

The result is sharp.

2 Initial Coefficient Bounds for Some Subclasses of Univalent Functions Involving Quasi-Subordination

In the present section, we obtain the initial coefficient bounds for few subclasses of analytic univalent functions that are defined using quasi-subordination.

$$\begin{aligned} \text{Let } f(z) &= z + a_2z^2 + a_3z^3 + \cdots, \\ \phi(z) &= 1 + B_1z + B_2z^2 + B_3z^3 + \cdots, B_1 > 0, \\ \text{and } \varphi(z) &= c_0 + c_1z + c_2z^2 + c_3z^3 + \cdots, |c_n| \leq 1. \end{aligned}$$

Theorem 1 *If f belongs to $\mathscr{R}^*_{p,q,Q}(\phi)$, then*

$$|a_2| \leq \frac{B_1}{p+q}, \; |a_3| \leq \frac{B_1}{p^2+pq+q^2}\left(1 + \max\left\{1, \frac{|B_2|}{B_1}\right\}\right). \tag{7}$$

Further, for any complex number μ,

$$|a_3 - \mu a_2^2| \leq \frac{B_1}{p^2+pq+q^2}\left(1 + \max\left\{1, \left|\mu\frac{B_1^2c_0(p^2+pq+q^2)}{(p+q)^2} - \frac{B_2}{B_1}\right|\right\}\right).$$

Proof Let $f \in \mathscr{R}^*_{p,q,Q}(\phi)$. Hence, there exist two analytic functions φ and ω, with $|\varphi(z)| \leq 1, \omega(0) = 0$ and $|\omega(z)| < 1$ such that

$$\partial_{p,q}(f(z)) - 1 \prec_Q = \varphi(z)(\phi(\omega(z)) - 1). \tag{8}$$

Since

$$\partial_{p,q}(f(z)) = 1 + [2]_{p,q}a_2z + [3]_{p,q}a_3z^2 + \cdots,$$

implies that

$$\partial_{p,q}(f(z)) - 1 = (p+q)\,a_2z + \left(p^2+pq+q^2\right)a_3z^2 + \cdots. \tag{9}$$

Also,

$$\varphi(z)(\phi(\omega(z)) - 1) = B_1c_0\omega_1z + \left(B_1c_1\omega_1 + c_0\left(B_1\omega_2 + B_2\omega_1{}^2\right)\right)z^2 + \cdots. \tag{10}$$

Upon substituting (9) and (10) in (8), we get

$$a_2 = \frac{B_1c_0\omega_1}{p+q}, \tag{11}$$

$$a_3 = \frac{1}{p^2+pq+q^2}\left(B_1c_1\omega_1 + B_1c_0\omega_2 + c_0B_2\omega_1^2\right). \tag{12}$$

As the function φ is analytic and bounded in the unit disk $\mathbb{D}$,

$$|c_n| \leq 1 - |c_0|^2 \leq 1 \quad (n > 0). \tag{13}$$

By virtue of the above result and the known inequality, $|\omega_1| \leq 1$, by using (11) and (12), we get,

$$|a_3 - \mu a_2^2| =$$

$$\frac{1}{p^2+pq+q^2}\left(B_1c_1\omega_1 + c_0\left(B_1\omega_2 + \left[B_2 - \mu\frac{B_1^2c_0(p^2+pq+q^2)}{(p+q)^2}\right]\omega_1^2\right)\right). \tag{14}$$

Taking modulus on both sides and simplifying, we get

$$|a_3 - \mu a_2^2| = \frac{B_1}{p^2+pq+q^2}\left(|B_1c_1\omega_1| + |c_0|\left(\omega_2 - t\,\omega_1^2\right)\right) \tag{15}$$

where $t = \mu\frac{B_1^2c_0(p^2+pq+q^2)}{(p+q)^2} - \frac{B_2}{B_1}$. By applying the inequalities $|c_n| \leq 1$ and $|\omega_1| \leq 1$ and applying Lemma 1 to the modulus term $\left|\omega_2 - t\omega_1^2\right|$, we get

$$|a_3 - \mu a_2^2| \leq \frac{B_1}{|p^2+pq+q^2|}\left(1 + \max\left\{1, \left|\mu\frac{B_1^2c_0(p^2+pq+q^2)}{(p+q)^2} - \frac{B_2}{B_1}\right|\right\}\right). \tag{16}$$

This completes the proof of Theorem 1.

If $p \longrightarrow 1^-$ and $q \longrightarrow 1^-$, Theorem 2 reduces to the following result obtained by Haji Mohd and Darus [5].

Corollary 1 *If f belongs to $\mathcal{R}_Q^*(\phi)$, then $|a_2| \leq \frac{B_1}{2}$,*
$|a_3| \leq \frac{B_1}{3}\left(1 + \max\left\{1, \left|-\frac{B_2}{B_1} - B_1^2c_0\right|\right\}\right)$, *and for any $\mu \in \mathbb{C}$,*

$$|a_3 - \mu a_2^2| \leq \frac{B_1}{3}\left(1 + \max\left\{1, \left|\mu\frac{3B_1^2}{4} - \frac{B_2}{B_1}\right|\right\}\right).$$

Corollary 2 *For the special case $\varphi(z) = 1$, we have $c_0 = 1$ and all other c_i's are 0. Under these assumptions, Theorem 1 reduces to the special case of the result obtained by Ali [1] for the univalent case (Theorem 3) and a particular estimate (Theorem 2.3) of [2] when $k = 1$.*

Theorem 2 *If f belongs to $\mathscr{S}^*_{p,q,Q}(\phi)$, then*

$$|a_2| \leq \frac{B_1}{|p+q-1|}, \tag{17}$$

$$|a_3| \leq \frac{B_1}{|p^2+pq+q^2-1|}\left(1+\max\left\{1, \left|-\frac{B_2}{B_1}-\frac{B_1^2 c_0}{(p+q-1)}\right|\right\}\right) \tag{18}$$

Further, for any complex number μ,

$$|a_3-\mu a_2^2| \leq \frac{B_1}{|p^2+pq+q^2-1|}$$
$$\times\left(1+\max\left\{1, \left|\mu\frac{B_1^2 c_0(p^2+pq+q^2-1)}{(p+q-1)^2}-\frac{B_2}{B_1}-\frac{B_1^2 c_0}{(p+q-1)}\right|\right\}\right).$$

Proof If $f \in \mathscr{S}^*_q(\phi)$, then there exist analytic functions φ and ω, with $|\varphi(z)| \leq 1, \omega(0)=0$, and $|\omega(z)|<1$ such that

$$\frac{z\partial_{p,q}(f(z))}{f(z)}-1 \prec_Q = \varphi(z)(\phi(\omega(z))-1). \tag{19}$$

Since

$$\frac{z\partial_{p,q}(f(z))}{f(z)}-1 = \left([2]_{p,q}-1\right)a_2 z+\left(\left([3]_{p,q}-1\right)a_3-\left([2]_{p,q}-1\right)a_2^2\right)z^2+\cdots,$$

which implies

$$\frac{z\partial_{p,q}(f(z))}{f(z)}-1 = (p+q-1)a_2 z$$
$$+\left(\left(p^2+pq+q^2-1\right)a_3-(p+q-1)a_2^2\right)z^2+\cdots. \tag{20}$$

Further,

$$\varphi(z)(\phi(\omega(z))-1) = B_1(k)c_0\omega_1 z+\left(B_1 c_1\omega_1+c_0\left(B_1\omega_2+B_2{\omega_1}^2\right)\right)z^2+\cdots. \tag{21}$$

Upon substituting (20) and (21) in (19), we get

$$a_2 = \frac{B_1 c_0\omega_1}{|p+q-1|}, \tag{22}$$

$$a_3 = \frac{1}{p^2+pq+q^2-1}\left(B_1 c_1\omega_1+B_1 c_0\omega_2+c_0\left(B_2+\frac{B_1^2 c_0}{p+q-1}\right){\omega_1}^2\right). \tag{23}$$

As the function φ is analytic and bounded in the unit disk $\mathbb{D}$,

$$|c_n| \leq 1 - |c_0|^2 \leq 1 \quad (n > 0). \tag{24}$$

By virtue of the above result and the known inequality, $|\omega_1| \leq 1$, by using (22) and (23), we get

$$a_3 - \mu a_2^2 = \frac{1}{p^2 + pq + q^2 - 1}$$
$$\times \left(B_1 c_1 \omega_1 + c_0 \left(B_1 \omega_2 + \left[B_2 + \frac{B_1^2 c_0}{(p+q-1)} - \mu \frac{B_1^2 c_0 (p^2 + pq + q^2 - 1)}{(p+q-1)^2} \right] \omega_1^2 \right) \right). \tag{25}$$

Taking modulus on both sides and simplifying, we get

$$|a_3 - \mu a_2^2| = \frac{B_1}{|p^2 + pq + q^2 - 1|} \left(|B_1 c_1 \omega_1| + \left| c_0 \left(\omega_2 - t \omega_1^2 \right) \right| \right). \tag{26}$$

The above term can be rearranged as

$$|a_3 - \mu a_2^2| = \frac{B_1}{|p^2 + pq + q^2 - 1|} \left(|B_1 c_1 \omega_1| + |c_0| \left| \omega_2 - t \omega_1^2 \right| \right) \tag{27}$$

where

$$t = \mu \frac{B_1^2 c_0 (p^2 + pq + q^2 - 1)}{(p + q - 1)^2} - \frac{B_2}{B_1} - \frac{B_1^2 c_0}{(p + q - 1)}.$$

$$|a_3 - \mu a_2^2| = \frac{B_1}{|p^2 + pq + q^2 - 1|} \left(|B_1 c_1 \omega_1| + \left| c_0 \left(\omega_2 - t \omega_1^2 \right) \right| \right). \tag{28}$$

By applying the inequalities $|c_n| \leq 1$ and $|\omega_1| \leq 1$, we have Applying Lemma 1 to the modulus term $|\omega_2 - t\omega_1^2|$, we get,

$$|a_3 - \mu a_2^2| \leq \frac{B_1}{|p^2 + pq + q^2 - 1|}$$
$$\times \left(1 + \max \left\{ 1, \left| \mu \frac{B_1^2 c_0 (p^2 + pq + q^2 - 1)}{(p + q - 1)^2} - \frac{B_2}{B_1} - \frac{B_1^2 c_0}{(p + q - 1)} \right| \right\} \right) \tag{29}$$

This completes the proof of Theorem 2.

If $p \longrightarrow 1^-$ and $q \longrightarrow 1^-$, Theorem 2 reduces to a result obtained by Haji Mohd and Darus [5].

Corollary 3 *For the special case* $\varphi(z) = 1$, *we have* $c_0 = 1$ *and all other* c_i*'s are* 0. *Under these assumptions, Theorem 2 reduces to the special case of the result obtained by Ali [1] for the univalent case (Theorem 1) and a particular estimate (Theorem 2.1) of [2] when* $k = 1$. *Further, for the special case* $\phi(z) = \left(\frac{1+z}{1-z}\right)^{\alpha}$, $B_1 = 2\alpha$ *and* $B_2 = 2\alpha^2, 0 < \alpha \leq 1$, *we have the following corollary. For the special case* $\phi(z) = \left(\frac{1+z}{1-z}\right)^{\alpha}$, $B_1 = 2\alpha$ *and* $B_2 = 2\alpha^2, 0 < \alpha \leq 1$, *and for* $\varphi(z) = 1$, *we have* $c_0 = 1$ *and all other* c_i*'s are* 0. *Under these assumptions, Theorem 2 reduces to the result obtained by Ali [2].*

Theorem 3 *If* f *belongs to* $\mathscr{C}^*_{p,q,Q}(\phi)$, *then*

$$|a_2| \leq \frac{B_1}{(p+q)^2|p+q-1|}, \tag{30}$$

$$|a_3| \leq \frac{B_1}{|p^2+pq+q^2-1|}\left(1+\max\left\{1, \left|-\frac{B_2}{B_1} - \frac{B_1^2 c_0}{(p+q-1)}\right|\right\}\right). \tag{31}$$

Further, for any complex number μ,

$$|a_3 - \mu a_2^2| \leq \frac{B_1}{|p^2+pq+q^2-1|}$$
$$\left(1+\max\left\{1, \left|\mu\frac{B_1^2 c_0(p^2+pq+q^2-1)}{(p+q-1)^2} - \frac{B_2}{B_1} - \frac{B_1^2 c_0}{(p+q-1)}\right|\right\}\right)$$

Proof If $f \in \mathscr{C}_q(\phi)$, then there exist analytic functions φ and ω, with $|\varphi(z)| \leq 1, \omega(0) = 0$ and $|\omega(z)| < 1$ such that

$$\frac{\partial_{p,q}\left(z\partial_{p,q}(f(z))\right)}{\partial_{p,q}(f(z))} - 1 \prec_Q \phi(z) - 1 = \varphi(z)(p_k(\omega(z)) - 1). \tag{32}$$

Continuing as in Theorems 1 and 2, we get the estimates as stated in Theorem 3.

Acknowledgements The work of the first author is supported by a grant from SDNB Vaishnav College for Women under Minor Research Project scheme. The work was completed when the first author was visiting VIT Vellore Campus for a research discussion with Prof. G.Murugusundaramoorthy during the second week of November 2017.

Conflicts of Interest The authors declare that they have no conflicts of interest regarding the publication of this paper.

References

1. R. M. Ali, V. Ravichandran, and N. Seenivasagan, Coefficient bounds for p-valent functions, Applied Mathematics and Computation, **187(1)**, 2007, 35–46.
2. R. M. Ali, S. K. Lee, V. Ravichandran, and S. Supramaniam, The Fekete-Szego coefficient functional for transforms of analytic functions, Bulletin of the Iranian Mathematical Society, **35(2)**, 2009, 119–142.
3. S. Araci, U. Duran, M. Acikgoz and H. M. Srivastava, *A certain (p, q)-derivative operator and associated divided differences*, J. Inequal. Appl., (2016), 2016:301.
4. R. Chakrabarti and R. Jagannathan, *A (p, q)-oscillator realization of two-parameter quantum algebras*, J. Phys. A **24**(13) (1991), L711–L718.
5. M. Haji Mohd and M. Darus, Fekete-Szegő problems for quasi-subordination classes, Abstr. Appl. Anal. **2012**, Art. ID 192956, 14 pp.
6. W. Ma and D. Minda,*A unified treatment of some special classes of univalent functions*, in Proceedings of the conference on complex Analysis, Z. Li, F. Ren, L. Lang and S. Zhang (Eds.), Int. Press (1994), 157–169.
7. M. Mursaleen, K. J. Ansari and A. Khan, *Some approximation results by (p, q)-analogue of Bernstein-Stancu operators*, Appl. Math. Comput. **264** (2015), 392–402.
8. F.R. Keogh, E.P. Merkes, A coefficient inequality for certain classes of analytic functions, *Proceedings of the American Mathematical Society*, **20** (1969), 171–180.
9. V. Sahai and S. Yadav, *Representations of two parameter quantum algebras and p, q-special functions*, J. Math. Anal. Appl. **335** (2007), 268–279.
10. M. S. Robertson, Quasi-subordination and coefficient conjectures, Bull. Amer. Math. Soc. **76** (1970), 1–9.

References

Part III
Operations Research

Sensitivity Analysis for Spanning Tree

K. Kavitha and D. Anuradha

Abstract This paper develops a heuristic procedure for finding the maximum increment and decrement of each edge weight separately without modifying the optimality of minimum spanning tree. The procedure of the proposed approach is illustrated by numerical example.

1 Introduction

The traveling salesman problem (TSP) is a class of combinatorial optimization problems. In TSP, the salesperson has to visit all the towns only one time, and he has to come back to the origin point of the town to complete the tour. The main objective of the problem is to find a tour of smallest distance in terms of time or cost on a completely connected graph. In traditional TSP, Hamiltonian cycles are generally called tours. Dantzig et al. [6] developed an algorithm to find the shortest route for the TSP. In literature, many researchers (Bhide et al. [3], Andreae [1], Bockenhauer et al. [5], Blaser et al. [4]) have developed numerous algorithms to solve the TSP. Anuradha and Bhavani [2] used various metrics to find the set of all efficient points for bi-criteria TSP. SA is one of the interesting domains in optimization, and it is to study the variations of the bounds in the optimization problems. Post-optimality analysis and parametric optimization techniques for integer programming are discussed by Geoffrion and Nauss [7]. Tarjan [16] studied the SA of shortest path trees and smallest spanning trees. An idea of SA for an approximate relaxation of the minimum Hamiltonian path has developed by Libura [10]. A procedure for finding lower bounds of the edge tolerances without changing the optimality of the minimum Hamiltonian path and TSP is discussed by Libura [11]. The cost sensitivity analysis in a transportation problem was discussed by Cabulea [9]. The smallest spanning tree is discussed by Pettie [14] with the help of an inverse-Ackermann

K. Kavitha · D. Anuradha (✉)
Department of Mathematics, School of Advanced Sciences, VIT University, Vellore, India
e-mail: kavinphd@gmail.com; k.kavitha@vit.ac.in; anuradhadhanapal1981@gmail.com; danuradha@vit.ac.in

V. Madhu et al. (eds.), *Advances in Algebra and Analysis*, Trends in Mathematics,
https://doi.org/10.1007/978-3-030-01120-8_26

type lower bound. The procedure for calculating the SA of a smallest spanning tree was presented by Pettie [15]. Moritz and Girard [12] investigated the application of TSP. The aim of this paper is to find the edge tolerances with respect to the optimal solution of spanning tree. A simple example is chosen to illustrate the proposed approach.

2 Travelling Salesman Problem

Suppose a salesperson has to trip towns. Starting from a specific town, he has to trip each and every town only one time and revisit to the beginning position. Our goal is to minimize the total traveling cost. Now, TSP can be modeled as given below LPP.

(P) Minimize Z = $\sum_{i=1}^{n}\sum_{j=1}^{n} c_{ij} y_{ij}$

Subject to

$$\sum_{i=1}^{n} y_{ij} = 1,\ j = 1, 2, \ldots, n \text{ and } j \neq i \tag{1}$$

$$\sum_{j=1}^{n} y_{ij} = 1,\ i = 1, 2, \ldots, n \text{ and } i \neq j \tag{2}$$

$$y_{ij} + y_{ji} \leq 1,\ 1 \leq i \neq j \leq n \tag{3}$$

$$y_{ip_1} + y_{p_1 p_2} + \ldots + y_{p_{(n-2)}i} \leq (n-2),\ 1 \leq i \neq p_1 \neq \ldots \neq p_{(n-2)} \leq n \tag{4}$$

$$y_{ij} = \begin{cases} 1; & \text{salesperson travels from town } i \text{ to } j \\ 0; & \text{otherwise} \end{cases} \tag{5}$$

where c_{ij} is the traveling cost from town i to j; y_{ij} is the connection from town i to j; (1) and (2) make assurance that each and every town is visit at one time; (3) is sub-tour elimination constraint and eliminates all two-town sub-tours; (4) eliminates all $(n-1)$-town sub-tours. A set $Y^{\circ} = \{y^{\circ}_{ij}, i = 1, 2, \ldots, n; j = 1, 2, \ldots, n\}$ is feasible to problem (P) if Y° satisfies from (1) to (5).

Traveling salesman problem is shown as graph by demonstrating the towns as nodes and the roads that connected the towns as edges. The costs are measured as weights (wij) that are allotted to the edges. Our aim is to identify a tour of minimal weight.

We want the basic definitions (graph, tree, spanning tree) that can be found in Deo [13].

Table 1 Cost matrix

Town→ ↓	1	2	3	4	5
1	–	25	40	10	12
2	25	–	20	23	11
3	40	20	–	23	33
4	10	23	23	–	20
5	12	11	33	20	–

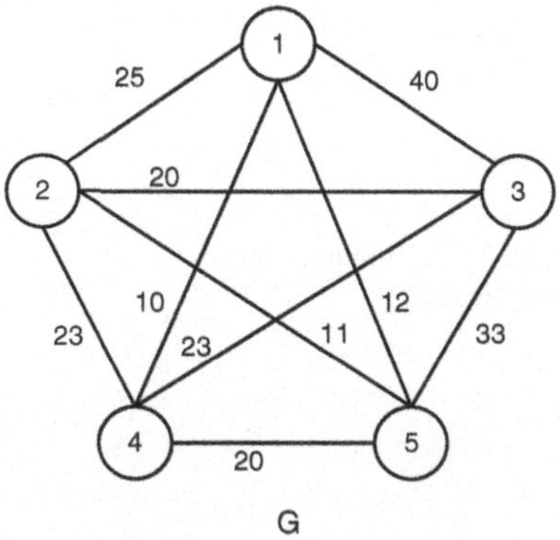

Fig. 1 Graphical TSP with weight

Table 2 Arrangement of edges and its weights

G			
Edge	Weight	Edge	Weight
(1,4)	10	(2,4)	23
(2,5)	11	(3,4)	23
(1,5)	12	(1,2)	25
(2,3)	20	(3,5)	33
(4,5)	20	(1,3)	40

3 Illustration

Consider TSP with five towns. Any salesperson has to visit all his towns starting from his origin point of the town and revisit to the same town. The cost ('00) between the towns are provided in the following Table 1.

In the above Table 1, the problem of traveling salesman can be designed as a graph G. The towns are expressed as nodes and the roads that connect the towns as edges. The costs are measured as weights allocated to the edges.

Complete graph shows in Fig. 1, as each pair of towns is linked by a road. Each edge of Fig. 1 is assigned with weights which are positive real numbers. For weighted graph, one minimal spanning tree or more than one minimal spanning trees can be generated. Generally, it is said that a spanning tree of weighted linked graph G is a minimal spanning tree if its total weight is less than or equal to any other spanning tree of G. In Table 2, using Kruskal's algorithm, we arranged the edges in the ascending manner of their weights for graph G.

Fig. 2 MST with weight

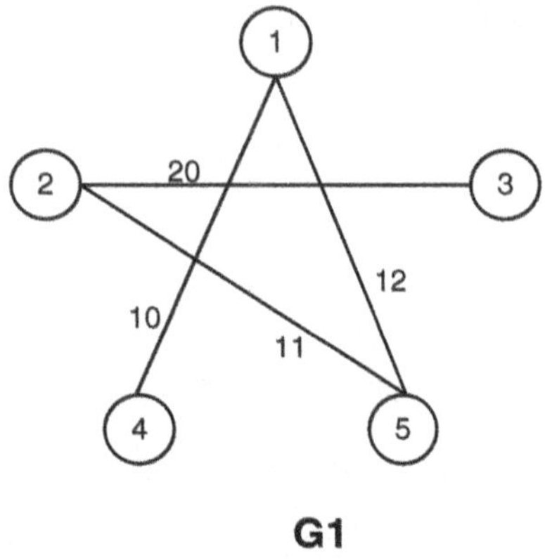

Fig. 3 Modi indices for graph G1

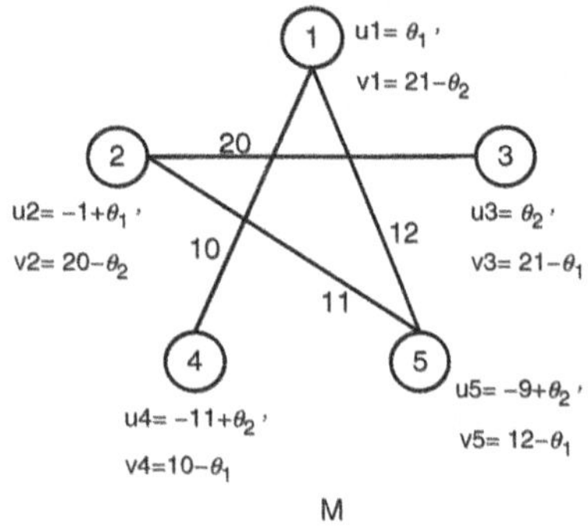

A structure of a minimal spanning tree (MST) is formed by choosing the edges from lesser to greater weights such that no edge forms a loop. Continue the selection procedure until all nodes are comprised in it. As an outcome minimal spanning tree G1 for G is obtained as follows: Now the graph $G1$ is the MST with weight 53('00).

Sensitivity analysis in MST tree is used to find the maximum incremented value and decremented value of each and every edge weight separately without changing the optimality of the solution. To find the sensitivity range for graph G1, we compute the Modi indices (Fig. 2).

Now, to calculate the relations among MODI indices parameters θ_i, $i = 1$, 2….., k and their intervals using the parametric MODI indices and the optimum condition $d_{ij} - (u_i + v_j) \geq 0$ for all unallocated cells (i,j) (Fig. 3).

Now, we find the relations for unallocated cells as $-\theta_1 + \theta_2 \geq -5$; $-\theta_1 + \theta_2 \leq 13$; $-\theta_1 + \theta_2 \leq 21$ *and* $-\theta_1 + \theta_2 \leq 19.\theta_1 - \theta_2$. From the relations, we found that $\theta_1 - \theta_2$ range from -13 to 5 (Fig. 4).

Modi indices M for graph G1 is minimum if and only if for every edge $T_{ij} = \max\{U_i, V_j\}$. Here U_i and V_j are the minimum values of the non-tree edges in M for G1. To verify the above said condition, we use the optimality conditions in [8].

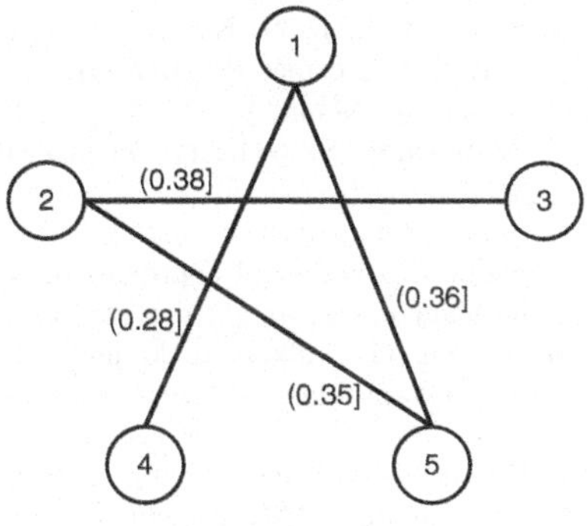

Fig. 4 Sensitivity range for graph G1

To compute the maximum incremented value and decremented value of the edge weight of (1,4), we choose the non-tree edges (2,4), (3,4),(5,4), (1,2), and (1,3). Using the optimality condition (i) and the range of $\theta_1 - \theta_2$, for the non-tree edges (2,4), (5,4), and (3,4), we get the following relations $23 - (-1 + \theta_1 + 10 - \theta_1 + \Delta w_{14}) \geq 0, 20-(-9+\theta_2+10-\theta_1+\Delta w_{14}) \geq 0$ and $23-(\theta_2+10-\theta_1+\Delta w_{14}) \geq 0$ which provides the minimum value as 14.

By using the optimality condition (ii) and the range of $\theta_1 - \theta_2$, for the non-tree edges (1,2) and (1,3), we get the relations as $25 - (\Delta w_{14} + \theta_1 + 20 - \theta_2) \geq 0$ and $40 - (\Delta w_{14} + \theta_1 + 21 - \theta_1) \geq 0$ which provides the minimum value as 18. Using the above condition $\mathrm{T_{ij}} = \max\left\{\mathrm{U_i}, \mathrm{V_j}\right\}$, the maximum of minimum value of (1,4) is 18.

Proceeding in this same manner, we can find the maximum of minimum value of (1,5) as 24; (2,3) as 18, and (2,5) as 24. Therefore, the re-optimization edge weight of (1,4) in M for G1 is 28 (replacing w_{ij} by $w_{ij} + \Delta w_{ij}$). Similarly, the re-optimization edge weight of (1,5) is 36, (2,3) is 38, and (2,5) is 35. Now, the sensitivity range for graph G1 is given below:

4 Conclusion

In this paper we obtained the maximum increment value and decrement value of each and every edge weight with respect to an optimal solution of spanning tree. The proposed approach allows the decision-maker to calculate the impact of variations on edge weight to create the problem sensible and valid.

References

1. Andreae, T. (2001) On the travelling salesman problem restricted to inputs satisfying a relaxed triangle inequality', *Networks*, Vol. 38, pp. 59–67.
2. Anuradha, D. and Bhavani, .S (2013), ?Multi Perspective Metrics for Finding All Efficient Solutions to Bi-Criteria Travelling Salesman Problem', *International Journal of Engineering and Technology*, Vol. 5, No. 2, pp. 1682–1687.

3. Bhide, S. John, N. Kabuka, M.R. (1993) ?A Boolean Neural Network Approach for the Traveling Salesman Problem', *IEEE Transactions on Computers*, Vol. 42, pp. 1271–1278.
4. Blaser, M., Manthey, B. and Sgall, J. (2006) ?An improved approximation algorithm for the asymmetric TSP with strengthened triangle inequality', *Journal of Discrete Algorithms*, Vol. 4, pp. 623–632.
5. Bockenhauer, H.J., Hromkovi, J., Klasing, R., Seibert, S. and Unger, W. (2002) 'Towards the notion of stability of approximation for hard optimization tasks and the travelling salesman problem', *Theoretical Computer Science*, Vol. 285, pp. 3–24.
6. Dantzig, G., Fulkerson, R. and Johnson, S. (1954) ?Solution of a large-scale Traveling-salesman problem', *Journal of the Operations Research Society of America*, Vol. 2, pp.393–410.
7. Geoffrion, A.M. and Nauss, R. (1977) ?Parametric and post optimality analysis in integer programming', *Management Sci.*, Vol. 23, pp. 453–466.
8. Kavitha, K. Anuradha, D. (2015) 'Heuristic Algorithm for finding Sensitivity Analysis of a More for Less Solution to Transportation Problems', *Global Journal of Pure and Applied Mathematics*, Vol.11, pp.479–485.
9. Lucia Cabulea, (2006) 'Sensitivity analysis of costs in a transportation problem', *ICTAMI, Alba Iulia, Romania*, Vol.11, pp. 39–46.
10. Marek Libura (1986) 'Sensitivity analysis of optimal solution for minimum Hamiltonian path', *Zeszyty Nauk. Politech. Slaskiej: Automatyka* , Vol. 84, pp. 131–139.
11. Marek Libura (1991), 'Sensitivity analysis for Hamiltonian path and travelling salesman problems, Discrete Applied Mathematics', Vol. 30, pp. 197–211.
12. Moritz Niendorf and Anouck R. Girard, (2016), 'Robustness of communication links for teams of unmanned aircraft by sensitivity analysis of minimum spanning trees' *American control conference*,pp.4623–4629.
13. Narsingh Deo, (2011) 'Graph Theory with Applications to Engineering and Computer Science', *Eastern Economy Edition.*
14. Pettie, S. (2006), 'An inverse-Ackermann type lower bound for online minimum spanning tree verification', *Combinatorica*, Vol.26, No. 2, pp. 207–230.
15. Pettie, S. (2015), 'Sensitivity analysis of minimum trees in sub inverse Ackermann time', *Journal of Graph algorithms and application*, Vol.19, No.1, pp. 375–391.
16. Tarjan, R.E. (1982), 'Sensitivity analysis of minimum spanning trees and shortest path trees', *Inform. Process. Lett.* Vol. 14, pp. 30–33.

On Solving Bi-objective Fuzzy Transportation Problem

V. E. Sobana and D. Anuradha

Abstract A fuzzy dripping method (FDM) has been proposed to find efficient solutions for the problem of bi-objective transportation under uncertainty. The procedure of the FDM is illustrated by numerical example.

1 Introduction

The transportation problem (TP) is a planning problem with the lowest costs for transporting goods from a set of sources to a set of destinations with the transporting cost from one position to another. In general, TPs can be designed more gainfully with the simultaneous consideration of multiple objectives, because a decision-maker of a transportation system generally chases multiple goals. The multi-objective functions include mean time of the goods, the reliability of transportation, and product deterioration and are thus generally considered in actual TPs. In [2], Biswal used fuzzy programming technique for solving multi-objective geometric programming problems. The efficiency of the solutions and the stability of MOTP in which parameters are imprecise this was studied by Ammar and Youness [3]. Bodkhe et al. [4] found bi-objective TP (BTP) as vector smallest problem by using the fuzzy programming approach with hyperbolic membership function. Pandian and Anuradha [6] proposed dripping method to find the set of all solutions which is efficient to the BTP. In [7], Zangiabadi and Maleki used a fuzzy goal programming technique with a special type of nonlinear membership function to obtaining an optimal compromise solution for the linear MOTP. Pandian and Natarajan [5] used fuzzy zero point method (ZPM) for finding the optimal solution to the FTP. SaruKumari and Priyamvada Singh [8] discussed the solution procedure for solving MOTP using fuzzy efficient interactive goal programming technique.

V. E. Sobana · D. Anuradha (✉)
Department of Mathematics, School of Advanced Sciences, VIT University, Vellore, Tamil Nadu, India
e-mail: murthyshobana1982@gmail.com; sobana.ve@vit.ac.in; anuradhadhanapal1981@gmail.com; danuradha@vit.ac.in

V. Madhu et al. (eds.), *Advances in Algebra and Analysis*, Trends in Mathematics,
https://doi.org/10.1007/978-3-030-01120-8_27

The primary hub of this paper is to find the set of efficient solutions to bi-objective FTP. Section 2 projects the mathematical formulation of BFTP and basic definitions. Section 3 presents FDM. Section 4 gives a numerical illustration for the proposed algorithm, and Sect. 5 concludes the paper.

2 Bi-objective Fuzzy Transportation Problem

Consider a mathematical model of a BFTP as shown as below:

$$(\text{P}) Minimize\ \tilde{z}_1 = \sum_{i=1}^{m} \sum_{j=1}^{n} \tilde{c}_{ij} \tilde{x}_{ij}$$

$$Minimize\ \tilde{z}_2 = \sum_{i=1}^{m} \sum_{j=1}^{n} \tilde{d}_{ij} \tilde{x}_{ij}$$

Subject to

$$\sum_{j=1}^{n} \tilde{x}_{ij} = \tilde{a}_i, i = 1, 2, \ldots, m \quad (2.1)$$

$$\sum_{i=1}^{m} \tilde{x}_{ij} = \tilde{b}_j, j = 1, 2, \ldots, n \quad (2.2)$$

$$\tilde{x}_{ij} \succeq 0, \text{ for all } i \text{ and } j \text{ are integers} \quad (2.3)$$

where $\tilde{c}_{ij}$ be the unit fuzzy transportation cost from origin i to the destination j, $\tilde{d}_{ij}$ is the unit fuzzy deterioration cost from ith origin to the jth destination, $\tilde{x}_{ij}$ is the fuzzy amount shipped from ith origin to jth destination, $\tilde{a}_i$ is the quantity available at origin point i, and $\tilde{b}_j$ is the demand at the jth destination.

The definitions of the following fuzzy set, fuzzy number, triangular fuzzy number, and arithmetic operations on the triangular fuzzy number can be found in [1, 9].

Definition 1 A set $\tilde{X}^0 = \{\tilde{x}_{ij}^0, i = 1, 2, \ldots, m; j = 1, 2, \ldots, n\}$ is said to be feasible to the problem (P) if $\tilde{X}^0$ satisfies the conditions (2.1) to (2.3).

Definition 2 A feasible solution $\tilde{X}^0$ would be an efficient solution to the problem (P) if there is no other feasible X of BFTP such that $\tilde{X}$ of BFTP such that $\tilde{Z}_1(\tilde{X}) \leq \tilde{Z}_1(\tilde{X}^0)$ and $\tilde{Z}_2(\tilde{X}) \leq \tilde{Z}_2(\tilde{X}^0)$ (or)$\tilde{Z}_2(\tilde{X}) < \tilde{Z}_2(\tilde{X}^0)$ and $\tilde{Z}_1(\tilde{X}) < \tilde{Z}_1(\tilde{X}^0)$; otherwise, it is called non-efficient solution to the problem (P). For simplicity, a pair $(\tilde{Z}_1(\tilde{X}^0), \tilde{Z}_2(\tilde{X}^0))$ is called an efficient solution (or) a non-efficient solution to the problem (P) if $\tilde{X}^0$ is efficient solution (or) non-efficient solution to the problem (P).

Definition 3 The percentage satisfaction level (PSL) of the objective at a solution $\tilde{U}$ to a problem is defined as follows.
PSL of the objective at $\tilde{U}$=$\left(1 - \frac{(\tilde{O}(\tilde{U}) - \tilde{O}_0)}{\tilde{O}_0} \times 100\right) = \left(\frac{(2\tilde{O}_0 - \tilde{O}(\tilde{U}))}{\tilde{O}_0} \times 100\right)$ where $\tilde{O}(\tilde{U})$ is the objective value at the solution $\tilde{U}$ and $\tilde{O}$ is the optimal objective value of the problem.

3 Fuzzy Dripping Method

We introduce the FDM for obtaining all the solutions of the BFTP (P).
The proposed procedure proceeds as follows:

Step 1: Construct FOFTP and SOFTP from the given BFTP and find an optimal solution to the FOFTP and SOFTP by the ZPM.
Step 2: Start with an optimal solution of FOFTP and consider the optimal solution as a feasible solution of SOFTP that will act as an efficient solution to BFTP.
Step 3: In the SOFTP after selecting the assigned cell (a,c) with the highest penalty, construct a rectangular loop that starts and ends at the assigned cell (a,c), and then join some of the assigned and unassigned cells together.
Step 4: Add and subtract from the transition cells of the loop so that the rim requirements are fulfilled. Then allocate a sequence of values to one by one, so that the assigned cell residue nonnegative. Then find the solution that should be feasible for SOFTP for each, and every value of which becomes an efficient solution and a non-efficient solution for BFTP.
Step 5: Check if the feasible solution for SOFTP is the optimum solution. Otherwise, repeat the Steps 3 and 4 until finding an optimum solution for SOFTP. If so, the process can be stopped and go to the next step.
Step 6: Start with an optimal solution of SOFTP and consider the optimal solution as a feasible solution of FOFTP that will act as an efficient solution to BFTP.
Step 7: Repeat Steps 3, 4, and 5 for the FOFTP.
Step 8: Combine all solutions (efficient/non-efficient) of BFTP obtained using the optimal solutions of FOFTP and SOFTP. From this, a set of efficient solutions and a set of non-efficient solutions to the BFTP can be obtained.

Now, the solution procedure of the FDM for solving a BFTP problem is demonstrated by means of the numerical example, shown below.

4 Numerical Example

Consider the following bi-objective fuzzy transportation problem:

	W1	W2	W3	W4	Supply
F1	((1, 2, 3), (7, 8, 9))	((3, 4, 5), (7, 8, 9))	((13, 14, 15), (5, 6, 7))	((13, 14, 15), (7, 8, 9))	(7, 8, 9)
F2	((1, 2, 3), (9, 10, 11))	((17, 18, 19), (15, 16, 17))	((5, 6, 7), (17, 18, 19))	((7, 8, 9), (19, 20, 21))	(18, 19, 20)
F3	((15,16,17), (11,12,13))	((17, 18, 19), (3, 4, 5))	((7, 8, 9), (9, 10, 11))	((11, 12, 13), (1, 2, 3))	(16, 17, 18)
Demand	(10, 11, 12)	(2, 3, 4)	(13, 14, 15)	(15, 16, 17)	

The aim is to find the set of all solutions for the BFTP.
Now, the FOFTP of BFTP is given below:

	W1	W2	W3	W4	Supply
F1	(1, 2, 3)	(3, 4, 5)	(13, 14, 15)	(13, 14, 15)	(7, 8, 9)
F2	(1, 2, 3)	(17, 18, 19)	(5, 6, 7)	(7, 8, 9)	(18, 19, 20)
F3	(15, 16, 17)	(17, 18, 19)	(7, 8, 9)	(11, 12, 13)	(16, 17, 18)
Demand	(10, 11, 12)	(2, 3, 4)	(13, 14, 15)	(15, 16, 17)	

Now, using the fuzzy ZPM, the optimal solution for the FOFTP is $\tilde{x}_{11}$=(3, 5, 7),$\tilde{x}_{12}$=(2, 3, 4),$\tilde{x}_{21}$=(3, 6, 9),$\tilde{x}_{24}$=(9, 13, 17),$\tilde{x}_{33}$=(13, 14, 15),$\tilde{x}_{34}$=(1, 3, 5) and the smallest fuzzy transportation cost is(242, 286, 330).
Now, the SOFTP of BFTP is given below:

	W1	W2	W3	W4	Supply
F1	(7, 8, 9)	(7, 8, 9)	(5, 6, 7)	(7, 8, 9)	(7, 8, 9)
F2	(9, 10, 11)	(15, 16, 17)	(17, 18, 19)	(19, 20, 21)	(18, 19, 20)
F3	(11, 12, 13)	(3, 4, 5)	(9, 10, 11)	(1, 2, 3)	(16, 17, 18)
Demand	(10, 11, 12)	(2, 3, 4)	(13, 14, 15)	(15, 16, 17)	

Now, using the fuzzy ZPM, the optimal solution for the SOFTP is $\tilde{x}_{13}$=(7, 8, 9), $\tilde{x}_{21}$=(10, 11, 12),$\tilde{x}_{22}$=(−2, 2, 6),$\tilde{x}_{23}$=(4, 6, 8),$\tilde{x}_{32}$=(−4, 1, 6),$\tilde{x}_{34}$=(15, 16, 17) and the smallest fuzzy transportation cost is(290, 334, 378).
Now, we consider the optimal FOFTP solution in the SOFTP as a feasible solution by using Step 2 in the below table.

	W1	W2	W3	W4	Supply
F1	(7, 8, 9) [(3, 5, 7)]	(7, 8, 9) [(2, 3, 4)]	(5, 6, 7)	(7, 8, 9)	(7, 8, 9)
F2	(9, 10, 11) [(3, 6, 9)]	(15, 16, 17)	(17, 18, 19)	(19, 20, 21) [(9, 13, 17)]	(18, 19, 20)
F3	(11, 12, 13)	(3, 4, 5)	(9, 10, 11) [(13, 14, 15)]	(1, 2, 3) [(1, 3, 5)]	(16, 17, 18)
Demand	(10, 11, 12)	(2, 3, 4)	(13, 14, 15)	(15, 16, 17)	

Therefore, the fuzzy bi-objective value (BOV) of BFTP is ((242, 286, 330), (382, 530, 678)), and the feasible solution is $\tilde{x}_{11}$=(3, 5, 7),$\tilde{x}_{12}$=(2, 3, 4),$\tilde{x}_{21}$=(3, 6, 9), $\tilde{x}_{24}$=(9, 13, 17),$\tilde{x}_{33}$=(13, 14, 15),$\tilde{x}_{34}$=(1, 3, 5).
According to Step 3, we form a rectangular loop (2, 4) − (2, 3) − (3, 3) − (3, 4) − (2, 4), and we get a reduced table by using Step 4.

	W1	W2	W3	W4	Supply
F1	(7, 8, 9) [(3, 5, 7)]	(7, 8, 9) [(2, 3, 4)]	(5, 6, 7)	(7, 8, 9)	(7, 8, 9)
F2	(9, 10, 11) [(3, 6, 9)]	(15, 16, 17)	(17, 18, 19) [(θ, θ, θ)]	(19, 20, 21) [$(9-\theta, 13-\theta, 17-\theta)$]	(18, 19, 20)
F3	(11, 12, 13)	(3, 4, 5)	(9, 10, 11) [$(13-\theta, 14-\theta, 15-\theta)$]	(1, 2, 3) [$(1+\theta, 3+\theta, 5+\theta)$]	(16, 17, 18)
Demand	(10, 11, 12)	(2, 3, 4)	(13, 14, 15)	(15, 16, 17)	

Now, for any value $\theta \in \{(0, 1, 2), (1, 2, 3), \ldots, (12, 13, 14)\}$, the deterioration cost of SOFTP is $(382 - 10\theta, 530 - 10\theta, 678 - 10\theta)$, and the transportation cost of FOFTP is $(242 + 2\theta, 286 + 2\theta, 330 + 2\theta)$. Therefore, the fuzzy BOV of BFTP is $((242 + 2\theta, 286 + 2\theta, 330 + 2\theta),(382 - 10\theta, 530 - 10\theta, 678 - 10\theta))$, and the feasible solution is $\tilde{x}_{11} = (3, 5, 7)$,$\tilde{x}_{12} = (2, 3, 4)$,$\tilde{x}_{21} = (3, 6, 9)$,$\tilde{x}_{23} = (\theta, \theta, \theta)$, $\tilde{x}_{24} = (9 - \theta, 13 - \theta, 17 - \theta)$,$\tilde{x}_{33} = (13 - \theta, 14 - \theta, 15 - \theta)$,$\tilde{x}_{34} = (1 + \theta, 3 + \theta, 5 + \theta)$. For the highest value of θ, that is, θ=(12, 13, 14), the deterioration cost of SOFTP is (262, 400, 538) and FOFTP is (268, 312, 356). Therefore, the fuzzy BOV of BFTP is ((268, 312, 356), (262, 400, 538)), and the feasible solution is $\tilde{x}_{11} = (3, 5, 7)$,$\tilde{x}_{12} = (2, 3, 4)$,$\tilde{x}_{21} = (3, 6, 9)$,$\tilde{x}_{23} = (13, 13, 13)$,$\tilde{x}_{33} = (0, 1, 2)$, $\tilde{x}_{34} = (13, 16, 18)$.
Now, the solution of SOFTP is not an optimal solution which we obtained here. But we will get a feasible solution that will be better than the previously obtained solution of SOFTP by repeating Steps 3 and 4.

	W1	W2	W3	W4	Supply
F1	(7, 8, 9) [$(3-\theta, 5-\theta, 7-\theta)$]	(7, 8, 9) [(2, 3, 4)]	(5, 6, 7) [(θ, θ, θ)]	(7, 8, 9)	(7, 8, 9)
F2	(9, 10, 11) [$(3+\theta, 6+\theta, 9+\theta)$]	(15, 16, 17)	(17, 18, 19) [$(13-\theta, 13-\theta, 13-\theta)$]	(19, 20, 21)	(18, 19, 20)
F3	(11, 12, 13)	(3, 4, 5)	(9, 10, 11) [(0, 1, 2)]	(1, 2, 3) [(13, 16, 18)]	(16, 17, 18)
Demand	(10, 11, 12)	(2, 3, 4)	(13, 14, 15)	(15, 16, 17)	

Now, for any value $\theta \in \{(0, 1, 2), (1, 2, 3), \ldots, (4, 5, 6)\}$, the deterioration cost of SOFTP is $(356 - 10\theta, 400 - 10\theta, 444 - 10\theta)$, and the transportation cost of FOFTP is $(268 + 8\theta, 312 + 8\theta, 356 + 8\theta)$. Therefore, the fuzzy BOV of BFTP is $((268 + 8\theta, 312 + 8\theta, 356 + 8\theta),(356 - 10\theta, 400 - 10\theta, 444 - 10\theta))$, and the feasible solution is $\tilde{x}_{11}$=$(3 - \theta, 5 - \theta, 7 - \theta)$,$\tilde{x}_{12}$=(2, 3, 4),$\tilde{x}_{13}$=$(\theta, \theta, \theta)$,$\tilde{x}_{21}$=$(3 + \theta, 6 + \theta, 9 + \theta)$,$\tilde{x}_{23}$=$(13 - \theta, 13 - \theta, 13 - \theta)$,$\tilde{x}_{33}$=(0, 1, 2),$\tilde{x}_{34}$=(13, 16, 18). For the highest value of θ, that is, θ=(4, 5, 6), the deterioration cost of

SOFTP is $(316, 350, 384)$, and FOFTP is $(300, 352, 404)$. Therefore, the fuzzy BOV of BFTP is $((300, 352, 404), (316, 350, 384))$, and the feasible solution is $\tilde{x}_{12}$=(2, 3, 4),$\tilde{x}_{13}$=(5, 5, 5),$\tilde{x}_{21}$=(8, 11, 14), $\tilde{x}_{23}$=(8, 8, 8),$\tilde{x}_{33}$=(0, 1, 2), $\tilde{x}_{34}$=(13, 16, 18).
Now, the solution of SOFTP is not an optimal solution which we obtained here. But we will get a feasible solution that will be better than the previously obtained solution of SOFTP by repeating Steps 3 and 4.

	W1	W2	W3	W4	Supply
F1	(7, 8, 9)	(7, 8, 9) $[(2-\theta, 3-\theta, 4-\theta)]$	(5, 6, 7) $[(5+\theta, 5+\theta, 5+\theta)]$	(7, 8, 9)	(7, 8, 9)
F2	(9, 10, 11) [(8, 11, 14)]	(15, 16, 17) $[(\theta, \theta, \theta)]$	(17, 18, 19) $[(8-\theta, 8-\theta, 8-\theta)]$	(19, 20, 21)	(18, 19, 20)
F3	(11, 12, 13)	(3, 4, 5)	(9, 10, 11) [(0, 1, 2)]	(1, 2, 3) [(13, 16, 18)]	(16, 17, 18)
Demand	(10, 11, 12)	(2, 3, 4)	(13, 14, 15)	(15, 16, 17)	

Now, for any value $\theta \in \{(0, 1, 2), (1, 2, 3), (2, 3, 4)\}$, the deterioration cost of SOFTP is $(306 - 4\theta, 350 - 4\theta, 394 - 4\theta)$, and the transportation cost of FOFTP is $(308 + 22\theta, 352 + 22\theta, 396 + 22\theta)$. Therefore, the fuzzy BOV of BFTP is $((308 + 22\theta, 352 + 22\theta, 396 + 22\theta),(306 - 4\theta, 350 - 4\theta, 394 - 4\theta))$, and the feasible solution is $\tilde{x}_{12}=(2-\theta, 3-\theta, 4-\theta)$,$\tilde{x}_{13}=(5+\theta, 5+\theta, 5+\theta)$,$\tilde{x}_{21}$=(8, 11, 14),$\tilde{x}_{22}=(\theta, \theta, \theta)$,$\tilde{x}_{23}=(8-\theta, 8-\theta, 8-\theta)$,$\tilde{x}_{33}$=(0, 1, 2),$\tilde{x}_{34}$=(13, 16, 18). For the highest value of θ, that is, θ=(2, 3, 4), the deterioration cost of SOFTP is $(298, 338, 378)$ and FOFTP is $(352, 418, 484)$. Therefore, the fuzzy BOV of BFTP is $((352, 418, 484), (298, 338, 378))$, and the feasible solution is $\tilde{x}_{13}$=(8, 8, 8),$\tilde{x}_{21}$=(8, 11, 14),$\tilde{x}_{22}$=(3, 3, 3),$\tilde{x}_{23}$=(5, 5, 5), $\tilde{x}_{33}$=(0, 1, 2),$\tilde{x}_{34}$=(13, 16, 18).
Now, the solution of SOFTP is not an optimal solution which we obtained here. But we will get a feasible solution that will be better than the previously obtained solution of SOFTP by repeating Steps 3 and 4.

	W1	W2	W3	W4	Supply
F1	(7, 8, 9)	(7, 8, 9)	(5, 6, 7) [(8, 8, 8)]	(7, 8, 9)	(7, 8, 9)
F2	(9, 10, 11) [(8, 11, 14)]	(15, 16, 17) $[(3-\theta, 3-\theta, 3-\theta)]$	(17, 18, 19) $[(5+\theta, 5+\theta, 5+\theta)]$	(19, 20, 21)	(18, 19, 20)
F3	(11, 12, 13)	(3, 4, 5) $[(\theta, \theta, \theta)]$	(9, 10, 11) $[(0-\theta, 1-\theta, (2-\theta)]$	(1, 2, 3) [(13, 16, 18)]	(16, 17, 18)
Demand	(10, 11, 12)	(2, 3, 4)	(13, 14, 15)	(15, 16, 17)	

Now, for the highest value of θ, that is, θ=(0, 1, 2), the deterioration cost of SOFTP is (275, 334, 393), and FOFTP is (372, 416, 458). Therefore, the fuzzy BOV of BFTP is ((372, 416, 458), (275, 334, 393)), and the feasible solution is $\tilde{x}_{13}$=(8, 8, 8), $\tilde{x}_{21}$ = (8, 11, 14), $\tilde{x}_{22}$ = (2, 2, 2), $\tilde{x}_{23}$ = (6, 6, 6), $\tilde{x}_{32}$ = (1, 1, 1), $\tilde{x}_{34}$ = (13, 16, 18).
Now, since (275, 334, 393) is the optimal value for the SOFTP, we stop the computations.
Therefore, the set of all S_1 solutions of the BFTP is found from FOFTP to SOFTP is given below:

The set of all solutions S_1 of the BFTP obtained from FOFTP to SOFTP

No		No	
1	((242, 286, 330), (382, 530, 678))	13	((264, 310, 356), (272, 410, 548))
2	((242, 288, 334), (382, 520, 658))	14	((266, 312, 358), (262, 400, 538))
3	((244, 290, 336), (372, 510, 648))	15	((268, 320, 372), (356, 390, 424))
4	((246, 292, 338), (362, 500, 638))	16	((276, 328, 380), (346, 380, 414))
5	((248, 294, 340), (352, 490, 628))	17	((284, 336, 388), (336, 370, 404))
6	((250, 296, 342), (342, 480, 618))	18	((292, 344, 396), (326, 360, 394))
7	((252, 298, 344), (332, 470, 608))	19	((300, 352, 404), (316, 350, 384))
8	((254, 300, 346), (322, 460, 598))	20	((308, 374, 440), (306, 346, 386))
9	((256, 302, 348), (312, 450, 588))	21	((330, 396, 462), (302, 342, 382))
10	((258, 304, 350), (302, 440, 578))	22	((352, 418, 484), (298, 338, 378))
11	((260, 306, 352), (292, 430, 568))	23	((372, 416, 458), (275, 334, 393))
12	((262, 308, 354), (282, 420, 558))		

Likewise, using the Steps 6 and 7, we find that the set of all S_2 solutions from SOFTP to FOFTP is shown below:

Iteration	θ	Solutions of BFTP	Fuzzy BOV
1	$\{(0,1,2),$ $(1,2,3)\}$	$\tilde{x}_{12}=(\theta,\theta,\theta)$, $\tilde{x}_{13}=(7-\theta,7-\theta,7-\theta)$, $\tilde{x}_{21}=(10,11,12)$, $\tilde{x}_{22}=(-2-\theta,2-\theta,6-\theta)$, $\tilde{x}_{23}=(4+\theta,6+\theta,8+\theta)$, $\tilde{x}_{32}=(-4,1,6)$, $\tilde{x}_{34}=(15,16,17)$,	$((381-22\theta,416-22\theta,$ $451-22\theta),$ $(290+4\theta,334+4\theta,$ $378+4\theta))$
2	$\{(0,1,2)\}$	$\tilde{x}_{12}=(2+\theta,2+\theta,2+\theta)$, $\tilde{x}_{13}=(5-\theta,6-\theta,7-\theta)$, $\tilde{x}_{21}=(10,11,12)$, $\tilde{x}_{23}=(6,8,10)$, $\tilde{x}_{32}=(-4-\theta,1-\theta,6-\theta)$, $\tilde{x}_{33}=(\theta,\theta,\theta)$, $\tilde{x}_{34}=(15,16,17)$,	((328, 352, 376), (298, 350, 402))

3	{(0, 1, 2), ..., (4, 5, 6)}	$\tilde{x}_{11}$=(θ, θ, θ), $\tilde{x}_{12}$=(3, 3, 3), $\tilde{x}_{13}$=$(4-\theta, 5-\theta, 6-\theta)$, $\tilde{x}_{21}$=$(10-\theta, 11-\theta, 12-\theta)$, $\tilde{x}_{23}$=$(6+\theta, 8+\theta, 10+\theta)$, $\tilde{x}_{33}$=(1, 1, 1), $\tilde{x}_{34}$=(15, 16, 17),	$((308-8\theta, 352-8\theta, 396-8\theta), (306+10\theta, 350+10\theta, 394+10\theta))$
4	{(0, 1, 2), ..., (12, 13, 14)}	$\tilde{x}_{11}$=(5, 5, 5), $\tilde{x}_{12}$=(3, 3, 3), $\tilde{x}_{13}$=(5, 6, 7), $\tilde{x}_{23}$=$(11-\theta, 13-\theta, 15-\theta)$, $\tilde{x}_{24}$=(θ, θ, θ), $\tilde{x}_{33}$=$(1+\theta, 1+\theta, 1+\theta)$, $\tilde{x}_{34}$=$(15-\theta, 16-\theta, 17-\theta)$,	$((268-2\theta, 312-2\theta, 356-2\theta)$ $(356+10\theta, 400+10\theta, 444+10\theta))$

Therefore, the set of all solutions S_2 of the BFTP obtained from SOFTP to FOFTP is shown below:

The set of all solutions S_2 of the BFTP obtained from FOFTP to SOFTP			
No		No	
1	((375, 416, 457), (290, 334, 378))	12	((264, 306, 348), (376, 430, 484))
2	((381, 394, 407), (290, 338, 386))	13	((262, 304, 346), (386, 440, 494))
3	((359, 372, 385), (294, 342, 390))	14	((260, 302, 344), (396, 450, 504))
4	((328, 352, 376), (298, 350, 402))	15	((258, 300, 342), (406, 460, 514))
5	((308, 344, 380), (306, 360, 414))	16	((256, 298, 340), (416, 470, 524))
6	((300, 336, 372), (316, 370, 424))	17	((254, 296, 338), (426, 480, 534))
7	((292, 328, 364), (326, 380, 434))	18	((252, 294, 336), (430, 490, 544))
8	((284, 320, 356), (336, 390, 444))	19	((250, 292, 334), (446, 500, 554))
9	((276, 312, 348), (346, 400, 454))	20	((248, 290, 332), (456, 510, 564))
10	((268, 310, 352), (356, 410, 464))	21	((246, 288, 330), (466, 520, 574))
11	((266, 308, 350), (366, 420, 474))	22	((244, 286, 328), (476, 530, 584))

Now the set of all S solutions of the BFTP is found from FOFTP to SOFTP and from SOFTP to FOFTP is given below:

$S = S_1 \cup S_2$			
No		No	
1	((244, 286, 328), (476, 530, 584))	14	((276, 312, 348), (346, 400, 454))
2	((246, 288, 330), (466, 520, 574))	15	((284, 320, 356), (336, 390, 444))
3	((248, 290, 332), (456, 510, 564))	16	((292, 328, 364), (326, 380, 434))
4	((250, 292, 334), (446, 500, 554))	17	((300, 336, 372), (316, 370, 424))
5	((252, 294, 336), (430, 490, 544))	18	((308, 344, 380), (306, 360, 414))
6	((254, 296, 338), (426, 480, 534))	19	((328, 352, 376), (298, 350, 402))
7	((256, 298, 340), (416, 470, 524))	20	((359, 372, 385), (294, 342, 390))
8	((258, 300, 342), (406, 460, 514))	21	((308, 374, 440), (306, 346, 386))
9	((260, 302, 344), (396, 450, 504))	22	((381, 394, 407), (290, 338, 386))
10	((262, 304, 346), (386, 440, 494))	23	((330, 396, 462), (302, 342, 382))
11	((264, 306, 348), (376, 430, 484))	24	((352, 418, 484), (298, 338, 378))
12	((266, 308, 350), (366, 420, 474))	25	((381, 416, 451), (290, 334, 378))
13	((268, 310, 352), (356, 410, 464))		

The table given below displays the satisfaction level of the objectives of the problem with each efficient solution.

		Satisfaction level	
No	Fuzzy BOV of BFTP	Objectives of FOFTP	Objective of SOFTP
1	((244, 286, 328), (476, 530, 584))	(100, 100, 100)	(35.86, 41.31, 45.50)
2	((246, 288, 330), (466, 520, 574))	(99.18, 99.3, 99.39)	(39.31, 44.31, 48.14)
3	((248, 290, 332), (456, 510, 564))	(98.38, 98.6, 98.7)	(42.75, 47.30, 50.79)
4	((250, 292, 334), (446, 500, 554))	(97.54, 97.9, 98.17)	(46.20, 50.29, 53.44)
5	((252, 294, 336), (436, 490, 544))	(96.72, 97.2, 97.56)	(49.65, 53.29, 56.08)
6	((254, 296, 338), (426, 480, 534))	(95.9, 96.5, 96.95)	(53.10, 56.28, 58.73)
7	((256, 298, 340), (416, 470, 524))	(95.08, 95.8, 96.34)	(56.55, 59.28, 61.37)
8	((258, 300, 342), (406, 460, 514))	(94.26, 95.10, 95.73)	(60, 62.27, 64.02)
9	((260, 302, 344), (396, 450, 504))	(93.44, 94.4, 95.12)	(63.44, 65.26, 66.67)
10	((262, 304, 346), (386, 440, 494))	(92.62, 93.7, 94.51)	(66.89, 68.26, 69.31)
11	((264, 306, 348), (376, 430, 484))	(91.80, 93.01, 93.90)	(70.34, 71.25, 71.95)
12	((266, 308, 350), (366, 420, 474))	(90.98, 92.30, 93.29)	(73.79, 74.25, 74.60)
13	((268, 310, 352), (356, 410, 464))	(90.16, 91.6, 92.68)	(77.24, 77.24, 77.24)
14	((276, 312, 348), (346, 400, 454))	(86.88, 90.90, 93.90)	(80.68, 80.23, 79.89)
15	((284, 320, 356), (336, 390, 444))	(83.60, 88.11, 91.46)	(84.1, 83.23, 82.53)
16	((292, 328, 364), (326, 380, 434))	(80.32, 85.31, 89.02)	(87.58, 86.22, 85.18)
17	((300, 336, 372), (316, 370, 424))	(77.04, 82.51, 86.58)	(91.03, 89.22, 87.83)
18	((308, 344, 380), (306, 360, 414))	(73.77, 79.72, 84.14)	(94.48, 92.21, 90.47)
19	((328, 352, 376), (298, 350, 402))	(65.57, 76.92, 83.26)	(97.24, 95.20, 93.65)
20	((359, 372, 385), (294, 342, 390))	(52.86, 69.93, 82.62)	(98.62, 97.60, 96.82)
21	((381, 394, 407), (290, 338, 386))	(43.85, 62.23, 75.91)	(100, 98.80, 97.88)
22	((381, 416, 451), (290, 334, 378))	(43.85, 54.55, 62.5)	(100, 100, 100)

The decision-makers can freely use the above satisfaction level table in selecting the appropriate solutions to BFTP according to their satisfaction level of objectives.

5 Conclusion

In this paper, we have presented a fuzzy dripping method for bi-objective transportation problem under imprecise environments. The proposed procedure gives the set of efficient solutions for BFTP. The FDM improves the decision makers to choose a suitable solution, depending on their economic location and their level of satisfaction of the objectives.

References

1. Zadeh,L.A.:Fuzzy sets. Inf contr.**8**,338–353(1965)
2. Biswal,M.P.:Fuzzy programming technique to solve multi-objective geometric programming problems.Fuzzy Set Syst.**51**,67–71(1992)
3. Ammar,E.E., Youness,E.A.:Study on multi objective transportation problem with fuzzy numbers. Appl. Math. Sci.**166**,241–253(2005)
4. Bodkhe,S.G., Bajaj, V.H., Dhaigude,R.M.:Fuzzy programming technique to solve bi-objective transportation problem.Int J Mach Intell. **2**,46–52(2010)
5. Pandian,P., Natarajan,G.: A new algorithm for finding a fuzzy optimal solution for fuzzy transportation problems. Appl. Math. Sci. **4**,79–90(2010)
6. Pandian,P., Anuradha,D.: A New Method for Solving Bi-Objective Transportation Problems. AJBAS.**5**,67–74(2011)
7. Zangiabadi,M., Maleki,H.R.: Fuzzy goal programming technique to solve multi objective transportation problems with some non-linear membership functions. IRAN J FUZZY SYST.**10**,61–74(2013)
8. SaruKumari, Priyamvada Singh,: Fuzzy efficient interactive goal programming approach for multi-objective transportation problems. J.Appl.Comput.Math.,1–21(2016)
9. Palanivel,K.: Fuzzy commercial traveler problem of trapezoidal membership functions within the sort of ? optimum solution using ranking technique. Afr. Mat.**27**,263–277(2016)

Nonlinear Programming Problem for an M-Design Multi-Skill Call Center with Impatience Based on Queueing Model Method

K. Banu Priya and P. Rajendran

Abstract A new method is proposed to obtain the state-transition rates and the service level using the Erlang-A queueing formula for an M-design multi-skill (MDMS) customer service center (CSC) with two kinds of calls and three service centers present to serve the different kinds of calls. The special case of this MDMS method is the third service center which has the ability to serve both kinds of calls. The main aim of customer service center is to minimize the customer's waiting time and cost. The calculated service-level values are applied in the nonlinear programming problem. The proposed method is illustrated with the help of numerical examples.

Keywords Quadratic programming · Queueing model · Multi-skill call center

1 Introduction

Customer service centers (CSC) have been used for decades to provide the customer service, support in Tele-marketing and in many other services for businesses. The main aim of the business is to satisfy the customers and to create a standard name for the company. Today one of the most popular ways in business is the use of customer service centers to provide technical support and customer service. CSC focused on customer's satisfaction of the service. To satisfy the customer, the company uses many ways like reducing the waiting time and saving the money by providing a better service to the customers. Nancy Marengo [7] mentioned the importance of call centers for many concerns like bank, insurance, and telephone companies and illustrated designs of call centers. Smith [2] shows steady state probability for M/M queues, queueing system with a large number of states. Lie [9] explained the M-design model with adequate numerical example by staffing problems.

K. Banu Priya · P. Rajendran (✉)
VIT, Vellore, India
e-mail: banupriya.k@vit.ac.in; prajendran@vit.ac.in

V. Madhu et al. (eds.), *Advances in Algebra and Analysis*, Trends in Mathematics,
https://doi.org/10.1007/978-3-030-01120-8_28

Li and Yue [10] have studied the multi-skill call center in N-design. To get more clear idea about telephone call center, staff hiring, queueing, Erlang A, Erlang B, Erlang C, and skill-based routing, refer to Gans et al. [4]. Koole et al. [3] applied queuing model on call center and assumed callers to be customers, waiting line to be a waiting queue for service, and call centers to be a service centers. Lerzan Ormeci [6] applied dynamic admission control in call center with one shared and two dedicated stations. Mamadou Thiongane et al. [8] found waiting time for multi-skill call center using N-design by applying regression splines and artificial neural network. Gans and Zhou [5] have taken queueing system which is mostly used in call center and gave type-H and type-L works, a new method is proposed to obtain the state-transition rates and the service level using the Erlang-A queueing formula for an M-design multi-skill (MDMS) customer service center (CSC) with two kinds of calls and three service centers present to serve the different kinds of calls. The special case of this MDMS model is the third service center which has the ability to serve both kinds of calls. In the Erlang-A queueing formula, the letter A stands for impatience and is given by the parameter. The main aim of customer service center is to minimize the customer's waiting time and cost. To begin with, we calculate the service level, and the calculated service-level values are applied in the nonlinear programming problem. The proposed method is illustrated with the help of numerical examples.

2 Preliminaries

The definitions of the terms M-design single skill, M-design multi-skill, service level, and M-design service center are found in [1] and Erlang-A queuing are found in [7].

3 M-Design Multi-Skill (MDMS) Model

We consider two types of calls following the Poisson processes, namely, call1 with arrival rate λ_1 and call2 with arrival rate λ_2. There is no restriction in the queueing process. The queue is of infinite in capacity. We provide the option to the customers to leave the queue due to their impatience with mean θ.
This MDMS model has three customer service centers (organizers), namely, N_1, N_2, and N_1N_2 . The service time of the service center N_1 is exponentially distributed with mean μ_1.The service time of the service center N_2 is exponentially distributed with mean μ_2. The service time of the service center N_1N_2 is exponentially distributed with mean μ_3. The special case of this MDMS model is the third service center which has the ability to serve both kinds of calls .The service center N_1 can serve only queue1 customers, and the service center N_2 can serve queue2 customers only, whereas we provide the option to the service center N_1N_2 which can serve both

queue1 and queue2 types of customers. It is to be noted that queue1 and queue2 are different and are not similar to each other. The model is shown in the following figure:

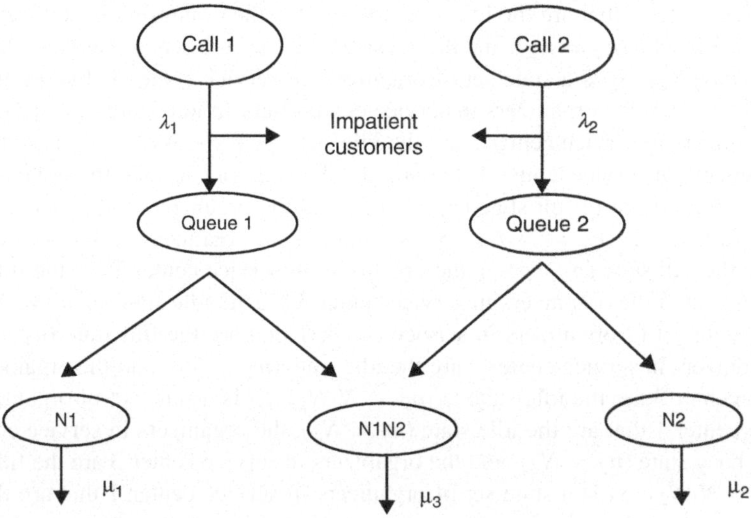

3.1 MDMS State Transition

We divide the state space by considering the relationship between the number of calls and the number of organizers in each group.

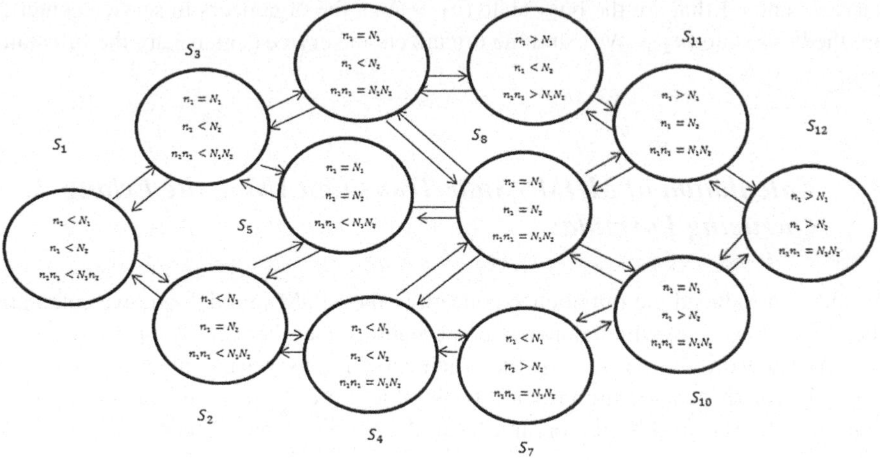

In this MDMS model, we have 12 states which are given by S_i,(i = 1,2,3,..12), where S_1 is a state set of organizers in service center 1 that are the idle state ($n_1 < N_1$), the organizers in service center 2 are the idle state ($n_2 < N_2$), and the organizers in service center 3 are also the idle state ($n_1n_2 < N_1N_2$). S_2 is a state set of organizers in service center 1 that are the idle state ($n_1 < N_1$), the organizers in service center 2 are the full state ($n_2 = N_2$), and the organizers in service center 3 are the idle state ($n_1n_2 < N_1N_2$). S_3 is a state set of organizers in service center 1 that are the full state ($n_1 = N_1$), the organizers in service center 2 are in idle state ($n_2 < N_2$), and the organizers in service center 3 are the idle state ($n_1n_2 < N_1N_2$). S_4 is a state set of organizers in service center 1 that are the idle state ($n_1 < N_1$), the organizers in service center 2 are in full state ($n_2 = N_2$), and the organizers in service center 3 are the full state ($n_1n_2 = N_1N_2$). S_5 is a state set of organizers in service center 1 that are the full state ($n_1 = N_1$), the organizers in service center 2 are the full state ($n_2 = N_2$), and the organizers in service center 3 are the idle state ($n_1n_2 < N_1N_2$). S_6 is a state set of organizers in service center 1 that are the full state ($n_1 = N_1$), the organizers in service center 2 are the idle state ($n_2 < N_2$), and the organizers in service center 3 are the idle state ($n_1n_2 < N_1N_2$). S_7 is a state set of organizers in service center 1 that are the idle state ($n_1 < N_1$), the organizers in service center 2 are the busy state ($n_2 > N_2$), and the organizers in service center 3 are the full state ($n_1n_2 = N_1N_2$). S_8 is a state set of organizers in service center 1 that are the full state ($n_1 = N_1$), the organizers in service center 2 are the full state ($n_2 = N_2$), and the organizers in service center 3 are the full state ($n_1n_2 = N_1N_2$). S_9 is a state set of organizers in service center 1 that are the busy state ($n_1 > N_1$), the organizers in service center 2 are the idle state ($n_2 < N_2$), and the organizers in service center 3 are the full state ($n_1n_2 = N_1N_2$). S_{10} is a state set of organizers in service center 1 that are the full state ($n_1 = N_1$), the organizers in service center 2 are the busy state ($n_1 > N_1$), and the organizers in service center 3 are the full state ($n_1n_2 = N_1N_2$). S_{11} is a state set of organizers in service center 1 that are the busy state ($n_1 > N_1$), the organizers in service center 2 are the full state ($n_2 = N_2$), and the organizers in service center 3 are the full state ($n_1n_2 = N_1N_2$). S_{12} is a state set of organizers in service center 1 that are the busy state ($n_1 > N_1$), the organizers in service center 2 are the busy state ($n_2 > N_2$), and the organizers in service center 3 are the full state ($n_1n_2 = N_1N_2$).

3.2 Calculation of MDMS State Transition Using the Erlang-A Queueing Formula

We have introduced the impatience concept in the MDMS model, and we calculate the $P(n_1 = N_1 + 1)$ which denotes the probability that there are $N_1 + 1$ customers of call1 needed to be serviced by the organizers in the service center1 and $P(n_2 = N_2 + 1)$ which denotes the probability that there are $N_2 + 1$ customers of call 2 needed to be serviced by the organizers in the service center 2 using the Erlang-A queuing formula with the condition that $n_1 < N_1$ and $n_2 > N_2$. Here it is interesting to note that the queue1 and queue2 are different and are not similar to each other. Now,

$$P(n_1 = N_1 + 1) = \left\{ \prod_{K=n+1} \frac{\lambda}{(N_1\mu_1 + (K-h)\theta)} \frac{\left(\frac{\lambda_1}{\mu_1}\right)^{N_1}}{N_1} P_0 \right\} \tag{1}$$

where

$$P_0 = \sum_{j=0}^{n} \frac{\left(\frac{\lambda_1}{\mu_1}\right)^j}{j!} + \sum_{j=n+1}^{\infty} \prod_{k=n+1}^{j} \left(\frac{\lambda_1 \left(\frac{\lambda_1}{\mu_1}\right)^{N_1}}{(N_1\mu_1 + (k-N_1)\theta)N_1!} \right)^{-1} \tag{2}$$

4 Calculation of the Steady-State Probability

Now, we consider the MDMS state transition, and we calculate the transition rates among different sets of states by means of figure MDMS state transition (See Sect. 4). Let P_i (for i = 1,2,...,12) denote the steady-state probability of each state, and $q(S_i - S_j)$, (i,j = 1,2,...,12) denote the state transition.
The equation for the steady-state probabilities of the system is given below:

$$P_1(q(S_1 - S_2) + q(S_1 - S_3)) = P_2 q(S_2 - S_1) + P_3 q (S_3 - S_1), \tag{3}$$

$$P_2(q(S_2 - S_1) + q(S_2 - S_4) + q(S_2 - S_5)) = P_1 q (S_1 - S_2) + P_4 q(S_4 - S_2) + P_5 q(S_5 - S_2), \tag{4}$$

$$P_3(q(S_3 - S_1) + q(S_3 - S_5) + q(S_3 - S_6)) = P_1 q (S_1 - S_3) + P_5 q(S_5 - S_3) + P_6 q(S_6 - S_3), \tag{5}$$

$$P_4(q(S_4 - S_2) + q(S_4 - S_7) + q(S_4 - S_8)) = P_2 q (S_2 - S_4) + P_7 q(S_7 - S_4) + P_8 q(S_8 - S_4), \tag{6}$$

$$P_5(q(S_5 - S_2) + q(S_5 - S_3) + q(S_5 - S_8)) = P_2 q (S_2 - S_5) + P_3 q(S_3 - S_5) + P_8 q(S_8 - S_5), \tag{7}$$

$$P_6(q(S_6 - S_3) + q(S_6 - S_8) + q(S_6 - S_9)) = P_3 q (S_3 - S_6) + P_8 q(S_8 - S_6) + P_9 q(S_9 - S_6), \tag{8}$$

$$P_7(q(S_7 - S_4) + q(S_7 - S_{10})) = P_4 q(S_4 - S_7) + P_{10} q(S_{10} - S_7) \tag{9}$$

$$P_8(q(S_8 - S_4) + q(S_8 - S_5) + q(S_8 - S_6) + q(S_8 - S_{10} + q(S_8 - S_{11})) = P_4q(S_4 - S_8) + P_5q(S_5 - S_8) + P_6q(S_6 - S_8) + P_{10}q(S_{10} - S_8) + P_{11}q(S_{11} - S_8), \tag{10}$$

$$P_9(q(S_9 - S_6) + q(S_9 - S_{11})) = P_6q(S_6 - S_9) + P_{11} q(S_{11} - S_9) \tag{11}$$

$$P_{10}(q(S_{10} - S_7) + q(S_{10} - S_8) + q(S_{10} - S_{12})) = P_7 q(S_7 - S_{10}) + P_8q(S_8 - S_{10}) + P_{12}q(S_{12} - S_{10}), \tag{12}$$

$$P_{11}(q(S_{11} - S_8) + q(S_{11} - S_9) + q(S_{11} - S_{12})) = P_8 q(S_8 - S_{11}) + P_9q(S_9 - S_{11}) + P_{12}q(S_{12} - S_{11}), \tag{13}$$

$$P_{12}(q(S_{12} - S_{10}) + q(S_{12} - S_{11})) = P_{10}q(S_{10} - S_{12}) + P_{11}q(S_{11} - S_{12}) \tag{14}$$

Subject to the condition that $\sum_{i=1}^{12} P_i = 1$
By solving equations from 3 to 14, we obtain the steady-state probabilities from P_1toP_{12}. These probabilities are calculated numerically by using MATLAB software.

5 Calculation of the Service Level

Here we consider the concept that the 80% of calls should get service within 20 s of waiting time. We got the service level using steady-state probability. Let P^1_{sl} be the probability of call1 providing in a fixed waiting time T_1, and let P^2_{sl} be the probability of call2 providing in a fixed waiting time T_2. We consider customers of call1, customers of call1 form a queue1, and the states of queue1 are S_9, S_{11}, S_{12}. The service rate of S_9 and S_11 of call 1 is $N_1\mu_1 + N_1N_2\mu_3$, and the service rate of $N_1\mu_1 + (N_1N_2\mu_3)/2$, so we have assumed that the probability of call1 that cannot be served in time T_1 is:

$$P^1_{ns} = P_9 \sum_{i=K_1}^{\infty} P_9(n_1 = i) + P_{11} \sum_{i=K_1}^{\infty} P_{11}(n_1 = i) + P_{12} \sum_{i=K_2}^{\infty} P_{12}(n_1 = i) \tag{15}$$

$$\begin{aligned} &where: \\ &K_1 = N_1 + N_1N_2 + [T_1(N_1\mu_1 + N_1N_2\mu_3)] \\ &K_2 = N_1 + 1/2N_1N_2 + [T_1(N_1\mu_1 + 1/2N_1N_2\mu_3)] \end{aligned}$$

$$P_9(n_1 = i) = \prod_{k=n+1}^{j} \left(\frac{1}{N_1\mu_1 + (K - N_1)\theta} \right) \frac{(\lambda_1)^{N_1+1}}{\mu_1 N_1!} P_0 \tag{16}$$

and:

$$P_0 = \sum_{j=0}^{n} \frac{(\frac{\lambda_1}{\mu_1})^j}{j!} + \sum_{j=n+1}^{\infty} \prod_{k=n+1}^{j} \left(\frac{\lambda_1 (\frac{\lambda_1}{\mu_1})^{N_1}}{(N_1\mu_1 + (k - N_1)\theta)N_1!} \right)^{-1} \tag{17}$$

$P_{11}(n_1 = i)$ and $P_{12}(n_1 = i)$ calculation is same as like $P_9(n_1 = i)$
We consider customers of call2, customers of call2 form a queue2, and the states of queue2 are S_7, S_{10}, S_{12}. The analysis method is the same as call1. We obtain the probability of call2 cannot be served in time T_2 is:

$$P_{ns}^2 = P_7 \sum_{i=K_1}^{\infty} P_7(n_2 = i) + P_{10} \sum_{i=K_1}^{\infty} P_{10}(n_2 = i) + P_{12} \sum_{i=K_2}^{\infty} P_{12}(n_2 = i) \tag{18}$$

where,
$K_3 = N_2 + N_1N_2 + [T_2(N_2\mu_2 + N_1N_2\mu_3)]$
$K_4 = N_2 + 1/2N_1N_2 + [T_2(N_2\mu_2 + 1/2N_1N_2\mu_3)]$

$$P_7(n_1 = i) = \prod_{k=n+1}^{j} \left(\frac{1}{N_1\mu_1 + (K - N_1)\theta} \right) \frac{(\lambda_1)^{N_1+1}}{\mu_1 N_1!} P_0 \tag{19}$$

and:

$$P_0 = \sum_{j=0}^{n} \frac{(\frac{\lambda_1}{\mu_1})^j}{j!} + \sum_{j=n+1}^{\infty} \prod_{k=n+1}^{j} \left(\frac{\lambda_1 (\frac{\lambda_1}{\mu_1})^{N_1}}{(N_1\mu_1 + (k - N_1)\theta)N_1!} \right)^{-1} \tag{20}$$

After finding the values of P_{ns}^1 and P_{ns}^2, now we find the values of ${P_{sl}}^1 = 1 - {P_{ns}}^1$ and ${P_{sl}}^2 = 1 - {P_{ns}}^2$ which are tabulated as follows:

N_1	N_2	N_3	P_{sl}^1	P_{sl}^2
11	16	9	0.8135	0.815
11	14	11	0.8187	0.82
11	15	10	0.805	0.8066

6 Staffing Problem

The problem which is worked out for a bunch of people or workers is known as the problem of workers or the staffing problem.
Let C_1 be the cost of the service center1, C_2 be the cost of the service center2, and C_3 be the cost of the service center3. The main aim of the customer service center

is to minimize the cost. We try to find the optimal number of servers N_1, N_1, and N_1N_2 which are utilized in the nonlinear programming problem.
The model is given below:

$$Solve : Minf(x) = -C_1N_1 - C_2N_2 - C_3N_1N_2$$

subject to constrain

$$P_{sl}^1 \leq \alpha_1$$

$$P_{sl}^2 \leq \alpha_2$$

$$N_1, N_1 \in Z^+$$

The constrains denote that the customer serviced in less than or equal to α_1,α_2, respectively. N_1,N_2 are integers. The parameter settings are $\alpha_1 = \alpha_2 = 0.8$, and then ${P_{sl}}^1$ and ${P_{sl}}^2$ are taken from service level.

7 Numerical Examples

We solve the MDMS model by using nonlinear programming problem.

C_1	C_2	C_3	N_1	N_2	N_1N_2	Z
10	10	10	1.1091	1.1092	1.2301	9.8808
10	60	10	6.1589	1.6136	9.9380	59.0248

7.1 Remark

We obtain the minimum cost using the nonlinear programming problem with impatience to be (9.8808,59.0248), but in [9], the without impatience is (320,1070).

8 Conclusion

The new method provides the state-transition rates and the service level using the Erlang queueing A formula for an M-design multi-skill (MDMS) customer service center (CSC) with two kinds of calls and three service centers present to serve the different kinds of calls. The special case of this MDMS method is the third service center which has the ability to serve both kinds of calls. This method is very easy to understand and apply and also will help the CSC to minimize the customers' waiting time and cost. Minimizing the customers' waiting time and cost will improve the name of the CSC to a good level, thereby increasing the income of the CSC to the maximum level.

References

1. Garnett,O.Mandelbaum,A. Reiman,M.: Designing a Call Center with Impatient Customers. Manufacturing Service Oper. Manag. **4**, 208–227(2002)
2. David,K.smith.: Calculation of steady-state probability of M/M Queues: Further approaches. J.App.Math. decision sci. **6**, 43–50(2002)
3. Koole,G. Mandelbaum,A.: Queueing Model of call centers : an introduction . Ann. Oper. Res. **113**, 41–59(2002)
4. Gans,N. Koole,G. Mandelbaum,A.: Telephone call centers: Tutorial, Review and Research prospects.Manufacturing Service Oper. Manag.**5**, 79–141(2003)
5. Noah Gans. Yong-Pin Zhou.: A call routing problem with service level constraints. Inst. Oper.Res. Management Sci. **51**, 255–271(2003)
6. Lerzan Ormeci.E.: Dynamic admission control in a call center with one shared and two dedicated service facilities. IEE trans. Automat. Contr. **49**, 1157–1161(2004)
7. Nancy Marengo.:Skill based routing in multi-skill call center. BMI (2014)
8. Mamadou Thiongane. WyeanChan. Pierre LE cuyer.: Waiting time predictor for multi-skill call center. WSC,3073–3084(2015)
9. Chun-Yan Li.: Performance analysis and the stuffing optimization for a multi- skill call center in M-design based on queuing model method. Computer Science, Technology and Application: chapter. **5**, 454–466(2016)
10. Chun-Yan Li.,De-Quan Yue.: The staffing problem of the N-design multi-skill call center based on queuing model. Advances in computer science research,3rd International Conference on Wireless Communication and Sensor Network. **44**, 427–432(2016)

Optimizing a Production Inventory Model with Exponential Demand Rate, Exponential Deterioration and Shortages

M. Dhivya Lakshmi and P. Pandian

Abstract A production inventory model having an exponential deterioration rate with an exponential demand rate and shortages is considered. In this model, the production rate is a function of the demand rate. The total inventory cost per cycle, the cycle length, the shortage and the production lengths are optimized. Numerical example of the proposed model is presented.

1 Introduction

Inventory means a physical stock of stored goods having some economic value to meet the anticipated demand. It can be in the form of physical resource, human resources, or financial resource. Inventory may be regarded as those goods which are acquired, accumulated and utilized for day-to-day functioning of a management smoothly and effectively. The study of the inventory system and the inventory control process helps us to economically manage the flow of materials, effectively make use of people and equipment, coordinate internal activities and communicate with customers. Demand and deterioration are two main factors in inventory system which have been of growing interest to researchers. In inventory models, five types of demand, namely, constant demand, time-dependent demand, probabilistic demand, stock-dependent demand and imprecise demand, are considered generally. Deterioration is referred as decay, damage, dryness and spoilage which acts vital role in the inventory model. Deterioration rate of the on-hand inventory may be a constant fraction or a function of time. The perishable inventory theory was developed in which products are to be deteriorated.

Harris [8] is the one who initially originated the economic order quantity (EOQ) model to assist organizations in minimizing total inventory costs. Wee [23], Niketa and Nita [12] and Rekha and Vikas [13] proposed a deterministic lot size inventory model for deteriorating items. Samanta and Roy [15] and Bhowmick

M. Dhivya Lakshmi (✉) · P. Pandian
Department of Mathematics, VIT, Vellore, Tamil Nadu, India
e-mail: dhivyalakshmi.m2015@vit.ac.in; ppandian@vit.ac.in

V. Madhu et al. (eds.), *Advances in Algebra and Analysis*, Trends in Mathematics,
https://doi.org/10.1007/978-3-030-01120-8_29

and Samanta [9] established a continuous production control inventory model for deteriorating items with shortages. Ouyang et al. [10], Jain et al. [11], Begum et al. [3] and Sheikh and Raman [19] developed an inventory model for deteriorating items with shortages. Gupta and Vrat [7] studied an inventory model in which demand is dependent on initial stock levels. Baker and Urban [1] presented an inventory model in which the on-hand inventory demand is in polynomial form. Begum et al. [2] proposed an inventory model with exponential demand rate and shortages. Sahoo and Sahoo [14] derived an inventory model with linear demand rate allowing shortages in the inventory. Whitin [24] initiated and studied inventory of deteriorating fashion goods at the end of prescribed storage period. Ghare and Schrader [6] developed and analysed an inventory model for exponentially decaying perishable items.

Sarkar and Chakrabarti [16] presented a production model for the lot-size inventory system with finite production rate, effect of decay and permissible delay in payments. An inventory model for deteriorating items having a time-dependent demand rate was studied and analysed by Shukla et al. [21], Bhanu et al. [4] and Geetha et al. [5]. Sunil and Pravin [22] constructed an inventory model for time varying holding cost and Weibull distribution for deterioration with fully backlogged shortages. Sharmila and Uthayakumar [18] presented the fuzzy inventory model for deteriorating items with shortages under fully backlogged conditions. A deteriorating items production inventory model with time- and price-dependent demand under inflation and trade credit period was determined by Shital [20]. Seyed et al. [17] presented and studied the production of a deteriorating item for a three-level supply chain.

In the present article, a production inventory model for deteriorating items with demand having exponential rate is proposed. In the proposed model, the production is finite and a function of the demand, deterioration is an exponential function, and shortages are permitted. We optimize the total inventory cost and length of the cycle in the developed model. Numerical example is presented for illustrating the developed inventory model.

In the proposed model, we assume that the demand rate is an exponential function, and the declining deterioration items follow exponential distribution. Because of the exponential nature, the demand is increasing smoothly which helps us to reduce the holding cost. The deterioration of the items is declining gradually because of exponential distribution which supports us to reduce the set-up cost. So, the developed model is applicable to all organizations having non-zero demand at the starting time of the production, smoothly increasing demand and gradually decreasing deterioration.

2 Assumptions and Notations and Assumptions

Now, we adopt the following variables and parameters in the developed model:

$D(t)$: the demand which is a function of t.
$P(t)$: the production which is a function of t.
$h(t)$: the rate of deterioration.

C_h : the holding cost per unit per unit time.
C_d : the deteriorating cost per unit per unit time.
C_o : the set-up cost per cycle.
C_s : the shortage cost per unit per unit time.
$I(t)$: the inventory level at time t.
Q : the maximum inventory level.
R : the maximum shortage level.
T_1 : the time at which the first production stops.
T_2 : the time at which the stock is fully consumed.
T_3 : the time at which the second production starts.
T : the production cycle time.
TC : the total inventory cost per cycle.

Now, the following conventions are assumed in the proposed model:

1. Demand function follows exponential distribution and is given by $D(t) = ae^{bt}$, $a > 0, b \in (0, 1)$. Also, the demand is not zero at t=0.
2. Production function is proportional to the demand function and is taken as $P(t) = \lambda D(t), \lambda > 1$.
3. Deterioration function follows exponential distribution $g(t)$ where $g(t) = \begin{cases} \theta e^{-\theta t}, t \geq 0 \\ 0, t < 0 \end{cases}$ The rate of deterioration, $h(t) = \theta$ a constant $(0 < \theta < 1)$. During a given cycle, repair or replacement of the deteriorated items does not take placc.
4. Shortages are allowed and there is no lead time.

3 Description of the Proposed Model

Based on the exponential market demand and production capacity of the firm, the model is constructed. With zero inventory, the production starts at $t = 0$. The demand exponentially changes time to time and is given by $D(t) = ae^{bt}$, $a > 0, 0 < b < 1$. Because of the market demand and limited shelf-life, deterioration occurs, during the time $t = 0$ to T_1. The inventory attains the level Q at $t = T_1$, and the production is stopped at $t = T_1$. Now, from T_1 to T_2, the inventory level reduces due to deterioration and demand, and it becomes zero at time $t = T_2$. Now, shortages occur from the time $t = T_2$ and accumulate to the level R at time $t = T_3$. Again, at $t = T_3$, the production starts with the same rate. The shortages are emptied at time $t = T$, and again the stock becomes empty. Then, after time T, the cycle repeats (Fig. 1).

Now, the governing equations with boundary conditions for the above developed model are given by

$$\frac{d}{dt}I(t) + \theta I(t) = (\lambda - 1)\, ae^{bt}, 0 \leq t \leq T_1; \tag{1}$$

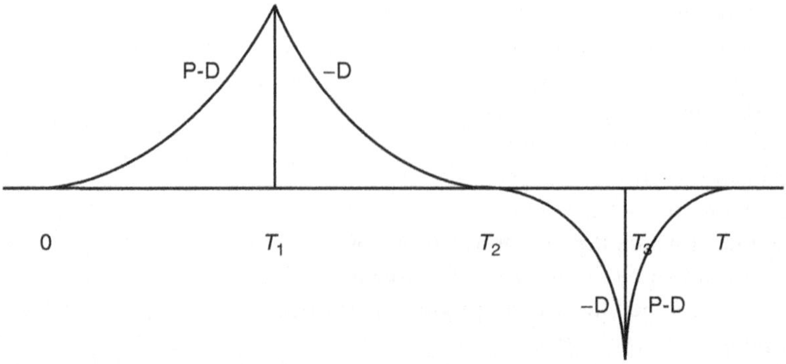

Fig. 1 The production inventory model

$$\frac{d}{dt}I(t) + \theta I(t) = -ae^{bt},\ T_1 \le t \le T_2; \tag{2}$$

$$\frac{d}{dt}I(t) = -ae^{bt},\ T_2 \le t \le T_3; \tag{3}$$

$$\frac{d}{dt}I(t) + \theta I(t) = (\lambda - 1)\,ae^{bt},\ T_3 \le t \le T; \tag{4}$$

$I(0) = 0$, $I(T_1) = Q$, $I(T_2) = 0$, $I(T_3) = -R$ and $I(T) = 0$.

Now, solving (1), (2), (3) and (4) with boundary conditions, we obtain

$$I(t) = \begin{cases} \frac{a(\lambda-1)}{b+\theta}\left(e^{bt} - e^{-\theta t}\right), 0 \le t \le T_1 \\ \frac{a}{b+\theta}\left(e^{(b+\theta)T_2-\theta t} - e^{bt}\right), T_1 \le t \le T_2 \\ \frac{a}{b}\left(e^{bT_2} - e^{bt}\right), T_2 \le t \le T_3 \\ \frac{a(\lambda-1)}{b+\theta}\left(e^{bt} - e^{(b+\theta)T-\theta t}\right), T_3 \le t \le T \end{cases}$$

Now, the holding cost, HC is given as follows:

$$HC = C_h\left\{\int_0^{T_2} I(t)dt\right\} = C_h\left\{\begin{array}{l} \frac{a(\lambda-1)}{b+\theta}\left(\frac{e^{bT_1}-1}{b}\right) + \frac{a(\lambda-1)}{b+\theta}\left(\frac{e^{-\theta T_1}-1}{\theta}\right) \\ +\frac{a}{b+\theta}\left(\frac{e^{bT_2+\theta(T_2-T_1)}-e^{bT_2}}{\theta}\right) - \frac{a}{b+\theta}\left(\frac{e^{bT_2}-e^{bT_1}}{b}\right) \end{array}\right\}.$$

Approximating the exponential functions by omitting the greater powers of b and θ, we have

$$HC = C_h\left\{\frac{a(\lambda-1)}{2}T_1^2 + \frac{a}{(b+\theta)}\{bT_2(T_2-T_1) + \frac{\theta}{2}(T_2-T_1)^2 - \frac{b}{2}(T_2^2 - T_1^2)\}\right\}. \tag{5}$$

Now, the deteriorating cost, DC is given as follows:

$$DC = C_d \left\{ \int_0^{T_1} \theta I(t)\,dt + \int_{T_1}^{T_2} \theta I(t)\,dt + \int_{T_3}^{T} \theta I(t)\,dt \right\}$$

$$= \theta C_d \left\{ \begin{array}{l} \frac{a(\lambda-1)}{b+\theta}\left(\frac{e^{bT_1}-1}{b}\right) + \frac{a(\lambda-1)}{b+\theta}\left(\frac{e^{-\theta T_1}-1}{\theta}\right) + \frac{a}{b+\theta}\left(\frac{e^{bT_2+\theta(T_2-T_1)}-e^{bT_2}}{\theta}\right) \\ -\frac{a}{b+\theta}\left(\frac{e^{bT_2}-e^{bT_1}}{b}\right) + \frac{a(\lambda-1)}{b+\theta}\left(\frac{e^{bT}-e^{bT_3}}{b}\right) + \frac{a(\lambda-1)}{b+\theta}\left(\frac{e^{bT}-e^{(b+\theta)T-\theta T_3}}{\theta}\right) \end{array} \right\}$$

Approximating the exponential functions by omitting the greater powers of b and θ, we obtain

$$DC = \theta C_d \left\{ \begin{array}{l} \frac{a(\lambda-1)}{2}T_1^2 + \frac{a}{(b+\theta)}\{bT_2(T_2-T_1) + \frac{\theta}{2}(T_2-T_1)^2 - \frac{b}{2}(T_2^2-T_1^2)\} \\ +\frac{a(\lambda-1)}{(b+\theta)}\{\frac{b}{2}(T^2-T_3^2) - bT(T-T_3) - \frac{\theta}{2}(T-T_3)^2\} \end{array} \right\}. \tag{6}$$

Now, the shortage cost, SC is given as follows:

$$SC = C_s \left\{ \int_{T_2}^{T} -I(t)\,dt \right\}$$

$$= C_s \left\{ \begin{array}{l} \frac{a}{b}\left(\frac{e^{bT_3}-e^{bT_2}}{b}\right) - \frac{a}{b}(T_3-T_2)e^{bT_2} + \frac{a(\lambda-1)}{b+\theta}\left(\frac{e^{bT+\theta(T-T_3)}-e^{bT}}{\theta}\right) \\ -\frac{a(\lambda-1)}{b+\theta}\left(\frac{e^{bT}-e^{bT_3}}{b}\right) \end{array} \right\}$$

Substituting $T_3 = T_2 + \mu(T - T_2)$, where $0 < \mu < 1$, we have

$$SC = C_s \left\{ \begin{array}{l} \frac{a}{b^2}\left[e^{bT_2+b\mu(T-T_2)} - e^{bT_2}\right] - \frac{a}{b}\mu(T-T_2)e^{bT_2} \\ +\frac{a(\lambda-1)}{b+\theta}\left(\frac{e^{bT+\theta(T-T_3)}-e^{bT}}{\theta}\right) - \frac{a(\lambda-1)}{b+\theta}\left(\frac{e^{bT}-e^{bT_3}}{b}\right) \end{array} \right\}$$

Approximating the exponential functions by omitting the greater powers of b and θ, we get

$$SC = C_s \left\{ \begin{array}{l} \frac{1}{2}a\mu\{\mu T^2 - 2\mu T T_2 + \mu T_2^2 - bT T_2^2 + bT_2^3\} \\ +\frac{a(\lambda-1)}{(b+\theta)}\{bT(T-T_3) + \frac{\theta}{2}(T-T_3)^2 - \frac{b}{2}(T^2-T_3^2)\} \end{array} \right\}. \tag{7}$$

Now, the total inventory cost per unit time, TC is given by

$$TC = \frac{1}{T}\{C_o + HC + DC + SC\}$$

$$TC = \frac{1}{T}\left[\begin{array}{l} C_o + C_h\left\{\frac{a(\lambda-1)}{2}T_1^2 + \frac{a}{(b+\theta)}\{bT_2(T_2-T_1) + \frac{\theta}{2}(T_2-T_1)^2 - \frac{b}{2}(T_2^2-T_1^2)\}\right\} \\ +\theta C_d\left\{\begin{array}{l} \frac{a(\lambda-1)}{2}T_1^2 + \frac{a}{(b+\theta)}\{bT_2(T_2-T_1) + \frac{\theta}{2}(T_2-T_1)^2 - \frac{b}{2}(T_2^2-T_1^2)\} \\ +\frac{a(\lambda-1)}{(b+\theta)}\left\{\frac{b}{2}\left(T^2-T_3^2\right) - bT(T-T_3) - \frac{\theta}{2}(T-T_3)^2\right\} \end{array}\right\} \\ +C_s\left\{\begin{array}{l} \frac{1}{2}a\mu\left\{\mu T^2 - 2\mu TT_2 + \mu T_2^2 - bTT_2^2 + bT_2^3\right\} \\ +\frac{a(\lambda-1)}{(b+\theta)}\left\{bT(T-T_3) + \frac{\theta}{2}(T-T_3)^2 - \frac{b}{2}(T^2-T_3^2)\right\} \end{array}\right\} \end{array}\right]. \quad (8)$$

Let us assume that $T_1 = pT_2$, and $T_3 = T_2 + \mu(T - T_2)$, where $0 < p < 1$ and $0 < \mu < 1$. Now, the Eq. (8) becomes

$$TC = \frac{1}{T}\left[\begin{array}{l} C_o + (C_h + \theta C_d)\left\{\begin{array}{l} \frac{a(\lambda-1)}{2}p^2T_2^2 \\ +\frac{a}{(b+\theta)}\left[bT_2^2(1-p) + \frac{\theta T_2^2}{2}(1-p)^2 - \frac{bT_2^2}{2}(1-p^2)\right] \end{array}\right\} \\ +(\theta C_d - C_s)\frac{a(\lambda-1)}{(b+\theta)}\left[\begin{array}{l} \frac{b}{2}\left(T^2\left(1-\mu^2\right) - T_2^2(1-\mu)^2 - 2\mu TT_2(1-\mu)\right) \\ -bT(1+\mu)(T-T_2) - \frac{\theta}{2}(1+\mu)^2(T-T_2)^2 \end{array}\right] \\ +C_s\left\{\frac{a\mu}{2}\left[\mu T^2 - 2\mu TT_2 + \mu T_2^2 - bTT_2^2 + bT_2^3\right]\right\} \end{array}\right] \quad (9)$$

For minimizing TC, TC should satisfy the conditions which are given below:

$$\frac{\partial(TC)}{\partial T} = 0; \ \frac{\partial(TC)}{\partial T_2} = 0; \ \left(\frac{\partial^2(TC)}{\partial T^2}\right)\left(\frac{\partial^2(TC)}{\partial T_2^2}\right) - \left(\frac{\partial^2(TC)}{\partial T\partial T_2}\right)^2 > 0 \text{ and}$$
$$\frac{\partial^2(TC)}{\partial T^2} \text{ or } \frac{\partial^2(TC)}{\partial T_2^2} > 0.$$

Since Eq. (9) is non-linear, it is solved by using MATLAB software. After knowing the optimal values of T, T_2 and TC, the optimal values of T_1 and T_3 can be determined using $T_1 = pT_2$, and $T_3 = T_2 + \mu(T - T_2)$, where $0 < p < 1$ and $0 < \mu < 1$.

4 Numerical Example

A numerical example is presented in this section to understand the proposed inventory model. Consider the inventory system with the following numerical data: $a = 100$, $b = 0.8$, $C_0 = 25$, $C_h = 30$, $C_s = 15$, $C_d = 20$, $\theta = 0.6$, $\lambda = 1.5$, $p = 0.4$ and $\mu = 0.6$.

Now, using MATLAB, we obtain the optimum length of the cycle value of T as $T^* = 0.2813$, the optimum value of T_1 as $T_1^* = 0.0375$, optimum value of T_2 as $T_2^* = 0.0937$, optimum value of T_3 as $T_3^* = 0.2063$ and the optimum total cost TC as $TC^* = 179.8331$.

5 Conclusion

In this article, a production inventory model for exponentially declining deterioration with an exponential demand rate and shortage is considered. In the proposed model, the production is a function of the demand. The optimal total inventory cost per cycle and the optimal cycle length and optimal production lengths are determined. A numerical example of the constructed model is shown.

References

1. Baker, R.C., Urban, T.L.: A deterministic inventory system with an inventory level dependent demand rate. J. Oper. Res. Soc. **39**(9), 823–831 (1988)
2. Begum, R., Sahu, S.K., Sahoo, R.R.: An inventory model with exponential demand rate, finite production rate and shortages. J. Sci. Res. **1**(3), 473–483 (2009)
3. Begum, R., Sahu, S.K., Sahoo, R.R.: An inventory model for deteriorating items with quadratic demand and partial backlogging. Br. J. Appl. Sci. Technol. **2**(2), 112–131 (2012)
4. Bhanu Priya Dash, Trailokyanath Singh, Hadibandhu Pattnayak.: An inventory model for deteriorating items with exponential declining demand and time-varying holding cost. Am. J. Oper. Res. **4**, 1–7 (2014)
5. Geetha, K., Anusheela, N., Raja, A.: An optimum inventory model for time dependent demand with shortages. Int. J. Math. Archive. **7**(10), 99–102 (2016)
6. Ghare, P., Schrader, G.: A model for an exponentially decaying inventory. J. Ind. Eng.**14**, 238–243 (1963)
7. Gupta, R., Vrat, P.: Inventory model for stock-dependent consumption rate. Opsearch. **23**, 19–24 (1986)
8. Harris, F.: Operations and cost. Chicago AW Shaw Co (1915)
9. Jhuma Bhowmick, Samanta, G.P.: A deterministic inventory model of deteriorating items with two rates of production, shortages and variable production cycle. International Scholarly Research Notices. ISRN Applied Mathematics. **2011**, (2011). https://doi.org/10.5402/2011/657464.
10. Liang-Yuh Ouyang, Kun-Shan WU, Mei-Chuan Cheng.: An inventory model for deteriorating items with exponential declining demand and partial backlogging. Yugosl. J. Oper. Res. **15**(2), 277–288 (2005)
11. Madhu Jain, Sharma, G.C., Shalini Rathore.: Economic production quantity models with shortage, price and stock-dependent demand for deteriorating items. IJE Transactions A: Basics. **20**(2), 159–168 (2007)
12. Niketa J. Mehta, Nita H. Shah.: An inventory model for deteriorating items with exponentially increasing demand and shortages under inflation and time discounting. Investigacao Operacional. **23**, 103–111 (2003)
13. Rekha Rani Chaudhary, Vikas Sharma.: Optimal inventory model with weibull deterioration with trapezoidal demand and shortages. Int. J. Eng. Res. Technol. **2**(3), 1–10 (2013)
14. Sahoo, C.K., Sahoo, S.K.: An Inventory Model with Linear Demand Rate, Finite Rate of Production with Shortages and Complete Backlogging. Proceedings of the 2010 International Conference on Industrial Engineering and Operations Management. Dhaka Bangladesh (2010)
15. Samanta, G.P., Ajanta Roy.: A production inventory model with deteriorating items and shortages. Yugosl. J. Oper. Res. **14**(2), 219–230 (2004)
16. Sarkar, S., Chakrabarti, T.: An EPQ model having weibull distribution deterioration with exponential demand and production with shortages under permissible delay in payments. Mathematical Theory and Modeling. **3**(1), 1–6 (2013)

17. Seyed Reza Moosavi Tabatabaei, Seyed Jafar Sadjadi, Ahmad Makui.: Optimal pricing and marketing planning for deteriorating items. PLOS ONE. **12**(3), pp. 1–21 (2017)
18. Sharmila, D., Uthayakumar, R.: Inventory model for deteriorating items involving fuzzy with shortages and exponential demand. Int. J. Supply.Oper. Manag. **2**(3), 888–904 (2015)
19. Sheikh, S.R., Raman Patel.: Production inventory model with different deterioration rates under shortages and linear demand. International Refereed Journal of Engineering and Science. **5**(3), 01–07 (2016)
20. Shital S. Patel.: Production inventory model for deteriorating items with different deterioration rates under stock and price dependent demand and shortages under inflation and permissible delay in payments. Global J. Pure. Appl. Math. **13**(7), 3687–3701 (2017)
21. Shukla, H.S., Vivek Shukla, Sushil Kumar Yadav.: EOQ model for deteriorating items with exponential demand rate and shortages. Uncertain Supply Chain Management. **1**, 67–76 (2013)
22. Sunil V. Kawale, Pravin B. Bansode.: An inventory model for time varying holding cost and weibull distribution for deterioration with fully backlogged shortages. Int. J. Math. Trends. Technol. **4**(10), 201–206 (2013)
23. Wee, H.M.: A deterministic lot-size inventory model for deteriorating items with shortages and a declining market. Comput. Oper. Res. **22**, 345–356 (1995)
24. Whitin, T.M.: The theory of inventory management. Princeton University Press. New Jersey USA (1957)

Analysis of Batch Arrival Bulk Service Queueing System with Breakdown, Different Vacation Policies, and Multiphase Repair

M. Thangaraj and P. Rajendran

Abstract In bulk queueing models, arrival comes in batches, and service provided to the customer in bulk with server breakdown, different vacation policies, and multiphase repair is considered. The queue size distribution and the performance measures of the developed queueing model are established. The particular cases of the proposed queuing model are also discussed. Also numerical example of the model is also discussed.

1 Introduction

Queueing theory was initiated by Erlang [1]. Ke et al. [5] have studied the queueing system with N-policy and at most j vacation. Lee and Kim [2] analyzed the queueing system with vacation interruption and single working vacation. Krishna Reddy et al. [4] analyzed bulk service queueing system with N-policy. Thangaraj and Rajendran [3] studied the batch arrival queueing model with two types of service pattern and two types of vacation. Ke [6] proposed the queueing system under two types of vacation policies. Balasubramanian and Arumuganathan [7] developed the bulk queueing model with modified M-vacation policy and variant arrival rate. Recently, Singh et al. [9] investigated the bulk service queueing system with unsatisfied customer, optional service, and multiphase repair. In multiphase repair process, the repaired server may not go under repair immediately due to unavailability of repair man, unavailability of spare parts, or other reasons. Similarly, we have discussed at most j vacation in this model. In at most j vacation, the server finds inadequate number of customer after completing first service; the server will take another vacation until required number of customer is in the system. Many of their application can be found in real life with bulk service such as CNC turning machines, soft flow dying machine, vegetable oil refinery, giant wheel, etc.

M. Thangaraj · P. Rajendran (✉)
VIT University, Vellore, India
e-mail: mthangaraj51@gmail.com; thangaraj.m2014phd0532@vit.ac.in; prajendran@vit.ac.in

V. Madhu et al. (eds.), *Advances in Algebra and Analysis*, Trends in Mathematics,
https://doi.org/10.1007/978-3-030-01120-8_30

2 Mathematical Model

A single server batch arrival queueing system with server breakdown, multiphase repair, and different vacation policies is considered. In batch service, the server provides service to the batch of customer (minimum of 'a' customers and maximum of 'b' customer). In this model, the server begins the service, when at least 'a' customers are waiting in the queue. If the queue length reaches the value 'a', the server begins the bulk service. After completing bulk service, if the queue length, Q, is greater than or equal to a, then the server will continue the bulk service according to Neuts [8] general bulk service rule. Whenever breakdown occurs in the main server, the failed server goes to the repair station. During the repair period, the server undergoes the k-different phases of repair. At the end of each phases of repair, the server either goes to next phase of repair or otherwise goes to the service station. After completion of repair process, the server either goes to the bulk service or the server goes to the different vacation policies (at most j vacation) according to the queue length. If the queue length is less than 'a', then the server goes to different vacation policies. After completing vacation if the queue length is less than 'a', then the server will be idle (dormant) until the queue length reaches 'a' and then provide bulk service. Otherwise, the server either goes setup time or then provides bulk service (Fig. 1).

3 Notation

λ, poisson arrival rate; Y, group size random variable of the arrival; g_k, probability that k customer arrive in a batch; $N_q(t)$, number of customers waiting for service at time t; $N_s(t)$, number of customers under the service at time t.

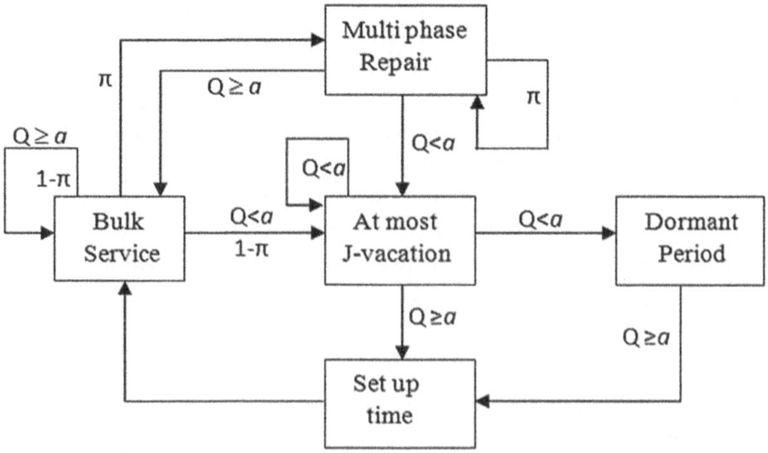

Fig. 1 Schematic representation of the model: Q, Queue length

Let L(x) [l(x)] $\{\tilde{L}(\theta)\}(L^0$(t)) denotes the cumulative distribution function (CDF) [probability density function(PDF)] {Laplace–Stieltjes transform (LST)}(remaining time) of the server repair. Let M(x) [m(x)] $\{\tilde{M}(\theta)\}(M^0$(t)) denotes the CDF [PDF] {LST} (remaining time) of batch service. Let N(x) [n(x)] $\{\tilde{N}(\theta)\}(N^0$(t)) denotes the CDF[PDF]{LST} (remaining time) of the server vacation. Let H(x) [h(x)] $\{\tilde{H}(\theta)\}(H^0$(t)) denotes the CDF [PDF]{LST} (remaining time) of the setup time.

$$C(t) = \begin{Bmatrix} [0] - \text{if the server is on single service} \\ [1] - \text{if the server is on batch service} \\ [2] - \text{if the server is on fast vacation} \\ [3] - \text{if the server is on slow vacation} \\ [4] - \text{if the server is on dormant period} \end{Bmatrix}$$

$$Z(t) = \begin{Bmatrix} [j] - \text{if the server is on j-th vacation} \\ [k] - \text{if the server is on k-th repair} \end{Bmatrix}$$

Now, the state probabilities are established as follows:

$$G_{ij}(x,t)\delta t = \Pr\{N_s(t) = i, N_q(t) = j, x \le M^0(t) \le x + \delta t, C(t) = 0\},$$
$$a \le i \le b, j \ge 0$$
$$T_{nj}(x,t)\delta t = \Pr\{, N_q(t) = j, x \le L^0(t) \le x + \delta t, C(t) = 1, \text{Z(t)} = \text{k}\},$$
$$n = 1, 2, 3, \ldots, k, j \ge 0$$
$$F_{ki}(x,t)\delta t = \Pr\{N_q(t) = j, x \le N^0(t) \le x + \delta t, C(t) = 2, \text{Z(t)} = \text{j}\},$$
$$k = 1, 2, 3, \ldots, j, 1 \le i \le a - 1$$
$$S_j(x,t)\delta t = \Pr\{N_q(t) = j, x \le H^0(t) \le x + \delta t, C(t) = 3\}, j \ge a$$
$$D_j(x,t)\delta t = \Pr\{N_q(t) = j, C(t) = 4\}, 0 \le j \le a - 1$$

4 Queue Size Distributions

Now, we obtain the following steady state system difference-differential equations for the proposed queueing model by using the above state probabilities.

$$-G_{i0}'(x) = -\lambda G_{i0}(x) + \sum_{n=1}^{k} T_{ni}(0)m(x) + S_i(0)m(x)$$
$$+(1-\pi)\sum_{m=a}^{b} G_{mi}(0)m(x), a \le i \le b \quad (1)$$

$$-G_{ij}'(x) = -\lambda G_{ij}(x) + \sum_{k=1}^{j} G_{ij-k}(x)\lambda g_k, a \le i \le b-1, j \ge 1 \quad (2)$$

$$-G_{bj}'(x) = -\lambda G_{bj}(x) + \sum_{n=1}^{k} T_{nb+j}(0)m(x) + S_{b+j}(0)m(x)$$
$$+(1-\pi)\sum_{m=a}^{b} G_{mb+j}(0)m(x) + \sum_{k=1}^{j} G_{bj-k}(x)\lambda g_k,\ j \geq 1 \tag{3}$$

$$T_{10}'(x) = -\lambda T_{10}(x) + (1-\pi)\sum_{m=c}^{b} G_{m0}(0)l(x) \tag{4}$$

$$-T_{1j}'(x) = -\lambda T_{1j}(x)) + \sum_{k=1}^{j} T_{1j-k}(x)\lambda g_k + (1-\pi)\sum_{m=a}^{b} G_{mj}(0)l(x),\ j \geq 1 \tag{5}$$

$$-T_{n0}'(x) = -\lambda T_{n0}(x) + \pi\sum_{n=2}^{k} T_{n-10}(0)l(x),\ n = 2, 3, .., k \tag{6}$$

$$-T_{kj}'(x) = -\lambda T_{kj}(x) + \pi\sum_{l=1}^{k}\sum_{n=2}^{k} T_{n-1,l-k}(x)\lambda g_k,\ j \geq 1 \tag{7}$$

$$-F_{01}'(x) = -\lambda F_{01}(x) + \sum_{n=1}^{k} T_{n0}(0)n(x) + (1-\pi)\sum_{m=a}^{b} G_{m0}(0)n(x) \tag{8}$$

$$-F_{i1}'(x) = -\lambda F_{i1}(x) + \sum_{k=1}^{n} F_{i-k,1}(x)\lambda g_k + (1-\pi)\sum_{m=a}^{b} G_{mi}(0)n(x)$$
$$+\pi\sum_{n=1}^{k} T_{ni}(0)n(x),\ 1 \leq i \leq a-1 \tag{9}$$

$$-F_{i1}'(x) = -\lambda F_{i1}(x) + \sum_{k=1}^{n} F_{i-k,1}(x)\lambda g_k,\ i \geq a \tag{10}$$

$$-F_{0k}'(x) = -\lambda F_{0k}(x) + \sum_{k=2}^{j} F_{0k-1}(0)n(x),\ k = 2, 3, \ldots, j \tag{11}$$

$$-F_{ik}'(x) = -\lambda F_{ik}(x) + \sum_{k=2}^{j} F_{ik-1}(0)n(x) + \sum_{n=1}^{j} F_{i-n,k}(x)\lambda g_n,$$
$$1 \leq i \leq a-1,\ k = 2, 3, \ldots, j \tag{12}$$

$$-F_{ik}^{'}(x) = -\lambda F_{ik}(x) + \sum_{n=1}^{j} F_{i-n,k}(x)\lambda g_n,\ i \geq a \tag{13}$$

$$-S_j^{'}(x) = -\lambda S_j(x) + \sum_{n=1}^{j} F_{nj}(0)h(x) + \sum_{k=0}^{a-1} D_k\lambda g_{j-k}(0),\ j \geq a \tag{14}$$

$$0 = -\lambda D_0 + \sum_{n=1}^{j} F_{n0}(0) \tag{15}$$

$$0 = -\lambda D_j + \sum_{n=1}^{j} F_{nj}(0) + \sum_{k=1}^{j} D_{j-k}\lambda g_k,\ 1 \leq j \leq a-1 \tag{16}$$

In order to find the system size distribution, we define the following PGF:

$$\left.\begin{array}{l}
\tilde{T}_n(z,\theta) = \sum_{j=0}^{\infty} \tilde{T}_{nj}(\theta)z^j \text{ and } T_n(z,0) = \sum_{j=0}^{\infty} T_{nj}(0)z^j, n = 1,2,\ldots,k \\
\tilde{G}_i(z,\theta) = \sum_{j=0}^{\infty} \tilde{G}_{ij}(\theta)z^j \text{ and } G_i(z,0) = \sum_{j=0}^{\infty} G_{ij}(0)z^j,\ a \leq i \leq b \\
\tilde{F}_k(z,\theta) = \sum_{i=0}^{\infty} \tilde{F}_{ik}(\theta)z^j \text{ and } F_k(z,0) = \sum_{i=0}^{\infty} F_{ik}(0)z^j, k = 1,2,\ldots,j \\
\tilde{S}(z,\theta) = \sum_{j=a}^{\infty} \tilde{S}_j(\theta)z^j \text{ and } S(z,0) = \sum_{j=a}^{\infty} S_j(0)z^j,\ D(z) = \sum_{j=0}^{a-1} D_j(0)z^j
\end{array}\right\} \tag{17}$$

5 PGF of the Queue Size

Now, the PGF of the queue size at an arbitrary time epoch is obtained as,

$$P(z) = \sum_{i=1}^{b-1} \tilde{G}_i(z,\theta) + \tilde{G}_b(z,\theta) + \sum_{n=1}^{k} \tilde{T}_n(z,\theta) + \sum_{k=1}^{j} \tilde{F}_k(z,\theta) + \tilde{S}(z,\theta) \tag{18}$$

$$
P(z) = \frac{\begin{array}{l}\left\{\left(\tilde{M}(\lambda - \lambda Y(z)) - 1\right) + (1-\pi)\tilde{M}(\lambda - \lambda Y(z))(\tilde{L}(\lambda - \lambda Y(z)) - 1)\right\} \sum_{j=a}^{b-1} (r_i + s_i)(z^b - z^j) \\ + \left\{z^b(1-\pi)\tilde{L}(\lambda - \lambda Y(z))\left(\tilde{L}(\lambda - \lambda Y(z)) - 1\right) + \tilde{L}(\lambda - \lambda Y(z))\left(\tilde{M}(\lambda - \lambda Y(z)) - 1\right)\right\} \\ \sum_{n=2}^{k} T_{n-1j}(0)z^j \\ + F(N,M,L)\sum_{i=0}^{a-1}\sum_{n=2}^{j} F_{in-1}(0)z^i + FD(H,M,L)\sum_{i=a}^{\infty}(f_i + \sum_{k=0}^{a-1} D_k \lambda g_{i-k}(0)) + R(N,M,L)\sum_{i=0}^{a-1} r_i\end{array}}{(-\lambda + \lambda Y(z))(z^b - (1-\pi)\tilde{M}(\lambda - \lambda Y(z))(1 + \tilde{L}(\lambda - \lambda Y(z)))} \tag{19}
$$

where

$$
\begin{aligned}
F(N,M,L) = {} & (\tilde{N}(\lambda - \lambda Y(Z)) - 1)[z^b - (1-\pi)\tilde{M}(\lambda - \lambda Y(Z)) \\
& -(1-\pi)\tilde{M}(\lambda - \lambda Y(Z))\tilde{N}(\lambda - \lambda Y(Z))] \\
& +(\tilde{M}(\lambda - \lambda Y(Z)) - 1)\tilde{L}(\lambda - \lambda Y(Z)) \\
FD(H,M,L) = {} & (\tilde{H}(\lambda - \lambda Y(Z)) - 1)[z^b - (1-\pi)\tilde{M}(\lambda - \lambda Y(Z)) \\
& -(1-\pi)\tilde{M}(\lambda - \lambda Y(Z))\tilde{L}(\lambda - \lambda Y(Z))] \\
& +(1-\pi)(\tilde{L}(\lambda - \lambda Y(Z)) - 1)\tilde{H}(\lambda - \lambda Y(Z)) \\
& \tilde{M}(\lambda - \lambda Y(Z))(1 + \tilde{L}(\lambda - \lambda Y(Z))) \\
& +(\tilde{M}(\lambda - \lambda Y(Z)) - 1)\tilde{H}(\lambda - \lambda Y(Z)) \\
R(N,M,L) = {} & (\tilde{N}(\lambda - \lambda Y(Z)) - 1)[z^b - (1-\pi)\tilde{M}(\lambda - \lambda Y(Z)) \\
& - (1-\pi)\tilde{M}(\lambda - \lambda Y(Z))\tilde{L}(\lambda - \lambda Y(Z))] \\
& - (1-\pi)(\tilde{L}(\lambda - \lambda Y(Z)) - 1) \\
& \tilde{M}(\lambda - \lambda Y(Z))(1 + \tilde{L}(\lambda - \lambda Y(Z)) \\
& - (\tilde{M}(\lambda - \lambda Y(Z)) - 1)
\end{aligned}
$$

6 Performance Measures

6.1 Expected Queue Length

The expected queue length (EQL), E(Q) is computed by differentiating the equation (15) with respect to z at z = 1.

$$
(2b\lambda E(Y) - 2M1^2)
$$

$$
\left\{
\begin{array}{l}
(3M2^2 + 3M2) \sum\limits_{j=c}^{b-1} (r_j + f_j + s_j + \lambda \sum\limits_{k=0}^{a-1} D_k g_{j-k})(b-j) \\
+3M1 \sum\limits_{j=c}^{b-1} (r_j + f_j + s_j + \lambda \sum\limits_{k=0}^{a-1} D_k g_{j-k}[b(b-1) - j(j-1)] \\
+[(3bL2^2 + 3bL2 + 3b(b-1)L1) \sum\limits_{j=a}^{c-1} (r_j + f_j + s_j \\
+\lambda \sum\limits_{k=0}^{a-1} D_k g_{j-k}) + 6bL1 \sum\limits_{j=a}^{c-1} (r_j + f_j + s_j + \lambda \sum\limits_{k=0}^{a-1} D_k g_{j-k}) j] \\
+[3bN2^2 + 3bN2 + 3b(b-1)N1]r_0 + [3bH2^2 \\
+3bH2 + 3b(b-1)H1] \sum\limits_{j=1}^{a-1} (r_j + f_j) + 6bH1 \sum\limits_{j=1}^{a-1} (r_j + f_j) j
\end{array}
\right\}
$$

$$
-2[M1 \sum_{j=c}^{b-1} (r_j + f_j + s_j + \lambda \sum_{k=0}^{a-1} D_k g_{j-k})(b-j)
$$

$$
+L1b \sum_{j=a}^{c-1} (r_j + f_j + s_j + \lambda \sum_{k=0}^{a-1} D_k g_{j-k})
$$

$$
E(Q) =
$$

$$
\frac{+bN1r_0 + bH1 \sum\limits_{j=1}^{a-1} (r_j + f_j)]\{3[b\lambda E(Y^2) + b(b-1)\lambda E(Y) - 2M1'' - M2'']\}}{4[b\lambda E(Y) - M1^2]^2}
$$

where

$\lambda E(M)E(Y) = M1$, $\lambda E(L)E(Y) = L1$, $\lambda E(N)E(Y) = N1$, $\lambda E(H)E(Y) = H1$, $\lambda^2 E(M^2)[E(Y)]^2 = M2^2$, $\lambda^2 E(L^2)[E(Y)]^2 = L2^2$, $\lambda^2 E(N^2)[E(Y)]^2 = N2^2$, $\lambda^2 E(H^2)[E(Y)]^2 = H2^2$, $\lambda E(M)E(Y^2) = M1'$, $\lambda E(L)E(Y^2) = L1'$, $\lambda E(N)E(Y^2) = N1'$, $\lambda E(H)E(Y^2) = H1'$, $\lambda^2 E(M)[E(Y)]^2 = M1^2$, $\lambda^3 E(M^2)[E(Y)]^3 = M2''$, $\lambda^2 E(M)E(Y)E(Y^2) = M1''$

6.2 Expected Busy Period

$$
E(B) = \frac{E(T)}{P(J=0)} = \frac{E(T)}{\sum\limits_{i=0}^{a-1} [(1-\pi)r_i + t_i]}
$$

6.3 Expected Length of Idle Period

$$E(I) = \frac{1}{\lambda} \sum_{j=0}^{a-1} D_j + E(N) \sum_{j=1}^{m} j[1 - \sum_{i=0}^{a-1} \sum_{n=0}^{i} (r_n + t_n)\alpha_{i-n}] + E(S)$$

7 Numerical Example

A numerical model is analyzed with the following assumptions: (1) Batch service time distribution is k-Erlang with k = 2. (2) Repair time, vacation time, and setup time follow exponential distribution. (3) Batch arrival size distribution is geometric with mean 2, π = 0.2. The expected queue size E(Q), expected length of idle period E(I), and expected length of busy period E(B) are computed. Repair rate η = 8, batch service rate μ , vacation rate α = 10, setup time rate γ = 5, threshold value 'a' = 2, maximum service capacity 'b' = 4 (Tables 1 and 2).

From the table and figures, the following observations are made: (1) As arrival rate increases, the mean queue size increases, expected length of idle period decreases and that of busy period increases. (2) Mean queue size decreases, when service rate increases for a particular arrival rate (Fig. 2).

8 Conclusion

The batch arrival queueing systems with two types of service pattern and with two types of vacation have been developed and studied in this paper. In the proposed model, the server will be repaired for all types of issue. We have derived the system

Table 1 Arrival rate versus various performance measures for μ=4

Arrival rate(λ)	E(Q)	E(B)	E(I)
3.1	3.0577	3.5638	0.2792
3.2	4.3421	4.4455	0.1805
3.3	6.2512	4.5399	0.1758
3.4	7.2311	5.6441	0.1341
3.5	9.0750	5.9481	0.1163

Table 2 Arrival rate versus E(Q) with various service rates

Arrival rate(λ)	E(Q1)	E(Q2)	E(Q3)
3.1	3.0577	2.8112	2.3204
3.2	4.3421	3.4034	2.8735
3.3	6.2512	5.3241	4.2423
3.4	7.2311	6.1132	5.5211
3.5	9.0750	8.2179	7.5241

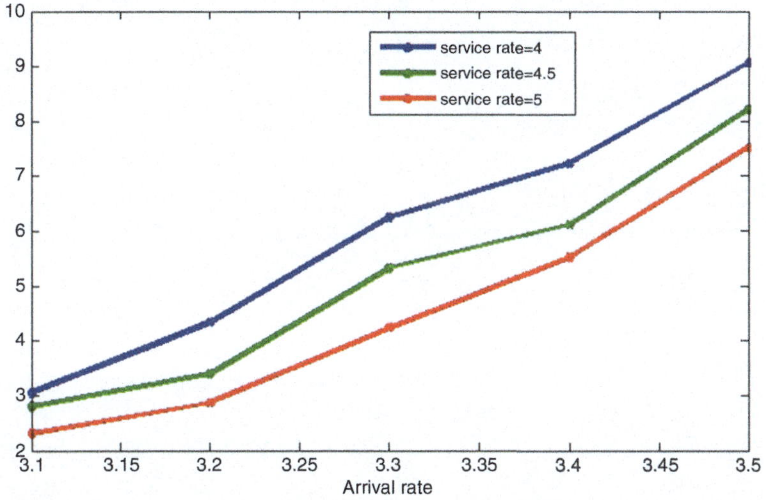

Fig. 2 Arrival rate versus various performance measures for μ=4

size distribution and the performance measures of the proposed queueing model by using PGF technique. Numerical example of the above model is also discussed.

References

1. Erlang AK (1909) The theory of probabilities and telephone conservations. Nyt Tidsskrift for Matematik 20: 33–39
2. Lee DH and Kim BK (2015) A note on the sojourn time distribution of an M/G/1 queue with a single working vacation and vacation interruption . Operations Research Perspectives 2:57–61
3. Thangaraj M and Rajendran P (2017) Analysis of Batch Arrival Queueing System with Two Types of Service and Two Types of Vacation. International Journal of Pure and Applied Mathematics 117: 263–272
4. Krishna Reddy GV, Arumuganathan R and Nadarajan R (1998) Analysis of bulk queue with N policy multiple vacations and setup times. Computers and Operations 25:957–967
5. Ke JC, Huang H and Chu Y (2010) Batch arrival queue with N-policy and at most J vacations. Applied Mathematical Modelling 34: 451–466
6. Ke JC (2003) The optimal control of an M/G/1 queueing system with server startup and two vacation types. Appl Math Model 27: 437–450
7. Balasubramanian M and Arumuganathan R (2011) Steady state analysis of a bulk arrival general bulk service queueing system with modified M-vacation policy and variant arrival rate. IntJ Oper Res 11: 383–407
8. Neuts MF (1967) A general class of bulk queues with Poisson input . Ann Math Stat 38: 759–770
9. Singh CJ, Jain M and Kaur S (2017) Performance analysis of bulk arrival queue with balking, optional service, delayed repair and multi-phase repair. Ain Shams Engineering Journal (2017)

An Improvement to One's BCM for the Balanced and Unbalanced Transshipment Problems by Using Fuzzy Numbers

Kirtiwant P. Ghadle, Priyanka A. Pathade, and Ahmed A. Hamoud

Abstract In this paper, we consider the pentagonal fuzzy number to solve the fuzzy transshipment problem. A new method namely, Ghadle and Pathade one's best candidate method (BCM), is proposed. BCM is for finding optimal solution to a transshipment problem. Proposed method in this paper gives the remarkable solutions on balanced and unbalanced fuzzy transshipment problem. The method has been illustrated with the help of an example.

Keywords Fuzzy transportation problem · Fuzzy transshipment problem · One's BCM · Fuzzy numbers · Optimal solution

Mathematics Subject Classification 03E72

1 Introduction

Transportation problem is nothing but a plan for transporting a commodity from a number of sources to a number of destinations. It is a determination of a minimum cost and maximum profit. In a transportation problem, shipments play an important role, because shipment of commodity takes place among source and destination, but instead of direct shipments to destinations, the commodity can be transported to a particular destination through one or more intermediate points. This is intermediate point which we call as transshipment point. A transshipment point is a point that can both receive goods from any other points and send goods to any other points. Transshipment problem is an extension of transportation problem with

K. P. Ghadle (✉) · P. A. Pathade · A. A. Hamoud
Department of Mathematics, Dr. Babasaheb Ambedkar Marathwada University, Aurangabad, Maharashtra, India
e-mail: drkp.ghadle@gmail.com; ghadle.maths@bamu.ac.in; priyankapathade88@gmail.com; drahmed985@yahoo.com

V. Madhu et al. (eds.), *Advances in Algebra and Analysis*, Trends in Mathematics,
https://doi.org/10.1007/978-3-030-01120-8_31

additional features. Transshipment problem is a sequence of points rather than direct connections from one origin to one of the destinations. Large amount of material can be assumed that shipped is available at each point like stockpile, which can be replenished. It is a function in any direction. Most of the time, transportation cost of transshipment problem is minimized. It is widely used in planning bulk distribution. Transportation problem is well known and has been studied long for minimizing the total cost. Orden [7] has extended the problem by including the case when transshipment is also allowed. The objective of the transshipment problem is to minimize the duration of transportation studied by Garg and Prakash [3]. They found the optimal routes of transportation problem from original point to destination. To deals with the problems with imprecision information, Zadeh [9] introduced the fuzzy set theory. Kaufmann et al. [5] introduced fuzzy numbers and investigate with arithmetic operations. Baskaran et al. [1] formulate the fuzziness in the goal programming formulation. They used unbalanced transshipment problem with budgetary constraints in which the demand and budget are specified imprecisely. Mohanpriya and Jayanthi [6] determined the efficient solutions for the large-scale fuzzy transshipment problem. They solved transshipment problem by VAM to find the efficient initial solution for the large-scale transshipment problem. Baskaran et al. [2] considered transit points, but these points have no demands. They convert transshipment problem as transportation problem. Problem solved by fuzzy cost deviation algorithm. Gani et al. [4] studied mixed constraint fuzzy transshipment problem. Rajarajendran and Pandian [8] proposed a newly splitting method to find optimal solution. This splitting method is extended to fully fuzzy transshipment problems.

2 Preliminary

In this section, we collect some basic definitions that will be important to us in the sequel.

Definition 1 A fuzzy set is characterized by a membership function mapping element of a domain, space, or the universe of discourse X to the unit interval $[0, 1]$, i.e, $A = \{(\mu_A(x); x \in X)\}$. Here $\mu_A(x) : X \longrightarrow [0, 1]$ is a mapping called the degree of membership value of $x \in X$ in the fuzzy set A. These membership grades are often represented by real numbers ranking from $[0, 1]$.

Definition 2 A fuzzy number f in the real line R is a fuzzy set $f : R \longrightarrow [0, 1]$ that satisfies the following properties (Fig. 1).

- f is piecewise continuous.
- There exists an $x \in R$ such that $f(x) = 1$.
- f is convex, i.e., if x_1, x_2 and $a \in [0, 1]$, then $f(\lambda x_1)+(1-\lambda x_2) \geq f(x_1) \wedge f(x_2)$

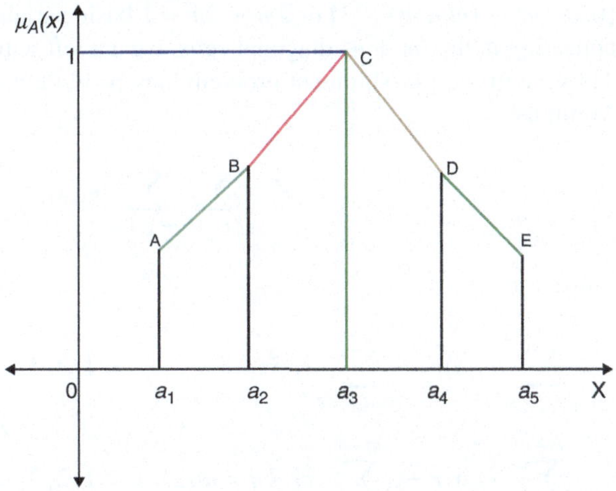

Fig. 1 Pentagonal fuzzy number

Definition 3 A fuzzy number A is defined to be a triangular fuzzy number if its membership function $\mu_A : R \longrightarrow [0, 1]$ is equal to

$$\mu_A(x) = \begin{cases} \frac{(x-a_1)}{(a_2-a_1)}, & \text{if} \quad x \in [a_1, a_2] \\ \frac{(a_3-x)}{(a_2-a_2)}, & \text{if} \quad x \in [a_2, a_3] \\ 0, & \quad \text{Otherwise.} \end{cases} \tag{1}$$

Definition 4 A fuzzy number $A = (a_1, a_2, a_3, a_4, a_5)$ is called a pentagonal fuzzy number when the membership function has the form,

$$\mu_A(x) = \begin{cases} 0, & x \leq a_1 \\ \frac{(x-a_1)}{(a_2-a_4)}, & a_1 \leq x \leq a_2 \\ 1, & a_2 \leq x \leq a_4 \\ \frac{a_4-x}{a_5-a_4}, & a_4 \leq x \leq a_5 \\ 0, & x > a_5. \end{cases} \tag{2}$$

3 Formulation of the General Fuzzy Transshipment Problem

The FTP assumes that direct routes exist from each source to each destination. There are situations in which units may be shipped from one source to another or to other destinations before reaching their final destinations. This situation is called a fuzzy transshipment problem. It dropped so that a transportation problem with m source and n destinations gives rise to a transshipment problem with $m + n$ source and $m + n$ destinations. The basic feasible solution(BFS) to such a problem will involve

$[(m+n)+(m+n)-1]$ or $2m+2n-1$ basic variables. If we omit the variables appearing in the $(m+n)$ diagonal cells, we are left with $m+n-1$ basic variables. Thus the fuzzy transshipment problem may be written as:
Minimize

$$\tilde{Z} = \sum_{i=1}^{m+n} \sum_{j=1, j\neq i}^{m+n} \tilde{c}_{ij}\tilde{x}_{ij}. \tag{3}$$

Subject to

$$\sum_{j=1, j\neq i}^{m+n} \tilde{c}_{ij}\tilde{x}_{ij} - \sum_{j=1, j\neq i}^{m+n} \tilde{c}_{ij}\tilde{x}_{ji} = a_i, \quad i = 1, 2, 3, \ldots m,$$

$$\sum_{j=1, j\neq i}^{m+n} \tilde{c}_{ij}\tilde{x}_{ij} - \sum_{j=1, j\neq i}^{m+n} \tilde{c}_{ij}\tilde{x}_{ji} = a_i, \quad i = 1, 2, 3, \ldots m,$$

$$\sum_{j=1, j\neq i}^{m+n} \tilde{c}_{ij}\tilde{x}_{ij} - \sum_{j=1, j\neq i}^{m+n} \tilde{c}_{ij}\tilde{x}_{ji} = \tilde{b_j}, \quad j = m+1, m+2, m+3, \ldots, m+n,$$

where

$$\tilde{x_i j} \geq 0, \quad i, j = 1, 2, 3, \ldots, m+n, \quad j \neq i$$

and

$$\sum_{i=1}^{m} \tilde{a_i} = \sum_{i=1}^{m} \tilde{a_j},$$

then the problem is balance otherwise unbalanced.
The above formulation shows fuzzy transshipment model; the transshipment model is reduced to transportation problem as:
Minimize

$$\tilde{Z} = \sum_{i=1}^{m+n} \sum_{j=1, j\neq i}^{m+n} \tilde{c}_{ij}\tilde{x}_{ij}. \tag{4}$$

Subject to:

$$\sum_{j=1}^{m+n} \tilde{x_{ij}} = \tilde{a_i} + T, \quad i = 1, 2, 3, \ldots, m,$$

$$\sum_{j=1}^{m+n} \tilde{x_{ij}} = T, \quad i = m+1, m+2, m+3, \ldots, m+n,$$

$$\sum_{j=1}^{m+n} \tilde{x_{ij}} = T, \quad j = 1, 2, 3 \ldots, m,$$

$$\sum_{j=1}^{m+n} \tilde{x_{ij}} = \tilde{b_i} + T, \quad j = m+1, m+2, m+3, \ldots, m+n,$$

where

$$\tilde{x_{ij}} \geq 0, \quad i, j = 1, 2, 3, \ldots, m+n, \quad j \neq i,$$

the above mathematical model represents a standard balanced transportation problem with $(m+n)$ origins and $(m+n)$ destinations. T is a buffer stock at each origin and each destination. Since we assume that any amount of goods can be transshipped at each point, T should be large enough to take care of all transshipments. It is clear that the volume of good transshipped at any point cannot exceed the amount produced or received; therefore we take

$$T = \sum_{i=1}^{m} \tilde{a_i}.$$

4 Ghadle and Pathade One's Best Candidate Method

Step 1: Matrix must be balanced.

Step 2: The one's best candidate method are selected by choosing minimum cost for minimization problem and maximum cost for maximization problems.

Step 3: Must be assign as much as possible to the cell with the smallest unit cost (or highest)in the whole tableau. If tie occurs, then choose arbitrarily.

Step 4: Check if each row and column has at least one best candidate. Assign one to all diagonally zero's.

Step 5: If smallest one occurs in entire tableau, then allocate first northwest corner and allocate it. After that allocate the demand and the supply as much as possible to the variable with the least unit cost in the selected row or column.

Step 6: Adjust supply and demand by crossing out the row/column to be then assigned to zero. If the row or column is not assigned to zero, then check the selected row if it has an element in the chosen combination, then we elect it.

Step 7: Elect the next least cost from the chosen combination and repeat step 5 until all column and rows are exhausted.

Numerical Example 1: Consider balanced fuzzy transshipment problem

	s_1	s_2
s_1	(0,0,0,0,0)	(1,2,3,4,5)
s_2	(0.5,1,1,2,3)	(0,0,0,0,0)

	d_1	d_2
s_1	(1,4,1,3,2)	(2,5,1,3,4)
s_2	(0.3,2.9,6,3.5,2)	(2,1,5,3,1)

	d_1	d_2
d_1	(0,0,0,0,0)	(0.5,1,1,1,2)
d_2	(0.7,2,1,2,1)	(0,0,0,0,0)

	s_1	s_2
d_1	(1,4,1,3,2)	(0.3,2.9,6,3.5,2)
d_2	(2,5,1,3,4)	(2,1,5,3,1)

We convert transshipment problem as transportation problem,

	s_1	s_2	d_1	d_2	Supply
s_1	(0,0,0,0,0)	(1,2,3,4,5)	(1,4,1,3,2)	(2,5,1,3,4)	(19,19, 15,12.5,11)
s_2	(0.5,1,1,2,3)	(0,0,0,0,0)	(0.3,2.9,6, 3.5,2)	(2,1,5,3,1)	(15,29,17, 18,17)
d_1	(1,4,1,3,2)	(0.3,2.9,6, 3.5,2)	(0,0,0,0,0)	(0.5,1,1,1,2)	(17,16,12, 19,14)
d_2	(2,5,1,3,4)	(2,1,5,3,1)	(0.7,2,1,2,1)	(0,0,0,0,0)	(15.5,15,14, 17,12.5)
Demand	(18,20,15, 17,10)	(14,24,16, 18,12)	(19,18,11, 20,13)	(15.5,17,16, 11.5,19.5)	

By using ranking method, we get crisp values as in the next table,

	s_1	s_2	d_1	d_2	Supply
s_1	0	3	2.2	3	15.3
s_2	1.5	0	2.94	2.4	19.2
d_1	2.2	2.94	0	1.1	15.6
d_2	3	2.4	1.34	0	14.8
Demand	16	16.8	16.2	15.9	

Now we illustrate the problem by using Ghadle and Pathade one's best candidate method,

	s_1	s_2	d_1	d_2	Supply
s_1	1	3	2.2	3	15.3
s_2	1.5	1	2.94	2.4	19.2
d_1	2.2	2.94	1	1.1	15.6
d_2	3	2.4	1.34	S1	14.8
Demand	16	16.8	16.2	15.9	

	s_1	s_2	d_1	d_2	Supply
s_1	$^{15.3}1$	3	2.2	3	15.3
s_2	1.5	1	$^{16.2}2.94$	$^{3}2.4$	19.2
d_1	$^{0.7}2.2$	$^{14.9}2.94$	1	1.1	15.6
d_2	3	$^{1.9}2.4$	1.34	$^{12.9}1$	14.8
Demand	16	16.8	16.2	15.9	

The optimal solution is given below:

$$= (15.3)(0) + (16.2)(2.94) + (3)(2.4) + (2.2)(0.7) + (14.9)(2.94)$$
$$+ (1.9)(2.4) + (12.9)(0)$$
$$= 104.72.$$

Numerical Example 2: Consider unbalanced fuzzy transshipment problem

	s_1	s_2
s_1	(0,0,0,0,0)	(2,3,5,6,4)
s_2	(1.1,2,2.5,4,3)	(0,0,0,0,0)

	d_1	d_2
s_1	(1,2,3,4,5)	(2,1,3,4,6)
s_2	(0.5,3,4,6,2)	(2,5,7,5,4)

	d_1	d_2
d_1	(0,0,0,0,0)	(1,4,5,6,7)
d_2	(2.5,3,3.5,4,4.5)	(0,0,0,0,0)

	s_1	s_2
d_1	(1,2,3,4,5)	(2.5,3,3.5,4,4.5)
d_2	(2,1,3,4,6)	(2,5,7,5,4)

We convert transshipment problem as transportation problem given below,

	s_1	s_2	d_1	d_2	Supply
s_1	(0,0,0,0,0)	(2,3,5,6,4)	(1,2,3,4,5)	(2,1,3,4,6)	(5,10,13,14,18)
s_2	(1.1,2,2.5,4,3)	(0,0,0,0,0)	(0.5,3,4,6,2)	(2,5,7,5,4)	(1,2,3,4,5)
d_1	(1,2,3,4,5)	(2.5,3,3.5,4,4.5)	(0,0,0,0,0)	(1,4,5,6,7)	(5,8,8,7,2)
d_2	(2,1,3,4,6)	(2,5,7,5,4)	(2.5,3,3.5,4,4.5)	(0,0,0,0,0)	(3,6,9,12,15)
Demand	(2,8,10,14,12)	(7,6,5,4,9)	(3,7,6,7,3)	(1.5,3,6,12,8)	

	s_1	s_2	d_1	d_2	Dummy	Supply
s_1	(0,0,0,0,0)	(2,3,5,6,4)	(1,2,3,4,5)	(2,1,3,4,6)	(0,0,0,0,0)	(5,10,13, 14,18)
s_2	(1.1,2, 2.5,4,3)	(0,0,0,0,0)	(0.5,3,4,6,2)	(2,5,7,5,4)	(0,0,0,0,0)	(1,2,3,4,5)
d_1	(1,2,3,4,5)	(2.5,3,3.5, 4,4.5)	(0,0,0,0,0)	(1,4,5,6,7)	(0,0,0,0,0)	(5,8,8,7,2)
d_2	(2,1,3,4,6)	(2,5,7,5,4)	(2.5,3,3.5, 4,4.5)	(0,0,0,0,0)	(0,0,0,0,0)	(3,6,9,12,15)
Demand	(2,8,10, 14,12)	(7,6,5,4,9)	(3,7,6,7,3)	(1.5,3,6, 12,8)	(0.5,2,6,0,8)	

	s_1	s_2	d_1	d_2	Dummy	Supply
s_1	0	4	3	3.2	0	12
s_2	12.6	0	15.5	19.8	0	3
d_1	3	3.5	0	4.6	0	6
d_2	3.2	4.6	3.5	0	0	9
Demand	9.2	6.2	5.2	6.1	3.3	

By using ranking method, we get crisp values which we see in above table,

	s_1	s_2	d_1	d_2	Dummy	Supply
s_1	$^{9.2}1$	4	$^{2.8}3$	3.2	0	12
s_2	12.6	$^{0.5}1$	15.5	$^{2.5}19.8$	0	3
d_1	3	3.5	$^{2.4}1$	$^{3.6}4.6$	0	6
d_2	3.2	$^{5.7}4.6$	3.5	$^{2.5}1$	$^{3.3}0$	9
Demand	9.2	6.2	5.2	6.1	3.3	

We used Ghadle and Pathade one's best candidate method, and the optimal solution is given below:

$$
\begin{aligned}
&= (0)(9.2) + (3)(2.8) + (0)(0.5) + (19.8)(2.5) + (0)(2.4) \\
&\quad + (3.6)(4.6) + (5.7)(4.6) + (0)(2.5) + (0)(3.3) \\
&= 106.08.
\end{aligned}
$$

5 Conclusion

In this paper, the transshipment with balanced and unbalanced pentagonal fuzzy numbers is taken as a problem. We have solved the problem by proposing one's best candidate method. Thus this method provides an applicable optimal solution which helps in handling real life transportation problem. The proposed method consumes less time as well as very easy to understand which is mathematically proved.

References

1. Baskaran, R., Dharmalingam, K.: Multi-objective fuzzy transshipment problem. Intern. J. Fuzzy Math. Archive. **10**, 161–167 (2016).
2. Baskaran, R., Dharmalingam, K., Mohamed S.: Fuzzy transshipment problem with transit points. Intern. J. Pure Appl. Math. (2016) https://doi.org/10.12732/ijpam.v107i4.22
3. Garg, R., Prakash, S.: Time minimizing transshipment problem. Indian J. Pure Appl. Math. **16**, 449–460 (1985).
4. Gani, A., Baskaran, R., Mohamed, S.: Mixed constraint fuzzy transshipment problem. Appl. Math. Sci. **6**, 2385–2394 (2012).
5. Kaufmann, A.: Introduction to the Theory of Fuzzy Sets. Academic Press, New York (1976).
6. Mohanpriya, S., Jeyanthi, V.: Modified procedure to solve fuzzy transshipment problem by using trapezoidal fuzzy number. Int. J. Math. and Stat. Inv. **4**, 30–34 (2016).
7. Orden, A.: The transshipment problem. Management Sci. **2**, 83–97 (1956).
8. Rajendran, P., Pandian, P.: Solving fully interval transshipment problems. Inter. Math. Forum. **7**, 2027–2035 (2012).
9. Zadeh, L.: Fuzzy sets. Inform. Contr. **8**, 338–353 (1965).

An Articulation Point-Based Approximation Algorithm for Minimum Vertex Cover Problem

Jayanth Kumar Thenepalle and Purusotham Singamsetty

Abstract The minimum vertex cover problem (MVCP) is a well-known NP complete combinatorial optimization problem. The aim of this paper is to present an approximation algorithm for minimum vertex cover problem (MVCP). The algorithm construction is based on articulation points/cut vertices and leaf vertices/pendant vertices. The proposed algorithm assures the near optimal or optimal solution for a given graph and can be solved in polynomial time. A numerical example is illustrated to describe the proposed algorithm. Comparative results show that the proposed algorithm is very competitive compared with other existing algorithms.

1 Introduction

Let $G = (V, E)$ be a simple, undirected, connected and unweighted graph, where V and E represent the vertex and edge set of G, respectively, such that $E = \{e = (u, v)/u, v \in V\}$. A vertex cover S of G is a subset of vertices, if and only if $\forall e = (u, v), u \in S \quad or \quad v \in S \quad or \quad u, v \in S$. The number of vertices in S is called cardinality of the vertex cover. The problem of finding least cardinality of S is called minimum vertex cover problem. Note that a minimum vertex cover is need not be unique. In a graph G, for any $u \in V$, then $N(u)$ denote the set of neighbours of u, and thus the *degree* of a vertex is equivalent to the cardinality of $N(u)$. A graph G is called connected, if every pair of vertices in G is connected. A maximal connected subgraph of the graph G is said to be *component*. A vertex in the graph G is said to be an *articulation point/cut vertex*, if and only if discarding the vertex makes the graph disconnected. In other words, let $W(G)$ denote the number of components of G. If $W(G \backslash v) > W(G)$, then the vertex v is called an articulation

J. K. Thenepalle · P. Singamsetty (✉)
Department of Mathematics, VIT, Vellore, TamilNadu, India
e-mail: jayanth.maths@gmail.com; thenepalle.jayanth@vit.ac.in; drpurusotham.or@gmail.com; purusotham.s@vit.ac.in

V. Madhu et al. (eds.), *Advances in Algebra and Analysis*, Trends in Mathematics,
https://doi.org/10.1007/978-3-030-01120-8_32

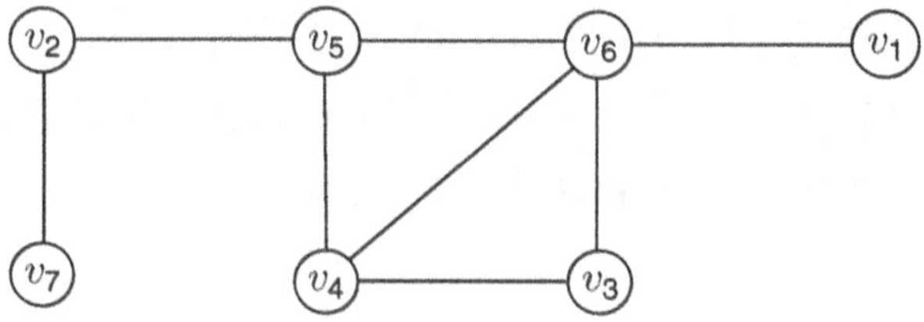

Fig. 1 An arbitrary graph G with 7 vertices

point. If a connected graph G does not contain any articulation point, then the graph is said to be *biconnected*. Any vertex in the graph with degree one is called *leaf vertex or pendant vertex* (Fig. 1). Note that G represents the simple graph.

Graph theory plays significant role in computer science in the context of data mining, clustering, networking, image processing, etc. The vertex cover problem (VCP) is one of the NP-complete conventional graph optimization problems [1]. Apart from the applications in graph theory, the VCP has numerous practical applications such as VLSI design [2], crew scheduling [3] and industrial machine assignments [4]. It is observed that there are some problems that do not have optimal solutions. However, we can have approximation algorithms that assure a solution, which will be near to the optimal solution. The calculation of minimum vertex cover can be interpreted into the computation of prime implicants of a Boolean function [5, 6]. Thus, one can find all the minimum vertex covers of a graph by means of Boolean operation [5]. As VCP is NP-complete, most of the existing works on VCP concerns approximation algorithms. Several solution techniques including direct, intelligent and parameterized algorithms have been developed for VCP and its allied problems.

Some of the direct algorithms include edge deletion (ED) approach [1], the ListLeft (LL) algorithm [7], the ListRight (LR) approach [8] and depth-first search (DFS) approach [9]. The intelligent algorithms such as a hybrid Genetic algorithm (GA) with Greedy approach [10], a revised Reactive Tabu Search (TS) algorithm with Simulated Annealing (SA) [11] were observed to find a minimum vertex cover on weighted graphs. Some of the parameterized algorithms includes an improved polynomial space algorithm [12] with a running time $O(1.286^k + kn)$ developed for VCP. A fixed-parameter approach [13] has been proposed for VCP_3 problem with time complexity $O(2^k k^{3.376} + n^4 m)$. In addition, some of the recent works including a revised approximation algorithm [14] for VCP has been presented with a time complexity $O(2(V + E))$. The Dijkstra algorithm-based approximation algorithm [15] for MVCP has been proposed with time complexity $O(n^3)$. A rough set-based approximation algorithm proposed for solving MVCP proved that computing an optimal vertex cover of a given graph is the same as of determining the best reduction of a decision table [16].

2 Proposed Approach

An approximation algorithm [14] was developed to solve MVCP, where it is based on two phases. In first phase, the articulation points are traced for the given graph using DFS algorithm, and then an approximation algorithm [17] is used in the second phase, on leftover graph to get minimum vertex cover. The proposed algorithm is an improved version of [14], where it is capable to solve biconnected graphs too.

The proposed algorithm is mainly based on articulation point and pendant vertices. Several algorithms have been developed to find articulation points. The DFS algorithm is a simple approach to find articulation points with time complexity $O(V + E)$. In the proposed Algorithm 2.1, first it verifies whether the given graph has articulation points or not. If the graph has no articulation points, then find the maximum degree vertex. If the maximum degree vertex is unique, then add it to the vertex cover *Vc* and remove the edges that are covered by that vertex; otherwise, if all the vertices are incident to each other, then add any one of the vertex arbitrarily to vertex cover *Vc*. Else, add a pair of vertices to *Vc* that have no common edge between them. Delete all the edges that are incident to the vertex or vertices. If the given or leftover graph has articulation points, then check whether pendant vertices exist or not. If it has, then add the vertices that are adjacent to the pendant vertices to the vertex cover *Vc* and delete all the edges that are covered by the added vertices. If the given or leftover graph does not have pendant vertices, then find all the articulation points. If there exists a single articulation point, then add it to vertex cover *Vc* and delete all the edges that are covered by that vertex. If there exists more than one articulation point, then find their degree, select the maximum degree articulation point and add it to vertex cover. If multiple articulation points have the same maximum degree, then choose a pair of articulation points in which no common edge between them; otherwise, choose any one of the articulation point randomly. Add that articulation point(s) to the vertex cover *Vc* and remove all the edges connected to that point(s). Remove the isolated vertices from the remaining graph wherever exists. If the remaining graph consists of more than one component, follow the same process until all the components get exhausted, or the graph will be null graph. Finally, the vertex cover *Vc* gives the minimum vertex cover for a given graph. The systematic procedure of proposed algorithm for MVCP is given in Algorithm 2.1.

3 Illustrative Example

The proposed algorithm is demonstrated with the help of an example given in Fig. 2. Figure 2 describes a graph with eight vertices and nine edges. To test the effectiveness of proposed algorithm over the existing algorithm [14], we applied both the algorithms to the example given in Fig. 2. The example is solved systematically as the algorithms discussed above.

Algorithm 2.1 Proposed algorithm for MVCP

Notations: $G \leftarrow$ A simple connected, undirected and unweighted graph, $Vc \leftarrow$ vertex cover set and $Deg \leftarrow$ degree of a vertex
Input: A graph G can be read as adjacency matrix
Output: Minimum vertex cover, i.e. Vc

1. $Vc \leftarrow \phi$, $E^* \leftarrow E$, and $G^* \leftarrow G$.
2. If (G^* has no articulation points)
3. $Deg \leftarrow$ Compute the degree of all vertices of G^*.
4. $D \leftarrow$ Vertices with maximum degree from Deg.
5. If (D contains a unique vertex), then add it to vertex cover Vc.
6. Else
   $D \leftarrow$ A pair of vertices that have no common edge between them, if exists. Else, add any one of the vertex arbitrarily from D.
7. $Vc \leftarrow Vc \cup D$, $E^* \leftarrow E^*$-{Set of edges covered by D} and $G^* \leftarrow G^*$ - D (It is understood that G^* is revised after removal of vertices of D and the edges which are incident to them).
8. End if
9. End if
10. If (G^* has any pendant vertices)
11. $L \leftarrow$ Set of all pendant vertices in G^*
12. $L^* \leftarrow$ Set of vertices that are adjacent to pendant vertices in L.
13. $Vc \leftarrow Vc \cup L^*$, $E^* \leftarrow E^*$ - {Set of edges covered by L^* } and $G^* \leftarrow G^*$- L^*.
14. End if
15. If (G^* has articulation points)
16. $A \leftarrow$ Set of articulation points of G^*.
17. If (A contains a unique articulation point), then add it to vertex cover Vc.
18. Else
    $Deg \leftarrow$ Compute the degree of all articulation points. $A \leftarrow$ The articulation point with maximum degree.
19. If (A contains a unique articulation point), then add it to vertex cover Vc.
20. Else
    $A \leftarrow$ A pair of articulation points having no common edge between them, if exists. Else, $A \leftarrow$ Add any one of the articulation point arbitrarily.
21. $Vc \leftarrow Vc \cup A$, $E^* \leftarrow E^*$ - {Set of edges covered by A } and $G^* \leftarrow G^*$- A.
22. End if
23. End if
24. End if
25. Remove isolated vertices in the remaining graph, if exists.
26. If (the revised graph G^* contains multiple connected components)
    Move to Step 2 and repeat the same process until all the components will be exhausted, or the graph will be empty.
27. End if
28. If ($E^* = \phi$), then print Vc and Stop
29. Else, move to Step 2 and repeat the process
30. End if

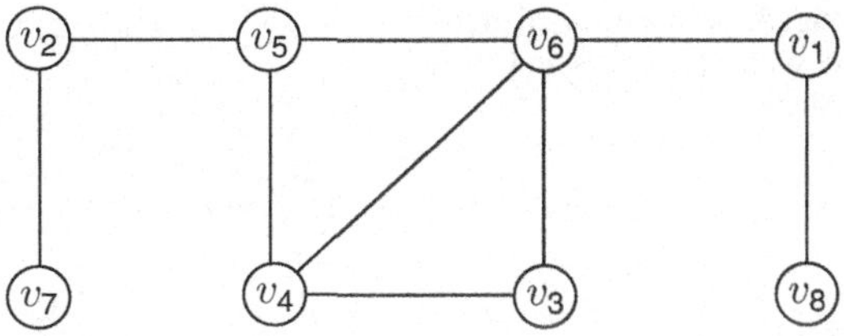

Fig. 2 A graph G with 8 vertices and 9 edges

Using existing algorithm [14]

1. $Vc \leftarrow \phi$ and $E^* \leftarrow \{(v_1, v_6),(v_1, v_8),(v_2, v_5),(v_2, v_7),(v_3, v_4),(v_3, v_6), (v_4, v_5),(v_4, v_6),(v_5, v_6)\}$.
2. $A \leftarrow$ Articulation Points, $\{v_1, v_2, v_5, v_6\}$. $Vc \leftarrow Vc \cup A$.
3. $Vc \leftarrow \{v_1, v_2, v_5, v_6\}$ and $E^* \leftarrow E^*$ - {Set of edges covered by A }.
4. $E^* \leftarrow \{(v_3, v_4)\}$
5. Take (v_3, v_4) as an arbitrary edge. Add either v_3 or v_4 to Vc.
6. $Vc \leftarrow \{v_1, v_2, v_3, v_5, v_6\}$.
7. $E^* \leftarrow \phi$, $Vc \leftarrow$ minimum vertex cover, $\{v_1, v_2, v_3, v_5, v_6\}$.

Finally, the vertex cover set Vc returns the minimum vertex cover, and it contains five elements.

Using proposed algorithm

1. $Vc \leftarrow \phi$, $E^* \leftarrow \{(v_1, v_6),(v_1, v_8),(v_2, v_5),(v_2, v_7),(v_3, v_4),(v_3, v_6),(v_4, v_5), (v_4, v_6),(v_5, v_6)\}$.
2. Given G^* has articulation points.
3. $L \leftarrow$ Pendant vertices in G^*, $\{v_7, v_8\}$.
4. $L^* \leftarrow \{v_1, v_2\}$, the vertices that are adjacent to pendant vertices in L.
5. $Vc \leftarrow \{v_1, v_2\}$, $E^* \leftarrow \{(v_3, v_4),(v_3, v_6),(v_4, v_5),(v_4, v_6),(v_5, v_6)\}$ and $G^* \leftarrow G^*$- L^*.
6. The leftover graph does not have pendant vertices and articulation points; thus it is biconnected.
7. $D \leftarrow$ Vertices v_4 and v_6 having same degree. Since, there is a common edge (v_4, v_6) between the vertices v_4 and v_6. Thus, add either vertex v_4 or vertex v_6 to the vertex cover set Vc.
8. $Vc \leftarrow \{v_1, v_2, v_6\}$, $E^* \leftarrow \{(v_3, v_4),(v_4, v_5)\}$ and $G^* \leftarrow G^*$- D.
9. The remaining graph has pendant vertices. $L \leftarrow \{v_3, v_5\}$, pendant vertices in G^*.
10. $L^* \leftarrow \{v_4\}$, the vertex adjacent to pendant vertex in L. $Vc \leftarrow \{v_1, v_2, v_4, v_6\}$, $E^* \leftarrow \phi$ and $G^* \leftarrow G^*$- L^*. Remove all isolated vertices in leftover graph, if exists.
11. $E^* \leftarrow \phi$, $Vc \leftarrow$ Minimum vertex cover set, $\{v_1, v_2, v_4, v_6\}$, Stop.

From the results, it is seen that existing algorithm found the feasible solution, but it is not optimum, while the proposed algorithm provides the least vertex cover. Moreover, it is noticed that the result produced by the existing approach includes all

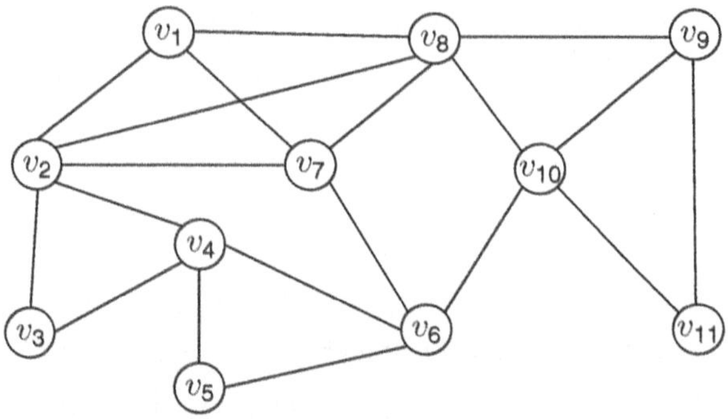

Fig. 3 A biconnected graph G with 11 vertices and 19 edges

the articulation points to the vertex cover. The solution found by proposed algorithm does not include all the articulation points in the minimum vertex cover.

In many cases, the given graph cannot have articulation points. The existing algorithm [14] does not focus on such graphs to find minimum vertex cover. The proposed algorithm is capable of finding minimum vertex cover including biconnected graphs. For ease of understanding, we considered an arbitrary simple biconnected graph with 11 vertices and 19 edges, given in Fig. 3. The example given in Fig. 3 is solved using proposed algorithm and is illustrated below.

1. $Vc \leftarrow \phi$, $E^* \leftarrow \{(v_1, v_2),(v_1, v_7),(v_1, v_8),(v_2, v_3),(v_2, v_4),(v_2, v_7),(v_2, v_8),(v_3, v_4),(v_4, v_5),(v_4, v_6),(v_5, v_6),(v_6, v_7),(v_6, v_{10}),(v_7, v_8),(v_8, v_9),(v_8, v_{10}),(v_9, v_{10}),(v_9, v_{11}),(v_{10}, v_{11})\}$.
2. Given G^* has no articulation point. $Deg \leftarrow$ Compute the degree of all vertices of G^*. Choose vertices v_2 and v_8 having same maximum degree 5 from Deg.
3. $D \leftarrow$ Vertices v_2 and v_8 having same degree. Since, there is a common edge (v_2, v_8) between the vertices v_2 and v_8. Thus, add either v_2 or v_8 to Vc. $Vc \leftarrow Vc \cup D$, $Vc \leftarrow \{v_2\}$, $E^* \leftarrow \{(v_1, v_7),(v_1, v_8),(v_3, v_4),(v_4, v_5),(v_4, v_6),(v_5, v_6),(v_6, v_7),(v_6, v_{10}),(v_7, v_8),(v_8, v_9),(v_8, v_{10}),(v_9, v_{10}),(v_9, v_{11}),(v_{10}, v_{11})\}$ and $G^* \leftarrow G^*\text{- } D$.
4. $L \leftarrow \{v_3\}$, pendant vertex of the leftover graph G^*. $L^* \leftarrow \{v_4\}$, the vertex adjacent to pendant vertex in L. $Vc \leftarrow Vc \cup L^*$.
5. $Vc \leftarrow \{v_2, v_4\}$, $E^* \leftarrow \{(v_1, v_7),(v_1, v_8),(v_5, v_6),(v_6, v_7),(v_6, v_{10}),(v_7, v_8),(v_8, v_9),(v_8, v_{10}),(v_9, v_{10}),(v_9, v_{11}),(v_{10}, v_{11})\}$ and $G^* \leftarrow G^*\text{- } L^*$.
6. $A \leftarrow$Set of articulation points for the graph G^*, $\{v_6\}$. $Vc \leftarrow Vc \cup A$.
7. $Vc \leftarrow \{v_2, v_4, v_6\}$, $E^* \leftarrow \{(v_1, v_7),(v_1, v_8),(v_7, v_8),(v_8, v_9),(v_8, v_{10}),(v_9, v_{10}),(v_9, v_{11}),(v_{10}, v_{11})\}$ and $G^* \leftarrow G^*\text{- } A$.
8. The remaining graph has articulation points, and it does not have pendant vertices.
9. $A \leftarrow$ Set of articulation points for the remaining graph, $\{v_8\}$. $Vc \leftarrow Vc \cup A$, $Vc \leftarrow \{v_2, v_4, v_6, v_8\}$, $E^* \leftarrow \{(v_1, v_7),(v_9, v_{10}),(v_9, v_{11}),(v_{10}, v_{11})\}$ and $G^* \leftarrow G^*\text{- } A$. Now the remaining graph is disconnected, and it contains two components.

10. The remaining graph has no articulation points. *Deg* ← Compute the degree of all vertices of G^*. Choose the vertices v_9,v_{10} and v_{11} having same maximum degree 2 from *Deg*.
11. D ← All the higher degree vertices have common edges with one another. Hence, pick either v_9, v_{10} or v_{11}. $Vc \leftarrow Vc \cup D$ and $G^* \leftarrow G^*\text{- } D$.
12. $Vc \leftarrow \{v_2, v_4, v_6, v_8, v_9\}$ and $E^* \leftarrow \{(v_1, v_7),(v_{10}, v_{11})\}$.
13. The remaining graph consists of two independent edges. Take any one vertex from each edge and add them to the vertex cover set Vc.
14. $Vc \leftarrow \{v_1, v_2, v_4, v_6, v_8, v_9, v_{11}\}$.
15. $E^* \leftarrow \phi, Vc \leftarrow$ Minimum vertex cover set,$\{v_1, v_2, v_4, v_6, v_8, v_9, v_{11}\}$. Stop.

Finally, the proposed algorithm provided the vertex cover with least cardinality 7.

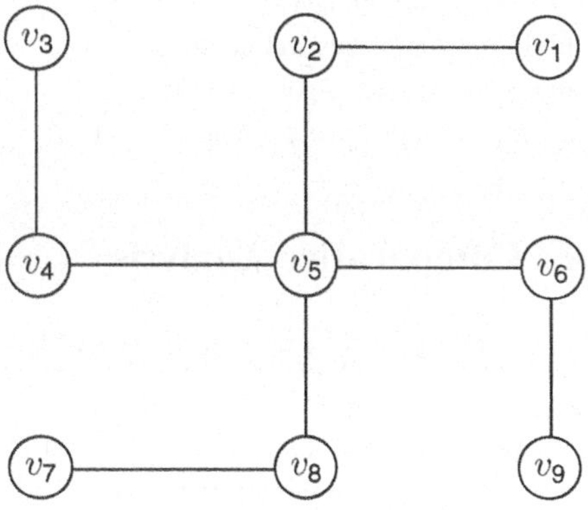

Fig. 4 A graph G with 9 vertices and 8 edges

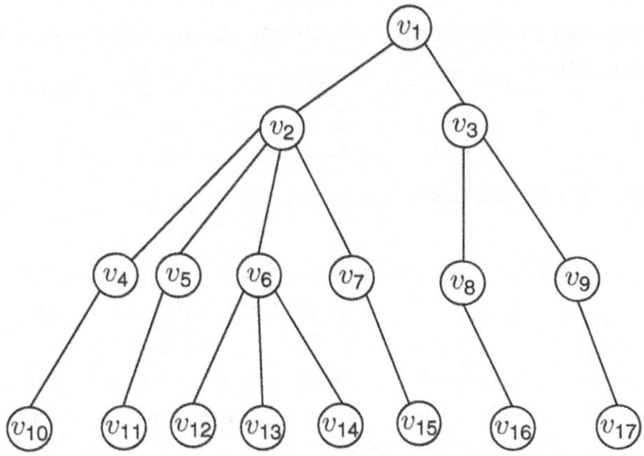

Fig. 5 A graph G with 17 vertices and 16 edges

Table 1 Comparative details of proposed algorithm with existing approaches

Dataset	Algorithm	$\mid N \mid$	$\mid E \mid$	$\mid Vc \mid$	Vertex cover set, Vc
Fig. 1	Approx.-Vertex-Cover Algorithm [19]	7	8	6	$\{v_2, v_3, v_4, v_5, v_6, v_7\}$
Fig. 1	Alom's algorithm [20]	7	8	3	$\{v_2, v_4, v_6\}$
Fig. 1	Approx. algorithm [14]	7	8	4	$\{v_2, v_4, v_5, v_6\}$
Fig. 1	*Proposed algorithm*	7	8	3	$\{v_2, v_4, v_6\}$
Fig. 4	Approx.-Vertex-Cover Algorithm [19]	9	8	7	$\{v_5, v_6, v_1, v_2, v_3, v_4, v_9\}$
Fig. 4	Alom's algorithm [20]	9	8	5	$\{v_5, v_1, v_3, v_6, v_8\}$
Fig. 4	Approx. algorithm [14]	9	8	5	$\{v_2, v_4, v_5, v_6, v_8\}$
Fig. 4	*Proposed algorithm*	9	8	4	$\{v_2, v_4, v_6, v_8\}$
Fig. 5	Approx.-Vertex-Cover Algorithm [19]	17	16	14	$\{v_1, v_2, v_4, v_{10}, v_5, v_{11}, v_6, v_{12}, v_7, v_{15}, v_8, v_{16}, v_9, v_{17}\}$
Fig. 5	Alom's algorithm [20]	17	16	8	$\{v_2, v_6, v_3, v_4, v_5, v_7, v_8, v_9\}$
Fig. 5	Approx. algorithm [14]	17	16	9	$\{v_1, v_2, v_3, v_4, v_5, v_6, v_7, v_8, v_9\}$
Fig. 5	*Proposed algorithm*	17	16	7	$\{v_1, v_4, v_5, v_6, v_7, v_8, v_9\}$

4 Comparative Analysis

To assess the capability, the proposed algorithm has been tested on three selected graphs [18], and the results are compared with those of existing algorithms. The comparative results are reported in Table 1. From the results, it is seen that for Fig. 1, the solution found by the proposed algorithm is better than two other methods, and it is same with Alom's algorithm solution. For rest of the cases, it is observed that our proposed algorithm is providing more efficient solutions than the existing algorithms.

5 Conclusions

In this study, an articulation point-based approximation algorithm is proposed to solve MVCP. The proposed algorithm is simple and easy to implement. The comparative results show that the proposed algorithm is quite efficient than the existing algorithms and guarantees in giving promising solutions for a given simple graphs. Further, the vertex cover problem finds numerous practical applications in communication networks.

References

1. Garry, M., Johnson, D.: Computers and Intractability: A User Guide to the Theory of NP Completeness. San Francisco (1979)
2. Hoo, C.S., Jeevan, K., Ganapathy, V., Ramiah, H.: Variable-order ant system for VLSI multiobjective floor planning. Appl. Soft Comput. **13** (7), 3285–3297 (2013)
3. Sherali, H.D., Rios, M.: An air force crew allocation and scheduling problem. J. Oper. Res. Soc. **35** (2), 91–103 (1984)
4. Woodyatt, L.R., Stott, K.L., Wolf, F.E., Vasko, F.J.: An application combining setcovering and fuzzy sets to optimally assign metallurgical grades to customer orders. Fuzzy Sets Syst. **53** (1), 15–25 (1993)
5. Eiter, T., Gottlob, G.: Identifying the minimal transversals of a hypergraph and related problems. SIAM J. Comput. **24** (6), 1278–1304 (1995)
6. Listrovoy, S., Minukhin, S.: The solution algorithms for problems on the minimal vertex cover in networks and the minimal cover in Boolean matrixes. IJCSI. **9** (5), 8–15 (2012)
7. Avis, D., Imamura, T.: A list heuristic for vertex cover. Oper. Res. Lett. **35** (2), 201–204 (2007)
8. Delbot, F. Laforest, C.: A better list heuristic for vertex cover. Inf. Process. Lett. **107** (3 -4), 125–127 (2008)
9. Savage, C.: Depth-first search and the vertex cover problem. Inf. Process. Lett. **14** (5), 233–235 (1982)
10. Singh, A., Gupta, A.K.: A hybrid heuristic for the minimum weight vertex cover problem. Asia-Pac. J. Oper. Res. **23**, 273–285 (2006)
11. Vob, S., Fink, A.: A hybridized tabu search approach for the minimum weight vertex cover problem. J. Heu. **18** (6), 869–876 (2012)
12. J. Chen, I.A. Kani, G. Xia, Improved upper bounds for vertex cover. Theor. Comput. Sci. **411**, 3736–3756 (2010)
13. Tu, J.: A fixed-parameter algorithm for the vertex cover P3 problem. Inf. Process. Lett. **115** (2), 96–99 (2015)
14. Shah, K., Reddy, P., Selvakumar, R.: Vertex Cover Problem - Revised Approximation Algorithm. In: Artificial Intelligence and Evolutionary Algorithms in Engineering Systems, pp. 9–16. Springer, New Delhi (2015)
15. Chen, J., Kou, L., Cui, X.: An Approximation Algorithm for the Minimum Vertex Cover Problem. Procedia Eng. **137**, 180–185 (2016)
16. Chen, J., Lin, Y., Li, J., Lin, G., Ma, Z., Tan, A.: A rough set method for the minimum vertex cover problem of graphs. Appl. Soft Comput. **42**, 360–367 (2016)
17. Hochbaum, D.S.: Approximation algorithms for the set covering and vertex cover problems. SIAM J. Comput. **11** (3), 555–556 (1982)
18. Dahiya, S.: A New Approximation Algorithm for Vertex Cover Problem. In: International Conference on Machine Intelligence and Research Advancement, pp. 472–478, IEEE (2013)
19. Cormen, T.H., Leiserson, C.E., Rivest, R.L., Stein, C.: Introduction to Algorithms. 3rd edn. MIT Press, Cambridge, MA (2009)
20. Alom, B.M., Das, S., Rouf, M.A.: Performance evaluation of vertex cover and set cover problem using optimal algorithm. DUET Journal. **1** (2), (2011)

On Bottleneck-Rough Cost Interval Integer Transportation Problems

A. Akilbasha, G. Natarajan, and P. Pandian

Abstract An innovative method, namely, level maintain method, is proposed for finding all efficient solutions to a bottleneck-rough cost interval integer transportation problem in which the unit transportation cost, supply, and demand parameters are rough interval integers and the transportation time parameter is an interval integer. The solving procedure of the suggested method is expressed and explained with a numerical example. The level maintain method will dispense the necessary determined support to decision-makers when they are handling time-related logistic problems in rough nature.

1 Introduction

Transportation problem (TP) is one of the most important and popular applications of the linear programming problem. Different types of efficient algorithms have been developed by various authors for solving TPs having deterministic parameters. The classical, interval, and fuzzy transportation problems have been formulated and discussed the methods for solution to the fuzzy TP by Chanas et al. [2]. Many researchers [6, 8, 12, 13, 15] have proposed various methods to solve interval and fuzzy TPs.

Pawlak [11] initiated the rough set theory. Then, it has been developed by many researchers both in theoretical and applied. Youness [16] introduced a rough programming problem; there he considered the decision set as a rough set. Some solid transportation models with crisp and rough costs were solved by Kundu et al. [4]. Subhakanta Dash et al. [14] proposed a compromise solution method to transportation problems; here they are considering the transportation cost as a

---

A. Akilbasha (✉) · G. Natarajan · P. Pandian
Department of Mathematics, Vellore Institute of Technology, Vellore, India
e-mail: bashaakil@gmail.com; akilbasha.a@vit.ac.in; gnatarajan@vit.ac.in; ppandian@vit.ac.in

V. Madhu et al. (eds.), *Advances in Algebra and Analysis*, Trends in Mathematics,
https://doi.org/10.1007/978-3-030-01120-8_33

rough interval integer. Various methods for solving interval integer transportation problems with rough nature are presented in Akilbasha et al. [1] and Pandian et al. [10].

The paper is constructed as follows: In Sect. 2 presents the basic results of rough sets. The bottleneck-rough cost interval integer transportation and its solution are discussed in the Sect. 3. Section 4 proposes an innovative method for finding an efficient solution to the given TP model, and a numerical example is shown. Finally Sect. 5 concludes the paper.

2 Preliminaries

The following are definitions we need, which can be found in [3, 5].

Let S denote the set of all rough intervals on the real line R. That is,

$$S = \{[[x_2, x_3], [x_1, x_4]],\ \ x_1 \le x_2 \le x_3 \le x_4 \text{ and } x_1,\ x_2,\ x_3 \text{ and } x_4 \text{ are in } R\}.$$

Note that,

- if $x_1 = x_2$ and $x_3 = x_4$ in S, then S becomes the set of all real intervals and
- if $x_1 = x_2 = x_3 = x_4$ in S, then S becomes the set of all real numbers.

Definition 1 Let $X = [[x_2, x_3], [x_1, x_4]]$ and $Y = [[y_2, y_3], [y_1, y_4]]$ be in S. Then,

- $X \oplus Y = [[x_2 + y_2, x_3 + y_3], [x_1 + y_1, x_4 + y_4]]$
- $kX = [[kx_2, kx_3], [kx_1, kx_4]]$ if k is a positive real interval and
- $X \otimes Y = [[x_2, x_3][y_2, y_3], [x_1, x_4][y_1, y_4]]$

Definition 2 Let $X = [[x_2, x_3], [x_1, x_4]]$ and $Y = [[y_2, y_3], [y_1, y_4]]$ be in S. Then,

- $X \le Y$ if $x_i \le y_i, i = 1, 2, 3, 4$
- $X \ge Y$ if $Y \le X$, that is, $x_i \ge y_i, i = 1, 2, 3, 4$ and
- $X = Y$ if $X \le Y$ and $Y \le X$, that is, $x_i = y_i,\ i = 1, 2, 3, 4$

Definition 3 Let $X = [[x_2, x_3], [x_1, x_4]]$ be in S. Then, X is said to be nonnegative, that is, $X \ge 0$ if $x_1 \ge 0$.

Remark 1 If $X = [[x_2, x_3], [x_1, x_4]]$ and $Y = [[y_2, y_3], [y_1, y_4]]$ in S are nonnegative, then $X \otimes Y = [[x_2 y_2, x_3 y_3], [x_1 y_1, x_4 y_4]]$.

Definition 4 Let $X = [[x_2, x_3], [x_1, x_4]]$ be in S. Then, X is said to be rough integer if $x_i,\ i = 1, 2, 3, 4$ are integers.

3 Bottleneck-Rough Cost Interval Integer Transportation Problems

Consider the following bottleneck-rough cost interval integer transportation problems:

(P) Minimize $[[z_2, z_3], [z_1, z_4]] = \sum_{i=1}^{m} \sum_{j=1}^{n} [[c_{ij}^2, c_{ij}^3], [c_{ij}^1, c_{ij}^4]] \otimes [[x_{ij}^2, x_{ij}^3], [x_{ij}^1, x_{ij}^4]]$

Minimize $[T_1, T_2]$ = [Maximize $[t_{ij}^1, t_{ij}^2] / [[x_{ij}^2, x_{ij}^3], [x_{ij}^1, x_{ij}^4]] > 0$]

Subject to

$$\sum_{j=1}^{n} [[x_{ij}^2, x_{ij}^3], [x_{ij}^1, x_{ij}^4]] = [[a_i^2, a_i^3], [a_i^1, a_i^4]], \ i \in I \quad (1)$$

$$\sum_{i=1}^{m} [[x_{ij}^2, x_{ij}^3], [x_{ij}^1, x_{ij}^4]] = [[b_j^2, b_j^3], [b_j^1, b_j^4]], \ j \in J \quad (2)$$

$$x_{ij}^1 \geq 0, \ i \in I \text{ and } j \in J \text{ and } x_{ij}^1, x_{ij}^2, x_{ij}^3 \text{ and } x_{ij}^4, \ i \in I \ \& \ j \in J \text{ are integers} \quad (3)$$

where $I = \{1, 2, 3, \ldots, m\}$ and $J = \{1, 2, 3, \ldots, n\}$, $c_{ij}^1, c_{ij}^2, c_{ij}^3$ & c_{ij}^4 are positive integers for all $i \in I$ and $j \in J$; t_{ij} = transporting time of goods from supply point i to demand point j; a_i^1, a_i^2, a_i^3 and a_i^4 are positive integers for all $i \in I$; and b_j^1, b_j^2, b_j^3 and b_j^4 are positive integers for all $j \in J$. The problem (P) is said to be balanced if total supply = total demand.

Definition 5 A set $\{([[x_{ij}^2, x_{ij}^3], [x_{ij}^1, x_{ij}^4]], [m_1, m_2])$, for all $i \in I$ and $j \in J\}$ where $[m_1, m_2]$ is a time interval is said to be a feasible solution to the problem (P) if the rough interval set $\{[[x_{ij}^2, x_{ij}^3], [x_{ij}^1, x_{ij}^4]]$ for all $i \in I$ and $j \in J\}$ satisfies the Eqs. (1), (2), and (3).

Definition 6 A set $\{([[x_{ij}^2, x_{ij}^3], [x_{ij}^1, x_{ij}^4]], [m_1, m_2])$, for all $i \in I$ and $j \in J\}$ is said to be an efficient solution of the problem (P) if there exists no other feasible $\{([[u_{ij}^2, u_{ij}^3], [u_{ij}^1, u_{ij}^4]], [n_1, n_2])$, for all $i \in I$ and $j \in J\}$ to (P) such that

$$\sum_{i=1}^{m} \sum_{j=1}^{n} [[c_{ij}^2, c_{ij}^3], [c_{ij}^1, c_{ij}^4]] \otimes [[x_{ij}^2, x_{ij}^3], [x_{ij}^1, x_{ij}^4]] \leq \sum_{i=1}^{m} \sum_{j=1}^{n} [[c_{ij}^2, c_{ij}^3], [c_{ij}^1, c_{ij}^4]] \otimes$$

$[[u_{ij}^2, u_{ij}^3], [u_{ij}^1, u_{ij}^4]]$ and $[m_1, m_2] < [n_1, n_2]$ or

$$\sum_{i=1}^{m} \sum_{j=1}^{n} [[c_{ij}^2, c_{ij}^3], [c_{ij}^1, c_{ij}^4]] \otimes [[x_{ij}^2, x_{ij}^3], [x_{ij}^1, x_{ij}^4]] < \sum_{i=1}^{m} \sum_{j=1}^{n} [[c_{ij}^2, c_{ij}^3], [c_{ij}^1, c_{ij}^4]] \otimes$$

$[[u_{ij}^2, u_{ij}^3], [u_{ij}^1, u_{ij}^4]]$ and $[m_1, m_2] \leq [n_1, n_2]$.

The problem (P) can be partitioned into four subproblems, namely, (P1), (P2), (P3), and (P4) as follows:

(P4) Minimize $z_4 = \sum_{i=1}^{m} \sum_{j=1}^{n} c_{ij}^4 \otimes x_{ij}^4$; Minimize T_2 = [Maximize $t_{ij}^2/x_{ij}^4 > 0$]

Subject to

$\sum_{j=1}^{n} x_{ij}^4 = a_i^4,\ i \in I;\quad \sum_{i=1}^{m} x_{ij}^4 = b_j^4,\ j \in J;\quad x_{ij}^4 \geq 0,\ i \in I\ \&\ j \in J$ are integers.

(P3) Minimize $z_3 = \sum_{i=1}^{m} \sum_{j=1}^{n} c_{ij}^3 \otimes x_{ij}^3$; Minimize T_1 = [Maximize $t_{ij}^1/x_{ij}^3 > 0$]

Subject to

$\sum_{j=1}^{n} x_{ij}^3 = a_i^3,\ i \in I;\quad \sum_{i=1}^{m} x_{ij}^3 = b_j^3,\ j \in J;\quad x_{ij}^3 \geq 0,\ i \in I\ \&\ j \in J$ are integers.

(P2) Minimize $z_2 = \sum_{i=1}^{m} \sum_{j=1}^{n} c_{ij}^2 \otimes x_{ij}^2$; Minimize T_1 = [Maximize $t_{ij}^1/x_{ij}^2 > 0$]

Subject to

$\sum_{j=1}^{n} x_{ij}^2 = a_i^2,\ i \in I;\quad \sum_{i=1}^{m} x_{ij}^2 = b_j^2,\ j \in J;\quad x_{ij}^2 \geq 0,\ i \in I\ \&\ j \in J$ are integers. and

(P1) Minimize $z_1 = \sum_{i=1}^{m} \sum_{j=1}^{n} c_{ij}^1 \otimes x_{ij}^1$; Minimize T_2 = [Maximize $t_{ij}^2/x_{ij}^1 > 0$]

Subject to

$\sum_{j=1}^{n} x_{ij}^1 = a_i^1,\ i \in I;\quad \sum_{i=1}^{m} x_{ij}^1 = b_j^1,\ j \in J;\quad x_{ij}^1 \geq 0,\ i \in I\ \&\ j \in J$ are integers.

We now prove the following theorem connecting the efficient solution of the problem (P) and the efficient solutions of the problems (P1), (P2), (P3), and (P4) which is used in the proposed method called level maintain method for finding an efficient solution for the problem (P).

Theorem 1 *If the set* $\{(\bar{x}_{ij}^4, T_2),\ \forall\ i \in I\ \&\ j \in J\}$ *is an efficient solution for the problem (P4) with objective value* $(\bar{z}_4, T_2)$*, the set* $\{(\bar{x}_{ij}^3, T_1),\ \forall\ i \in I\ \&\ j \in J\}$ *is an efficient solution for the problem (P3) with objective value* $(\bar{z}_3, T_1)$*, the set* $\{(\bar{x}_{ij}^2, T_1),\ \forall\ i \in I\ \&\ j \in J\}$ *is an efficient solution for the problem (P2) with objective value* $(\bar{z}_2, T_1)$*, and the set* $\{(\bar{x}_{ij}^1, T_2),\ \forall\ i \in I\ \&\ j \in J\}$ *is an efficient solution for the problem (P1) with objective value* $(\bar{z}_1, T_2)$*, then the set* $\{([[\bar{x}_{ij}^2, \bar{x}_{ij}^3], [\bar{x}_{ij}^1, \bar{x}_{ij}^4]], [T_1, T_2]),\ \forall\ i \in I\ \&\ j \in J\}$ *is an efficient solution for the problem (P) with objective value* $([[\bar{z}_{ij}^2, \bar{z}_{ij}^3], [\bar{z}_{ij}^1, \bar{z}_{ij}^4]], [T_1, T_2])$ *provided* $\bar{x}_{ij}^1 \leq \bar{x}_{ij}^2 \leq \bar{x}_{ij}^3 \leq \bar{x}_{ij}^4,\ \forall\ i \in I\ \&\ j \in J$ *and* $T_1 \leq T_2$.

Proof Now, since $\{(\bar{x}_{ij}^4, T_2),\ \forall\ i \in I\ \&\ j \in J\}$, $\{(\bar{x}_{ij}^3, T_1),\ \forall\ i \in I\ \&\ j \in J\}$, $\{(\bar{x}_{ij}^2, T_1),\ \forall\ i \in I\ \&\ j \in J\}$, and $\{(\bar{x}_{ij}^1, T_2),\ \forall\ i \in I\ \&\ j \in J\}$ are

efficient solutions for the problems (P4), (P3), (P2), and (P1), respectively, and $\bar{x}_{ij}^1 \leq \bar{x}_{ij}^2 \leq \bar{x}_{ij}^3 \leq \bar{x}_{ij}^4$, $\forall\, i \in I \;\&\; j \in J$ and $T_1 \leq T_2$, we can conclude that the set $\{([[\bar{x}_{ij}^2, \bar{x}_{ij}^3], [\bar{x}_{ij}^1, \bar{x}_{ij}^4]], [T_1, T_2]), \forall\, i \in I \;\&\; j \in J\}$ is a feasible solution to the problem (P).
Assume that the set $\{([[\bar{x}_{ij}^2, \bar{x}_{ij}^3], [\bar{x}_{ij}^1, \bar{x}_{ij}^4]], [T_1, T_2]), \forall\, i \in I \;\&\; j \in J\}$ is not an efficient solution to the problem (P).
Then, there exists other feasible $\{([[u_{ij}^2, u_{ij}^3], [u_{ij}^1, u_{ij}^4]], [n_1, n_2])\; \forall\, i \in I \;\&\; j \in J\}$ to (P) such that

$$\sum_{i=1}^{m}\sum_{j=1}^{n}[[c_{ij}^2, c_{ij}^3], [c_{ij}^1, c_{ij}^4]] \otimes [[x_{ij}^2, x_{ij}^3], [x_{ij}^1, x_{ij}^4]] \leq \sum_{i=1}^{m}\sum_{j=1}^{n}[[c_{ij}^2, c_{ij}^3], [c_{ij}^1, c_{ij}^4]] \otimes$$

$$[[u_{ij}^2, u_{ij}^3], [u_{ij}^1, u_{ij}^4]] \text{ and } [T_1, T_2] < [n_1, n_2] \text{ or}$$

$$\sum_{i=1}^{m}\sum_{j=1}^{n}[[c_{ij}^2, c_{ij}^3], [c_{ij}^1, c_{ij}^4]] \otimes [[x_{ij}^2, x_{ij}^3], [x_{ij}^1, x_{ij}^4]] < \sum_{i=1}^{m}\sum_{j=1}^{n}[[c_{ij}^2, c_{ij}^3], [c_{ij}^1, c_{ij}^4]] \otimes$$

$$[[u_{ij}^2, u_{ij}^3], [u_{ij}^1, u_{ij}^4]] \text{ and } [T_1, T_2] \leq [n_1, n_2]$$

This implies that the set $\{(\bar{x}_{ij}^4, T_2), \forall\, i \in I \;\&\; j \in J\}$ is not an efficient solution for the problem (P4) and the set $\{(\bar{x}_{ij}^1, T_2), \forall\, i \in I \;\&\; j \in J\}$ is not an efficient solution for the problem (P1) or the set $\{(\bar{x}_{ij}^3, T_1), \forall\, i \in I \;\&\; j \in J\}$ is not an efficient solution for the problem (P3) and the set $\{(\bar{x}_{ij}^2, T_1), \forall\, i \in I \;\&\; j \in J\}$ is not an efficient solution for the problem (P2).
This is a contradiction to our assumption. Therefore, the set $\{([[\bar{x}_{ij}^2, \bar{x}_{ij}^3], [\bar{x}_{ij}^1, \bar{x}_{ij}^4]], [T_1, T_2]), \forall\, i \in I \;\&\; j \in J\}$ is an efficient solution for the problem (P) with objective value $([[\bar{z}_{ij}^2, \bar{z}_{ij}^3], [\bar{z}_{ij}^1, \bar{z}_{ij}^4]], [T_1, T_2])$.
Hence, the theorem is proved. □

4 Level Maintain Method

We propose a new method, namely, level maintain method for solving the bottleneck-rough cost interval integer TP, (P).
The proposed method proceeds as follows.

Step 1: Check if the given problem (P) is balanced. If not, make it into balanced.
Step 2: Construct four problems (P1), (P2), (P3), and (P4) from the given problem (P).
Step 3: Solve the problem (P4) by using the blocking zero point method [9] (the blocking zero point method is based on the zero point method [7]). Let $\{(\bar{x}_{ij}^4, T_2), \forall\, i \in I \;\&\; j \in J\}$ be an efficient solution of the (P4) problem with objective value $(\bar{z}_4, T_2)$.

Step 4: Solve the problem (P3) by using the blocking zero point method [9] with the upper bound constraints $x_{ij}^3 \leq \bar{x}_{ij}^4$ and $T_1 \leq T_2$. Let $\{(\bar{x}_{ij}^3, T_1), \forall i \in I \;\&\; j \in J\}$ be an efficient solution of the (P3) problem with objective value $(\bar{z}_3, T_1)$.

Step 5: Solve the problem (P2) by using the blocking zero point method [9] with the upper bound constraints $x_{ij}^2 \leq \bar{x}_{ij}^3$ and $T_1 \leq T_2$. Let $\{(\bar{x}_{ij}^2, T_1), \forall i \in I \;\&\; j \in J\}$ be an efficient solution of the (P2) problem with objective value $(\bar{z}_2, T_1)$.

Step 6: Solve the problem (P1) by using the blocking zero point method [9] with the upper bound constraints $x_{ij}^1 \leq \bar{x}_{ij}^2$ and $T_1 \leq T_2$. Let $\{(\bar{x}_{ij}^1, T_2), \forall i \in I \;\&\; j \in J\}$ be an efficient solution of the (P1) problem with objective value $(\bar{z}_1, T_2)$.

Step 7: The set $\{([[\bar{x}_{ij}^2, \bar{x}_{ij}^3], [\bar{x}_{ij}^1, \bar{x}_{ij}^4]],[T_1, T_2]), \forall i \in I \;\&\; j \in J\}$ is an optimal solution of the problem (P) with objective value $([[\bar{z}_2, \bar{z}_3], [\bar{z}_1, \bar{z}_4]],[T_1, T_2])$. (by Theorem 1).

The solution procedure of the proposed method is solved with the given numerical example.

Example 1 Consider the below bottleneck-rough cost interval integer TP:

	D1	D2	D3	D4	Supply
F1	[[4,5], [3,6]] [6,8]	[[4,6], [2,8]] [66,68]	[[9,10], [7,11]] [69,71]	[[9,11], [7,13]] [49,51]	[[6,8], [4,10]]
F2	[[4,6], [2,8]] [64,66]	[[6,7], [5,8]] [91,93]	[[10,12], [8,14]] [28,30]	[[13,14], [12,15]] [17,19]	[[17,19], [15,21]]
F3	[[13,14], [12,15]] [93,95]	[[9,11], [7,13]] [61,63]	[[8,9], [7,10]] [15,17]	[[5,7], [3,9]] [21,23]	[[15,17], [13,19]]
Demand	[[10,11], [9,12]]	[[2,3], [1,4]]	[[12,14], [10,16]]	[[14,16], [12,18]]	

Now, since the total supply = the total demand = [[38,44],[32,50]], the given problem is a balanced one. Now, using the Steps 2 and 3, the efficient solutions of the problem (P4) with their objective values are obtained as follows:

S. No.	(P4) Efficient solution	(P4)Objective value
1	$x_{11} = 7; x_{14} = 3; x_{21} = 5; x_{23} = 16; x_{32} = 4$ and $x_{34} = 15$ with time 66.	(532, 66)
2	$x_{11} = 6; x_{12} = 4; x_{21} = 6; x_{23} = 15; x_{33} = 1$ and $x_{34} = 18$ with time 68.	(498, 68)
3	$x_{12} = 4; x_{13} = 6; x_{21} = 12; x_{23} = 9; x_{33} = 1$ and $x_{34} = 18$ with time 71.	(492, 71)
4	$x_{13} = 10; x_{21} = 12; x_{22} = 4; x_{23} = 5; x_{33} = 1$ and $x_{34} = 18$ with time 93.	(480, 93)

Now, using the Steps 2 and 4, the efficient solutions of the problem (P3) with their objective values are tabulated below:

S. No.	(P3) Efficient solution	(P3)Objective value
1	$x_{11} = 6; x_{14} = 2; x_{21} = 5; x_{23} = 14; x_{32} = 3$ and $x_{34} = 14$ with time 64.	(381, 64)
2	$x_{11} = 5; x_{12} = 3; x_{21} = 6; x_{23} = 13; x_{33} = 1$ and $x_{34} = 16$ with time 66.	(356, 66)
3	$x_{12} = 3; x_{13} = 5; x_{21} = 11; x_{23} = 8; x_{33} = 1$ and $x_{34} = 16$ with time 69.	(351, 69)
4	$x_{13} = 8; x_{21} = 11; x_{22} = 3; x_{23} = 5; x_{33} = 1$ and $x_{34} = 16$ with time 91.	(348, 91)

Now, by the Steps 2 and 5, the efficient solutions of the problem (P2) with their objective values are shown below:

S. No.	(P2) Efficient solution	(P2)Objective value
1	$x_{11} = 5; x_{14} = 1; x_{21} = 5; x_{23} = 12; x_{32} = 2$ and $x_{34} = 13$ with time 64.	(252, 64)
2	$x_{11} = 4; x_{12} = 2; x_{21} = 6; x_{23} = 11; x_{33} = 1$ and $x_{34} = 14$ with time 66.	(236, 66)
3	$x_{12} = 2; x_{13} = 4; x_{21} = 10; x_{23} = 7; x_{33} = 1$ and $x_{34} = 14$ with time 69.	(232, 69)
4	$x_{13} = 6; x_{21} = 10; x_{22} = 2; x_{23} = 5; x_{33} = 1$ and $x_{34} = 14$ with time 91.	(234, 91)

Now, using the Steps 2 and 6, the efficient solutions of the problem (P1) with their objective values are obtained and tabulated as given below:

S. No.	(P1) Efficient solution	(P1) Objective value
1	$x_{11} = 4; x_{14} = 0; x_{21} = 5; x_{23} = 10; x_{32} = 1$ and $x_{34} = 12$ with time 66.	(145, 66)
2	$x_{11} = 3; x_{12} = 1; x_{21} = 6; x_{23} = 9; x_{33} = 1$ and $x_{34} = 12$ with time 68.	(138, 68)
3	$x_{12} = 1; x_{13} = 3; x_{21} = 9; x_{23} = 6; x_{33} = 1$ and $x_{34} = 12$ with time 71.	(132, 71)
4	$x_{13} = 4; x_{21} = 9; x_{22} = 1; x_{23} = 5; x_{33} = 1$ and $x_{34} = 12$ with time 93.	(134, 93)

Now, by the Step 7, the efficient solutions of the problem (P) with their objective values are given below:

S. No.	Efficient solution of (P)	Objective value of (P)
1	$x_{11} = [[5,6],[4,7]]$; $x_{14} = [[1,2],[0,3]]$; $x_{21} = [[5,5],[5,5]]$; $x_{23} = [[12,14],[10,16]]$; $x_{32} = [[2,3],[1,4]]$ and $x_{34} = [[13,14],[12,15]]$ with time [64,66].	([[252,381],[145,532]], [64,66])
2	$x_{11} = [[4,5],[3,6]]$; $x_{12} = [[2,3],[1,4]]$; $x_{21} = [[6,6],[6,6]]$; $x_{23} = [[11,13],[9,15]]$; $x_{33} = [[1,1],[1,1]]$ and $x_{34} = [[14,16],[12,18]]$ with time [66,68].	([[236,356],[138,498]], [66,68])
3	$x_{12} = [[2,3],[1,4]]$; $x_{13} = [[4,5],[3,6]]$; $x_{21} = [[10,11],[9,12]]$; $x_{23} = [[7,8],[6,9]]$; $x_{33} = [[1,1],[1,1]]$ and $x_{34} = [[14,16],[12,18]]$ with time [69,71].	([[232,351],[132,492]], [69,71])
4	$x_{13} = [[6,8],[4,10]]$; $x_{21} = [[10,11],[9,12]]$; $x_{22} = [[2,3],[1,4]]$; $x_{23} = [[5,5],[5,5]]$; $x_{33} = [[1,1],[1,1]]$ and $x_{34} = [[14,16],[12,18]]$ with time [91,93].	([[234,348],[134,480]], [91,93])

5 Conclusion

In this paper, a TP having bi-objective functions, namely, cost objective and time objective functions, is discussed; here the parameters of transportation cost, supply, and demand are in rough interval integers, and transportation time parameter is an interval integer. For solving considering problem, a new method, namely, level maintain method, has been developed, and its solution procedure is presented. A numerical example is illustrated for understanding the proposed method. The proposed method gives a set of transportation schedules to bottleneck-rough cost interval integer transportation problems which can help the decision-makers to select the suitable transportation schedule, depending on their financial level and the extent of bottleneck that they can afford.

References

1. Akilbasha, A., Natarajan, G., Pandian, P.: Finding an optimal solution of the interval integer transportation problems with rough nature by split and separation method. International Journal of Pure and Applied Mathematics. **106**, 1–8 (2016)
2. Chanas, S., Delgado, M., Verdegayand, J.L., Vila, M.A.: Interval and fuzzy extensions of classical transportation problems. TranspPlann Technol. **17**, 203–218 (1993)
3. Hongwei Lu., Guohe Huang., Li He.: An inexact rough-interval fuzzy linear programming method for generating conjunctive water-allocation strategies to agricultural irrigation systems. Applied Mathematical Modelling. **35**, 4330–4340 (2011)

4. Kundu, P., Kar, S., Maiti, M.: Some solid transportation model with crisp and rough costs. World Academy of Science, Engineering and Technology. **73**, 185–192 (2013)
5. Moore, R.E.: Method and applications of interval analysis. SLAM, Philadelphia, PA (1979)
6. Pandian, P., Natarajan, G.: A new algorithm for finding a fuzzy optimal solution for fuzzy transportation problems. Appl. Math. Sci. **4**, 79–90 (2010)
7. Pandian, P., Natarajan, G.: A new method for finding an optimal solution for transportation problems. International Journal of Math. Sci. & Engg. Appls. **4**, 59–65 (2010)
8. Pandian, P., Natarajan, G.: A new method for finding an optimal solution of fully interval integer transportation problems. Appl. Math. Sci. **4**, 1819–1830 (2010)
9. Pandian, P., Natarajan, G.: A new method for solving bottleneck-cost transportation problems. International Mathematical Forum. **6**, 451–460 (2011)
10. Pandian, P., Natarajan, G., Akilbasha, A.: Fully rough integer interval transportation problems. International Journal Of Pharmacy and Technology. **8**, 13866–13876 (2016)
11. Pawlak, Z.: Rough sets. International Journal of Information and Computer Science. **11**, 341–356 (1981)
12. Sengupta, A., Pal, T.K.: Interval-valued transportation problem with multiple penalty factors. VU Journal of Physical Sciences. **9**, 71–81 (2003)
13. Shanmugasundari, M., Ganesan, K.: A novel approach for the fuzzy optimal solution of fuzzy transportation problem. Int. J. Eng. Res. Appl. **3**, 1416–1421 (2013)
14. Subhakanta Dash., Mohanty, S.P.: Transportation programming under uncertain environment. International Journal of Engineering Research and Development. **7**, 22–28 (2013)
15. Sudhagar, S., Ganesan, K.: A fuzzy approach to transport optimization problem. Optimization and Engineering. **17**, 965–980 (2016)
16. Youness, E.: Characterizing solutions of rough programming problems. European Journal of Operational Research. **168**, 1019–1029 (2006)

Direct Solving Method of Fully Fuzzy Linear Programming Problems with Equality Constraints Having Positive Fuzzy Numbers

C. Muralidaran and B. Venkateswarlu

Abstract In the process of solving fully fuzzy linear programming (FLP) problems, many methods have been investigated. To find the fuzzy optimal solution of fully FLP problems, there is a need to convert the fully FLP problems into crisp linear programming (CLP) problems. There is no method which can be used directly to find the fuzzy optimal solution of the fully FLP problems without converting it into CLP problems. In this paper we investigate fully FLP problems with equality constrains in which all the parameters and variables are positive triangular fuzzy numbers. This approach can be used directly to find the fuzzy optimal solution by the simplex method(Big-M).

Keywords Fully fuzzy linear programming problem · Triangular fuzzy numbers

1 Introduction

Many decision-making problems have been solved using one of the most important technique, linear programming in operation research. Every linear programming model representing real-world situations involves various parameters. The values of those parameters are assigned by experts which are not precise in many situations. Therefore, it is best to consider the parameters are in fuzzy.

Finding the best solution is the main objective of a fuzzy linear programming (FLP) problem in the environment of uncertain, imprecise, vague, or incomplete information. When Zadeh and Bellman [1] initiated the concept of decision-making in fuzzy environment, the research on fuzzy linear programming attains the tremendous growth. The first formulation of the fuzzy linear programming was introduced by Zimmermann [2] in 1978. FLP varies depending upon the vagueness in the situations of the problem.

C. Muralidaran · B. Venkateswarlu (✉)
Department of Mathematics, School of Advanced Sciences, Vellore Institute of Technology, Vellore, India
e-mail: muralidaran.c@vit.ac.in; venkatesh.reddy@vit.ac.in

V. Madhu et al. (eds.), *Advances in Algebra and Analysis*, Trends in Mathematics,
https://doi.org/10.1007/978-3-030-01120-8_34

In 2000 Buckley [4] and Feuring introduced a method to find a solution for fully FLP problem. A computational method was proposed by Dehghan et al. [9] to find the exact solution of a fully FLP problem. Allahviranloo et al. in 2008 [5] proposed a kind of defuzzification method for solving fully FLP problem. Lotfi et al. [6] discussed fully FLP problem in which all the variables and parameters were symmetric triangular fuzzy numbers.

A method was proposed by Kumar et al. [3, 7] for finding the fuzzy optimal solution of fully FLP problem with the inequality constraints. Whereas Nasseri et al. [8] proposed a new method to find the fuzzy solution of fully FLP problems with inequality constraints, there is no method in the literature for finding the fuzzy optimal solution of a fully FLP problem without converting it into crisp linear programming (CLP) problem.

2 Preliminaries

2.1 *Mathematical Form of a Linear Programming Problem*

The linear programming problem or model involving n decision variables and m constraints can be written in the form of

Maximize (or) minimize $Z = c_1x_1 + c_1x_2 + \cdots + c_nx_n$.

Here, $x_1, x_2, \ldots, x_n$ are decision variables, and $c_1, c_1, \ldots, c_n$ are the unit contributions of the decision-making variables $x_1, x_2, \ldots, x_n$, respectively.

Subject to the constraints,

$$a_{11}x_1 + a_{12}x_2 + \cdots + a_{1n}x_n (\leq or = or \geq) b_1$$

$$a_{21}x_1 + a_{22}x_2 + \cdots + a_{2n}x_n (\leq or = or \geq) b_2$$

$$\cdots\cdots\cdots\cdots\cdots\cdots\cdots\cdots\cdots\cdots\cdots\cdots$$

$$\cdots\cdots\cdots\cdots\cdots\cdots\cdots\cdots\cdots\cdots\cdots\cdots$$

$$a_{m1}x_1 + a_{m2}x_2 + \cdots + a_{mn}x_n (\leq or = or \geq) b_m$$

and the non-negativity condition is

$$x_1, x_2, \ldots, x_n \geq 0$$

Solution: A set of values $(x_1, x_2, \ldots, x_n)$ is said to be a solution when it satisfies the constraints of the given linear programming problem.

Feasible Solution: Any solution to a linear programming problem is said to be a feasible solution when it satisfies the non-negativity conditions of the same linear programming problem.

Optimal Solution: Any feasible solution of a linear programming problem is said to be optimal solution when it optimizes (maximizes or minimizes) the objective function of the same linear programming problem.

2.2 *Fuzzy Set*

If X is a collection of objects (denoted generally by x), then a fuzzy set $\tilde{A} \in X$ is a set of ordered pairs:

$$\tilde{A} = \{(x, \mu_{\tilde{A}}(x))|x \in X\}$$

where $\mu_{\tilde{A}}(x)$ is the membership function (or grade of membership) with respect to x in $\tilde{A}$.

2.3 *Fuzzy Number*

$\tilde{A} = (a_1, a_2, a_3, \ldots, a_n)$ is a real fuzzy number, and it is a fuzzy subset of the real number $\mathbb{R}$ with membership function $\mu_{\tilde{A}}$ that satisfies below conditions:

- $\mu_{\tilde{A}}$ is continuous from $\mathbb{R}$ to [0, 1]
- for odd n

$$= \begin{cases} \mu_{\tilde{A}} \text{ is strictly increasing and continuous on } [a_1, a_{\frac{n+1}{2}}] \\ \mu_{\tilde{A}} \text{ is strictly decreasing and continuous on } [a_{\frac{n+1}{2}}, a_n] \end{cases}$$

- for even n

$$= \begin{cases} \mu_{\tilde{A}} \text{ is strictly increasing and continuous on } [a_1, a_{\frac{n}{2}}] \\ \mu_{\tilde{A}} \text{ is strictly decreasing and continuous on } [a_{\frac{n}{2}+1}, a_n] \end{cases}$$

2.4 *Positive Fuzzy Number*

$\tilde{A} = (a_1, a_2, a_3, \ldots, a_n)$ is a fuzzy number, and it is said to be positive fuzzy number (nonnegative fuzzy number) if and only if $a_1 \geq 0$. Otherwise, $\tilde{A}$ is said to be nonpositive fuzzy number.

2.5 Operations on Fuzzy Number

If $\tilde{A} = (a_1, a_2, a_3, \ldots, a_n)$ and $\tilde{B} = (b_1, b_2, b_3, \ldots, b_n)$ are positive fuzzy numbers, then the following four operations can be performed as follows:

- Addition: $\tilde{A} \oplus \tilde{B} = (a_1 + b_1, a_2 + b_2, a_3 + b_3, \ldots, a_n + b_n)$
- Subtraction: $\tilde{A} \ominus \tilde{B} = (a_1 - b_1, a_2 - b_2, a_3 - b_3, \ldots, a_n - b_n)$
- Multiplication: $\tilde{A} \otimes \tilde{B} = (a_1 \times b_1, a_2 \times b_2, a_3 \times b_3, \ldots, a_n \times b_n)$
- Division: $\frac{\tilde{A}}{\tilde{B}} = (\frac{a_1}{b_1}, \frac{a_2}{b_2}, \frac{a_3}{b_3}, \ldots, \frac{a_n}{b_n})$

2.6 Triangular Fuzzy Number

$\tilde{A} = (a_1, a_2, a_3)$ is a real triangular fuzzy number (Fig. 1), and it is a fuzzy subset of the real number $\mathbb{R}$ with membership function $\mu_{\tilde{A}}$ that satisfies below conditions:

- $\mu_{\tilde{A}}$ is continuous from $\mathbb{R}$ to [0, 1]
- $\mu_{\tilde{A}}$ is strictly increasing and continuous on $[a_1, a_2]$
- $\mu_{\tilde{A}}$ is strictly decreasing and continuous on $[a_2, a_3]$

Membership function is

$$\mu_{\tilde{A}}(x) = \begin{cases} \frac{(x-a_1)}{(a_2-a_1)} & if \; a_1 \leq x < a_2 \\ 1 & if \quad x = a_2 \\ \frac{(x-a_3)}{(a_2-a_3)} & if \; a_2 < x \leq a_3 \\ 0 & otherwise \end{cases}$$

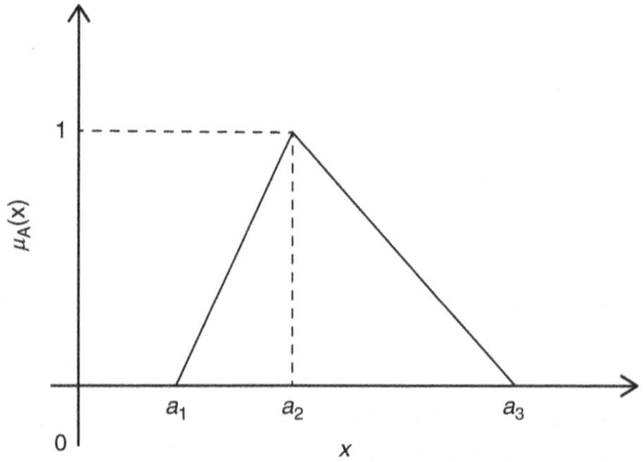

Fig. 1 Graphical representation of a triangular fuzzy number

3 Fully Fuzzy Linear Programming Problem Having Positive Fuzzy Numbers

The formulation can also be expressed in a compact form using summation sign. Maximize (or) minimize $Z = \sum_{j=1}^{n} \tilde{a}_j \tilde{x}_j$ is the generalized fuzzy objective function of a linear programming problem.
Subject to the linear constraints

$$\sum_{j=1}^{n} \tilde{a}_{ij}\tilde{x}_j (\preceq or \approx or \succeq) \tilde{b}_i;\ i = 1, 2, 3, \ldots, m$$

$$\tilde{x}_j \succeq 0;\ \ j = 1, 2, 3, \ldots, n \text{ (non-negativity condition)}$$

where $\tilde{a}_j$,$\tilde{a}_{ij}$,$\tilde{b}_i$ $(i = 1, 2, 3, \ldots, m\ and\ j = 1, 2, 3, \ldots, n)$ are positive fuzzy numbers.

4 Proposed Method of Solving

This proposed method can be used to find fuzzy optimal solution for fully FLP problems with equality constrains having positive fuzzy variables. The steps of proposed method are:

Step 1: Formulate the fully FLP problem with positive variables.
Step 2: Solve the formulated problems using simplex method(Big-M).
Note: In Big-M method, we have to use proposed operations on positive fuzzy numbers.
Using these simple steps, we can get fuzzy optimal solution without converting them to crisp.

5 Numeral Example

Consider the following numerical example [10].

Maximize $\tilde{Z} \approx (0, 1, 4) \otimes \tilde{x_1} \oplus (2, 4, 5) \otimes \tilde{x_2}$
Subject to

$$(2, 3, 7) \otimes \tilde{x_1} \oplus (2, 4, 5) \otimes \tilde{x_2} \approx (6, 18, 46)$$

$$(0, 2, 4) \otimes \tilde{x_1} \oplus (3, 5, 8) \otimes \tilde{x_2} \approx (6, 19, 52)$$

$$\tilde{x_1}, \tilde{x_2} \succeq 0$$

Here, decision variables, unit contributions of the decision-making variables, coefficient of the decision variable, and right-hand side in the constrains of the problem all are positive fuzzy numbers.

Note: $-\tilde{M} = (-M, -M, -M), \tilde{0} = (0, 0, 0)$ and $\tilde{1} = (1, 1, 1)$
From the first iteration (Table 1), the most negative value of $\tilde{Z}_j - \tilde{C}_j$ is $(-5M - 2, -9M - 4, -13M - 5)$, and the minimum ratio is (2, 19/5, 13/2). Therefore, $\tilde{x_2}$ is incoming variable, and $\tilde{A}_2$ is leaving from the basic variable.

From the second iteration, the most negative value of $\tilde{Z}_j - \tilde{C}_j$ is $(-2M, -\frac{7M}{5} + \frac{3}{5}, -\frac{9M}{2} - \frac{3}{2})$, and the minimum ratio is (1, 2, 3). Therefore, $\tilde{x_1}$ is incoming variable, and $\tilde{A}_1$ is leaving from the basic variable (Table 2).

In third iteration (Table 3), all the values of $\tilde{Z}_j - \tilde{C}_j$ is positive. Therefore, the optimal solution is $\tilde{x_1} \approx (1, 2, 3)$, $\tilde{x_2} \approx (2, 3, 5)$ and the maximized $\tilde{Z} \approx (4, 14, 37)$.

Table 1 First iteration

	$\tilde{C}_j$	(0, 1, 4)	(2, 4, 5)	$-\tilde{M}$	$-\tilde{M}$		
	B.V	$\tilde{x_1}$	$\tilde{x_2}$	$\tilde{A}_1$	$\tilde{A}_2$	RHS	Min. ratio
$-\tilde{M}$	$\tilde{A}_1$	(2, 3, 7)	(2, 4, 5)	$\tilde{1}$	$\tilde{0}$	(6, 18, 46)	(3, 9/2, 46/5)
$-\tilde{M}$	$\tilde{A}_2$	(0, 2, 4)	(3, 5, 8)	$\tilde{0}$	$\tilde{1}$	(6, 19, 52)	(2, 19/5, 13/2)
	$\tilde{Z}_j$	$(-2M, -5M, -11M)$	$(-5M, -9M, -13M)$	$-\tilde{M}$	$-\tilde{M}$		
	$\tilde{Z}_j - \tilde{C}_j$	$(-2M, -5M -1, -11M -4)$	$(-5M -2, -9M -4, -13M -5)$	$\tilde{0}$	$\tilde{0}$		

Table 2 Second iteration

	$\tilde{C}_j$	(0, 1, 4)	(2, 4, 5)	$-\tilde{M}$		
	B.V	$\tilde{x_1}$	$\tilde{x_2}$	$\tilde{A}_1$	RHS	Min. ratio
$-\tilde{M}$	$\tilde{A}_1$	(2, 7/5, 9/2)	(0, 0, 0)	$\tilde{1}$	(2, 14/5, 27/2)	(1, 2, 3)
(2, 4, 5)	$\tilde{x_2}$	(0, 2/5, 1/2)	(1, 1, 1)	$\tilde{0}$	(2, 19/5, 13/2)	Not defined
	$\tilde{Z}_j$	$(-2M, -\frac{7M}{5} + \frac{8}{5}, -\frac{9M}{2} + \frac{5}{2})$	(2, 4, 5)	$-\tilde{M}$		
	$\tilde{Z}_j - \tilde{C}_j$	$(-2M, -\frac{7M}{5} + \frac{3}{5}, -\frac{9M}{2} - \frac{3}{2})$	(0, 0, 0)	$\tilde{0}$		

Table 3 Third iteration

	$\tilde{C}_j$	(0, 1, 4)	(2, 4, 5)	
	B.V	$\tilde{x_1}$	$\tilde{x_2}$	RHS
(0, 1, 4)	$\tilde{x_1}$	(1, 1, 1)	(0, 0, 0)	(1, 2, 3)
(2, 4, 5)	$\tilde{x_2}$	(0, 0, 0)	(1, 1, 1)	(2, 3, 5)
	$\tilde{Z}_j$	(0, 1, 4)	(2, 4, 5)	
	$\tilde{Z}_j - \tilde{C}_j$	(0, 0, 0)	(0, 0, 0)	

6 Comparison and Advantages

The optimal solution that we got from proposed method is $\tilde{x_1} \approx (1, 2, 3)$,$\tilde{x_2} \approx (2, 3, 5)$, maximize $\tilde{Z} \approx (4, 14, 37)$ same as solution from Amith Kumar's method. Previously, there is no direct method for solving fully FLP problem without converting them to crisp. Here we proposed a method for solving fully FLP problems with equality constrains having positive fuzzy numbers.

7 Conclusion

In this paper, we have proposed a direct method for solving a fully FLP problem with equality constrains having positive fuzzy numbers which never been done before. In this method there is no need to convert fully FLP problem to crisp. The problem will be solved by the traditional way of simplex method(Big-M). The numerical example is illustrated. From this, it is guaranteed that the proposed method provides the optimal solution for a fully FLP problem with equality constrains having positive fuzzy numbers.

References

1. Zadeh and Bellman: Decision making in a fuzzy environment, Management science. **17**, PP.141–164 (1970)
2. H.J.Zimmerman: Fuzzy programming and linear programming with several objective functions, Fuzzy sets and systems. **1**, PP.45–55 (1978)
3. Amit Kumar and Jagdeep Kaur: A New method for solving fuzzy linear programs with trapezoidal fuzzy numbers,International scientific publications and consulting services (ISPACS). (2011) doi:10.5899
4. J. Buckley and T. Feuring:Evolutionary Algorithm Solution to Fuzzy Problems: Fuzzy Linear Programming, Fuzzy Sets and Systems. **109**, PP.35–53 (2000)
5. T. Allahviranloo, F. H. Lotfi, M. K. Kiasary, N. A. Kiani and L. Alizadeh: Solving Fully Fuzzy Linear Programming Problem by the Ranking Function, Applied Matematical Sciences. **2**, PP.19–32 (2008)

6. F. H. Lotfi, A. Ebrahimnejad, S. H. Nasseri and M. Soltanifar: A Primal Dual Method for Linear Programming Problems with Fuzzy Variables, European Journal of Industrial Engineering. **4**, PP.189–209 (2010)
7. Amit Kumar, Jagdeep Kaur, Pushpinder Singh: Fuzzy optimal solution of fully fuzzy linear programming problems with inequality constraints, International Journal of Mathematical and Computer Sciences. **6**, PP.37–41 (2010)
8. S. H. Nasseri et al: Fully fuzzy linear programming with inequality constraints, International Journal of Industrial Mathematics. **5**, Article ID IJIM-00280, 8 pages (2013)
9. M. Dehghan, B. Hashemi and M. Ghatee: Computational methods for solving fully fuzzy linear systems, Applied Mathematics and Computation. **179**, PP.328–343 (2006)
10. Kumar A, Kaur J and Singh P: A new method for solving fully fuzzy linear programming problems, Appl. Math. Modell. **35**, PP.817–823 (2011)

An Optimal Deterministic Two-Warehouse Inventory Model for Deteriorating Items

K. Rangarajan and K. Karthikeyan

Abstract In this paper, an optimal deterministic two-warehouse inventory model for non-instantaneous deteriorating items under the dispatching policy of last in, first out (LIFO) has been developed. The demand rate and deterioration rate of an item are considered as ramp type and time dependent. Time dependent holding cost, inflation and shortages in inventory are allowed, which is partially backlogged are the various factors considered in this model. The main objective of this model is to develop an optimal policy, which minimizes the total average inventory cost. Finally, this model is illustrated, thoroughly by numerical examples.

1 Introduction

In today's wholesale business market, we can see various situations like price discounts given by the supplier, customer's high demand for the product, product's storage cost that is low, and seasonal products that make the retailer to buy more than the owned warehouse capacity. Then the excess products are kept in other storage places. Such place is called as rented warehouse. Products start to deteriorate after a time lag, as the retailer received it from the supplier or from the factory without using any preservation technology. Those types of products are called as non-instantaneous deteriorating items. In that time lag, there is no deterioration occurring for the products. In 1963, an inventory model for exponentially decaying inventory is first studied by Ghare and Schrader [1]. Many researchers developed the two-warehouse inventory models with various features like inflation, permissible delay in payments, allowing shortages or not, and so on. Skouri et al. [3] considered a two-warehouse model with ramp-type demand. Sanni et al. [4] considered an EOQ model for items with three-parameter Weibull distribution deterioration, ramp-type demand, and shortages. Kumar et al. [5] considered a partial backlogging inventory

K. Rangarajan · K. Karthikeyan (✉)
Department of Mathematics, SAS, Vellore Institute Of Technology, Vellore, India
e-mail: rangarajancheyyaru@gmail.com; kkarthikeyan67@yahoo.co.in; k.karthikeyan@vit.ac.in

V. Madhu et al. (eds.), *Advances in Algebra and Analysis*, Trends in Mathematics,
https://doi.org/10.1007/978-3-030-01120-8_35

model deteriorating items with ramp-type demand. Jaggi et al. [2] developed the two-warehouse inventory model for deteriorating items with imperfect quality under the conditions of permissible delay in payments.

2 Assumptions and Notations

1. Demand rate of an item is a ramp-type function of time.

$$i.e., \quad D(t) = \begin{cases} at, & \text{if } t < \mu \\ a\mu, & \text{if } t \geq \mu \end{cases}$$

   where $t \geq 0$, $a > 0$ stands for initial demand and $\mu > 0$ is a fixed point in time.
2. Deterioration rate of an item is time dependent.
3. Shortages in inventory are allowed, which is partially backlogged and backlogging rate is constant.
4. Holding cost is a linear function of time.
5. Lead time is zero.
6. $t_d > t_r$
7. Replenishment rate is infinite and instantaneous.
8. Planning horizon is assumed as finite.
9. $I_r(t)$: Inventory level in rented warehouse at any time t, $0 \leq t \leq t_r$.
10. $I_o(t)$: Inventory level in owned warehouse at any time t, $0 \leq t \leq t_w$.
11. $I_s(t)$: Inventory level at any time t, $t_w \leq t \leq T$.
12. t_d : Length of time in which the product exhibits no deterioration.
13. t_r : Time at which the inventory level vanishes in rented warehouse.
14. t_w : Time at which the inventory level vanishes in owned warehouse.
15. Q : Order quantity
16. Q_1 : Maximum positive inventory level at time $t = 0$.
17. Q_2 : Maximum negative inventory level at time $t = T$.
18. W : Capacity of owned warehouse.
19. $Q_1 - W$: Capacity of rented warehouse.
20. C_1 : Unit purchasing cost of an item.
21. C_2 : Shortage cost of an item.
22. C_3 : Lost sale cost of an item.
23. A : Ordering cost per unit order is known and constant.
24. T : The time interval between two successive orders.

3 Mathematical Formulation and Solution of the Model : $0 < \mu < t_d < t_r < t_w < T$

Inventory levels at any time in the duration time $(0, T)$ are governed by the following differential equations:

$$\frac{dI_r(t)}{dt} = -at, \qquad 0 \le t \le \mu \quad with\ I_r(\mu-) = I_r(\mu+) \tag{1}$$

$$\frac{dI_r(t)}{dt} = -a\mu, \qquad \mu \le t \le t_r \quad with\ I_r(t_r) = 0 \tag{2}$$

$$\frac{dI_o(t)}{dt} = 0, \qquad 0 \le t \le t_r \quad with\ I_0(0) = I_0(t_r) = W \tag{3}$$

$$\frac{dI_o(t)}{dt} = -a\mu, \qquad t_r \le t \le t_d \quad with\ I_0(t_r) = W \tag{4}$$

$$\frac{dI_o(t)}{dt} + \theta_2 t I_o(t) = -a\mu, \qquad t_d \le t \le t_w \quad with\ I_0(t_w) = 0 \tag{5}$$

$$\frac{dI_s(t)}{dt} = -a\mu\delta, \qquad t_w \le t \le T \quad with\ I_s(t_w) = 0 \tag{6}$$

Now solving all the Eqs. (1)–(6) by using boundary conditions, we have

$$I_r(t) = -\frac{at^2}{2} - a\mu\left(t_r - \frac{\mu}{2}\right), \qquad 0 \le t \le \mu \tag{7}$$

$$I_r(t) = a\mu\,(t_r - t)\,, \qquad \mu \le t \le t_d \tag{8}$$

$$I_o(t) = W, \qquad 0 \le t \le t_r \tag{9}$$

$$I_o(t) = a\mu\,(t_r - t) + W, \qquad t_r \le t \le t_d \tag{10}$$

$$I_0(t) = a\mu\left[t_w - t + \frac{\theta_2}{6}\left(t_w^3 - 3t^2 t_w + 2t^3\right)\right], \qquad t_d \le t \le t_w \tag{11}$$

$$I_r(t) = a\mu\delta\,(t_w - t)\,, \qquad t_w \le t \le T \tag{12}$$

From Eqs. (10) and (11), we can obtain initial capacity of inventory level of OW as

$$W = a\mu\left[t_w - t_r + \frac{\theta_2}{6}\left(t_w^3 + 2t_d^3 - 3t_d^2 t_w\right)\right] \tag{13}$$

From Eq. (12) at t=0, we can obtain initial capacity of inventory level of RW as

$$Q_1 = W + a\mu\left[t_r - \frac{\mu}{2}\right] \tag{14}$$

With boundary condition $I_s(t) = -Q_2$, we can obtain the maximum capacity of negative inventory level as

$$I_s(t) = -Q_2 = -a\mu\delta\,(t_w - T) \tag{15}$$

Total inventory is given by $Q = Q_1 + Q_2$
Total inventory cost (TC) per cycle consists of the following costs:

1. Setup cost: $SC = \frac{A}{T}$
2. Total holding cost per cycle for rented warehouse:

$$HC_r = \tfrac{1}{T}\int_0^{t_r}(x_1 + y_1 t)I_r(t)\,e^{-Rt}dt$$

$$HC_r = \frac{1}{T}\begin{bmatrix} x_1\left\{-\frac{2a\mu^3}{3} + a\mu^2 t_r - R\left(-\frac{3a\mu^4}{8} + \frac{a\mu^3 t_r}{2}\right)\right\} \\ +y_1\left\{-\frac{3a\mu^4}{8} + \frac{a\mu^3 t_r}{2} - R\left(-\frac{4a\mu^5}{15} + \frac{a\mu^4 t_r}{3}\right)\right\} \\ +x_1 a\mu\left\{\frac{t_r^2}{2} - \mu t_r + \frac{\mu^2}{2} - R\left(\frac{t_r^3}{6} - \frac{\mu^2 t_r}{2} + \frac{\mu^3}{3}\right)\right\} \\ +y_1 a\mu\left\{\frac{t_r^3}{6} - \frac{\mu^2 t_r}{2} + \frac{\mu^3}{3} - R\left(\frac{t_r^4}{12} - \frac{\mu^3 t_r}{3} + \frac{\mu^4}{4}\right)\right\} \end{bmatrix}$$

3. Total holding cost per cycle for owned warehouse:

$$HC_o = \tfrac{1}{T}\int_0^{t_w}(x_2 + y_2 t)I_o(t)\,e^{-Rt}dt$$

$$HC_o = \frac{1}{T}\begin{bmatrix} x_2 W\left(t_r - \frac{Rt_r^2}{2}\right) + y_2 W\left(\frac{t_r^2}{2} - \frac{Rt_r^3}{3}\right) \\ +x_2 W(t_d - t_r) + x_2 a\mu\left(t_d t_r - \frac{t_r^2}{2} - \frac{t_d^2}{2}\right) \\ -x_2 R\left\{\frac{W(t_d^2 - t_r^2)}{2} - a\mu\left(\frac{t_d^2 t_r}{2} - \frac{t_r^3}{6} - \frac{t_d^3}{3}\right)\right\} \\ +y_2 W\frac{(t_d^2 - t_r^2)}{2} + y_2 a\mu\left(\frac{t_d^2 t_r}{2} - \frac{t_r^3}{6} - \frac{t_d^3}{3}\right) \\ -y_2 R\left\{\frac{W(t_d^3 - t_r^3)}{3} - a\mu\left(\frac{t_d^3 t_r}{3} - \frac{t_r^4}{12} - \frac{t_d^4}{4}\right)\right\} \\ +x_2 a\mu\left\{\begin{array}{c} \frac{t_w^2}{2} - t_d t_w + \frac{t_d^2}{2} + \frac{\theta_2}{6}\left(\frac{t_w^4}{2} - t_d t_w^3 - \frac{t_d^4}{2} + t_w t_d^3\right) \\ -R\left(\frac{t_w^3}{6} - \frac{t_d^2 t_w}{2} + \frac{t_d^3}{3} + \frac{\theta_2}{6}\left(\frac{3t_w^5}{20} - \frac{t_d^2 t_w^3}{2} - \frac{2t_d^5}{5} + \frac{3t_w t_d^4}{4}\right)\right) \end{array}\right\} \\ +y_2 a\mu\left\{\begin{array}{c} \frac{t_w^3}{6} - \frac{t_d^2 t_w}{2} + \frac{t_d^3}{3} + \frac{\theta_2}{6}\left(\frac{3t_w^5}{20} - \frac{t_d^2 t_w^3}{2} - \frac{2t_d^5}{5} + \frac{3t_w t_d^4}{4}\right) \\ -R\left(\frac{t_w^4}{12} - \frac{t_d^3 t_w}{3} + \frac{t_d^4}{4} + \frac{\theta_2}{6}\left(\frac{t_w^6}{15} - \frac{t_d^3 t_w^3}{3} - \frac{2t_d^6}{6} + \frac{3t_w t_d^5}{5}\right)\right) \end{array}\right\} \end{bmatrix}$$

4. Total deterioration cost per cycle for rented warehouse: $DC_r = 0$
5. Total deterioration cost per cycle for owned warehouse: $DC_w = \frac{C_1}{T}\int_{t_d}^{t_w} \theta_2 t I_0(t)$
$e^{-Rt}dt = \frac{C_1\theta_2 a\mu}{T}\left[\frac{t_w^3}{6} - \frac{t_d^2 t_w}{2} + \frac{t_d^3}{3} - R\left(\frac{t_w^4}{12} - \frac{t_d^3 t_w}{3} + \frac{t_d^4}{4}\right)\right]$
6. Shortage cost per cycle:
$SC = \frac{C_2}{T}\int_{t_w}^{T} I_s(t)e^{-Rt}dt = -\frac{C_2 a\mu\delta}{T}\left[Tt_w - \frac{t_w^2}{2} - \frac{T^2}{2} - R\left(\frac{T^2 t_w}{2} - \frac{t_w^3}{6} - \frac{T^3}{3}\right)\right]$
7. Cost due to lost sales per cycle :
$CLS = \frac{C_3}{T}\int_{t_w}^{T} D(t)(1-\delta)e^{-Rt}dt = \frac{C_3 a\mu(\delta-1)}{T}\left[T - t_w - R\left(\frac{T^2 - t_w^2}{2}\right)\right]$

The total average inventory cost (TC) per cycle is given by
$TC = OC + HC_r + HC_o + DC_r + DC_o + SC + CLS$
Our main objective of this model is to minimize the total average inventory cost per cycle.
Necessary condition for a total average inventory cost to be minimized are
$(i)\ \frac{d(TC)}{dt_w} = 0\ and\ \ (ii)\ \frac{d^2(TC)}{dt_w^2} > 0$

4 Numerical Examples

Using MATLAB software, the following examples are solved, and the optimal solutions are found.

Example 1 (Model I—Partial Backlogging Model) Let $A = Rs.300$, $x_1 = 3$, $y_1 = 0.3$, $x_2 = 1$, $y_2 = 0.1$, $a = 150$, $W = 100$, $R = 0.1$, $\mu = 1.5$ *weeks*, $t_d = 5$ *weeks*, $t_r = 2$ *weeks*, $T = 10$ *weeks*, $\theta_1 = 0.5$, $\theta_2 = 0.5$, $\delta = 0.5$, $C_1 = 5$, $C_2 = 7$ $C_3 = 3$.
Optimal solutions are $t_w^* = 6.4543$, *and* $TC^* = Rs.716.1$

Example 2 (Model II—Complete Backlogging Model) Let $A = Rs.300$, $x_1 = 3$, $y_1 = 0.3$, $x_2 = 1$, $y_2 = 0.1$, $a = 150$, $W = 100$, $R = 0.1$, $\mu = 1.5$ *weeks*, $t_d = 5$ *weeks*, $t_r = 2$ *weeks*, $T = 10$ *weeks*, $\theta_1 = 0.5$, $\theta_2 = 0.5$, $\delta = 1$, $C_1 = 5$, $C_2 = 7$ $C_3 = 3$.
Optimal solutions are $t_w^* = 7.2148$, *and* $TC^* = Rs.1{,}121$

5 Observations

1. The total optimal inventory cost (TC^*) in Model I is less than the total optimal inventory cost in Model II.
2. The total optimal time (t_w^*) in Model I is less than the total optimal time in Model II.

6 Conclusion

An optimal deterministic two-warehouse inventory model is developed under last in, first out dispatching policy for non-instantaneous deteriorating items with ramp-type demand rate, time-varying deterioration rate, time-varying holding cost, and inflation. Shortages in inventory are also considered in this model. This model is applicable to retailers to minimize the total cost for maintaining the inventory in various situations like price discounts given by supplier; customer's high demand for the product; product's storage cost that is low; and some new brand of cosmetic products, electronic items, seasonal products, etc. which are entered in the business market, in which the demand for those products is increasing at the beginning up to a particular time and then remains constant for the remaining period. An optimal policy, which minimizes the total inventory cost, is developed. Numerical examples for each case are given to explain the developed model.

References

1. Ghare, P.M., Schrader, G.F.: A model for exponentially decaying inventories. Journal of indusrial engineering. **14**, 238–243 (1963).
2. Jaggi, C.K., Cardenos Barron,L., Tiwari,S., Sha, A.A. : Two-warehouse inventory model for deteriorating items with imperfect quality under the conditions of permissible delay in payments. Scientia Iranica. **24**, 390–412 (2017).
3. Skouri, K., Konstantaras, I. : Two-warehouse inventory models for deteriorating products with ramp type demand rate. Journal of Industrial and Management Optimization. **9**, 855–883 (2013).
4. Sanni, S.S., Chukwu, W.I.E. : An Economic order quantity model for items with three parameter Weibull distribution deterioration, ramp type demand and shortages. Applied Mathematical Modelling. **37**, 9698–9706 (2013).
5. Kumar, S., Rajput, U.S. : A partially backlogging inventory model for deteriorating items with ramp type demand rate. American Journal of Operational Research. **5**, 39–46 (2015).

Analysis on Time to Recruitment in a Three-Grade Marketing Organization Having Classified Sources of Depletion of Two Types with an Extended Threshold and Shortage in Manpower Forms Geometric Process

S. Poornima and B. Esther Clara

Abstract Shortage in manpower owing to classified sources of depletion takes place in any marketing organizations. As frequent recruitment involves cost and time, it is not advisable to recruit as when the shortage in manpower occurs. Since the shortage in manpower and the inter-decision times are probabilistic, the organization requires appropriate recruitment policy for recruiting personnel. In this paper, the problem of time to recruitment based on shock model approach is studied by considering classified policy and transfer decisions when the shortage in manpower due to policy decisions forms geometric process and the extended threshold gives a better allowable cumulative shortage in manpower in the organization. Analytical expressions, namely, mean and variance for the time to recruitment, are obtained, and the results are numerically illustrated, and conclusions are drawn.

Keywords Three-grade marketing organization · Two sources of shortage in manpower · Classified policy and transfer decisions · Extended threshold · Geometric process · Univariate CUM policy of recruitment · Shock model approach

MSC Classification codes: Primary: 90B70, Secondary: 60H30, 60K05

S. Poornima (✉) · B. Esther Clara
PG and Research Department of Mathematics, Bishop Heber College, Trichy, India
e-mail: poornimasridharan3@gmail.com; poornima.res@bhc.edu.in; jecigrace@gmail.com; estherclara@bhc.edu.in

V. Madhu et al. (eds.), *Advances in Algebra and Analysis*, Trends in Mathematics,
https://doi.org/10.1007/978-3-030-01120-8_36

1 Introduction

Attrition of manpower is usual in all organizations and that will lead to decline in the total strength of marketing personnel which affects the organization, if it is not planned properly. For balancing this unpredictable shortage in manpower, suitable recruitment policy has to be designed. Several researchers have studied the problem of time to recruitment for two-grade manpower system using shock model approach. In this context, the authors in [1, 2] and [3] have given statistical approach for the stochastic models in manpower planning and manpower models in social processes. In [7], the authors have obtained a stochastic model for two-grade manpower system with wastage as a geometric process which is extended for three-grade manpower system in more general setting. The authors in [8] have obtained the mean and variance of the time to recruitment for an organization consisting of two grades (two-grade manpower system) by assuming that the distribution of shortage in manpower in different decisions and that of the inter-decision times as exponential according as the threshold for the shortage in manpower in the organization is maximum (minimum) of the exponential thresholds in the two grades. Recently, in [4] and [5], the authors have obtained the time to recruitment for stochastic model under two sources of depletion of manpower using univariate policy of recruitment by considering various assumptions for breakdown thresholds, shortage in manpower, and the inter-policy decisions. This paper analyzes the research work by taking into account a realistic aspect of policy decisions and transfer decisions having high or low intensity of attrition for inter-policy and inter-transfer decisions. An attempt has been made to study the problem of time to recruitment for a three-grade manpower system with an extended threshold and loss of manpower due to policy decisions forms geometric process. An extended threshold is introduced to give a better allowable cumulative shortage in manpower in the manpower system. It is assumed that the inter-policy decisions times for the three grades form the same ordinary renewal process; the inter-transfer decisions times for the three grades form the same ordinary renewal process which is different from that of inter-policy decisions. The conventional breakdown threshold used in all the earlier studies is identified as the level of alertness in the present paper. If the organization is not alert when the cumulative shortage in manpower exceeds this level of alertness, an allowable shortage in manpower of magnitude D is permitted. However, recruitment has to be done when the cumulative shortage in manpower exceeds this extended threshold. A univariate recruitment policy, usually known as univariate CUM policy of recruitment in the literature, is based on the replacement policy associated with the shock model approach in reliability theory and is stated as follows: **Recruitment is made whenever the cumulative shortage in manpower exceeds the extended threshold**. Analytical results related to time to recruitment are derived, and relevant conclusions are made with the help of numerical illustrations.

2 Model Description

Let us consider three-grade marketing organizations having policy and transfer decisions at random epochs in $(0, \infty)$. The policy decisions are classified into two types depending upon their intensity of attrition. Let X_{A_i}, X_{B_i} and X_{C_i}, $i = 1, 2, 3, \ldots$ be a sequence of exponential random variables representing the shortage in manpower due to ith policy decision in grade A, B, and C, respectively. Here X_{A_i},X_{B_i} and X_{C_i}, $i = 1, 2, 3, \ldots$ follows geometric process with positive constants $a_1, a_2, a_3 > 0$, respectively. Let $\widetilde{X_m}$ be the cumulative shortage in manpower for the three grades in the first m policy decisions. Let Y_{A_j},Y_{B_j}, and Y_{C_j} be independent and identically distributed exponential random variables representing the loss of manpower in the organization due to jth transfer decision with mean $\frac{1}{\alpha_{2A}}$,$\frac{1}{\alpha_{2B}}$, and $\frac{1}{\alpha_{2C}}$, respectively, $\alpha_{2A}, \alpha_{2B}, \alpha_{2C} > 0$. Let $\widetilde{Y_n}$ be the cumulative loss of manpower for the three grades in the first n transfer decisions. For $i = 1, 2, \ldots,$ let U_i be independent and identically distributed hyper-exponential random variable representing the time between $(i - 1)$th and ith policy decisions with mean $\frac{p_1}{\lambda_1} + \frac{(1-p_1)}{\lambda_2}$, $0 < p_1 < 1$,where p_1 is the proportion of policy decisions having high attrition rate $\lambda_1 > 0$ and $(1 - p_1)$ is the proportion of policy decisions having low attrition rate $\lambda_2 > 0$. For $j = 1, 2, \ldots,$ let V_j be independent and identically distributed hyper-exponential random variable representing the time between $(j - 1)$th and jth transfer decisions with mean $\frac{p_2}{\lambda_3} + \frac{(1-p_2)}{\lambda_4}$, $0 < p_2 < 1$, where p_2 is the proportion of transfer decisions having high attrition rate $\lambda_3 > 0$ and $(1 - p_2)$ is the proportion of transfer decisions having low attrition rate $\lambda_4 > 0$. Let $N_p(t)$ and $N_{Tr}(t)$ be the number of policy and transfer decisions taken in the organization during the period of recruitment $(0, t]$. Let $\widetilde{X_{N_p(t)}}$ and $\widetilde{Y_{N_{Tr}(t)}}$ be the total shortage in manpower in $N_p(t)$ decisions and $N_{Tr}(t)$ decisions. Let the cumulative distribution function of the random variable K be $W_K(.)$ (density function $w_K(.)$, and the Laplace transform of $w_K(.)$ be $\overline{w}_K(.)$). Assume that Z_A, Z_B, and Z_C represent the threshold levels for the cumulative shortage in manpower in grade A, B, and C with mean $\frac{1}{\theta_A}$,$\frac{1}{\theta_B}$, and $\frac{1}{\theta_C}$, respectively, where $\theta_A, \theta_B, \theta_C > 0$. Let Z be the threshold level for the cumulative shortage in manpower for the entire organization and D represent the extended threshold with mean $\frac{1}{\theta_D}$, respectively, where $\theta_D > 0$. Let T be the time to recruitment for the entire organization. Here, X_i and Y_j are linear and cumulative, and Z , X_i , and Y_j are statistically independent.

2.1 *Analytical Results*

The event of time to recruitment is defined as follows: Recruitment occurs beyond t($t > 0$) if and only if the total shortage in manpower up to $N_p(t)$ policy decisions and $N_{Tr}(t)$ transfer decisions does not exceed the breakdown threshold of the organization, and it is given by

$$\{T > t\} \Longleftrightarrow \left\{\widetilde{X_{N_p}(t)} + \widetilde{Y_{N_{Tr}(t)}} < Z\right\} \tag{1}$$

Hence the probability of occurrences of these two events is equal.

$$P(T > t) = P(\widetilde{X_{N_p}(t)} + \widetilde{Y_{N_{Tr}(t)}} < Z) \tag{2}$$

Invoking the law of total probability and the result of renewal theory [6], the survival function of time to recruitment is determined. The rth moment for the time to recruitment is determined by taking the rth derivative of the Laplace transform of density function for the random variable with respect to s and at s=0. Using this result, the fundamental performance measures like mean and variance of time to recruitment are determined.

Let $Z = (Z_A + Z_B + Z_C) + D$. Conditioning upon D, we get the distribution of the threshold as

$$P(Z > t) = C_1[e^{-\theta_D t} - e^{-\theta_A t}] - C_2[e^{-\theta_D t} - e^{-\theta_B t}] + C_3[e^{-\theta_D t} - e^{-\theta_C t}] \tag{3}$$

Taking derivative for the Laplace transform of T at s=0 gives the mean time to recruitment.

$$E(T) = (C_1 - C_2 + C_3)T_D - C_1T_1 + C_2T_2 - C_3T_3 \tag{4}$$

Here, $C_1 = \frac{\theta_B\theta_C\theta_D(\theta_B-\theta_C)}{(\theta_A-\theta_B)(\theta_B-\theta_C)(\theta_A-\theta_C)(\theta_A-\theta_D)}$; $C_2 = \frac{\theta_A\theta_C\theta_D(\theta_A-\theta_C)}{(\theta_A-\theta_B)(\theta_B-\theta_C)(\theta_A-\theta_C)(\theta_B-\theta_D)}$; $C_3 = \frac{\theta_A\theta_B\theta_D(\theta_A-\theta_B)}{(\theta_A-\theta_B)(\theta_B-\theta_C)(\theta_A-\theta_C)(\theta_C-\theta_D)}$. Similarly, the second moment for time to recruitment is determined by taking second derivative for the Laplace transform of T at s=0. Thus, from the two results, variance of time to recruitment can be obtained. For all the above cases, the following notations are used

$$E = \prod_{r=1}^{m} \overline{w}_U^r\left(\frac{\theta_D}{a_1^{r-1}}\right)\prod_{r=1}^{m} \overline{w}_U^r\left(\frac{\theta_D}{a_2^{r-1}}\right)\prod_{r=1}^{m} \overline{w}_U^r\left(\frac{\theta_D}{a_3^{r-1}}\right);$$

$$A_1 = \prod_{r=1}^{m} \overline{w}_U^r\left(\frac{\theta_A}{a_1^{r-1}}\right)\prod_{r=1}^{m} \overline{w}_U^r\left(\frac{\theta_A}{a_2^{r-1}}\right)\prod_{r=1}^{m} \overline{w}_U^r\left(\frac{\theta_A}{a_3^{r-1}}\right);$$

$$A_2 = \prod_{r=1}^{m} \overline{w}_U^r\left(\frac{\theta_B}{a_1^{r-1}}\right)\prod_{r=1}^{m} \overline{w}_U^r\left(\frac{\theta_B}{a_2^{r-1}}\right)\prod_{r=1}^{m} \overline{w}_U^r\left(\frac{\theta_B}{a_3^{r-1}}\right);$$

$$A_3 = \prod_{r=1}^{m} \overline{w}_U^r\left(\frac{\theta_C}{a_1^{r-1}}\right)\prod_{r=1}^{m} \overline{w}_U^r\left(\frac{\theta_C}{a_2^{r-1}}\right)\prod_{r=1}^{m} \overline{w}_U^r\left(\frac{\theta_C}{a_3^{r-1}}\right);$$

$$F = \overline{w}_{Y_A}(\theta_D)\overline{w}_{Y_B}(\theta_D)\overline{w}_{Y_C}(\theta_D);\quad B_1 = \overline{w}_{Y_A}(\theta_A)\overline{w}_{Y_B}(\theta_A)\overline{w}_{Y_C}(\theta_A);$$

$$B_2 = \overline{w}_{Y_A}(\theta_B)\overline{w}_{Y_B}(\theta_B)\overline{w}_{Y_C}(\theta_B);\quad B_3 = \overline{w}_{Y_A}(\theta_C)\overline{w}_{Y_B}(\theta_C)\overline{w}_{Y_C}(\theta_C);$$

$$f_1 = d_1 = d_3 = d_5 = p_2\lambda_3\lambda_4 + (1 - p_2)\lambda_3\lambda_4;$$

$$f_2 = d_2 = d_4 = d_6 = p_2\lambda_3 + (1 - p_2)\lambda_4;$$

$$f_1{}^* = \lambda_3 + \lambda_4 - Fp_2\lambda_3 - F(1 - p_2)\lambda_4;\ f_2{}^* = \lambda_3\lambda_4 - Fp_2\lambda_3\lambda_4 - F(1 - p_2)\lambda_3\lambda_4;$$

$$\psi_1 = \frac{f_1{}^* - \sqrt{f_1{}^{*2} - 4f_2{}^*}}{2};\ \psi_2 = \frac{f_1{}^* + \sqrt{f_1{}^{*2} - 4f_2{}^*}}{2};$$

For $i = 1, 2, 3$ and $j = 1, 3, 5$

$$\gamma_j = \frac{d_j{}^* - \sqrt{d_j{}^{*2} - 4d_{j+1}{}^*}}{2};\ \gamma_{j+1} = \frac{d_j{}^* + \sqrt{d_j{}^{*2} - 4d_{j+1}{}^*}}{2};$$

$$d_j{}^* = \lambda_3 + \lambda_4 - B_i p_2\lambda_3 - B_i(1 - p_2)\lambda_4;\ d_{j+1}{}^* = \lambda_3\lambda_4 - B_i p_2\lambda_3\lambda_4 - B_i(1 - p_2)\lambda_3\lambda_4;$$

$$T_i = \left[\begin{array}{c} \left[\dfrac{\gamma_j\gamma_{j+1} - (1 - B_i)d_j}{\gamma_j\gamma_{j+1}}\right]\displaystyle\sum_{m=0}^{\infty}\left[m\overline{w}_U^{m-1}(0)\overline{w}'_U(0) - (m+1)\overline{w}_U^m(0)\overline{w}'_U(0)\right]A_i \\ -\left[\dfrac{(1 - B_i)(\gamma_j d_{j+1} - d_j)}{\gamma_{j+1} - \gamma_j}\right]\displaystyle\sum_{m=0}^{\infty}\dfrac{\overline{w}_U^m(\gamma_j)}{\gamma_j^2}A_i - \left[\dfrac{(1 - B_i)(d_j - d_{j+1}\gamma_{j+1})}{\gamma_{j+1} - \gamma_j}\right]\displaystyle\sum_{m=0}^{\infty}\dfrac{\overline{w}_U^m(\gamma_{j+1})}{\gamma_{j+1}^2}A_i \\ +\left[\dfrac{(1 - B_i)(\gamma_j d_{j+1} - d_j)}{\gamma_{j+1} - \gamma_j}\right]\displaystyle\sum_{m=0}^{\infty}\dfrac{\overline{w}_U^{m+1}(\gamma_j)}{\gamma_j^2}A_i + \left[\dfrac{(1 - B_i)(d_j - d_{j+1}\gamma_{j+1})}{\gamma_{j+1} - \gamma_j}\right]\displaystyle\sum_{m=0}^{\infty}\dfrac{\overline{w}_U^{m+1}(\gamma_{j+1})}{\gamma_{j+1}^2}A_i \end{array}\right]$$

$$T_D = \left[\begin{array}{c} \left[\dfrac{\psi_1\psi_2 - (1 - F)f_1}{\psi_1\psi_2}\right]\displaystyle\sum_{m=0}^{\infty}\left[m\overline{w}_U^{m-1}(0)\overline{w}'_U(0) - (m+1)\overline{w}_U^m(0)\overline{w}'_U(0)\right]E \\ -\left[\dfrac{(1 - F)(\psi_1 f_2 - f_1)}{\psi_2 - \psi_1}\right]\displaystyle\sum_{m=0}^{\infty}\dfrac{\overline{w}_U^m(\psi_1)}{\psi_1^2}E - \left[\dfrac{(1 - F)(f_1 - f_2\psi_2)}{\psi_2 - \psi_1}\right]\displaystyle\sum_{m=0}^{\infty}\dfrac{\overline{w}_U^m(\psi_2)}{\psi_2^2}E \\ +\left[\dfrac{(1 - F)(\psi_1 f_2 - f_1)}{\psi_2 - \psi_1}\right]\displaystyle\sum_{m=0}^{\infty}\dfrac{\overline{w}_U^{m+1}(\psi_1)}{\psi_1^2}E + \left[\dfrac{(1 - F)(f_1 - f_2\psi_2)}{\psi_2 - \psi_1}\right]\displaystyle\sum_{m=0}^{\infty}\dfrac{\overline{w}_U^{m+1}(\psi_2)}{\psi_2^2}E \end{array}\right]$$

Here $\overline{w}_U(s) = \frac{p_1\lambda_1}{\lambda_1+s} + \frac{(1-p_1)\lambda_2}{\lambda_2+s};\ \overline{w}_U(0) = 1;$

$$\overline{w}'_U(s) = -\frac{p_1\lambda_1}{(\lambda_1 + s)^2} - \frac{(1 - p_1)\lambda_2}{(\lambda_2 + s)^2};\ \ \overline{w}'_U(0) = -\frac{p_1}{\lambda_1} - \frac{(1 - p_1)}{\lambda_2};$$

$$\overline{w}''_U(s) = \frac{2p_1\lambda_1}{(\lambda_1 + s)^3} + \frac{2(1 - p_1)\lambda_2}{(\lambda_2 + s)^3};\ \ \overline{w}''_U(0) = \frac{2p_1}{\lambda_1^2} + \frac{2(1 - p_1)}{\lambda_2^2};$$

Note 1 If the organization does not want to allow transfer of personnel within their sister organizations, then the breakdown threshold for the organizations is given by $Z = min(Z_A, Z_B, Z_C) + D$, and similarly if the organization allows transfer of personnel within their sister organizations, then $Z = max(Z_A, Z_B, Z_C) + D$.

3 Numerical Illustration

The numerical results for this model are determined by varying the rates of geometric processes which are assumed for the shortage in manpower and fixing the other parameters for three grades.

$$(\alpha_{1A} = 0.5; \alpha_{1B} = 0.6; \alpha_{1C} = 0.7; \alpha_{2A} = 0.2; \alpha_{2B} = 0.6; \alpha_{2C} = 0.7; p_1 = 0.4;$$
$$p_2 = 0.3; \theta_A = 0.4; \theta_B = 0.48; \theta_C = 0.52; \theta_D = 0.5; \lambda_1 = 0.3; \lambda_2 = 0.3;$$
$$\lambda_3 = 0.6; \lambda_4 = 0.2)$$

Table 1 Mean and variance of time to recruitment

a_1	a_2	a_3	$E(T)$	$V(T)$
0.3	0.2	0.6	0.8939	606.8457
0.4	0.2	0.6	0.9927	602.0091
0.5	0.2	0.6	1.0655	597.8405
3	2	6	1.7303	476.2401
4	2	6	1.7383	471.9973
5	2	6	1.7418	469.1710
0.2	0.5	0.3	0.8660	608.5488
0.2	0.6	0.3	0.8910	604.6453
0.2	0.7	0.3	0.9105	601.1701
2	5	3	1.7173	475.9127
2	6	3	1.7204	473.7059
2	7	3	1.7226	472.0575
0.5	0.7	0.2	1.2120	590.6229
0.5	0.7	0.3	1.2202	586.3936
0.5	0.7	0.4	1.2298	582.3832
5	7	2	1.7391	464.5564
5	7	3	1.7563	457.8993
5	7	4	1.7661	453.9912

4 Conclusions

In Table 1, increasing the rates of geometric processes which are assumed for the shortage in manpower for the three-grade marketing organizations, the average time to recruitment increases, and the variance decreases which are found to be realistic.

The above work contributes to the existing literature in such a way that the model in this paper is new by considering (1) the classified policy and transfer decisions (which are recurrent and nonrecurrent) and (2) the extended threshold. It is contemplated to study the present work for the different set of recruitment policies. One may study this paper when the organization allows backup resource after exceeding the breakdown threshold, i.e., recruitment time.

References

1. Bartholomew D.J.: The statistical approach to manpower planning. Statistician. **20**, 3–26, (1971)
2. Bartholomew D.J.: Stochastic Model for Social Processes. 2nd Edition, John Wiley, New York (1973)
3. Bartholomew D.J.: and Forbes A.F.: Statistical Techniques for Manpower Planning. John Wiley, New York (1979)
4. Dhivya S., Srinivasan A.: Estimation of variance of time to recruitment for a two grade manpower system with two sources of depletion and two types of policy decisions. Proceedings of the International Conference on Mathematics and its Applications. Anna University. 1230–1241, (2014)
5. Dhivya S., Srinivasan A.: Stochastic model for time to recruitment under two sources of depletion of manpower using univariate policy of recruitment, to be appear in International Journal of Multidisciplinary Research and Advances in Engineering. **5(4)** (2013)
6. Karlin Samuel, Taylor M. Haward.: A First Course in Stochastic Processes. Second Edition, Academic Press, New York (1975)
7. Saranya P., Srinivasan A.: Stochastic models for a two grade manpower system with wastage as a geometric process. International Journal of Innovative Research in Science, Engineering and Technology. **5(3)**, 3552–3559, (2016)
8. Sathiyamoorthi R., Parthasarathy S.: On the expected time to recruitment in a two graded marketing organization. Indian Association for Productivity Quality and Reliability. **27(1)**, 77–81, (2002)

Neutrosophic Assignment Problem via BnB Algorithm

S. Krishna Prabha and S. Vimala

Abstract This paper attempts to commence branch and bound technique to unravel the triangular fuzzy neutrosophic assignment problem (TFNAP). So far there are many researches based on fuzzy and intuitionistic fuzzy assignment problems; this is the first paper to deal with TFNAP which have been introduced as a simplification of crisp sets and intuitionistic fuzzy sets to indicate vague, imperfect, unsure, and incoherent notification about the existent world problem. Here a real-life agricultural problem where the farmer's objective is to locate the optimal assignment of paddocks to crops in such comportment that the total fertilizer cost becomes least is worked out to illustrate the efficiency of the branch and bound (BnB) algorithm in neutroshopic approach.

Keywords Triangular fuzzy neutroshopic assignment problem · Agricultural problem · Branch and bound algorithm

1 Introduction

To simplify the idea of fuzzy sets and intuitionistic fuzzy sets, Smarandache in 1998 [7] projected the perception of neutrosophic set and neutrosophic logic for managing problems concerning vague, imperfect, uncertain, and incoherent information which may not be handled using fuzzy sets and intuitionistic fuzzy sets. Three different membership degrees explicitly truth-membership degree (T), indeterminacy-membership degree (I), and falsity-membership degree (F), which

S. Krishna Prabha (✉)
Department of Mathematics, Mother Teresa Women's Universiy, Kodaikannal, India

Department of Mathematics, PSNA College of Engineering and Technology, Dindigul, India
e-mail: jvprbh1@gmail.com; krishna_praba@psnacet.edu.in.

S. Vimala
Department of Mathematics, Mother Teresa Women's Universiy, Kodaikannal, India
e-mail: tvimss@gmail.com

V. Madhu et al. (eds.), *Advances in Algebra and Analysis*, Trends in Mathematics,
https://doi.org/10.1007/978-3-030-01120-8_37

elongate out flanked through nonstandard unit interval]− 0, 1+[, are the categorization of the perception of neutrosophic set. Smarandache [7] and Wang et al. projected a subclass of the neutrosophic sets named single-valued neutrosophic sets (SVNS). By merging triangular fuzzy numbers (TFNs) and single-valued neutrosophic set (SVNS), Biswas et al. [4] introduced the idea of triangular fuzzy neutrosophic sets (TFNS).Trapezoidal fuzzy neutrosophic set was proposed by Ye [17], and he urbanized weighted arithmetic and geometric averaging for TFNS. Assignment problem (AP) is applied universal in unraveling genuine real tribulations. Among the premeditated optimization tribulations in administration discipline, the assignment problem has been widely enforced in both mechanized and repair systems. The intent of assignment problem is to consign n tasks to n machines at a least cost. As conservative traditional assignment problems cannot be effectively dealt with situations concerning imprecision in the data, the concept of fuzziness proposed by Zadeh [18] is employed. Researchers like Chen [6], Chen Liang-Hsuan et al. [5], and Long-Sheng Huang et al. [10] have explored various concepts for solving assignment problems. Ones assignment method for unraveling assignment problems was put forth by Hadi Basirzadeh [8]. Yager [16] has introduced a new method to rank the fuzzy subsets of unit interval. Various ranking methods have been proposed by Abbasbandy et al. [1] and Nagarajan et al. [12] to defuzzify fuzzy numbers. Srinivas and Ganesan [14] have applied branch and bound method for unraveling assignment problems. Transportation problems under neutrosophic domain were resolved by Thamaraiselvi et al. [15] and Akansha Singh et al. [2]. With linguistic variables, Anil and Khot [3] resolved fuzzy assignment problem through BnB method. Broumi et al. [13] have discussed the shortest path problem in neutrosophic domain. An application in agriculture by intuitionistic fuzzy assignment problem has been derived by Lone et al. [9]. Aggravated by the works done by the above researchers, in this paper branch and bound method is projected for solving triangular neutroshopic fuzzy assignment problems (TFNAP).

2 Preliminaries

Some important results regarding neutrosophic sets, single-valued neutrosophic sets, and triangular fuzzy neutrosophic sets have been referred from [13, 17].
The concept of score function S and accuracy function H are proposed by Ye[17].

Definition 2.1 Let F [0, 1] be the set of all triangular fuzzy numbers on [0, 1] and X be a finite universe of discourse. A triangular fuzzy neutrosophic set (TFNS) in X is given by $\tilde{A} = \{\langle x : \tilde{T}_A(x), \tilde{I}_A(x), \tilde{F}_A(x)\rangle, x\epsilon X\} where \tilde{T}_A(x) : X \longrightarrow F[0, 1]$, $\tilde{I}_A(x) : X \longrightarrow F[0, 1] and \tilde{F}_A(x) : X \longrightarrow F[0, 1]$,
The triangular fuzzy numbers $\tilde{T}_A(x) = (T_A^1(X), T_A^2(X), T_A^3(X))$,
$\tilde{I}_A(x) = (I_A^1(X), I_A^2(X), I_A^3(X))$,
$\tilde{F}_A(x) = (F_A^1(X), F_A^2(X), F_A^3(X))$,

respectively, denote the truth-membership, indeterminacy- membership, and falsity-membership degree of x in Ã
and for every $x \epsilon X, 0 \leq T_A^3(X) + I_A^3(X) + F_A^3(X) \leq 3$.
For notational convinence the triangular fuzzy neutrosophic value TFNV is denoted by $(T_A^1(X), T_A^2(X), T_A^3(X)) = (a, b, c)$,
$(I_A^1(X), I_A^2(X), I_A^3(X)) = (e, f, g), (F_A^1(X), F_A^2(X), F_A^3(X)) = (r, s, t)$

Definition 2.2 Let $\tilde{A} = \langle (a, b, c), (e, f, g), (r, s, t) \rangle$ be a TFNV; then score function $S(\tilde{A}_1)$ and accuracy function $H(\tilde{A}_1)$ are defined as follows:
$S(\tilde{A}_1) = \frac{1}{12}[8 + (a_1 + 2b_1 + c_1) - (e_1 + 2f_1 + g_1) - (r_1 + 2s_1 + t_1)]$
$H(\tilde{A}_1) = \frac{1}{4}[(a_1 + 2b_1 + c_1) - (r_1 + 2s_1 + t_1)]$

3 Assignment Method [6, 8, 9, 11, 12]

Mathematically a TFNAP is formulated as minimize z= $\sum_{i=1}^n \sum_{j=1}^n (\tilde{C}_{ij})^I$
$where\ i = 1, 2, 3, ., ., ., .n,\ j = 1, 2, 3, ., ., ., .n$
Subject to $\sum_{i=1}^n x_{ij} = 1, i = 1, 2, 3 \ldots .n$
$\sum_{j=1}^n x_{ij} = 1,\ j = 1, 2, 3 \ldots .n,, x_{ij} \epsilon \{0, 1\}$

$$\text{where } x_{ij} = \begin{cases} 1 if the i^{th} crop\ is\ assigned\ to\ the j^{th} paddock \\ 0\ otherwise \end{cases}$$

$\tilde{C}_{ij} = ((C_{ij}^1, C_{ij}^2 c_{ij}^3)(C_{ij}^{1^I}, C_{ij}^{2^I}, C_{ij}^{3^I}))$

$\tilde{C}_{ij}^I$ is the cost of allotting the ith crop to the jth paddock. The goal is to reduce the total cost of allotting all crops to the paddocks (One crop to one paddock). If the costs of $\tilde{c}_i j^I$ are TNF costs, then the TFNAP becomes
$Y(\tilde{Z}) = \sum_{i=1}^n \sum_{j=1}^n Y(\tilde{C}_{ij})^I x_{ij}$
subject to the same conditions.
For an unbalanced problem, add dummy rows/ columns and then follow the same procedure.

4 Branch and Bound Technique [3, 11, 14]

1. Presume that the source node is 0 by taking the level number as δ and allotment number as β in the present node of a branching tree.
2. $\P_\beta^\delta$ be an allotment at level δ of the branching tree. The set of assigned cells be $\S_A$ up to the node$\P_\beta^\delta$ as of the root node (set of i, j values regarding the alloted cells up to the node as of the source node), assuming the upper bound of the partial allotment up to$\P_\beta^\delta$ be V_ϵ such that $V_\epsilon = \Sigma_{i,j \epsilon X} C_{ij} + \Sigma_{i \epsilon x} \Sigma'_{j \epsilon y}$ $max C_{ij}$

The cell entry of the profit matrix with respect to the ıth row and jth column is denoted as C_{ij}. Presume X as the set of rows that are not eliminated up to the node $¶^{\delta}_{\beta}$ from the node in the branching node.

4.1 Branching Methodology

1. The column noted as δ of the AP will be allotted with the prime row of the AP at level δ .
2. The end node at the upmost level is to be taken into account for further branching, if there is a tie on the upper bound.
3. The optimality is obtained only if the greatest upper bound occurs to be at any one of the end nodes at the (n-1)th level. The optimal solution will be the allotment on the trail from the source node together with the omitted pair of row/column combination.

5 Numerical Example

Consider a TFNAP, where a farmer intends to plant four disparate crops in each of four equal-sized paddocks with rows instead of four different crops C1, C2, C3, C4 and columns instead of four equal sized paddocks like P1, P2, P3 and P4. The nutrient requirements required for different crops vary, and the paddocks vary in soil fertility. Thus the cost of the fertilizers which must be applied depends on which crop is grown in which paddock. The cost matrix be $[\tilde{C}_{ij}]$ whose components are given as TFNS. The farmer's aim is to locate the best allotment of paddocks to crops in such a manner that the entire fertilizer price becomes least (Table 1).

By using score function formula, TFNSV are converted to crisp values as follows:
$S(\tilde{A}_1) = \frac{1}{12}[8 + (a_1 + 2b_1 + c_1) - (e_1 + 2f_1 + g_1) - (r_1 + 2s_1 + t_1)]$

$\tilde{C}_{11} = 0.25, \tilde{C}_{12} = 0.33, \tilde{C}_{13} = 1.33, \tilde{C}_{14} = 3.33,$
$\tilde{C}_{21} = 0.33, \tilde{C}_{22} = 0.083, \tilde{C}_{23} = 1.58, \tilde{C}_{24} = 2.92,$
$\tilde{C}_{31} = 1.67, \tilde{C}_{32} = 3.33, \tilde{C}_{33} = 0.583, \tilde{C}_{34} = 0.583,$
$\tilde{C}_{41} = 2.92, \tilde{C}_{42} = 0.33, \tilde{C}_{43} = 0.583, \tilde{C}_{44} = 0.583,$

$$\begin{pmatrix} 0.25 & 0.33 & 1.33 & 3.33 \\ 0.33 & 0.083 & 1.58 & 2.92 \\ 1.67 & 3.33 & 0.083 & 0.25 \\ 2.92 & 0.33 & 0.583 & 0.583 \end{pmatrix}$$

Table 1 Assignment of paddocks to crops

ine CP	P_1	P_2	P_3	P_4
ine C_1	$\langle$ (3,4,5)(2,3,5)(1,2,3) $\rangle$	$\langle$ (8,12,16)(4,6,8)(6,7,8) $\rangle$	$\langle$ (20,22,24)(7,9,11)(9,11,13) $\rangle$	$\langle$ (34,38,40)(10,12,14)(12,14,16) $\rangle$
ine C_2	$\langle$ (8,12,16)(4,6,8)(6,7,8) $\rangle$	$\langle$ (2,3,5)(1,2,3)(2,3,4) $\rangle$	$\langle$ (23,26,28)(10,11,12)(11,12,13) $\rangle$	$\langle$ (27,30,32)(10,11,12)(11,12,13) $\rangle$
ine C_3	$\langle$ (13,15,17)(6,7,8)(3,5,7) $\rangle$	$\langle$ (34,38,40)(10,12,14)(12,14,16) $\rangle$	$\langle$ (2,3,5)(1,2,3)(2,3,4) $\rangle$	$\langle$ (3,4,5)(2,3,5)(1,2,3) $\rangle$
ine C_4	$\langle$ (27,30,32)(10,11,2)(11,12,13) $\rangle$	$\langle$ (8,12,16)(4,6,8)(6,7,8) $\rangle$	$\langle$ (19,22,24)(10,12,14)(8,10,12) $\rangle$	$\langle$ (19,22,24)(10,12,14)(8,10,12) $\rangle$
ine				

Farmer desires to sow four disparate crops in each of four equal-sized paddocks. The cost matrix whose elements are given as TFNS

At first no crops are allotted to any paddocks, so the allotment σ at the source (level 0) of the branching tree is the empty set, and the subsequent lower bound is also 0 for all. $P_{11}^{I} = 0.25 + [2.92 + 3.33 + 0.583] = 7.083$,

$P_{21}^{I} = 0.33 + [3.33 + 3.33 + 0.583] = 7.573$,

$P_{31}^{I} = 1.67 + [3.33 + 2.92 + 0.583] = 8.503$,

$P_{41}^{I} = 2.92 + [3.33 + 1.58 + 3.33] = 11.16$,

Further branching is done from the terminal node which has the greatest upper bound. P_{11}^{I}, P_{21}^{I}, P_{31}^{I}, and P_{41}^{I} are the terminal nodes. The node P_{41}^{I} has the greatest upper bound .Eliminate fourth row and first column. Hence further branching from this node is shown as follows:

$V_{22} = C_{41} + C_{12} + \Sigma_{i \in x} \Sigma_{j \in y} maxC_{ij}$

$$\begin{pmatrix} 0.33 & 1.33 & 3.33 \\ 0.083 & 1.58 & 2.92 \\ 3.33 & 0.083 & 0.25 \end{pmatrix}$$

$P_{12}^{I} = 0.33 + [2.92 + 0.25] = 3.5$, $P_{22}^{I} = 0.083 + [3.33 + 0.25] = 3.663$, $P_{32}^{I} = 3.33 + [3.33 + 2.92] = 9.58$,

At this stage the nodes P_{41}^{I}, P_{12}^{I}, P_{22}^{I}, and P_{32}^{I} are the terminal nodes. Among these nodes P_{32}^{I} is the upper bound. By considering end nodes at the uppermost for further branching, eliminate 3rd row and 2nd column. $V_{33} = C_{41} + C_{32} + \Sigma_{i \in 4} \Sigma_{j \in 4} maxC_{ij}$

$$\begin{pmatrix} 1.33 & 3.33 \\ 1.58 & 2.92 \end{pmatrix}$$

$P_{13}^{I} = [1.33 + 2.92] = 4.25$, $P_{23}^{I} = [1.58 + 3.33] = 4.91$; eliminate 2nd row and 3rd column. $P_{41}^{I} = 3.33$. The optimal Assignment is given by $P_{14}^{I} + P_{32}^{I} + P_{23}^{I} + P_{41}^{I} = 3.33 + 1.58 + 3.33 + 2.92 = 11.16$ (Fig. 1).

6 Conclusion

The assignment cost has been measured as vague numbers narrated by TFNS in this manuscript. The TFNAP has been defuzzified into crisp AP by score value, and BnB technique has been implied to derive an optimal result for the first time in neutroshopic assignment problems. Mathematical instance has been exposed that the allocation acquired is best. The optimal assignment of paddocks to crops is found in such a manner to satisfy the farmer's objective of making the total fertilizer cost becomes minimum. In the future the problem can be unraveled by the following methods: reduced matrix method, ones assignment method, approximation method, and best candidate method. BnB technique can be applied in solving integer programming, nearest neighbor search, knapsack problem, bin packing, and MAX-SAT problems.

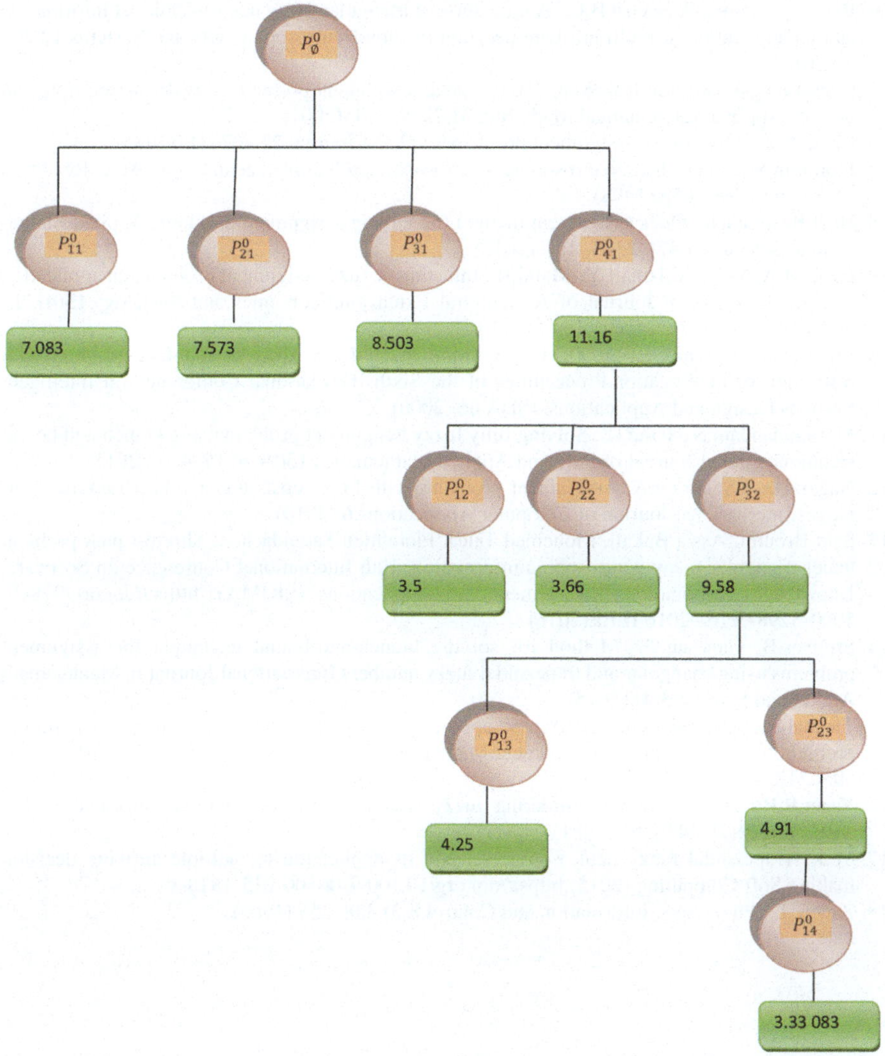

Fig. 1 Optimal solution for type-4 triangular neutroshopic assignment problem

References

1. Abbasbandy,S.,Hajjari,T.: A new approach for ranking of trapezoidal fuzzy numbers. Computers and Mathematics with Applications. **57**,413–419(2009)
2. Akansha Singh,Amit Kumar, Apppadoo,S.S.: Modified approach for optimization of real life transportation problem in neutrosophic environment.Mathematical Problems in Engineering.(2017) Article id: 2139791
3. Anil Gotmare,D., Khot, P.G.: Solution of fuzzy assignment problem by using branch and bound technique with application of lingustic variable.International Journals of Computer and Technology.**15(4)** (2016)

4. Biswas,P. Pramanik,S. Giri,B.C.: Aggregation of triangular fuzzy neutrosophic set information and its application to multi attribute decision making.Neutrosophic Sets and Systems.**12**,20–40(2016)
5. Chen Liang-Hsuan, Lu Hai-Wen.: An extended assignment problem considering multiple and outputs.Applied Mathematical Modelling.**31**,2239–2248(2007)
6. Chen, M.S.: On a fuzzy assignment problem.Tamkang Journal.**22**,407–411(1985)
7. Florentin Smarandache.: Neutrosophic ,neutrosophic probability set and logic.Amer.Res Press .Rehoboth.USA.105p,(1998)
8. Hadi Basirzadeh.: Ones assignment method for solving assignment problems.Applied Mathematical Sciences.**6(47)**, 2345–2355(2012)
9. Lone,M.A.,Mir,S.A.,Ismail,Y., Majid,R.: Intustinistic fuzy assignment problem, an application in agriculture .Asian Journal of Agricultural Extension,Economics and Socialogy.**15(4)**, 1–6(2017)
10. Long-Sheng Huang., Li-pu Zhang.: Solution method for fuzzy assignment problem with restriction of qualification.Proceedings of the Sixth International Conference on Intelligent Systems Design and Applications. ISDA'06 (2006)
11. Muruganandam,S.,Hema,K.: Solving fully fuzzy assignment problem using branch and bound technique.Global Journal of Pure and Applied Mathematics.**13(9)**, 4515–4522(2017)
12. Nagarajan,R.Solairaju,A.:Assignment problems with fuzzy costs under robust ranking techniques.International Journal of Computer Applications.**6**,(2010)
13. Said Broumi, Assia Bakali, Mohemed Talea, Florentien Sarandache .: Shortest path problem under triangular fuzzy neutrosophic information. 10th International Conference on Software, Knowledge, Information Management and Applications (SKIMA). https://doi.org/978-1-5090-3298-3/16©2016 IEEE(2016)
14. Srinivas,B. Ganesan,G.: Method for solving branch-and-bound technique for assignment problems using triangular and trapezoidal fuzzy numbers.International Journal In Management And Social Science.**3(3)**,(2015)
15. Thamaraiselvi,A. Shanthi ,R.: A new approach for optimization of real life transportation problem in neutrosophic environment.Mathematical Problems in Engineering.(2016) doi: 5950747.
16. Yager,R,R.: A procedure for ordering fuzzy subsets of the unit interval.Information Sciences.vol. **24(2)**, 143–161 (1981)
17. Ye,J.: Trapezoidal fuzzy neutrosophic set and its application to multiple attribute decision making.Soft Computing.(2015) https://doi.org/10.1007/s00500-015-1818-y
18. Zadeh, L.: Fuzzy sets, Information and Control.**8(3)**,338–353 (1965)

Part IV
Statistics

An Approach to Segment the Hippocampus from $T2$-Weighted MRI of Human Head Scans for the Diagnosis of Alzheimer's Disease Using Fuzzy C-Means Clustering

T. Genish, K. Prathapchandran, and S. P. Gayathri

Abstract The human brain plays a key role in memory-related functions such as encoding, storage, and retrieval of information. A defect in the brain results in memory impairment such as Alzheimer's disease (AD). Atrophy in the volume of hippocampus (Hc) is the earlier symptom of AD. Therefore, to study the Hc, one needs to segment it from the magnetic resonance imaging (MRI) slice. In this paper, a semiautomatic method is proposed to segment the Hc from MRI of human head scans. The proposed method uses geometric mean filter for image smoothing. The fuzzy C-means clustering is applied to convert the filtered image into three distinct regions. From those regions, the image is classified into region of interest (ROI) pixels and non-ROI pixels. The proposed method is applied to five volumes of human brain MRI. The Jaccard (J) and Dice (D) indices are used to quantify the performance of the proposed method. The results show that the proposed method works better than the existing method. The average value of Jaccard and Dice is obtained as 0.9530 and 0.9744, respectively, for the five volumes.

Keywords Segmentation · Hippocampus · Alzheimer's disease · Post-mortem MRI · Fuzzy clustering · ITK-SNAP

T. Genish (✉)
Department of Computer Science, Karpagam Academy of Higher Education, Coimbatore, Tamil Nadu, India
e-mail: th.genish@gmail.com; genish.t@kahedu.edu.in

K. Prathapchandran
Department of Computer Science and Applications, Karpagam Academy of Higher Education, Coimbatore, Tamil Nadu, India
e-mail: kpratapchandran@gmail.com; Prathapchandran.k@kahedu.edu.in

S. P. Gayathri
Department of Commerce (CA), PSGR Krishnammal College for Women, Coimbatore, Tamil Nadu, India
e-mail: gayathrisp12@gmail.com; gayathri@psgrkc.ac.in

V. Madhu et al. (eds.), *Advances in Algebra and Analysis*, Trends in Mathematics,
https://doi.org/10.1007/978-3-030-01120-8_38

1 Introduction

The hippocampus (Hc) is a paired structure. One Hc is located at the right hemisphere of the brain and the other in the left hemisphere. Atrophy in hippocampal volume leads to Alzheimer's disease. Estimation of volume of Hc is a biomarker to study memory-related diseases like AD. The cells in the hippocampus degenerate in the early stages of AD and the memory begins to fade. For treating and diagnosing AD, the image of hippocampus is needed for a neurologist. A large number of medical imaging modalities are available to image the brain and its substructure. One of the most powerful imaging modalities used to image human brain is magnetic resonance imaging (MRI). In earlier days, the Hc was segmented manually by a clinical expert. The manual segmentation is done by drawing desired borders directly onto the raw image. The expert then picks intensity by pointing to a pixel that seems to be on the border of a structure or the intensity by hand. Manual segmentation is labor-intensive and has high inter-rater and intra-rater variability.

To overcome these problems, several automatic and semiautomatic methods were reported. Some of the popular methods are adaptive-focus deformable model (AFDM) [1], automatic nonlinear image matching and anatomical labeling (ANIMAL) [2], FreeSurfer (FS) [3], and statistical parametric mapping (SPM) [4]. Few recent methods that are classified into different categories are shape-based [5], atlas-based [6], graph-cut-based [7], etc. Each of these methods has their own merits and demerits. Some methods worked well for a specific type of images, few worked well for certain conditions, few of them are time-consuming, and few methods have no consistency in segmenting Hc. But fully automated brain segmentation methods have not been widely adopted for clinical use because of issues related to reliability, accuracy, and limitations of delineation protocol.

To address few of these problems, the authors have proposed a simple semiautomatic method to segment hippocampus from MRI of the human brain. The proposed method provides good solution for hippocampus segmentation problem as they consider priori knowledge of hippocampal location, anatomical boundaries, and shape in the segmentation process. The fuzzy C-means (FCM) clustering technique is used to divide the input image into three distinct regions. From the three clusters, ROI and non-ROI regions are obtained, and these regions are separated into two regions using binarization. The largest connected component (LCC) [8] is then applied on the binary image to get the Hc mask.

2 Proposed Method

2.1 Applying Geometric Mean (GM) Filter for Image Smoothing

The original image I_{orig} is applied with geometric mean filter of size 3×3 to reduce noise in an image. The geometric mean filter achieves similar smoothing to the arithmetic mean but tends to lose less image details [9]. It is defined as the nth root of the product of the values of n times in a given series. The geometric mean (GM) [10] is calculated as $GM = \sqrt[n]{\prod x_i} = \sqrt[n]{x_1 \cdot x_2 \cdots x_n}$, where n is the number of items, x_i is the ith value in the list x, and Π is the conventional product notation. The image obtained after applying GM filter is I_{GM}.

2.2 *Image Binarization Using Fuzzy C-Means*

The fuzzy C-means (FCM) [11] is applied to divide the image I_{GM} into three distinct regions. The pixel value of hippocampus is depending upon the type of images taken for experimentation. Hence, the clusters 140, 180, and 220 are initialized by trial. Let R be a whole image and R_1, R_2, and R_3 are the clusters formed by FCM; R can be expressed as:

$$R = \bigcup_{i=1}^{3} R_i \tag{1}$$

Among the three clusters, it is found that hippocampus is visible in the clusters R_2 and R_3. Hence, these clusters are considered as a region of interest (ROI) and the remaining cluster R_1 as non-ROI. The ROI and non-ROI can be calculated as:

$$Non - ROI = R_1 \tag{2}$$

From Eqs. (1) and (2), it is divided as ROI pixels in one group and Non-ROI as another group by:

$$Bin_{GM} = \begin{cases} 1, & if\ I_{GM} \in ROI \\ 0, & otherwise \end{cases} \tag{3}$$

The individual regions of the image are labeled using labeling method. The run length identification scheme for region labeling described by Sonka et al. [12] is used to find the LCC among the regions as:

$$R_{LCC} = R\,(arg\ \max\ R_A(i)) \tag{4}$$

where the area $R_A(i)$ of ith region $R(i)$ is the total number of pixels in that region. This labeling helps to identify the several connected regions. The HC mask is retrieved by using the appropriate label. Image filling is used to remove the unwanted regions that are connected with HC. The HC region is extracted by

Table 1 Total number of slices used in the proposed method for experimentation

Vol-ID	Slices used in the proposed method
1R	22 (Penn001R_01059-01080)
2L	27 (Penn002L_01070-01096)
2R	16 (Penn002R_01024-01139)
3L	16 (Penn003L_01035-01150)
3R	11 (Penn003R_01071-01181)

mapping the mask and the input slice. The mask contains pixel values as 1 in the hippocampus region and other pixels are 0, and the extraction is defined as:

$$Hc_{Seg}(i,j) = \begin{cases} I_{orig}(i,j), & if\ Hc_{mask}(i,j) = 1 \\ 0, & otherwise \end{cases} \quad (5)$$

where, Hc_{mask} is the mask, I_{orig} is the input slice and Hc_{Seg} is the segmented hippocampus.

3 Materials Used

The data set used for the proposed method is obtained from Penn Hippocampus Atlas (PHA) [13]. The total number of slices in each volume used for experiments is given in Table 1.

4 Results and Discussions

The authors performed experiments by applying the proposed method on the images given in the material pool. The segmentation of hippocampus in some sample slices from volume 1R is shown in Fig. 1. From Fig. 1, it is noted that the proposed method segments hippocampus more clearly than the semiautomatic tool ITK-SNAP [14]. But the proposed method gives under segmentation results for the slices 20, 21, and 22. This is because the proposed method failed to detect the boundary in those slices. But ITK-SNAP 2.4.0 produced poor results for most of the slices in the volume. The performance of method for the volume 1R is evaluated by computing the similarity indices Jaccard coefficient (J) [15] and dice coefficient (D) [16], sensitivity (S) [17], specificity (Sp), predictive accuracy (PA), false-positive rate (FPR), and false-negative rate (FNR) and is given in Table 2.

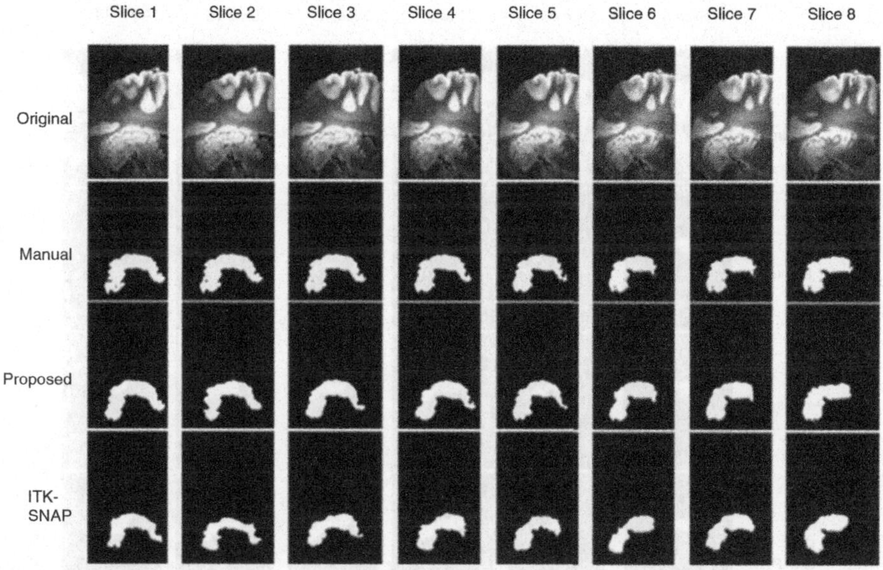

Fig. 1 Hippocampus segmentation results for slices 1–8. Row 1 shows original slices. Row 2 shows manual segmentation. Row 3 shows hippocampus segmentation by the proposed method. Row 4 shows segmentation by ITK-SNAP 2.4.0

Table 2 Average values of J, D, S, Sp, PA, FPR and FNR for volume 1R

Method	J	D	S	Sp	PA	FPR	FNR
ITK-SNAP 2.4.0	0.7274	0.7910	0.7289	0.9897	96.1321	0.1654	0.0875
Proposed	**0.8965**	**0.9410**	**0.9606**	**0.9994**	**98.2205**	**0.0547**	**0.0415**

The extraction of hippocampus from volume 2L is shown in Fig. 2. Slices 3, 6, 9, 12, 15, and 18 are selected at regular intervals from volume 2L. Figure 2 shows that the proposed method segmented the portion of hippocampus which is very closer to the manual segmentation. The average values of J, D, S, Sp, PA, FPR, and FNR for the proposed method and ITK-SNAP 2.4.0 against the manual segmentation are given in Table 3. From Table 3, it is noted that the proposed method produced an average value of 0.9605 for J and 0.9798 for D. The segmentation of hippocampus from volume 2R is shown in Fig. 3.

Slices 2, 4, 6, 8, 10, and 12 are chosen at regular intervals from volume 2R. From Fig. 3, it is observed that the proposed method gives better results than ITK-SNAP 2.4.0. The average values of J, D, S, Sp, PA, FPR, and FNR for the

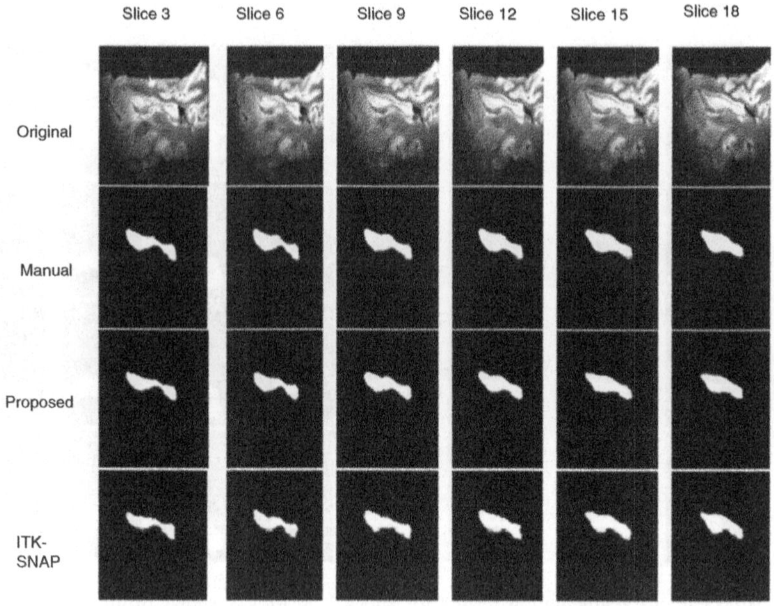

Fig. 2 Hippocampus segmentation results for volume 2L. Row 1 shows original slices. Row 2 shows manual segmentation. Row 3 shows hippocampus segmentation by the proposed method. Row 4 shows segmentation by ITK-SNAP 2.4.0

Table 3 Average values of J, D, S, Sp, PA, FPR and FNR for volume 2L

Method	J	D	S	Sp	PA	FPR	FNR
ITK-SNAP 2.4.0	0.8601	0.9247	0.8601	0.9645	96.2875	0.0789	0.0564
Proposed	**0.9605**	**0.9798**	**0.9605**	**0.9987**	**99.5405**	**0.0347**	**0.0215**

proposed method and ITK-SNAP 2.4.0 are given in Table 4. From Table 4, it is observed that the proposed method gives an average value of 0.9670 for J and 0.9832 for D. The results of hippocampus segmentation from volume 3L are shown in Fig. 4.

Slices 2, 4, 6, 8, and 10 are chosen at regular intervals from volume 3L. Figure 4 showed that the performance of the proposed method is good than that of ITK-SNAP 2.4.0. The average values of J, D, S, Sp, PA, FPR, and FNR for the proposed method and ITK-SNAP 2.4.0 are given in Table 5. From Table 5, it is noted that the proposed method gives an average value of 0.9730 for J and 0.9863 for D. The hippocampus segmentation in images of volume 3R is shown in Fig. 5.

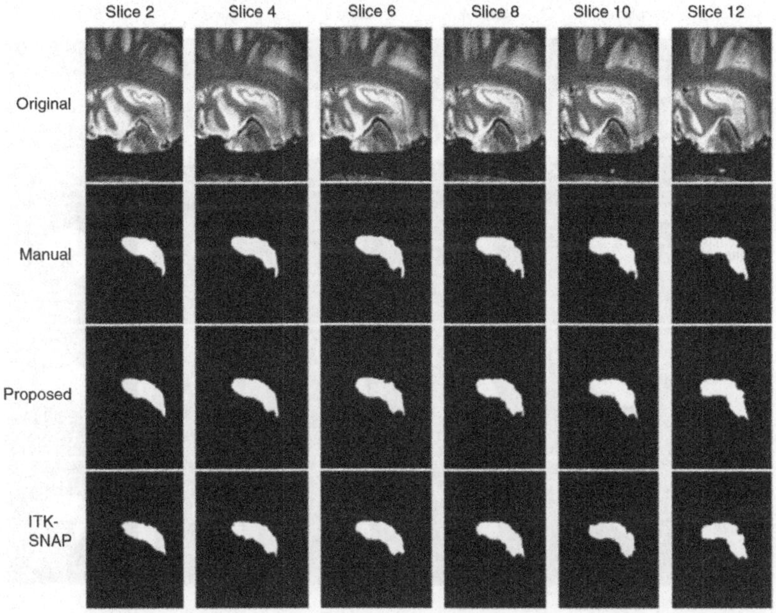

Fig. 3 Hippocampus segmentation results for volume 2R. Row 1 shows original slices. Row 2 shows manual segmentation. Row 3 shows hippocampus segmentation by the proposed method. Row 4 shows segmentation by ITK-SNAP 2.4.0

Table 4 Average values of J, D, S, Sp, PA, FPR and FNR for volume 2R

Method	J	D	S	Sp	PA	FPR	FNR
ITK-SNAP 2.4.0	0.8300	0.8637	0.9398	0.9622	95.7256	0.0409	0.0388
Proposed	**0.9670**	**0.9832**	**0.9670**	**0.9932**	**99.3210**	**0.0322**	**0.0285**

From Fig. 5, it is observed that the proposed method gives better results for all the slices in the volume. ITK-SNAP 2.4.0 gives poor results because the hippocampal pixels are removed in most of the slices. The average values of J, D, S, Sp, PA, FPR, and FNR for the proposed method are given in Table 6. From the results of all the volumes, it is observed that the performance of the proposed method is closer to the manual segmentation. The proposed method clearly identifies hippocampal and non-hippocampal pixels. The semiautomatic method ITK-SNAP 2.4.0 is unable to detect the edges of hippocampus. It removed hippocampal pixels in the slices and hence failed to segment structure of hippocampus as whole.

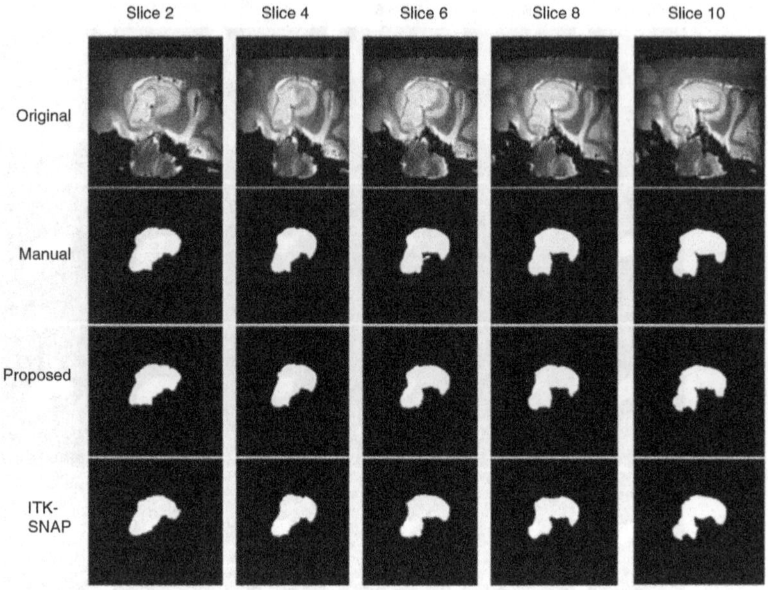

Fig. 4 Hippocampus segmentation results for volume 3L. Row 1 shows original slices. Row 2 shows manual segmentation. Row 3 shows hippocampus segmentation by the proposed method. Row 4 shows segmentation by ITK-SNAP 2.4.0

Table 5 Average values of J, D, S, Sp, PA, FPR, and FNR for volume 3L

Method	J	D	S	Sp	PA	FPR	FNR
ITK-SNAP 2.4.0	0.8267	0.8700	0.9354	0.9535	96.99	0.0601	0.0312
Proposed	**0.9730**	**0.9863**	**0.9889**	**0.9932**	**99.55**	**0.0109**	**0.0231**

5 Conclusion

In this paper, the authors have proposed a semiautomatic method to segment hippocampus from five volumes of postmortem MRI slices available at PHA. The geometric mean filter is applied for smoothing the original image. From the filtered image, the binary image is generated using fuzzy C-means technique. Then the mask of hippocampus is obtained from the binary image using connected component analysis. The quantitative results show that the proposed method produced results which are closer to the manual segmentation. The proposed method is also a simple method compared to ITK-SNAP 2.4.0.

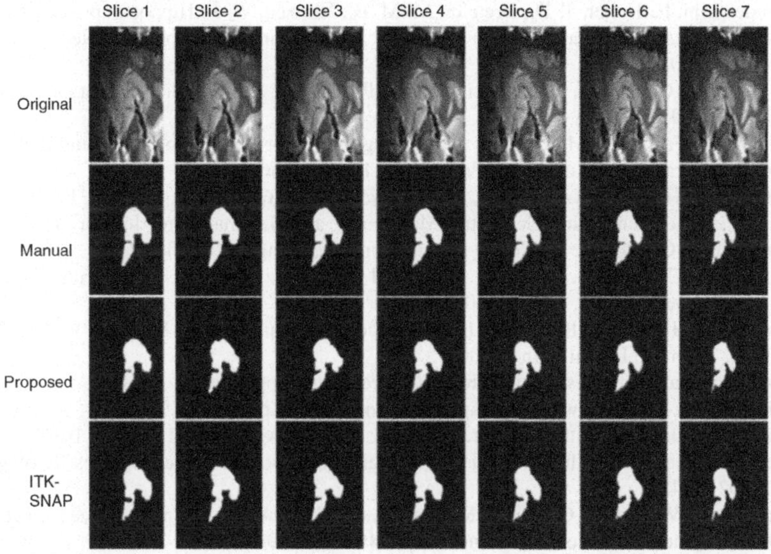

Fig. 5 Segmentation of hippocampus for volume 3R. Row 1 shows original slices. Row 2 shows manual segmentation. Row 3 shows hippocampus segmentation by the proposed method. Row 4 shows segmentation by ITK-SNAP 2.4.0

Table 6 Average values of J, D, S, Sp, PA, FPR, and FNR for volume 3R

Method	J	D	S	Sp	PA	FPR	FNR
ITK-SNAP 2.4.0	0.7987	0.8270	0.9572	0.9711	96.8759	0.0401	0.0522
Proposed	**0.9677**	**0.9821**	**0.9900**	**0.9918**	**99.2104**	**0.0200**	**0.0287**

References

1. Shen, D., Moffat, S., Resnick, S. M., Davatzikos, C.: Measuring Size and Shape of the Hippocampus in MR Images Using a Deformable Shape Model. NeuroImage. **15**, 422–434 (2002).
2. Kim, H., Chupin, M., Colliot, O., Bernhardt, B. C., Bernasconi, N., Bernasconi, A.: Automatic hippocampal segmentation in temporal lobe epilepsy: Impact of developmental abnormalities. NeuroImage. **59**, 3178–3186 (2012).
3. Morey, R. A., Petty, C. M., Xu, Y., Hayes, J. P., Wagner, H. R., Lewis, D. V., LaBar, K. S., Styner, M., McCarthy, G.: A comparison of automated segmentation and manual tracing for quantifying hippocampal and amygdala volumes. NeuroImage. **45**, 855–866 (2009).
4. Chupin, M., Hammers, A., Liu, R. S., Colliot, O., Burdett. J., Bardinet, E., Duncan, J. S, Garnero, L., Lemieux, L.: Automatic segmentation of the hippocampus and the amygdala driven by hybrid constraints: Method and validation. NeuroImage. **46**, 749–761 (2009).
5. Somasundaram, K., Genish, T.: An atlas based approach to segment the hippocampus from MRI of human head scans for the diagnosis of Alzheimers disease. International Journal of Computational Intelligence and Informatics. **5**, 7–13 (2015).
6. Kim, M., Wu, G., Li, W., Wang, L., Don Son, Y., Cho, Z. H., Shen, D.: Automatic hippocampus segmentation of 7.0 Tesla MR images by combining multiple atlases and auto-context models. NeuroImage. **83**, 335–345 (2013).

7. Van der Lijn, F., Heijer, T. D., Breteler, M. M. B., Niessen, W. J.: Hippocampus segmentation in MR images using atlas registration, voxel classification, and graph cuts. NeuroImage. **43**, 708–720 (2008).
8. Gonzelez, R. C., Woods, R. E.: Digital Image Processing, Second edition. Pearson Education. 117–118 (1992).
9. Takeda, H., Farsiu, S., Milanfar, P.: Kernel Regression for Image Processing and Reconstruction. IEEE Transactions on Image Processing. **16**, 349–366 (2007).
10. Suman, S., Hussin, F. A., Malik, A. S., Walter, N., Goh, K. L., Hilmi, I., Ho, S. H.: Image Enhancement Using Geometric Mean Filter and Gamma Correction for WCE Images. International Conference on Neural Information Processing, 276–283 (2014).
11. Bezdek, J. C., Ehrlich, R., Full, W.: FCM The Fuzzy c-Means Clustering Algorithm. Computers Geosciences. **10**, 191–203 (1984).
12. Sonka, M., Hlavac, V., Boyle, R.: Image Processing, Analysis and Machine Vision, Second Edition, Thomson Learning Inc. (2007).
13. Penn Hippocampus Atlas, www.nitrc.org/projects/pennhippoatlas/
14. ITK-SNAP 2.4.0, http://www.itksnap.org/pmwiki/pmviki.php/
15. Jaccard, P.: The Distribution of Flora in Alpine Zone. New Phytol. **11**, 37–50 (1912).
16. Dice, L.: Measures of the Amount of Ecologic Association between Species. Ecology. **26**, 297–302 (1945).
17. Shattuck, D. W., Prasad, G., Mirza, M., Narr, K. L., Toga, A. W.: Online resource for validation of brain segmentation methods. NeuroImage. **45**, 431–439 (2009).

Analysis of $M^{[X]}/G_k/1$ Retrial Queueing Model and Standby

J. Radha, K. Indhira, and V. M. Chandrasekaran

Abstract A batch arrival retrial queueing model with k optional stages of service is studied. Any arriving batch of customer finds the server free, one of the customers from the batch enters into the first stage of service, and the rest of them join into the orbit. After completion of the ith stage of service, the customer may have the option to choose (i+1)th stage of service with probability θ_i or may leave the system with probability $q_i = \begin{Bmatrix} 1-\theta_i, i = 1, 2, \ldots k-1 \\ 1, i = k \end{Bmatrix}$. Busy server may get to breakdown, and the standby server provides service only during the repair times. At the completion of service, the server remains idle to provide the service. By using the supplementary variable method, steady-state probability generating function for system size, some system performance measures are discussed. Simulation results are given using MATLAB.

Keywords Retrial · k-optional service · Standby · Supplementary variable technique

MSC Classification codes: 60J10, 90B18, 90B22

1 Introduction

There is an extensive literature on the retrial queues [7, 13]. We refer the works by Falin and Templeton [6] and Artalejo [2], Krishnakumar et al. [8] and Maraghi et al. [9] as a few. The multistage service in queues is studied by Artalejo and Choudhury [1], Wang and Li [12], Choudhury and Deka [3], Salehirad and Badmachizadeh

J. Radha · K. Indhira · V. M. Chandrasekaran (✉)
School of Advanced Sciences, VIT, Vellore, India
e-mail: jradha100@yahoo.com; radha.j2013@vit.ac.in; kindhira@vit.ac.in; vmcsn@vit.ac.in

V. Madhu et al. (eds.), *Advances in Algebra and Analysis*, Trends in Mathematics,
https://doi.org/10.1007/978-3-030-01120-8_39

[11], and Radha et al. [10]. Authors like Wang and Li [12] and Choudhury et al. [4, 5] discussed about the retrial queuing systems with the concept of breakdown and repair.

Practical Application of the Model

The proxy server is a server, in which the HTTP requests (customers) arrive according to the Poisson process. One of the requests is selected for service (such as a file, connection, web page or other resource etc.,), and other requests will join in the buffer (orbit). In the buffer each requests wait for some time and requires the service again (retrial). HAProxy is able to handle a lot of traffic (repair). HAProxy is standby for load balancing.

2 Model Description

Customers arrive in batches according to a compound Poisson process with arrival rate λ. Let X_l denote the number of customers in the lth arrival batch, where X_l, $l = 1, 2, 3, \ldots$ are with a common distribution $\Pr[X_l = n] = \chi_n$, $n = 1, 2, 3 \ldots X(z)$ denotes the probability generating function of X. The first and second moments are $E(X)$ and $E(X(X-1))$. Assume that there is no waiting space, and therefore if an arriving batch of customers finds the server free, one of the arrivals from the batch begins his service, and the rest of them join into the pool of blocked customers called an orbit. If an arriving batch finds the server either busy or on vacation or breakdown, then the batch joins into an orbit. Here inter-retrial times form an arbitrary distribution $R(t)$ with corresponding Laplace-Stieltjes transform (LST) $R^*(\varphi)$.

The server provides k stages of service in succession. The First Stage Service (FSS) is followed by i stages of service. The service time for all the stages has a general distribution. It is denoted by the random variable S_i with distribution function $S_i(t)$ having LST$S_i^*(\varphi)$, and first and second moments are $E(S_i)$ and $E(S_i^2)$, $(i = 1, 2, \ldots k)$.

From this model, the service time or the time required by the customer to complete the service cycle is a random variable S which is given by $S = \sum_{i=1}^{k} \Theta_{i-1} S_i$ having the LST $S^*(\varphi) = \prod_{i=1}^{k} \Theta_{i-1} S_i^*(\varphi)$, and the expected value is $E(S) = \sum_{i=1}^{k} \Theta_{i-1} E(S_i)$, where $\Theta_i = \theta_1\theta_2 \ldots \theta_i$ and $\Theta_0 = 1$.

While the server is working with any phase of service, it may breakdown at any time. As soon as breakdown occurs, the server is sent for repair; during that time, a standby server provides service to the customer. Assume that the standby service time distribution follows an exponential distribution with service rate h_i for ith stage, respectively, for $(i = 1, 2, \ldots k)$. The repair time (denoted by G_i) distributions of the server for i stages are assumed to be arbitrarily distributed with d.f. $G_i(t)$ and LST $G_i^*(\varphi)$ for $(i = 1, 2, \ldots k)$.

Various stochastic processes involved in the system are assumed to be mutually exclusive. The Markov process $\{C(t), N(t); t \geq 0\}$ describes the system state, where the server state.

$$C(t) = \begin{cases} 0, & \text{if the server is idle at time } t, \\ 1, & \text{if the server is busy on } i^{\text{th}} \text{ stage at time } t, \\ 2, & \text{if the server is repair on } i^{\text{th}} \text{ stage at time } t. \end{cases}$$

$N(t)$- is the number in orbit at time t; the functions $a(x)$, $\mu_i(x)$ and $\xi_i(x)$ are the conditional completion rates for retrial, service, and repair, respectively, $(1 \leq i \leq k)$:

$$a(x)dx = \frac{dR(x)}{1 - R(x)}, \quad \mu_i(x)dx = \frac{dS_i(x)}{1 - S_i(x)} \text{ and } \xi_i(x)dx = \frac{dG_i(x)}{1 - G_i(x)}.$$

Then define $B_i^* = S_1^* S_2^* \ldots S_i^*$ and $B_0^* = 1$. The first moment M_{1i} and second moment M_{2i} of B_i^* are given by

$$M_{1i} = \lim_{z \to 1} dB_i^*[A_i(z)] \Big/ dz = \sum_{j=1}^{i} -\lambda X^{(1)} S_i^{*'}(\alpha_i)$$

$$M_{2i} = \lim_{z \to 1} d^2 B_i^*[A_i(z)] \Big/ dz^2 = \sum_{j=1}^{i} \left[\left(\lambda X^{(1)}\right)^2 S_i^{*''}(\alpha_i) - \lambda X^{(2)} S_i^{*'}(\alpha_i) \right]$$

where $A_i(z) = b(z) + h_i - \frac{h_i}{z}$, $a(z) = \alpha_i + b(z)$ and $b(z) = \lambda(1 - X(z))$

Let $\{t_n;\ n = 1,2,\ldots\}$ be the sequence of time either a service period or repair period ends. In this system, $Z_n = \{C(t_n+), N(t_n+)\}$ forms an embedded Markov chain. The embedded Markov chain $\{Z_n;\ n \in N\}$ is ergodic $\Leftrightarrow \rho < 1$, where $\rho = \tau_1 + \frac{\alpha_i}{\lambda X^{(1)}} \left(1 - \tau - L + L_1 g^{(1)} \left[h_i - \lambda X^{(1)}\right]\right) - \frac{\alpha_i}{\lambda} (1 - R^*(\lambda))$.

3 Steady-State Distribution

The following probabilities are used in the sequential sections. At time t, $P_0(t)$ is the probability that the system is empty, $P_n(x, t)$ is the probability that an elapsed retrial time x of the retrial customers, $\Pi_{i,n}(x, t)$ is the probability that an elapsed service time x on ith stage of the customer under service, and $R_{i,n}(x, t)$ is the probability that an elapsed repair time x on ith stage of the server.

For the process $\{N(t),\ t \geq 0\}$, we define the probability $P_0(t) = P\{C(t) = 0, N(t) = 0\}$, and the following probability density functions:

$$P_n(x,t)dx| = P\left\{C(t) = 0, N(t) = n,\ x \leq R^0(t) < x + dx\right\},$$
$$\text{for } t, x \geq 0 \text{ and } n \geq 1,$$
$$\Pi_{i,n}(x,t)dx = P\left\{C(t) = 1,\ N(t) = n,\ x \leq S_i^0(t) < x + dx\right\},$$
$$\text{for } t, x, n \geq 0, (1 \leq i \leq k)$$
$$R_{i,n}(x,t)dx = P\left\{C(t) = 2,\ N(t) = n,\ x \leq G_i^0(t) < x + dx\right\},$$
$$\text{for } t, x, n \geq 0, (1 \leq i \leq k)$$

The stability condition exists for t, x, y and $n \geq 0$.

$$P_0 = \lim_{t\to\infty} P_0(t),\ P_n(x) = \lim_{t\to\infty} P_n(x,t),\ \Pi_{i,n}(x) = \lim_{t\to\infty} \Pi_{i,n}(x,t),$$
$$R_{i,n}(x) = \lim_{t\to\infty} R_{i,n}(x,t).$$

Steady-State Equations

By the method of supplementary variable, the following governing equations are obtained for $(i = 1, 2, \ldots k)$:

$$\lambda P_0 = \sum_{i=1}^{k} q_i \int_0^\infty \Pi_{i,0}(x,t)\mu_i(x)dx + \int_0^\infty R_{i,0}(x)\zeta_i(x)dx \tag{1}$$

$$\frac{dP_n(x)}{dx} + [\lambda + a(x)]P_n(x) = 0,\ n \geq 1 \tag{2}$$

$$\frac{d\Pi_{i,0}(x)}{dx} + [\lambda + \alpha_i + \mu_i(x)]\Pi_{i,0}(x) = 0, n = 0, \tag{3}$$

$$\frac{d\Pi_{i,n}(x)}{dx} + [\lambda + \alpha_i + \mu_i(x)]\Pi_{i,n}(x) = \lambda \sum_{l=1}^{n} \chi_l \Pi_{i,n-l}(x), n \geq 1 \tag{4}$$

$$\frac{dR_{i,0}(x)}{dx} + [\lambda + h_i + \xi_i(x)]R_{i,0}(x) = h_i R_{i,1}(x), n = 0, \tag{5}$$

$$\frac{dR_{i,n}(x)}{dx} + [\lambda + \xi_i(x) + h_i]R_{i,n}(x) = \lambda \sum_{l=1}^{n} \chi_l R_{i,n-l}(x) + h_i R_{i,k+1}(x),\ n \geq 1 \tag{6}$$

The steady-state boundary conditions at $x = 0$ and $y = 0$ are

$$P_n(0) = \left\{ \sum_{i=1}^{k} q_i \int_0^\infty \Pi_{i,0}(x,t)\mu_i(x)dx + \int_0^\infty R_{i,0}(x)\zeta_i(x)dx \right\},\ n \geq 1 \quad (7)$$

$$\Pi_{1,n}(0) = \left\{ \int_0^\infty P_{n+1}(x)a(x)dx + \lambda \sum_{l=1}^{n} \chi_l \int_0^\infty P_{n-l+1}(x)dx + \lambda \chi_{n+1} P_0 \right\},\ n \geq 1 \quad (8)$$

$$\Pi_{i,n}(0) = \theta_{i-1} \int_0^\infty \Pi_{i-1,n}(x)\mu_{i-1}(x)dx,\ n \geq 1,\ (2 \leq i \leq k) \quad (9)$$

$$R_{i,n}(x,0) = \alpha_i \int_0^\infty \Pi_{i,n}(x),\ n \geq 1, \text{for } (1 \leq i \leq k) \quad (10)$$

The normalizing condition is

$$P_0 + \sum_{n=1}^{\infty} \int_0^\infty P_n(x)dx + \sum_{n=0}^{\infty} \left(\sum_{i=1}^{k} \left(\int_0^\infty \prod_{i,n}(x)dx + \int_0^\infty R_{i,n}(x)dx \right) \right) = 1 \quad (11)$$

Under $\rho < 1$, probability generating function of the system size K(z) and orbit size H(z) distribution at stationary point of time is

$$K(z) = P_0 \left(\frac{N1(z)}{Dr(z)} + \frac{zN2(z)}{A_i(z)Dr(z)} \right)$$

where,

$$N1(z) = \Bigg(zA_i(z)\left[1 - (1 - R^*(\lambda))\right] - \left[R^*(\lambda) + X(z)\left(1 - R^*(\lambda)\right)\right] \begin{bmatrix} a_i(z) \sum_{i=1}^{k} q_i \Theta_{i-1} \left(B_i^* \left[a_i(z)\right]\right) \\ +\alpha_i\, \Theta_{i-1} B_{i-1}^* \left[a_{i-1}(z)\right] G_i^*(A_i(z))[1 - S_i^*(a_i(z))] \end{bmatrix} \Bigg)$$

$$N2(z) = \lambda \sum_{i=1}^{k} \Theta_{i-1} \left(B_{i-1}^* \left[a_{i-1}(z)\right]\right) \left[1 - S_i^*(a_i(z))\right] \left(X(z) - \left[R^*(\lambda) + X(z)\left(1 - R^*(\lambda)\right)\right]\right) \left[A_i(z) + \alpha_i (1 - G_i^*(A_i(z)))\right]$$

and $$Dr = za_i(z) - \left(\left[R^*(\lambda) + X(z)\left(1 - R^*(\lambda)\right)\right]\right) \begin{bmatrix} a_i(z) \sum_{i=1}^{k} q_i \Theta_{i-1} \left(B_i^* \left[a_i(z)\right]\right) \\ +\alpha_i\, \Theta_{i-1} B_{i-1}^* \left[a_{i-1}(z)\right] \\ G_i^*(A_i(z))[1 - S_i^*(a_i(z))] \end{bmatrix}$$

Table 1 The effect of $X^{(1)}$ on W_q, Lq, and Ls

Average batch size	Exponential			Erlang—2 stage			Hyper—exponential		
$X^{(1)}$	W_q	Lq	Ls	W_q	Lq	Ls	W_q	Lq	Ls
0.10	1.5214	1.2682	1.4354	0.0768	0.0165	0.2551	4.1003	0.6902	1.2111
0.20	1.8787	1.5357	1.6301	0.5864	0.0373	0.5923	6.3123	1.5001	2.1945
0.30	2.1771	1.8443	1.9905	1.1145	0.2036	1.0083	8.4232	2.2540	3.3009
0.40	2.6017	2.2005	2.4132	1.5999	0.4120	1.4485	10.4443	3.3365	4.0082
0.50	2.9459	2.4311	2.6723	2.0171	0.6450	1.9241	12.4321	4.5409	5.5737

Table 2 The effect of of (h_1) on P_0, Lq and Wq

Standby service rate	Exponential			Erlang—2 stage			Hyper—Exponential		
h_1	P_0	Lq	Wq	P_0	Lq	Wq	P_0	Lq	Wq
1.00	5.2286	2.0915	0.5752	8.1559	7.6624	0.5300	12.3052	8.7221	0.5982
2.00	3.9051	1.1620	0.5772	7.1866	4.2746	0.5379	10.5912	7.2365	0.5993
3.00	2.2701	0.7081	0.5792	6.5418	2.0167	0.5459	7.5949	5.7380	0.6004
4.00	1.0129	0.6052	0.5812	3.0840	0.6336	0.5538	6.8116	3.1236	0.6014
5.00	0.3563	0.1425	0.6832	0.5889	0.3356	0.5617	3.0890	2.9026	0.6025

$$H(z) = P_0 \left(\frac{N1(z)}{Dr(z)} + \frac{N2(z)}{A_i(z)Dr(z)} \right)$$

Table 1 shows the effect of $X^{(1)}$ on W_q, L_q, and L_s and Table 2 shows the effect of (h_1) on P_0, L_q and W_q.

4 Performance Measures

If the system satisfies $\rho < 1$, then the following probabilities of the server state, that is, the server is idle during the retrial, busy during ith stage, and under repair on ith stage, respectively, are obtained.

$$P = \frac{P_0 \overline{\alpha} \left(1 - R^*(\lambda)\right) \left[Dr' + \alpha_i X^{(1)}\right]}{Dr'},$$

$$\Pi_i = \sum_{i=1}^{k} \frac{\lambda P_0 L_1 X^{(1)} \left(1 - \overline{\alpha} \left(1 - R^*(\lambda)\right)\right)}{Dr'},$$

$$R_i = \sum_{i=1}^{k} \frac{\lambda P_0 \alpha_i L_1 X^{(1)} g^{(1)} \left(1 - \overline{\alpha} \left(1 - R^*(\lambda)\right)\right)}{Dr'}$$

$$Dr' = -\lambda X^{(1)}\left(1 - \tau_1 - \frac{\alpha_i}{\lambda X^{(1)}}\left(1 - \tau - L + L_1 g^{(1)}\left[h_i - \lambda X^{(1)}\right]\right)\right.$$
$$\left. + \frac{\alpha_i}{\lambda}\left(1 - R^*(\lambda)\right)\right)$$

Let L_s, L_q, W_s, and W_q be the average system size, average orbit size, average waiting time in the system, and average waiting time in the orbit, respectively, and then under $\rho < 1$:

$$L_q = \frac{Nr(z)}{Dr(z)} = \lim_{z\to 1}\frac{d}{dz}H(z) = H'(1) = P_0\left[\frac{Nr_q'''(1)DR_q''(1) - Dr_q'''(1)Nr_q''(1)}{3\left(DR_q''(1)\right)^2}\right]$$

$$Nr_q''(1) = 2\left[h_i - \lambda X^{(1)}\right]\left(R^*(\lambda)\right)\left\{-\lambda X^{(1)}L_1 + \alpha_i\left(1 - \tau - L - L_1 g^{(1)}\right.\right.$$
$$\left.\left.\left[h_i - \lambda X^{(1)}\right]\right) + \lambda X^{(1)}\sum_{i=1}^{k} L_1(1 + \alpha_i g^{(1)})\right\}$$

$$Nr_q'''(1) = \begin{pmatrix} 3\left[h_i - \lambda X^{(1)}\right]\left\{-2\lambda X^{(1)}(1-\tau) - \alpha_i\left(\omega + N + L_1\left(-\lambda X^{(2)} + \right.\right.\right. \\ \left.\left.\left. g^{(1)}\left[\lambda X^{(2)} + 2h_i\right] + g^{(2)}\left[h_i - \lambda X^{(1)}\right]^2\right) - 2L_1 g^{(1)}\left[h_i - \lambda X^{(1)}\right]\right)\right\} \\ -3\left[\lambda X^{(2)} + 2h_i\right]\left(L_1\left(-\lambda X^{(1)} + g^{(1)}\left[h_i - \lambda X^{(1)}\right]\right) + \alpha_i(1 - \tau - L)\right) \end{pmatrix}$$
$$+ \left(R^*(\lambda)\right)\sum_{i=1}^{k}\left\{(1 + \alpha_i g^{(1)})\left(3L_1\left(X^{(2)}\left[h_i - \lambda X^{(1)}\right]\right.\right.\right.$$
$$\left.\left.\left. - X^{(1)}\left[\lambda X^{(2)} + 2h_i\right]\right) + 4LX^{(1)}\left[h_i - \lambda X^{(1)}\right]\right) - 3L_1 X^{(1)} g^{(2)}\left[h_i - \lambda X^{(1)}\right]^2\right\}$$

$$Dr_q'''(1) = 3\left\{\begin{matrix} \left[h_i - \lambda X^{(1)}\right]\begin{pmatrix} -L_1\lambda X^{(1)} - 2\lambda X^{(1)}(1-\tau) - \alpha_i\omega - \left(\overline{\alpha}R^*(\lambda) + \alpha\right) \\ \left(-2\tau_1\lambda(X^{(1)})^2 + \alpha_i\left\{X^{(2)}\right.\right. \\ \left.\left. + X^{(1)}\left[\tau + L - L_1 g^{(1)}\left[h_i - \lambda X^{(1)}\right]\right]\right\}\right) \end{pmatrix} \\ + \lambda X^{(1)}(1-\rho)\left[\lambda X^{(2)} + 2h_i\right] \end{matrix}\right\}$$

$$DR_q''(1) = -2\lambda X^{(1)}\left[h_i - \lambda X^{(1)}\right](1-\rho)$$

here $\tau = \sum_{i=1}^{k}\Theta_{i-1}M_{1i} - \sum_{i=1}^{k-1}\Theta_{i-1}M_{1i}$, $\quad L = \sum_{i=1}^{k}\Theta_{i-1}(M_{1i-1} - M_{1i})$,

$\omega = \sum_{i=1}^{k}\Theta_{i-1}M_{2i} - \sum_{i=1}^{k-1}\Theta_{i-1}M_{2i}$ $\quad N = \sum_{i=1}^{k}\Theta_{i-1}(M_{2i-1} - M_{2i})$,

$\tau_1 = \sum_{i=1}^{k} q_i\Theta_{i-1}B_i^*(\alpha_i)$, *and* $L_1 = \Theta_{i-1}(B_{i-1}^*(\alpha_{i-1}) - B_i^*(\alpha_i))$.

$$L_s = \frac{Nr(z)}{Dr(z)} = \lim_{z\to 1}\frac{d}{dz}K(z) = K'(1) = P_0\left[\frac{Nr_s'''(1)DR_q''(1) - Dr_q'''(1)Nr_q''(1)}{3\left(DR_q''(1)\right)^2}\right]$$

$$Nr_s''(1) = Nr_q''(1)+4\left(R^*(\lambda)\right)X^{(1)}\sum_{i=1}^{k}L_1(1+\alpha_i g^{(1)})\left[h_i - \lambda X^{(1)}\right]$$

$$W_s = \frac{L_s}{\lambda X^{(1)}} \text{ and } W_q = \frac{L_q}{\lambda X^{(1)}}$$

5 Numerical Illustration

Here, some numerical examples are given using MATLAB. And assume the arbitrary values to the parameters satisfy $\rho < 1$. The following tables give the computed values of P_0, P, Π_i, for $(i = 1, 2, \ldots k)$, respectively. For the effect of a, h_1, are retrial rate and standby probability, respectively, graphs are drawn in Figs. 1, 2, and 3.

6 Conclusion

In this paper, we have studied a batch arrival retrial queue with multistage service, where the server is subject to server breakdowns and standby server during repair. The mean number of customers in the system/orbit, the average waiting time of

Fig. 1 L_q versus λ

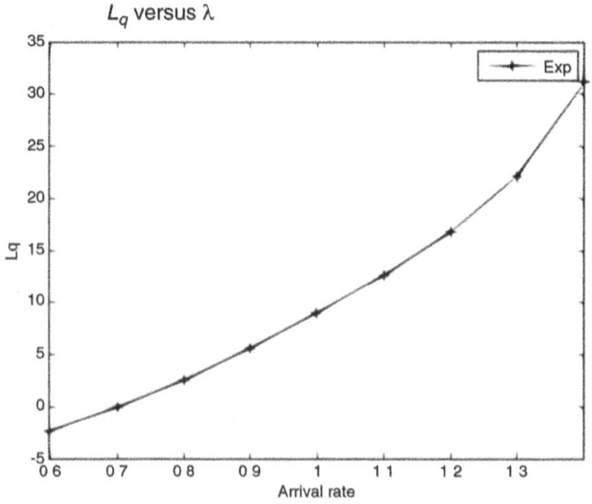

Fig. 2 L_q versus h_i

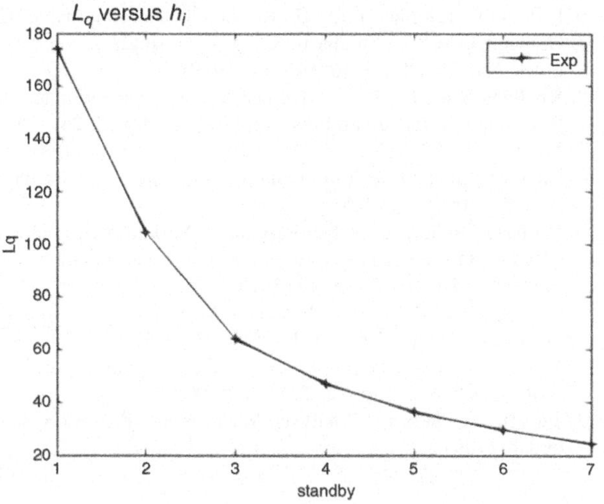

Fig. 3 L_q versus h_1 and a

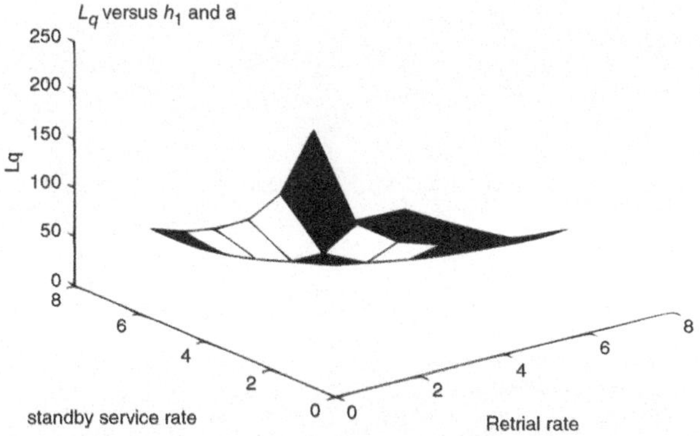

customer in the system/orbit, and some system probabilities were obtained. The analytical results are validated with the help of numerical illustrations.

References

1. Artalejo, J.R., Choudhury, G.: Steady state analysis of an M/G/1 queue with repeated attempts and two-phase service. Qual. Tec. & Quant. Mana., **1**, 189–199 (2004)
2. Artalejo, J.R.: A classified bibliography of research on retrial queues. Top 7, 187–211 (1990–1999)
3. Choudhury, G., Deka, K.: A single server queueing system with two phases of service subject to server breakdown and Bernoulli vacation. App. Mat. Mode. **36**, 6050–6060 (2012)
4. Choudhury, G., Deka, K.: An M/G/1 retrial queueing system with two phases ofservice subject to the server breakdown and repair. Per. Eval. **65**, 714–724 (2008)
5. Choudhury, G., Tadj, L., Deka, K.: A Batch arrival retrial queueing system with two phases of service and service interruption. Comp. & Mat. with Appl. **59**, 437–450 (2010)

6. Falin, G.I.: Templeton, J.C.G.: Retrial Queues.Chapman & Hall-London (1997)
7. Gomez-Correl, A.: Stochastic analysis of single server retrial queue with the general retrial times. Nav. Res. Logi. **46**, 561–81 (1999)
8. Krishnakumar, B., Pavai Madheswari, S., Vijayakumar, A.: The M/G/1 retrial queue with feedback and starting failures. Appl. Math. Modell. **26**, 1057–1075 (2002)
9. Maraghi, F. A., Madan, K.C., Darby-Dowman, K.: Bernoulli schedule vacation queue with batch arrivals and random breakdowns having general repair time distributions. Int. J of Ope. Rese.**7**, 240–256 (2010)
10. Radha, J., Indhira, K., Chandrasekaran, V.M.: An unreliable feedback retrial queue with multi optional stages of services under atmost J vacation and non-persistent customer. Int. J of App. Eng. Rese. **10**, 36435–36449 (2015)
11. Salehurad, M.R., Badamchizadeh, A.: On the multi-phase M/G/1 queueing system with random feedback. Cen. Eur. J of Ope. Rese. **17**, 131–139 (2009)
12. Wang , J. Li, Q.: A single server retrial queue with general retrial times and two phase service. J. of sys. sci. & Comp. **22**, 291–302 (2009)
13. Takagi, H.: Queueing Analysis, Vacation and Priority Systems Elsevier-North Holland Amsterdam (1991)

μ-Statistically Convergent Multiple Sequences in Probabilistic Normed Spaces

Rupam Haloi and Mausumi Sen

Abstract In this article, we introduce the notions of μ-statistically convergent and μ-statistically Cauchy multiple sequences in probabilistic normed spaces (in short PN-spaces). We also give a suitable characterization for μ-statistically convergent multiple sequences in PN-spaces. Moreover, we introduce the notion of μ-statistical limit points for multiple sequences in PN-spaces, and we give a relation between μ-statistical limit points and limit points of multiple sequences in PN-spaces.

Keywords Probabilistic normed space · μ-statistical convergence · Multiple sequence · Two-valued measure

1 Introduction

The notion of PN-space was first introduced by Šerstnev [20] in 1963. In this theory, it has been observed that these spaces are nothing but real linear spaces where the norm of a vector is a distribution function rather than just a number. Later this theory was generalized by many authors [1, 12]. The concept of statistical convergence was first developed by Steinhaus [23] as well as by Fast [8] in 1951. Later on, this theory has been investigated by many authors in recent papers [3, 5, 9–11]. Karakus [14] has extended the concept of statistical convergence to the probabilistic normed space in 2007. In the recent past, sequence spaces have been studied by various authors [21, 26, 27] from different point of view. Moreover, Tripathy et al. [28] have studied the concepts of I-limit inferior and I-limit superior of sequences in PN-space. The notion of convergence for a sequence is also considered in measure theory. In [4], Connor has extended the concept of statistical convergence, by replacing the asymptotic density with a finitely additive two-valued measure μ. Some more work can be found in [22].

R. Haloi · M. Sen (✉)
Department of Mathematics, NIT Silchar, Silchar, Assam, India
e-mail: rupam.haloi15@gmail.com; rupam@rs.math.student.nits.ac.in; senmausumi@gmail.com; mausumi@math.nits.ac.in

V. Madhu et al. (eds.), *Advances in Algebra and Analysis*, Trends in Mathematics,
https://doi.org/10.1007/978-3-030-01120-8_40

The concepts of sequence space had been extended to double sequence by Pringsheim [17] in 1900. Then Hardy [13] introduced the concept of regular convergence for double sequence in 1917. In [14], Karakus has investigated the concept of statistical convergence in PN-spaces for single sequences. Similar concept for double sequences has been developed by Karakus and Demirci [15]. More works on statistically convergent double sequences in PN-spaces can be found in [16, 18] from different aspects. The notion of statistically convergent triple sequences defined by Orlicz function has been investigated by Datta et al. [6]. Later on, Esi and Sharma [7] have studied some paranormed sequence spaces defined by Musielak-Orlicz functions over n-normed spaces. Recently, Tripathy and Goswami [24] have introduced the notion of multiple sequences in PN-spaces, and then they have studied the statistical convergence for the same in [25]. In this paper, we investigate this concept from measure theoretic aspects.

2 Preliminaries

Throughout the paper, $\mathbb{N}$, $\mathbb{R}$, and $\mathbb{R}^+$ denote the sets of natural, real, and nonnegative real numbers, respectively. Moreover, μ denotes a complete $\{0, 1\}$-valued finitely additive measure defined on a field Γ of all finite subsets of $\mathbb{N}$ and suppose that $\mu(B) = 0$, if $|B| < \infty$; if $B \subset A$ and $\mu(A) = 0$, then $\mu(B) = 0$; and $\mu(\mathbb{N}) = 1$.

The definitions of distribution function and continuous t-norm can be found in [19]. Let Δ denotes the set of all distribution functions. For the definition and example of a PN-space, one may refer to [1, 2].

Definition 1 ([24]) Let $(Y, M, *)$ be a PN-space. Then, we say a multiple sequence $y = (y_{k_1k_2\ldots k_n})$ is convergent to $\xi \in Y$ in terms of probabilistic norm M, if for every $\delta > 0$ and $\gamma \in (0, 1)$, there is an $n_0 \in \mathbb{N}$ such that $M_{y_{k_1k_2\ldots k_n}-\xi}(\delta) > 1 - \gamma$, for all $k_i \geq n_0$, for $i = 1, 2, \ldots, n$. It is denoted by $M - \lim y_{k_1k_2\ldots k_n} = \xi$.

Definition 2 ([24]) Let $(Y, M, *)$ be a PN-space. Then, we say a multiple sequence $y = (y_{k_1k_2\ldots k_n})$ is Cauchy in terms of probabilistic norm M, if for every $\delta > 0$ and $\gamma \in (0, 1)$, there is an $n_0 \in \mathbb{N}$ such that $M_{y_{k_1k_2\ldots k_n}-y_{m_1m_2\ldots m_n}}(\delta) > 1 - \gamma$, for all $k_i \geq n_0$ and $m_i \geq n_0$, for $i = 1, 2, \ldots, n$.

3 μ-Statistically Convergent Multiple Sequences in PN-Space

In this section, we introduce the following definitions and give some useful characterizations for μ-statistical convergence of multiple sequence in PN-spaces.

Definition 3 A multiple sequence $y = (y_{k_1k_2...k_n})$ in a PN-space $(Y, M, *)$ is said to be μ-statistically null in terms of the probabilistic norm M, if for every $\delta > 0$ and $\gamma \in (0, 1)$, we have

$$\mu\left(\left\{(k_1, k_2, \ldots, k_n) \in \mathbb{N}^n : M_{y_{k_1k_2...k_n}}(\delta) \leq 1 - \gamma\right\}\right) = 0.$$

Definition 4 A multiple sequence $y = (y_{k_1k_2...k_n})$ in a PN-space $(Y, M, *)$ is said to be μ-statistically bounded in terms of probabilistic norm M, if there exists an $\delta > 0$ such that

$$\mu\left(\left\{(k_1, k_2, \ldots, k_n) \in \mathbb{N}^n : M_{y_{k_1k_2...k_n}}(\delta) \leq 1 - \gamma\right\}\right) = 0, \text{ for every } \gamma \in (0, 1).$$

Definition 5 A multiple sequence $y = (y_{k_1k_2...k_n})$ in a PN-space $(Y, M, *)$ is said to be μ-statistically convergent to $\xi \in Y$ in terms of the probabilistic norm M, if for every $\delta > 0$ and $\gamma \in (0, 1)$, we have

$$\mu\left(\left\{(k_1, k_2, \ldots, k_n) \in \mathbb{N}^n : M_{y_{k_1k_2...k_n}-\xi}(\delta) \leq 1 - \gamma\right\}\right) = 0,$$

and we write as $\mu - stat_M - \lim y_{k_1k_2...k_n} = \xi$.

Definition 6 A multiple sequence $y = (y_{k_1k_2...k_n})$ in a PN-space $(Y, M, *)$ is called μ-statistically Cauchy in terms of probabilistic norm M, if for every $\delta > 0$ and $\gamma \in (0, 1)$, there is an $n_0 \in \mathbb{N}$ such that

$$\mu\left(\left\{(k_1, k_2, \ldots, k_n) \in \mathbb{N}^n : M_{y_{k_1k_2...k_n}-y_{m_1m_2...m_n}}(\delta) \leq 1 - \gamma\right\}\right) = 0.$$

From the above definitions, we have the following two results. The proofs are obvious, so omitted.

Theorem 1 *Let $(Y, M, *)$ be a probabilistic normed space. Then, for every $\gamma \in (0, 1)$ and $\delta > 0$, the following statements are equivalent:*

1. $\mu - stat_M - \lim y_{k_1k_2...k_n} = \xi$.
2. $\mu\left(\left\{(k_1, k_2, \ldots, k_n) \in \mathbb{N}^n : M_{y_{k_1k_2...k_n}-\xi}(\delta) \leq 1 - \gamma\right\}\right) = 0$.
3. $\mu\left(\left\{(k_1, k_2, \ldots, k_n) \in \mathbb{N}^n : M_{y_{k_1k_2...k_n}-\xi}(\delta) > 1 - \gamma\right\}\right) = 1$.
4. $\mu - stat - \lim M_{y_{k_1k_2...k_k}-\xi}(\delta) = 1$.

Corollary 1 *Let $(Y, M, *)$ be a PN-space. If a multiple sequence $y = (y_{k_1k_2...k_n})$ in $(Y, M, *)$ is μ-statistically convergent in terms of probabilistic norm M, then $\mu - stat_M - \lim y$ is unique.*

Corollary 2 *Let $(Y, M, *)$ be a probabilistic normed space. If $M - \lim y_{k_1k_2...k_n} = \xi$, then $\mu - stat_M - \lim y_{k_1k_2...k_n} = \xi$.*

Proof Suppose $y = (y_{k_1k_2...k_n})$ converges to ξ in terms of probabilistic norm M. Then, for every $\delta > 0$ and $\gamma \in (0, 1)$, there exists an $n_0 \in \mathbb{N}$ such that

$$M_{y_{k_1k_2...k_n}-\xi}(\delta) > 1 - \gamma, \quad \text{for all } k_i \geq n_0,\ i = 1, 2, \ldots, n.$$

Then, the set $\left\{(k_1, k_2, \ldots, k_n) \in \mathbb{N}^n : M_{y_{k_1k_2...k_n}-\xi}(\delta) \leq 1 - \gamma\right\}$ contains at most finite numbers of terms, and so we have

$$\mu\left(\left\{(k_1, k_2, \ldots, k_n) \in \mathbb{N}^n : M_{y_{k_1k_2...k_n}-\xi}(\delta) \leq 1 - \gamma\right\}\right) = 0.$$

Consequently, $\mu - stat_M - \lim y_{k_1k_2...k_n} = \xi$.

The converse of the Corollary 2 does not hold, in general.

Example 1 Suppose $(\mathbb{R}, ||\cdot||)$ is the space of all real numbers with the standard norm. Let $a_1 * a_2 = a_1a_2$ and $M_y(s) = \frac{s}{s+||y||}$, where $y \in R$ and $s \geq 0$. Then, we see that $(\mathbb{R}, M, *)$ is a probabilistic normed space. Let $K \subset \mathbb{N}^n$ be such that $\mu(K) = 0$. We define a sequence $y = (y_{k_1k_2...k_n})$ as follows:

$$y_{k_1k_2...k_n} = \begin{cases} k_1k_2 \ldots k_n, & \text{if } (k_1, k_2, \ldots, k_n) \in K \\ 0, & \text{otherwise.} \end{cases} \tag{1}$$

Then, one can easily verify that $y = (y_{k_1k_2...k_n})$ is μ-statistically convergent in terms of the probabilistic norm M. However, the sequence $y = (y_{k_1k_2...k_n})$ defined by (1) is not convergent in the space $(\mathbb{R}, ||\cdot||)$, thus we conclude that y is also not convergent in terms of the probabilistic norm M.

Theorem 2 *Suppose that $y = (y_{k_1k_2...k_n})$ is a multiple sequence in a probabilistic normed space $(Y, M, *)$. Then $\mu - stat_M - \lim y_{k_1k_2...k_n} = \xi$ if and only if there is an index subset $A = \{(n_{k_1}, n_{k_2}, \ldots, n_{k_n}) : n_{k_i} \in \mathbb{N}\}$ of $\mathbb{N}^n$ such that $\mu(A) = 1$ and*

$$M - \lim_{(k_1,k_2,\ldots,k_n)\in A} y_{k_1k_2...k_n} = \xi.$$

Proof First, suppose that $\mu - stat_M - \lim y_{k_1k_2...k_n} = \xi$. Then, for every $\delta > 0$ and $s \in \mathbb{N}$, we define the following two sets:

$$A(s, \delta) = \left\{(k_1, k_2, \ldots, k_n) \in \mathbb{N}^n : M_{y_{k_1k_2...k_n}-\xi}(\delta) \leq 1 - \frac{1}{s}\right\} \tag{2}$$

$$B(s, \delta) = \left\{(k_1, k_2, \ldots, k_n) \in \mathbb{N}^n : M_{y_{k_1k_2...k_n}-\xi}(\delta) > 1 - \frac{1}{s}\right\}. \tag{3}$$

Then, we have $\mu\,(A(s, \delta)) = 0$ and

$$B(1, \delta) \supset B(2, \delta) \supset \cdots \supset B(j, \delta) \supset B(j + 1, \delta) \supset \ldots \tag{4}$$

$$\mu(B(s, \delta)) = 1, \quad \text{for } s = 1, 2, \ldots. \tag{5}$$

Now, we need to show that, the sequence $y = (y_{k_1k_2\ldots k_n})$ is convergent to ξ in terms of probabilistic norm M, for $(k_1, k_2, \ldots, k_n) \in B(s, \delta)$. If possible, suppose that $y = (y_{k_1k_2\ldots k_n})$ is not convergent to ξ in terms of the probabilistic norm M. Then, there exists $\gamma > 0$ such that the set

$$\left\{(k_1, k_2, \ldots, k_n) \in \mathbb{N}^n : M_{y_{k_1k_2\ldots k_n}-\xi}(\delta) \leq 1 - \gamma\right\}$$

contains infinite number of terms. Let

$$B(\gamma, \delta) = \left\{(k_1, k_2, \ldots, k_n) \in \mathbb{N}^n : M_{y_{k_1k_2\ldots k_n}-\xi}(\delta) > 1 - \gamma\right\},$$

where $\gamma > \frac{1}{s}$, for $s = 1, 2, \ldots$. Then $\mu(B(\gamma, \delta)) = 0$. But from (4), we have $B(s, \delta) \subset B(\gamma, \delta)$. Thus, we obtain $\mu(B(s, \delta)) = 0$, which is a contradiction to (5). Hence $y = (y_{k_1k_2\ldots k_n})$ is convergent to ξ in terms of the probabilistic norm M.

Conversely, we assume that there is an index subset $A = \{(k_1, k_2, \ldots, k_n) : k_i \in \mathbb{N}\} \subset \mathbb{N}^n$ such that $\mu(A) = 1$ and

$$N - \lim_{(k_1,k_2,\ldots,k_n)\in A} y_{k_1k_2\ldots k_n} = \xi.$$

Then, for every $\delta > 0$ and $\gamma \in (0, 1)$, there is an $m_0 \in \mathbb{N}$ such that

$$M_{y_{k_1k_2\ldots k_n}-\xi}(\delta) > 1 - \gamma, \quad \text{for } k_i \geq m_0,\ i = 1, 2, \ldots, n.$$

Now, we see that

$$\{(k_1, k_2, \ldots, k_n) \in \mathbb{N}^n : M_{y_{k_1k_2\ldots k_n}-\xi}(\delta) \leq 1 - \gamma\}$$
$$\subset \mathbb{N}^n - \{(k_{1(m_0+1)}, \ldots, k_{n(m_0+1)}), (k_{1(m_0+2)}, \ldots, k_{n(m_0+2)}), \ldots\}.$$

Therefore, we have $\mu\left(\left\{(k_1, k_2, \ldots, k_n) \in \mathbb{N}^n : M_{y_{k_1k_2\ldots k_n}-\xi}(\delta) \leq 1 - \gamma\right\}\right) \leq 1 - 1 = 0$. Consequently, we have $\mu - stat_M - \lim y_{k_1k_2\ldots k_n} = \xi$.

Theorem 3 *Let $y = (y_{k_1k_2\ldots k_n})$ be a multiple sequence in a PN-space $(Y, M, *)$. Then the following statements are equivalent:*

1. *y is a μ-statistically Cauchy sequence in terms of probabilistic norm M.*
2. *There is an index subset $A = \left\{(m_{k_1}, m_{k_2}, \ldots, m_{k_n}) \in \mathbb{N}^n : m_{k_i} \in \mathbb{N}\right\} \subset \mathbb{N}^n$ such that $\mu(A) = 1$ and the subsequence $\left\{y_{m_{k_1}m_{k_2}\ldots m_{k_n}}\right\}_{(m_{k_1},m_{k_2},\ldots,m_{k_n})\in A}$ is a Cauchy sequence in terms of the probabilistic norm M.*

Proof The proof is easy and so omitted.

We now give some arithmetical properties of μ-statistical convergence for a multiple sequence on PN-space.

Theorem 4 *Let $(Y, M, *)$ be a probabilistic normed space. Then*

1. *If $\mu - stat_M - \lim x_{k_1k_2...k_n} = \alpha$ and $\mu - stat_M - \lim y_{k_1k_2...k_n} = \beta$, then $\mu - stat_M - \lim(x_{k_1k_2...k_n} + y_{k_1k_2...k_n}) = \alpha + \beta$.*
2. *If $\mu - stat_M - \lim x_{k_1k_2...k_n} = \alpha$ and $a \in \mathbb{R}$, then $\mu - stat_M - \lim ax_{k_1k_2...k_n} = a\alpha$.*
3. *If $\mu - stat_M - \lim x_{k_1k_2...k_n} = \alpha$ and $\mu - stat_M - \lim y_{k_1k_2...k_n} = \beta$, then $\mu - stat_M - \lim(x_{k_1k_2...k_n} - y_{k_1k_2...k_n}) = \alpha - \beta$.*

Proof The proof follows from the definition of μ-statistical convergence of a multiple sequence in PN-space itself.

4 μ-Statistical Limit Points for Multiple Sequences in PN-Space

In this section, we introduce the concepts of μ-statistical limit points of multiple sequences in PN-spaces and investigate their relation with limit points of multiple sequences in PN-spaces.

Definition 7 ([24]) Let $(Y, M, *)$ be a probabilistic normed space, and let $y = (y_{k_1k_2...k_n})$ be a multiple sequence. We say that $\xi \in Y$ is a limit point of y in terms of the probabilistic norm M, if there is a subsequence of y that converge to ξ in terms of the probabilistic norm M. Let $L_M(y)$ denotes the set of all limit points of the multiple sequence $y = (y_{k_1k_2...k_n})$.

Definition 8 Let $(Y, M, *)$ be a probabilistic normed space, and let $y = (y_{k_1k_2...k_n})$ be a multiple sequence. We say that $\eta \in Y$ is a μ-statistical limit point of the multiple sequence y in terms of the probabilistic norm M, if there is a set

$$A=\{(k_1(i), k_2(i), \dots, k_n(i)) : k_j(1)<k_j(2)<k_j(3)< \dots, \text{ for } j=1, 2, \dots, n\} \subset \mathbb{N}^n$$

such that $\mu(A) \neq 0$ and $M - \lim y_{k_1(i)k_2(i)...k_n(i)} = \eta$. Let $\Lambda_M^{\mu}(y)$ denote the set of all $\mu - stat_M - limit$ points of the multiple sequence $y = (y_{k_1k_2...k_n})$.

Theorem 5 *Suppose $y = (y_{k_1k_2...k_n})$ is a multiple sequence in a PN-space $(Y, M, *)$. If $\mu - stat_M - \lim y = L_1$, then $\Lambda_M^{\mu}(y) = \{L_1\}$.*

Proof If possible, suppose that $\Lambda_M^{\mu}(y) = \{L_1, L_2\}$ such that $L_1 \neq L_2$. Then there exists two sets:

$$A=\{(k_1(i), k_2(i), \dots, k_n(i)) : k_j(1)<k_j(2)<k_j(3)< \dots, \text{ for } j=1, 2, \dots, n\} \subset \mathbb{N}^n$$

$$B=\{(u_1(i), u_2(i), \dots, u_n(i)) : u_j(1)<u_j(2)<u_j(3)< \dots, \text{ for } j=1, 2, \dots, n\}\subset\mathbb{N}^n$$

such that $\mu(A) \neq 0$, $\mu(B) \neq 0$ and $M - \lim y_{k_1(i)k_2(i)...k_n(i)} = L_1$, $M - \lim y_{u_1(i)u_2(i)...u_n(i)} = L_2$. Since $M - \lim y_{u_1(i)u_2(i)...u_n(i)} = L_2$, so for every $\delta > 0$ and $\gamma \in (0, 1)$, we have

$$\mu\left(\left\{(u_1(i), u_2(i), \dots, u_n(i)) \in \mathbb{N}^n : M_{y_{u_1(i)u_2(i)...u_n(i)} - L_2}(\delta) \leq 1 - \gamma\right\}\right) = 0.$$

Now, we see that

$$\{(u_1(i), u_2(i), \ldots, u_n(i)) \in \mathbb{N}^n : i \in \mathbb{N}\}$$
$$= \left\{(u_1(i), u_2(i), \ldots, u_n(i)) \in \mathbb{N}^n : M_{y_{u_1(i)u_2(i)\ldots u_n(i)}-L_2}(\delta) > 1-\gamma\right\}$$
$$\cup \left\{(u_1(i), u_2(i), \ldots, u_n(i)) \in \mathbb{N}^n : M_{y_{u_1(i)u_2(i)\ldots u_n(i)}-L_2}(\delta) \le 1-\gamma\right\}$$

which implies that

$$\mu\left(\left\{(u_1(i), u_2(i), \ldots, u_n(i)) \in \mathbb{N}^n : M_{y_{u_1(i)u_2(i)\ldots u_n(i)}-L_2}(\delta) > 1-\gamma\right\}\right) \neq 0. \tag{6}$$

However $\mu - stat_M - \lim y = L_1$ implies that for every $\delta > 0$,

$$\mu\left(\left\{(k_1, k_2, \ldots, k_n) \in \mathbb{N}^n : M_{y_{k_1k_2\ldots k_n}-L_1}(\delta) \le 1-\gamma\right\}\right) = 0. \tag{7}$$

Thus, we can write $\mu\left(\left\{(k_1, k_2, \ldots, k_n) \in \mathbb{N}^n : M_{y_{k_1k_2\ldots k_n}-L_1}(\delta) > 1-\gamma\right\}\right) \neq 0$. Now, for every $L_1 \neq L_2$, we have

$$\left\{(u_1(i), u_2(i), \ldots, u_n(i)) \in \mathbb{N}^n : M_{y_{u_1(i)u_2(i)\ldots u_n(i)}-L_2}(\delta) > 1-\gamma\right\}$$
$$\cap \left\{(k_1, k_2, \ldots, k_n) \in \mathbb{N}^n : M_{y_{k_1k_2\ldots k_n}-L_1}(\delta) > 1-\gamma\right\} = \phi.$$

Therefore

$$\left\{(u_1(i), u_2(i), \ldots, u_n(i)) \in \mathbb{N}^n : M_{y_{u_1(i)u_2(i)\ldots u_n(i)}-L_2}(\delta) > 1-\gamma\right\}$$
$$\subseteq \left\{(k_1, k_2, \ldots, k_n) \in \mathbb{N}^n : M_{y_{k_1k_2\ldots k_n}-L_1}(\delta) \le 1-\gamma\right\},$$

which implies that $\mu\left(\left\{(u_1(i), u_2(i), \ldots, u_n(i)) \in \mathbb{N}^n : M_{y_{u_1(i)u_2(i)\ldots u_n(i)}-L_2}(\delta) > 1-\gamma\right\}\right) = 0$. This contradicts the Eq. (6). Hence, we must have $\Lambda^{\mu}_M(y) = \{L_1\}$.

Acknowledgements The work of the first author has been supported by the Research Project SB/S4/MS:887/14 of SERB - Department of Science and Technology, Govt. of India.

References

1. Alsina, C., Schweizer, B., Sklar, A.: On the definition of a probabilistic normed space. Aequationes Math. **46**, 91–98 (1993)
2. Asadollah, A., Nourouzi, K.: Convex sets in probabilistic normed spaces. Chaos, Solutions & Fractals. **36**, 322–328 (2008)

3. Connor, J.: The statistical and strong p-Cesàro convergence of sequences. Analysis. **8**, 47–63 (1988)
4. Connor, J.: Two valued measure and summability. Analysis. **10**, 373–385 (1990)
5. Connor, J.: R-type summability methods, Cauchy criterion, P-sets and statistical convergence. Proc. Amer. Math. Soc. **115**, 319–327 (1992)
6. Datta, A.J., Esi, A., Tripathy, B.C.: Statistically convergent triple sequence spaces defined by Orlicz function. Journal of Mathematical Analysis. **4** (2), 16–22 (2013)
7. Esi, A., Sharma, S.K.: Some paranormed sequence spaces defined by a Musielak-Orlicz function over n-normed spaces. Konuralp Journal of Mathematics. **3** (1), 16–28 (2015)
8. Fast, H.: Sur la convergence statistique. Colloq. Math. **2**, 241–244 (1951)
9. Fridy, J.A.: On Statistical convergence. Analysis. **5**, 301–313 (1985)
10. Fridy, J.A., Orhan, C.: Lacunary Statistical convergence. Pacific J. Math. **160**, 43–51 (1993)
11. Fridy, J.A., Orhan, C.: Lacunary statistical summability. J. Math. Anal. Appl. **173**, 497–503 (1993)
12. Guillén, B., Lallena, J., Sempi, C.: Some classes of probabilistic normed spaces. Rend. Math. **17** (7), 237–252 (1997)
13. Hardy, G.H.: On the Convergence of Certain Multiple Series. Proceedings of the Cambridge Philosophical Society. **19** (3), 86–95 (1917)
14. Karakus, S.: Statistical Convergence on PN-spaces. Mathematical Communications. **12**, 11–23 (2007)
15. Karakus, S., Demirci, K.: Statistical Convergence of Double Sequences on Probabilistic Normed Spaces. International Journal of Mathematics and Mathematical Sciences. (2007) https://doi.org/10.1155/2007/14737
16. Mohiuddine, S.A., Savaş, E.:, Lacunary statistically convergent double sequences in probabilistic normed spaces. Ann Univ Ferrara. **58** (2), 331–339 (2012)
17. Pringsheim, A.: Zur Theorie der zweifach unendlichen Zahlenfolgen. Mathematische Annalen. **53** (3), 289–321 (1900)
18. Savaş, E., Mohiuddine, S.A.: $\overline{\lambda}$-statistically convergent double sequences in probabilistic normed spaces. Mathematica Slovaca. **62** (1), 99–108 (2012)
19. Schweizer, B., Sklar, A.: Statistical metric spaces. Pacific J. Math. **10**, 313–334 (1960)
20. Šerstnev, A.N.: On the notion of a random normed space. Dokl. Akad. Nauk. SSSR. **142** (2), 280–283 (1963)
21. Sharma, S.K., Esi, A.: Some $\mathcal{I}$-convergent sequence spaces defined by using sequence of moduli and n-normed space. Journal of the Egyptian Mathematical Society. **21**, 29–33 (2013)
22. Sharma, S.K., Esi, A.: μ-statistical convergent double lacunary sequence spaces. Afrika Matematika. **26** (7–8), 1467–1481 (2015)
23. Steinhaus, H.: Sur la convergence ordinaire et la convergence asymptotique. Colloq. Math. **2**, 73–74 (1951)
24. Tripathy, B.C., Goswami, R.: Multiple sequences in probabilistic normed spaces. Afr. Mat. **26**, 753–760 (2015)
25. Tripathy, B.C., Goswami, R.: Statistically Convergent Multiple Sequences in Probabilistic Normed Spaces. U.P.B. Sci. Bull. Series A. **78** (4), 83–94 (2016)
26. Tripathy, B.C., Sen, M., Nath, S.: I-convergence in probabilistic n-normed spaces. Soft Computing. **16** (6), 1021–1027 (2012)
27. Tripathy, B.C., Sen, M., Nath, S.: Lacunary I-convergence in probabilistic n-normed spaces. IMBIC 6th International Conference on Mathematical Sciences for Advancement of Science and Technology (MSAST 2012), December 21–23, Salt Lake City, Kolkata, India
28. Tripathy, B.C., Sen, M., Nath, S.: I-Limit Superior and I-Limit Inferior of Sequences in Probabilistic Normed Space. International Journal of Modern Mathematical Sciences. **7** (1), 1–11 (2013)

A Retrial Queuing Model with Unreliable Server in K Policy

M. Seenivasan and M. Indumathi

Abstract The retrial queue with unreliable server with provision of temporary server has been studied. A temporary server is installed when the primary server is over loaded. It means that a fixed queue length of K-policy customers including the customer with the primary server has been build up. The primary server may breakdown while rendering service to the customers; it is sent for the repair. This type of queuing system has been investigated using matrix geometric method and obtains the probabilities of the system steady state. From the probabilities, we found some performance measures.

Keywords Retrial queue · Retrial rate · Stationary distribution · Server breakdown · Matrix geometric method

AMS Subject Classification 60K25, 60K30 and 90B22

1 Introduction

It is observed in daily routine activities that the provision of temporary servers in case when the server load increases can work a significant role to improve the system capacity. The provision of additional temporary server is done to reduce the workload on a single server; this may also be useful in reducing the waiting time of the customers. In real-life congestion problems, the concept of installing temporary server finds several applications such as telecommunication systems, computer protocols, web servers, admission counters, message transmission, dispensaries, and many other types of situations.

M. Seenivasan (✉) · M. Indumathi
Department of Mathematics, Annamalai University, Annamalainagar, Chidambaram, India
e-mail: emseeni@rediffmail.com; ms08318@annamalaiuniversity.ac.in

V. Madhu et al. (eds.), *Advances in Algebra and Analysis*, Trends in Mathematics,
https://doi.org/10.1007/978-3-030-01120-8_41

From the application point of view, heavy traffic analysis has been a challenging topic of investigation to the queuing theorists. The analysis of heavy traffic of controlled queuing and communication networks was introduced by Kushner [11]. The number of customers of a parallel system of two queues, operating under heavy traffic, by formulating reflected stochastic differential equation, was studied by Leite and Fragoso [12]. Also the theoretical works have been appeared on the queues with more than one server with respect to heavy traffic. Normally the secondary server is installed with an aim to reduce the waiting time of the customers and to increase the efficiency of the system in terms of faster service rendered. Retrial queues with a finite source and identical multiple server in parallel were studied by Alfa and Isotupa [2]. A two-server Markovian queue was studied by Krishna Kumar and Madheswari [9] by using matrix geometric method. Queues associated with reattempt are a common phenomenon of day-to-day congestion situations. These queues are associated with the fact that a customer, when deprived of service, is forced to join the virtual queue of the customers called orbit from where he can try again and again to get served. Falin and Templeton [7] and Artalejo and Corral [3] found in the detailed account of the retrial queues. The retrial queues with server subject to breakdowns were studied by Kulkarni and Choi [10]. Aissani and Artalejo [1] analyzed the single server retrial queue subject to breakdown. Recently, Bhagat and Jain [4] and Wu Lian [15] presented retrial queues with unreliable server. Dimitrious [6] analyzed a non-Markovian queue with multi-optional services and vacation. He computed stability condition for the unreliable server retrial queue with priority and negative customers. Unreliable retrial queue with Bernoulli vacation and obtained various performance measures was analyzed by Choudhury and Ke [5].

Several queue theorists developed repetitive matrix block-structured model to obtain solutions of various queuing problems using matrix geometric approach which was introduced by Neuts [13, 14]. Kalyanaraman and Seenivasan [8] analyzed a multi-server retrial queuing system with unreliable server.

The content of this paper is the retrial queue with the provision of additional temporary server and a fixed queue length of K-policy customers including the customer with the primary server may breakdown while rendering service to the customers; it is sent for the repair. These types of model have been investigated. The rest of the paper is organized as follows: The model description and governing equations are presented in Sect. 2. Section 3 contains the analysis of the queuing model and the various performance measures. The model analyzed using the numerical examples to the given particular values of parameters in Sect. 4. The last section contains a brief conclusion.

2 The Model and Governing Equations

We consider retrial queue with unreliable server which is the primary server. The system has the provision of installing a second temporary server which is turned on when the number of customer with the first server threshold level. The various description studied in the model are discussed in the following subsections.

2.1 Model Description

The retrial queuing model under consideration has the provision of two servers, out of which second temporary sever is activated only when the workload with the primary server crosses a pre-specified level. The various type of assumptions are introducing are as follows:

2.1.1 Arrival Process

The customer in the system follows Poisson arrival with rate λ. There is a provision of two servers; the first is primary server, and second one is temporary server. The second temporary server is installed only if K-policy customers are already queued up before the primary server including the one in the service. If an arriving customer finds less than K-policy customers with the primary server, then either customer wait for its turn in the queue with the primary server or customer may join the orbit. But if on arrival, the primary server's buffer is fully occupied with K-policy customers, then the new arrival has no other option rather than to join the buffer of secondary server.

2.1.2 Retrial Process

The customers accumulated in the orbit and retry with rate γ and fulfill for the service with primary customer as soon as they find the server is idle.

2.1.3 Service Process

The customers are served following exponential distribution with rate μ_i, if queued before ith server (i = 1 for primary server and i = 2 for secondary server). The maximum number of customers joining the primary sever is K-policy, i.e., a buffer of fixed capacity K is provided for the primary server. However the number of customer joining the secondary server is unlimited. Both the servers have their own independent queues, but the formation of second queue takes place when the buffer of primary server is full. Queue shifting is not permitted to the customers once they join it.

2.1.4 Breakdown and Repair Process

The primary server is unreliable and may breakdown while serving the customer; the broken down server is sent for repair immediately, and after repair, it becomes as good as before failure. However, temporary second server is considered as reliable server. The inter-failure time of the primary server follows exponential distribution with rate α_1. The repair time of the primary server follows exponential distribution with rate β_1. The Markov process is$\{X(t) = (S(t), C_1(t), C_2(t)); t \geq 0\}$
To describe the state of the system at any instant, we consider the following three random variables that describe the system completely.

1. where $S(t)$ denote the server state at time t, $S(t) = 0$ if the sever is idle, $S(t) = 1$ if the sever is busy and $S(t) = 2$ if the sever is breakdown,
2. where C_1(t) denote the number of customers with the first server, such that C_1(t) = i, (0 $\leq$ i $\leq$ K).
3. where C_2(t) denote the number of customers with the second server, such that $C_2(t) = j, j \geq 0$.

The state space of the process is $\{0, 1, 2\} \times \{0, 1, 2, \ldots\ldots K\} \times \{0, 1, 2, \ldots\ldots\}$.

2.2 *Governing Equation*

The Chapman–Kolmogrov equations corresponding to different system states are formulated as:

2.2.1 Retrial State

$$(\lambda + \gamma)P_{0,1,0} = \lambda\ P_{0,0,0} \tag{1}$$

$$(\lambda + \gamma)P_{0,i,0} = \lambda\ P_{0,i-1,0}, \qquad (1 \leq i \leq K-1) \tag{2}$$

$$(\lambda + \gamma)P_{0,K,0} = \lambda\ P_{0,K-1,0} \tag{3}$$

2.2.2 Idle State

$$0 = \lambda\ P_{0,0,0} + \mu_1\ P_{1,0,0} \tag{4}$$

$$0 = \lambda\ P_{0,i-1,0} + \mu_1\ P_{1,i-1,0} \tag{5}$$

$$(\lambda + \mu_1)P_{0,K,0} = \lambda\ P_{0,K-1,0} \tag{6}$$

$$0 = \lambda\ P_{0,K,0} + \mu_2\ P_{1,K,0} \tag{7}$$

2.2.3 Busy State

$$(\lambda+\alpha_1)P_{1,0,0}=\mu_1\,P_{1,1,0}+\beta_1\,P_{2,0,0}+\mu_2\,P_{1,0,1} \tag{8}$$

$$(\lambda+\alpha_1+\mu_1)P_{1,i,0}=\mu_1\,P_{1,i+1,0}+\beta_1\,P_{2,i,0}+\mu_2\,P_{1,i,1}+\lambda\,P_{1,i-1,0}+\gamma\,P_{0,i,0},\ (1\le \mathrm{i}\le K-1) \tag{9}$$

$$(\lambda+\alpha_1+\mu_1)P_{1,K,0}=\mu_2\,P_{1,K,1}+\beta_1\,P_{2,K,0}+\lambda\,P_{1,K-1,0}+\gamma_1\,P_{0,K,0} \tag{10}$$

$$(\lambda+\alpha_1+\mu_2)P_{1,0,j}=\mu_2\,P_{1,0,j+1}+\beta_1\,P_{2,0,j}+\lambda\,P_{1,0,j-1}+\mu_1\,P_{1,1,j},\quad \mathrm{j}\ \ge\ 1 \tag{11}$$

$$(\lambda+\alpha_1+\mu_1+\mu_2)P_{1,i,j}=\mu_1\,P_{1,i+1,j}+\beta_1\,P_{2,i,j}+\mu_2\,P_{1,i,j+1}+\lambda\,P_{1,i-1,j}+\lambda\,P_{1,i,j-1},\ (1\ \le \mathrm{i}\ \le\ \mathrm{K}-1),\ \ \mathrm{j}\ge 1 \tag{12}$$

$$(\lambda+\alpha_1+\mu_1+\mu_2)P_{1,K,j}=\mu_2\,P_{1,K,j+1}+\beta_1\,P_{2,K,j}+\lambda\,P_{1,K-1,j}+\lambda\,P_{1,K,j-1},\ \mathrm{j}\ge 1 \tag{13}$$

2.2.4 Repair State

$$(\lambda+\beta_1)P_{2,0,j}=\alpha_1\,P_{1,0,j},\qquad \mathrm{j}\ \ge\ 1 \tag{14}$$

$$(\lambda+\beta_1)P_{2,i,j}=\alpha_1\,P_{1,i,j}+\lambda\,P_{2,i-1,j},\qquad (1\ \le\ \mathrm{i}\ \le\ K-1),\quad \mathrm{j}\ \ge\ 1 \tag{15}$$

$$(\lambda+\beta_1)P_{2,K,j}=\alpha_1\,P_{1,K,j}+\lambda\,P_{2,K,j-1}+\lambda\,P_{2,K-1,j},\qquad \mathrm{j}\ \ge\ 1 \tag{16}$$

In order to determine the solution of Eqs. (1) to (16), we shall employ matrix geometric method as explained in Sect. 3.

3 The Analysis

The matrix geometric method (cf. Neuts [14]) can be used to solve the stationary state probabilities for the vector space Markov process with repetitive structure. In order to find the solution for the system of equations constructed in Sect. 2.2, we consider this technique to determine the associated state probability vectors.

3.1 Matrix Geometric Method

The above set of Eqs. (1)–(16) can be written in matrix form as πQ=0, where Q is the infinitesimal generator of the continuous time Markov chain. Also, let $\pi = (\pi_0, \pi_1, \pi_2, \pi_3, \ldots)$ be the vector defining the steady-state probabilities of all the governing states of the retrial queuing system under consideration.

The matrix Q can be given in partition form as

$$Q = \begin{pmatrix} F_0 & F_1 & 0 & 0 & 0 & 0 & 0 & \ldots \\ F_2 & F_3 & F_4 & 0 & 0 & 0 & 0 & \ldots \\ 0 & F_5 & F_3 & F_4 & 0 & 0 & 0 & \ldots \\ 0 & 0 & F_5 & F_3 & F_4 & 0 & 0 & \ldots \\ 0 & 0 & 0 & F_5 & F_3 & F_4 & 0 & \ldots \\ M & M & 0 & 0 & 0 & 0 & 0 & \ldots \end{pmatrix} \tag{17}$$

The block sub matrices of Q are

$$F_0 = \begin{pmatrix} A_0 & B_0 \\ 0 & A_1 \end{pmatrix}_{(2n+1)\times(2n+1)}; \quad F_1 = \begin{pmatrix} 0 & 0 \\ B_1 & C_1 \end{pmatrix}_{(2n+2)\times(2n+2)};$$

$$F_2 = \begin{pmatrix} 0 & D_1 \\ 0 & G_1 \end{pmatrix}_{(2n+2)\times(2n+1)};$$

$$F_3 = \begin{pmatrix} E_1 & 0 \\ 0 & H_1 \end{pmatrix}_{(2n+2)\times(2n+2)}; \quad F_4 = \begin{pmatrix} C_1 & 0 \\ B_1 & 0 \end{pmatrix}_{(2n+2)\times(2n+2)};$$

$$F_5 = \begin{pmatrix} 0 & D_1 \\ 0 & G_1 \end{pmatrix}_{(2n+2)\times(2n+2)};$$

$$B_1 = diag.\left(\alpha_1\right)_{(n+1)\times(n+1)}; \quad D_1 = diag.\left(\beta_1\right)_{(n+1)\times(n+1)};$$

$$G_1 = diag.\left(\mu_2\right)_{(n+1)\times(n+1)};$$

$$A_0 = \begin{pmatrix} -(\lambda+\gamma) & \lambda & 0 & 0 & \ldots & 0 \\ 0 & -(\lambda+\gamma) & \lambda & 0 & \ldots & 0 \\ 0 & 0 & -(\lambda+\gamma) & \lambda & \ldots & 0 \\ M & 0 & 0 & -(\lambda+\gamma) & \lambda & 0 \\ 0 & 0 & 0 & 0 & -(\lambda+\gamma) & \lambda \\ 0 & 0 & 0 & 0 & 0 & -\gamma \end{pmatrix}_{n\times n};$$

$$G_1 = \begin{pmatrix} \mu_2 & 0 & 0 & 0 & \ldots & 0 \\ 0 & \mu_2 & 0 & 0 & \ldots & 0 \\ 0 & 0 & \mu_2 & 0 & \ldots & 0 \\ M & 0 & 0 & \mu_2 & 0 & 0 \\ 0 & 0 & 0 & 0 & \mu_2 & 0 \\ 0 & 0 & 0 & 0 & 0 & \mu_2 \end{pmatrix}_{(n+1)\times(n+1)}$$

$$A_1 = \begin{pmatrix} -(\lambda+\alpha_1) & \lambda & 0 & 0 & \ldots & 0 \\ \mu_1 & -(\lambda+\alpha_1+\mu_1) & \lambda & 0 & \ldots & 0 \\ 0 & \mu_1 & -(\lambda+\alpha_1+\mu_1) & \lambda & \ldots & 0 \\ M & 0 & \mu_1 & -(\lambda+\alpha_1+\mu_1) & \lambda & 0 \\ 0 & 0 & 0 & \mu_1 & -(\lambda+\alpha_1+\mu_1) & \lambda \\ 0 & 0 & 0 & 0 & \mu_1 & -(\lambda+\alpha_1+\mu_1) \end{pmatrix}_{(n+1)\times(n+1)};$$

$$B_0 = \begin{pmatrix} 0 & \gamma & 0 & 0 & 0 & \ldots & 0 \\ 0 & 0 & \gamma & 0 & 0 & \ldots & 0 \\ 0 & 0 & 0 & \gamma & 0 & \ldots & 0 \\ M & 0 & 0 & 0 & \gamma & 0 & 0 \\ 0 & 0 & 0 & 0 & 0 & \gamma & 0 \\ 0 & 0 & 0 & 0 & 0 & 0 & \gamma \end{pmatrix}_{(n)\times(n+1)}$$

$$E_1 = \begin{pmatrix} -(\lambda+\beta_1) & \lambda & 0 & 0 & \dots & 0 \\ 0 & -(\lambda+\beta_1) & \lambda & 0 & \dots & 0 \\ 0 & 0 & -(\lambda+\beta_1) & \lambda & \dots & 0 \\ M & 0 & 0 & -(\lambda+\beta_1) & \lambda & 0 \\ 0 & 0 & 0 & 0 & -(\lambda+\beta_1) & \lambda \\ 0 & 0 & 0 & 0 & 0 & -(\lambda+\beta_1) \end{pmatrix}_{(n+1)\times(n+1)};$$

$$C_1 = \begin{pmatrix} 0 & 0 & 0 & 0 & \dots & 0 \\ 0 & 0 & 0 & 0 & \dots & 0 \\ 0 & 0 & 0 & 0 & \dots & 0 \\ M & 0 & 0 & 0 & 0 & 0 \\ 0 & 0 & 0 & 0 & 0 & 0 \\ 0 & 0 & 0 & 0 & 0 & \lambda \end{pmatrix}_{(n+1)\times(n+1)}$$

$$B_1 = \begin{pmatrix} \alpha_1 & 0 & 0 & 0 & \dots & 0 \\ 0 & \alpha_1 & 0 & 0 & \dots & 0 \\ 0 & 0 & \alpha_1 & 0 & \dots & 0 \\ M & 0 & 0 & \alpha_1 & 0 & 0 \\ 0 & 0 & 0 & 0 & \alpha_1 & 0 \\ 0 & 0 & 0 & 0 & 0 & \alpha_1 \end{pmatrix}_{(n+1)\times(n+1)}; D_1 = \begin{pmatrix} \beta_1 & 0 & 0 & 0 & \dots & 0 \\ 0 & \beta_1 & 0 & 0 & \dots & 0 \\ 0 & 0 & \beta_1 & 0 & \dots & 0 \\ M & 0 & 0 & \beta_1 & 0 & 0 \\ 0 & 0 & 0 & 0 & \beta_1 & 0 \\ 0 & 0 & 0 & 0 & 0 & \beta_1 \end{pmatrix}_{(n+1)\times(n+1)};$$

The normalizing condition is represented by $\Pi e = 1$, where "e" is a column vector of suitable dimension with all its entries as 1. In order to determine the probability vector, we partition vector π conformably with the block of matrix Q as

$$\pi_0 = (P_{0,0,0}, P_{1,0,0}; P_{0,1,0}, P_{1,1,0}; \dots\dots; P_{0,k,0}, P_{1,k,0}); P_{0,0,0} = 0 \tag{18}$$

$$\pi_1 = (P_{2,0,0}, P_{1,0,1}; P_{2,1,0}, P_{1,1,1}; \dots\dots; P_{2,k,0}, P_{1,k,1})$$

$$\pi_2 = (P_{2,0,1}, P_{1,0,2}; P_{2,1,1}, P_{1,1,2}; \dots\dots; P_{2,k,1}, P_{1,k,2})$$

$$\vdots$$

$$\pi_j = (P_{2,0,j-1}, P_{1,0,j}; P_{2,1,j-1}, P_{1,1,j}; \dots\dots; P_{2,k,j-1}, P_{1,k,j}); \qquad j \geq 1 \tag{19}$$

Using matrix geometric approach (cf. [14]), we have

$$\pi_j = \pi_1 R^{j-1}, \qquad (j \geq 2) \tag{20}$$

where, R is the minimal nonnegative matrix known as rate matrix.

Balance equation for repeating states $\pi_{j-1} F_4 + \pi_j F_3 + \pi_{j+1} F_5 = 0$; $j = 2, 3, 4, \dots$. (21)

The value π_j, $(j \geq 2)$ is a probability function of the transition between the states with $j-1$ queued customer and states with j queued customers using Eqs. (20) and (21), we have

$$F_4 + RF_3 + R^2 F_5 = 0 \tag{22}$$

On solving Eq. (22), we get the rate matrix R.

$$R(n+1) = -[F_4 + R^2 F_5] F_3^{-1}, \qquad n \geq 0$$

3.2 Performance Measures

Performance measures calculated in terms of steady-state probabilities are as follows:

3.2.1 Server State Probabilities

The probabilities of the server being present in different states are expressed as:

1. Probability of the primary server in retrial state is framed as: $P_r = \sum_{n=0}^{k} P_{0,n,0}$
2. Probability of primary server being busy: $P_{B1} = \sum_{n=1}^{k} P_{1,n,0}$
3. Probability when both server is busy in servicing = $P_{B2} = \sum_{j=1}^{\infty} \sum_{n=1}^{k} P_{1,n,j}$
4. Probability of the primary server being in broken down state: $P_D = \sum_{j=0}^{\infty} \sum_{n=0}^{q} P_{2,n,j}$

3.2.2 Queue Length

1. The expected number of customers in the retrial orbit is $E[N_r] = \sum_{n=1}^{k} nP_{0,n,0}$
2. The expected number of customer in the busy state when primary server is:
$$E[N_1] = \sum_{n=1}^{k} nP_{1,n,0}$$
3. The expected number of customer in the busy state when both the servers are busy in rendering service to the customers $=E[N_2] = \sum_{j=1}^{\infty} \sum_{n=1}^{K} nP_{1,n,j} + \sum_{j=1}^{\infty} jP_{1,k,j}, j \geq 1$
4. The expected number of customer when primary server is in broken state is
$$E[N_d] = \sum_{n=0}^{k} nP_{2,n,j}, j \geq 0$$
5. Expected number of customer in system $E[N] = E[N_r] + E[N_1] + E[N_2] + E[N_d]$

3.2.3 Throughput

Throughput gives the number of effective services rendered by the servers in the system. $TP = \mu_1 \sum_{n=0}^{k} P_{1,n,0} + (\mu_1 + \mu_2) \sum_{j=1}^{\infty} \sum_{n=0}^{k} P_{1,n,j}$

3.2.4 Expected Delay

The expected delay experienced by the customer in the system is $E[D] = \frac{E[N]}{TP}$

3.2.5 Waiting Time

The customer needs to wait in the system so as to get served either due to the unavailability of the server or due to busy behavior of the server. $E[W] = \frac{E[N]}{\lambda}$

4 Numerical Study

We now present numerical results related to the model discussed in the above section. We take the parameters are $\lambda = 0.4$, $\mu_1 = 3$, $\mu_2 = 5$, $\gamma = 0.4$, $\alpha_1 = 0.1$, $\beta_1 = 1$, K = 5. Using different parameters, we obtained various submatrices F_0, F_1, F_2, F_3, F_4, and F_5, and rate matrix **R** is computed:

$$R = \begin{pmatrix}
0 & 0 & 0 & 0 & 0 & 0 & 0 & 0 & 0 & 0 \\
0 & 0 & 0 & 0 & 0 & 0 & 0 & 0 & 0 & 0 \\
0 & 0 & 0 & 0 & 0 & 0 & 0 & 0 & 0 & 0 \\
0 & 0 & 0 & 0 & 0 & 0 & 0 & 0 & 0 & 0 \\
0.482 & 0.722 & 0.158 & 0.024 & 0.049 & 0.068 & 0.097 & 0.168 & 0.056 & 0.056 \\
0.009 & 0.045 & 0.469 & 0.059 & 0.068 & 0.086 & 0.153 & 0.162 & 0.068 & 0.078 \\
0.046 & 0.041 & 0.023 & 0.012 & 0.009 & 0.003 & 0.001 & 0.004 & 0.002 & 0.001 \\
0 & 0 & 0 & 0 & 0 & 0 & 0 & 0 & 0 & 0 \\
0 & 0 & 0 & 0 & 0 & 0 & 0 & 0 & 0 & 0 \\
0 & 0 & 0 & 0 & 0 & 0 & 0 & 0 & 0 & 0 \\
0 & 0 & 0 & 0 & 0 & 0 & 0 & 0 & 0 & 0 \\
0 & 0 & 0 & 0 & 0 & 0 & 0 & 0 & 0 & 0 \\
0 & 0 & 0 & 0 & 0 & 0 & 0 & 0 & 0 & 0
\end{pmatrix}$$

By using the above R matrix, the probability vectors are calculated. Our objective is to demonstrate the effect of the parameters on the probabilities by varying λ_i's, from the value 0.1 to 1 which is given in Table 1. Also all the performance measures are calculated and presented in Table 2.

Table 1 Performance measures

λ	0.1	0.2	0.3	0.4	0.5	0.6	0.7	0.8	0.9	1
p_{000}	0.321	0.359	0.368	0.383	0.388	0.393	0.398	0.421	0.433	0.441
p_{010}	0.122	0.086	0.057	0.048	0.041	0.033	0.024	0.021	0.019	0.017
p_{020}	0.005	0.009	0.007	0.008	0.006	0.005	0.002	0.001	0.001	0.000
p_{110}	0.123	0.147	0.152	0.159	0.164	0.183	0.189	0.196	0.221	0.232
p_{120}	0.056	0.072	0.062	0.061	0.058	0.045	0.022	0.018	0.016	0.013
p_{130}	0.012	0.007	0.009	0.010	0.008	0.006	0.004	0.003	0.002	0.001
p_{111}	0.098	0.091	0.086	0.083	0.068	0.051	0.043	0.038	0.031	0.024
p_{112}	0.009	0.015	0.012	0.011	0.009	0.005	0.003	0.002	0.001	0.000
p_{113}	0.006	0.005	0.004	0.006	0.008	0.012	0.016	0.021	0.027	0.033
p_{121}	0.026	0.022	0.017	0.014	0.012	0.008	0.007	0.005	0.003	0.002
p_{122}	0.007	0.006	0.009	0.008	0.006	0.005	0.004	0.003	0.002	0.001
p_{123}	0.006	0.005	0.005	0.006	0.005	0.002	0.002	0.001	0.001	0.000
p_{131}	0.017	0.012	0.028	0.034	0.043	0.048	0.059	0.062	0.069	0.073
p_{132}	0.002	0.002	0.008	0.006	0.004	0.003	0.002	0.004	0.002	0.001
p_{133}	0.001	0.001	0.006	0.004	0.002	0.002	0.005	0.006	0.003	0.002
p_{151}	0.012	0.018	0.012	0.011	0.008	0.006	0.004	0.002	0.001	0.000
p_{152}	0.005	0.008	0.007	0.008	0.009	0.014	0.025	0.029	0.036	0.041
p_{153}	0.003	0.004	0.003	0.002	0.001	0.001	0.001	0.001	0.000	0.000
p_{200}	0.023	0.022	0.018	0.016	0.013	0.011	0.009	0.007	0.005	0.003
p_{201}	0.002	0.013	0.010	0.009	0.006	0.005	0.003	0.002	0.001	0.000
p_{202}	0.000	0.008	0.011	0.014	0.019	0.028	0.035	0.039	0.042	0.046
p_{210}	0.021	0.018	0.016	0.014	0.011	0.008	0.006	0.005	0.004	0.002
p_{211}	0.008	0.007	0.006	0.005	0.004	0.003	0.002	0.001	0.001	0.000
p_{212}	0.003	0.002	0.004	0.006	0.009	0.011	0.013	0.016	0.018	0.021
p_{220}	0.025	0.021	0.019	0.016	0.012	0.009	0.007	0.005	0.003	0.001
p_{221}	0.006	0.004	0.011	0.023	0.034	0.046	0.052	0.056	0.059	0.062
p_{222}	0.002	0.001	0.008	0.009	0.013	0.016	0.020	0.024	0.026	0.029
p_{240}	0.018	0.017	0.013	0.011	0.009	0.007	0.005	0.004	0.003	0.002
p_{241}	0.006	0.005	0.008	0.013	0.017	0.023	0.026	0.029	0.031	0.033
p_{242}	0.004	0.004	0.006	0.008	0.012	0.017	0.021	0.026	0.029	0.031

5 Conclusion

In this paper, we have considered a multi-server retrial queuing system with unreliable server. We have obtained the steady-state probability vector by applying

Table 2 Performance measures

λ	0.1	0.2	0.3	0.4	0.5	0.6	0.7	0.8	0.9	1
P_r	0.448	0.454	0.432	0.439	0.435	0.431	0.424	0.443	0.453	0.458
P_{B1}	0.191	0.226	0.223	0.23	0.23	0.234	0.213	0.217	0.239	0.246
P_{B2}	0.192	0.189	0.197	0.193	0.175	0.157	0.171	0.174	0.176	0.177
P_D	0.118	0.122	0.130	0.144	0.159	0.184	0.199	0.214	0.222	0.230
$E[N_r]$	0.132	0.104	0.071	0.064	0.026	0.043	0.028	0.023	0.021	0.017
$E[N_1]$	0.271	0.312	0.303	0.311	0.304	0.291	0.245	0.241	0.259	0.261
$E[N_2]$	0.382	0.418	0.435	0.426	0.397	0.399	0.493	0.518	0.545	0.578
$j = 0$	0.143	0.128	0.106	0.09	0.071	0.054	0.04	0.031	0.022	0.012
E$[N_d]j = 1$	0.044	0.035	0.06	0.103	0.14	0.187	0.21	0.229	0.246	0.256
$j = 2$	0.023	0.02	0.044	0.056	0.083	0.111	0.137	0.168	0.186	0.203
$E[N]$	0.995	0.017	1.019	1.05	1.021	1.085	1.153	1.21	1.279	1.327
TP	0.454	0.488	0.514	0.526	0.483	0.487	0.529	0.563	0.627	0.664
$E[D]$	2.189	2.083	1.984	1.995	2.114	2.228	2.178	2.148	2.040	1.999
$E[W]$	9.95	5.085	3.396	2.625	2.042	1.808	1.647	1.513	1.421	1.327

matrix geometric method. Furthermore, we have performed numerical analysis by assuming particular values to the parameter. Various performance measures are computed to analyze the system behavior in a better server. It is verified that the total probability is ≈ 1.

References

1. Aissani, A., and Artalejo, J.R., On the single server retrial queue subject to breakdowns.Queuing systems, 30:307–321,(1998).
2. Alfa, A.S. and Isotupa, K.P.S., An M/PH/K retrial queue with finite number of sources, Comput. Oper. Res., Vol.31, pp. 1455–1464, (2004).
3. Artalejo, J.R. and Corral, A.G., Retrial Queuing Systems: A Computational Approach,Springer, (2008).
4. Bhagat, A. and Jain, M., Unreliable M^x/G/1 retrial queue with multi-optional services and impatient customers, Int.J.Oper. Res., Vol.17, pp.248–273, (2013).
5. Choudhury, G. and Ke, J.C., A batch arrival retrial queue with general retrial times under Bernoulli vacation schedule for unreliable server and delaying repair, Appl.Math. Model., Vol. 36,pp.255–269,(2012).
6. Dimitrious, I., A mixed priority retrial queue with negative arrival, unreliable server and multiple vacations.Appl.Math.Model.,Vol. 37,pp. 1295–1309. (2013).
7. Falin, G. I.and templeton, J.G.C., Retrial Queues, Chapman and Hall, (1997).
8. Kalyanaraman, R. and Seenivasan, M., A multi-server retrial queuing system with unreliable server, International Journal of Computational Cognation. Vol.8,NO.3,September(2010).
9. Krishna Kumar, B. and Madheswari, S.P., An M/M/2 queuing system with heterogeneous servers and multiple vacations, Math. Comput. Model., Vol.41,pp. 1415–1429, (2005).
10. Kulkarni, V.G., and Choi, B.D., Retrial queues with server subject to breakdowns. Queuing Systems,7:191–208, (1990).

11. Kushner, H.J., Heavy traffic analysis of controlled Queuing and Communication Networks Springer-Verlag, New York, (2001).
12. Leite, S.C. and Fragoso, M.D., Heavy traffic analysis of state-dependent parallel queues with trigger and an application to web search systems Perf.Eval., Vol.67, pp.913–928,(2010).
13. Neuts, M.F., Markov chains with applications queueing theory, which have a matrix geometric invariant probability vector, Adv. Appl. Prob., Vol.10, pp.185–212, (1978).
14. Neuts, M.F., Matrix-Geometric Solutions in Stochastic Models: An Algorithmic Approach. The Johns Hopkins University Press, Baltimore, 1981.
15. Wu, J. and Lian, Z. A single-server retrial G-queue with priority and unreliable server under Bernoulli vacation schedule.Comp.Ind.Engg.,Vol.64,pp.84–93,(2013).

Two-Level Control Policy of an Unreliable Queueing System with Queue Size-Dependent Vacation and Vacation Disruption

S. P. Niranjan, V. M. Chandrasekaran, and K. Indhira

Abstract The objective of the paper is to analyse two-level control policy of an $M^X/G(a,b)/1$ queueing system with fast and slow vacation rates and vacation disruption. In the service completion epoch, if the queue length is less than 'a', then the server leaves for a vacation. In this model depending upon the queue length, the server is allowed to take two types of vacation called fast vacation and slow vacation. Addressing this in the service completion epoch, if the queue length ψ(say) is less than β where $\beta < a - 1$, then the server leaves for slow vacation. On the other hand, if $\psi > \zeta$, where $a - 1 \geq \zeta > \beta$ during service completion, then the server leaves for fast vacation. During slow vacation if the queue length reaches the value ζ, then the server breaks the slow vacation and switches over to fast vacation. Also if the queue length attains the threshold value 'a' during fast vacation, then the server breaks the fast vacation too and moves to tune-up process to start the service. After tune-up process service will be initiated only if $\psi \geq N (N > b)$. For the designed queueing system probability, generating function of the queue size at an arbitrary time epoch is obtained by using supplementary variable technique. Various performance characteristics will also be derived with suitable numerical illustrations. Cost-effective analysis is also carried out in the paper.

1 Introduction

In server vacation models the server is used to do some supplementary jobs, during its idle time. These types of additional work may increase the quality of service. Many researchers have modelled certain type of queueing systems with the intention of effective utilization of an idle time of the server. Lee et al. [3] have analysed fixed batch service queueing system with vacations. They

S. P. Niranjan (✉) · V. M. Chandrasekaran · K. Indhira
Department of Mathematics, School of Advanced Sciences, VIT, Vellore, India
e-mail: niran.jayan092@gmail.com; niranjan.sp@vit.ac.in; vmcsn@vit.ac.in; kindhira@vit.ac.in

V. Madhu et al. (eds.), *Advances in Algebra and Analysis*, Trends in Mathematics,
https://doi.org/10.1007/978-3-030-01120-8_42

used decomposition technique to derive queue length distribution of the system. Neuts [4] introduced general bulk service rule for batch service queueing system. Ke [2] introduced the concept 'T-vacation policy' for an unreliable queueing system with setup times. In this paper, the server takes the vacation with constant time length 'T'. Haridass and Nithya [1] have studied $M^X/G(a,b)/1$ queueing system with server breakdown and vacation break-off. Singh and Kumar [6] have discussed maximum entropy analysis of $M^X/G/1$ queueing system with threshold, Bernoulli scheduled vacation and m-optional services. Wu et al. [7] studied non-Markovian queueing system with threshold, vacation, server failure and changeable repair facility. Niranjan et al. [5] have analysed performance measures of batch service queueing system with state-dependent service, server failure and vacation interruption.

In all the batch service queueing models, the server continues the vacation period though the queue length reaches the value 'a'. But this type of assumption will increase the waiting time of the customer. By considering the above situation, the authors introduced queue size-dependent vacation (fast vacation and slow vacation) for $M^X/G(a,b)/1$ queueing system with vacation disruption.

2 Model Description

In this paper, two-level control policy of an $M^X/G(a,b)/1$ queueing system with fast and slow vacation and vacation disruption is considered. Customers are arriving into the system in bulk according to the Poisson process with rate λ. Arriving customers are served in batches according to general bulk service rule. Depending upon the queue length, the server takes two types of vacation called fast vacation and slow vacation. In the service completion epoch, if the queue length ψ(say) is less than β where $\beta < a-1$, then the server decides to take slow vacation. On the other hand if the queue length ranges from $\beta < \psi \leq a-1$, then the server leaves for slow vacation. The duration of secondary job at slow vacation is high when compared to fast vacation.

An identification of server failure or proper maintenance of the server is called renewal of service station. After completing a batch of service, if the server is not reliable, then the renewal of service station will be considered with probability δ. If the server is reliable after the service completion with probability $1-\delta$ and the queue length is less than 'a', then depending upon the queue length, the server leaves for either fast vacation or slow vacation. During slow vacation if $\psi > \beta$, then the server breaks the slow vacation and switches over to fast vacation.

Tune-up time is defined as time needs to start the service after vacation or idle period of the server. Also at the time of fast vacation if the queue length reaches the threshold value 'a' , then the server breaks the fast vacation too and switches over to do tune-up process. If the queue length is still less than 'a' even after slow vacation

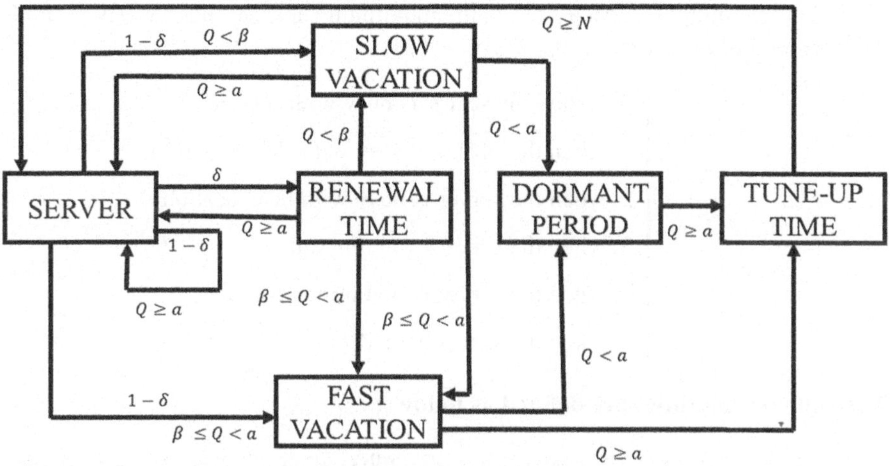

Fig. 1 Schematic representation of the model: Q-queue length

completion, then the server becomes dormant(idle) until the queue length reaches the value 'a'. Though the server completes tune-up process, service will be initiated only if the queue length is at least 'N' ($N > b$). The schematic representation of the designed queueing system is depicted below (Fig. 1).

2.1 Notations

Let γ be the Poisson arrival rate, X be the group size random variable of the arrival, g_k be the probability that k customers arrive in a batch, $X(z)$ be the probability-generating function(PGF) of X, $N_q(t)$ be the number of customers waiting for service at time t and $N_s(t)$ be the number of customers under the service at time t. Let $P(x)(p(x))\{\tilde{P}(\theta)\}[P^0(x)]$ be the cumulative distribution function (probability density function) {Laplace-Stieltjes transform} [remaining service time] of service. Let $Q(x)(q(x)\{\tilde{Q}(\theta)\}[Q^0(x)]$ be the cumulative distribution function (probability density function) {Laplace-Stieltjes transform} [remaining vacation time] of fast vacation. Let $S(x)(s(x)\{\tilde{S}(\theta)\}[S^0(x)]$ be the cumulative distribution function (probability density function) {Laplace-Stieltjes transform} [remaining vacation time] of slow vacation. Let $R(x)(r(x))\{\tilde{R}(\theta)\}[R^0(x)]$ be the cumulative distribution function (probability density function){Laplace-Stieltjes transform}[remaining renewal time] of renewal of service station. Let $U(x)(u(x))\{\tilde{U}(\theta)\}[U^0(x)]$ be the cumulative distribution function (probability density function){Laplace-Stieltjes transform}[remaining tune-up time] of tune-up time.

'a' is the minimum capacity, 'b' is the maximum capacity and 'N' $(N > B)$ is the threshold of the server

$$Y(t) = \begin{cases} 0, & \text{When the server is busy with service} \\ 1, & \text{when the server is busy with slow vacation} \\ 2, & \text{when the server is busy with fast vacation} \\ 3, & \text{when the server is on renewal} \\ 4, & \text{when the server is on tuning process} \\ 5, & \text{when the server is on dormant period} \end{cases}$$

The state probabilities are defined as follows:

$$\begin{aligned}
A_{ij}(x,t)dt &= Pr\{N_s(t)=i,\ N_q(t)=j,\ x \le P^0(t) \le x+dt,\ Y(t)=0\};\ a \le i \le b;\ j \ge 1 \\
B_j(x,t)dt &= Pr\{N_q(t)=j, x \le Q^0(t) \le x+dt, Y(t)=1\};\ j \ge \beta, \beta < a-1 \\
C_j(x,t)dt &= Pr\{N_q(t)=j, x \le S^0(t) \le x+dt, Y(t)=2\};\ \beta < j \le a-1 \\
R_n(x,t)dt &= Pr\{N_q(t)=n, x \le R^0(t) \le x+dt, Y(t)=3\};\ n \ge 0 \\
U_n(x,t)dt &= Pr\{N_q(t)=n, x \le U^0(t) \le x+dt, Y(t)=4\};\ n \ge a \\
I_n(t)dt &= Pr\{N_q(t)=n, Y(t)=5\};\ 0 \le n \le a-1
\end{aligned}$$

3 Steady-State Queue Size Distribution

$$-A'_{i0}(x) = -\gamma A_{i0}(x) + \Big(\delta \sum_{m=a}^{b} A_{mi}(0) + R_i(0) + B_i(0)\Big)p(x) \quad a \le i \le b \tag{1}$$

$$-A'_{ij}(x) = -\gamma A_{ij}(x) + \sum_{k=1}^{j} A_{ij-k}(x)\gamma g_k \quad a \le i \le b-1;\ j \ge 1 \tag{2}$$

$$-A'_{bj}(x) = -\gamma A_{bj}(x) + \Big(\delta \sum_{m=a}^{b} A_{mb+j}(0) + R_{b+j}(0) + B_{b+j}(0)\Big)p(x) \tag{3}$$

$$+ \sum_{k=1}^{j} A_{bj-k}(x)\gamma g_k \qquad 1 \le j \le N-b-1$$

$$-A_{bj}^{'}(x) = -\gamma A_{bj}(x) + \big(\delta \sum_{m=a}^{b} A_{mb+j}(0) + R_{b+j}(0) + B_{b+j}(0) + U_{b+j}(0)\big)p(x) \tag{4}$$

$$+ \sum_{k=1}^{j} A_{bj-k}(x)\gamma g_k \qquad j \geq N - b$$

$$-B_0^{'}(x) = -\gamma B_0(x) + \big((1-\delta)\sum_{m=a}^{b} A_{m0}(0) + R_0(0)\big)q(x) \tag{5}$$

$$-B_n^{'}(x) = -\gamma B_n(x) + \big((1-\delta)\sum_{n=1}^{\beta}\sum_{m=a}^{b} A_{mn}(0) + R_n(0)\big)q(x) \tag{6}$$

$$+ \sum_{n=1}^{k} B_{n-k}(x)\gamma g_k \qquad \beta < a - 1$$

$$-C_n^{'}(x) = -\gamma C_n(x) + \big((1-\delta)\sum_{m=a}^{b} A_{mn}(0) + R_n(0)\big)s(x) \tag{7}$$

$$+ \sum_{n=1}^{k} C_{n-k}(x)\gamma g_k + \sum_{k=0}^{\beta} B_k(x)\gamma g_{n-k}s(x) \qquad \beta < n \leq a - 1$$

$$-R_0^{'}(x) = -\gamma R_0(x) + \delta \sum_{m=a}^{b} A_{m0}(0)r(x) \tag{8}$$

$$-R_n^{'}(x) = -\gamma R_n(x) + \delta \sum_{m=a}^{b} A_{mn}(0)r(x) + \sum_{k=1}^{n} R_{n-k}(x)\gamma g_k \qquad n \geq 1 \tag{9}$$

$$-U_n^{'}(x) = -\gamma U_n(x) + \sum_{k=1} U_{n-k}(x)\gamma g_k + \sum_{k=0}^{a-1} C_k(x)\gamma g_{n-k}u(x) \tag{10}$$

$$+ \sum_{k=0}^{a-1} I_k \gamma g_{n-k}u(x) \tag{11}$$

$$0 = -\gamma I_0 + B_0(0) + C_0(0) \tag{12}$$

$$0 = -\gamma I_n + C_n(0) + B_n(0) + \sum_{k=1}^{n} I_{n-k}\gamma g_k \qquad 1 \leq n \leq a - 1 \tag{13}$$

4 Probability-Generating Function (PGF) of the Queue Size

The PGF of the queue size at an arbitrary time epoch is obtained by the methodology used in Niranjan et al. [5]

$$P(z)=\frac{\begin{array}{c}F(\gamma,z)\sum_{n=a}^{b-1}(A_n+B_n+R_n)-\sum_{n=0}^{b-1}(A_n+B_n+R_n)z^nW_3+M_1\sum_{n=1}^{\beta}(A_n+R_n)z^n\\+W_2\left(\left(\widetilde{U}(\gamma-\gamma X(z))-1\right)W_1+\widetilde{U}(\gamma-\gamma X(z))W_3\right)+G(\gamma,z)+I(z)W_1(-\gamma+\gamma X(z))\end{array}}{W_1(-\gamma+\gamma X(z))}$$

where

$$\begin{aligned}F(\gamma,z)&=\left(W_1+\widetilde{P}(\gamma-\gamma X(z))\left((1-\delta)\right)+\delta\widetilde{R}(\gamma-\gamma X(z))\right)\\&\quad\times W_3\left\{W_1+(1-\delta)\widetilde{P}(\gamma-\gamma X(z))+\delta\widetilde{R}(\gamma-\gamma X(z))\right\}\end{aligned}$$

$$\begin{aligned}G(\gamma,z)&=\left(\widetilde{S}(\gamma-\gamma X(z))-1\right)W_1\left(\sum_{i=\beta}^{a-1}(A_i+R_i)z^i\right.\\&\quad\left.+\gamma\left(\sum_{i=0}^{\beta}\widetilde{B}_i(\gamma-\gamma X(z))z^i\left(X(z)-\sum_{j=0}^{a-i-1}g_jz^j\right)\right)\right)\end{aligned}$$

$$W_1=z^b-(1-\delta)\widetilde{P}(\gamma-\gamma X(z))-\delta\widetilde{P}(\gamma-\gamma X(z))\widetilde{R}(\gamma-\gamma X(z))$$

$$\begin{aligned}W_2&=\gamma\left(\sum_{i=0}^{a-1}\widetilde{S}_i(\gamma-\gamma X(z))z^i\left(X(z)-\sum_{j=0}^{b-i-1}g_jz^j\right)\right.\\&\quad\left.+\sum_{m=0}^{a-1}I_mz^m\left(X(z)-\sum_{j=0}^{b-m-1}g_jz^j\right)\right)\end{aligned}$$

$$W_3=\left(\left(\widetilde{P}(\gamma-\gamma X(z))-1\right)+\delta\left(\widetilde{R}(\gamma-\gamma X(z))-1\right)\widetilde{P}(\gamma-\gamma X(z))\right)$$

$$M_1=\left(\widetilde{Q}(\gamma-\gamma X(z))-1\right)W_1+\widetilde{Q}(\gamma-\gamma X(z))W_3$$

5 Performance Measures

5.1 Expected Length of Busy Period

Let Y be the busy period random variable and T be the residence time that the server is rendering service or under repair. Then $E(T) = E(A) + \delta E(R)$ where $E(A)$ is the expected service time, $E(R)$ is the expected renewal time of the server, then

$$E(Y) = \frac{E(T)}{(1-\delta)\sum_{m=a}^{b}\sum_{i=a}^{b}\left(A_{mi}(0) + R_i\right)}$$

5.2 Expected Length of Idle Period

$$E(D) = \frac{1}{\gamma}\sum_{n=0}^{a-1}\alpha_j$$

where $\frac{1}{\gamma}$ is the expected staying time in the state 'n' during an idle period.

5.3 Expected Queue Length

$$E(Q) = \frac{2\psi_2''\left(\tau_1'''+\tau_2'''+\tau_3'''\right) - 2\psi_2'''\left(\tau_1''+\tau_2''+\tau_3''\right) - Y_1 - 3\psi_2^{iv}\left(\tau_1'+\tau_2'+\tau_3'\right)}{24\left(\lambda E(X)(b - \lambda E(A)E(X) - \delta E(R)E(X))\right)^2}$$

where

$$\tau_1 = F(\gamma, z)\sum_{n=a}^{b-1}(A_n + B_n + R_n) - \sum_{n=0}^{b-1}(A_n + B_n + R_n)\, z^n W_3 + M_1\sum_{n=1}^{\beta}(A_n + R_n)\, z^n$$

$$\tau_2 = \left(\widetilde{S}(\gamma - \gamma X(z)) - 1\right) W_1\left(\sum_{i=\beta}^{a-1}(A_i + R_i)z^i + \gamma\left(\sum_{i=0}^{\beta}\widetilde{B}_i(\gamma - \gamma X(z))z^i \times\left(X(z) - \sum_{j=0}^{a-i-1} g_j z^j\right)\right)\right)$$

$$\tau_3 = W_2\left(\left(\widetilde{U}(\gamma-\gamma X(z))-1\right)W_1+\widetilde{U}(\gamma-\gamma X(z))W_3\right)+I(z)W_1(-\gamma+\gamma X(z))$$

$$\psi_2 = (-\gamma+\gamma X(z)))\left(z^b-(1-\delta)\widetilde{P}(\gamma-\gamma X(z))-\delta\widetilde{P}(\gamma-\gamma X(z))\widetilde{R}(\gamma-\gamma X(z))\right)$$

6 Cost Model

Optimum cost analysis of queueing systems is much useful in minimizing the total average cost. The following assumptions are made to obtain the total average cost of the system.

A_h : Holding cost per customer	A_o : Operating cost per unit time
A_s : Tune-up cost per cycle	A_r : Reward cost per cycle due to vacation

$$\text{Total average cost(TAC)} = [A_s - A_r E(I)]\frac{1}{E(B+E(I))} + A_h E(Q) + A_o\left(\frac{\gamma E(X)(E(P)+\delta E(R))}{b}\right)$$

The simple direct search method is used to find the threshold value 'a^*' to minimize the TAC of the system is presented below.

step 1: Fix the threshold value 'N' $(N > b)$
step 2: Select the value of 'a' which satisfies the above condition

$$TAC(a^*) \leq TAC(a), \quad 1 \leq a \leq N$$

step 3: The value a^* is optimum, because it gives minimum TAC.

Some numerical example is presented in the next section to illustrate the above direct search method.

7 Numerical Illustrations

In this section, the derived analytical results are justified numerically with the following assumptions. Service time distribution is 4-Erlang with parameter μ; batch size distribution of the arrival is geometric with mean 2; fast vacation time, slow vacation time, renewal time and tune-up time follow exponential with parameters $\varphi_1, \varphi_2, \eta$ and ω, respectively (Table 1).

Start-up cost	Rs. 4	Holding cost	Rs. 0.50	Operating cost	Rs. 5
Reward cost	Rs. 1	Renewal cost	Rs. 0.4		

Table 1 Arrival rate versus performance measures ($\mu = 2.0, a = 3, b = 7, N = 11, \varphi_1 = 8, \varphi_2 = 6, \eta = 4, \omega = 5$)

γ	E(Q)	E(Y)	E(D)	E(W)
2.0	0.5469	1.6283	0.1473	0.3281
2.5	0.7742	1.7140	0.1085	0.3276
3.0	1.2973	1.8375	0.0517	0.3362
3.5	1.3715	2.1269	0.0193	0.3481
4.0	1.4928	2.2761	0.0158	0.3499
4.5	1.7193	2.5083	0.0132	0.3516
5.0	2.0481	2.7740	0.0113	0.3530

Table 2 Threshold value 'a' versus performance measures ($\gamma = 3, \mu = 2.0, b = 7, N = 11, \varphi_1 = 7, \varphi_2 = 5, \eta = 6, \omega = 4$)

a	E(Q)	E(Y)	E(D)	TAC
1	0.3741	2.5247	1.1749	0.7183
2	0.4379	3.3879	2.1693	0.5490
3	0.7962	5.8041	3.3150	0.4062
4	0.9617	5.5882	4.6241	0.4229
5	1.2438	10.0155	5.5907	0.4588
6	1.4962	15.8138	7.2321	0.5371
7	1.7709	19.2751	8.0171	0.5867
8	2.2310	17.6084	9.1982	0.7196
9	2.6019	15.1530	10.4763	0.9342

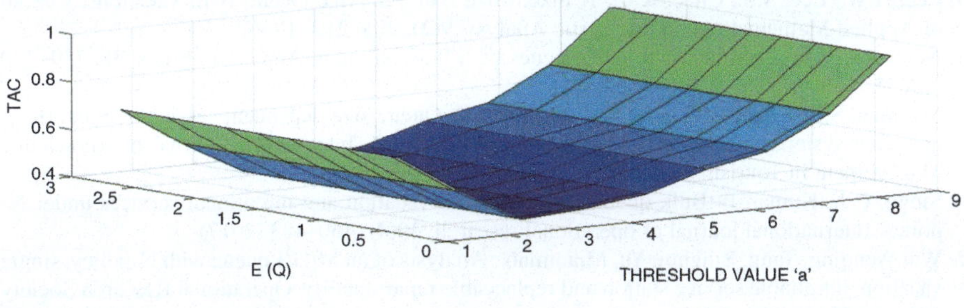

Fig. 2 Threshold value 'a' versus total average cost

7.1 *Optimal Value of 'a'*

The effects of the threshold value 'a' on the total average cost with $N = 11$ are given in Table 2 and Fig. 2. From that table and figure, it can be seen that to minimize the total average cost of the system, the management has to set the threshold value 'a' as 4. Similarly, the management has to fix the threshold value 'a' to minimize the total average cost for various arrival rates, service rates, renewal rates and vacation rates.

8 Conclusion

In this paper bulk arrival and batch service queueing system with fast vacation, slow vacation and vacation disruption are considered. The proposed queueing system is unique because fast and slow vacations are introduced for batch service queueing system. The PGF of the queue size at an arbitrary time epoch is obtained by using supplementary variable technique. Various performance characteristics are also presented with appropriate numerical illustrations. Additionally cost-effective model is presented for the queueing system.

Acknowledgements This work was supported by the "NBHM DAE, Government of India" and "Ref.No.2/48(6)/2015/$NBHM(R.P)/R\&D$11/14129".

References

1. Haridass, M., Nithya, R.P.: Analysis of a bulk queueing system with server breakdown and vacation interruption. International Journal of Operations Research.**12(3)**, 069–090 (2015)
2. Ke, J.C.: Modified T Vacation Policy for an M/G/1 Queueing System with an Unreliable Server and start-up. Mathematical and Computer Modelling.**41**, 1267–1277 (2005)
3. Lee, H.W., Lee, S.S., Chae, K.C.: A Fixed-Size Batch Service Queue With Vacations. Journal of Applied Mathematics and Stochastic Analysis.**9(2)**, 205–219 (1996)
4. Neuts, M.F.: A general class of bulk queues with Poisson input. Ann.Math. Statis. **38**, 759–770 (1967).
5. Niranjan, S.P., Chandrasekaran, V.M., Indhira, K.:Queue size dependent service in bulk arrival queueing system with server loss and vacation break-off. International journal of Knowledge Management in Tourism and Hospitality.**1(2)**, 176–207 (2017)
6. Singh, C.J., Kumar, B.:Bulk queue with Bernoulli vacation and m-optional services under N-policy. International journal of operational research. **30(4)**, 460–483 (2017)
7. Wu, Wenqing, Tang, Yinghui, Yu, Miaomiao.: Analysis of an M/G/1 queue with N-policy, single vacation, unreliable service station and replaceable repair facility. Operational Research Society of India. **52**, 670691 (2015)

Analysis of M/G/1 Priority Retrial G-Queue with Bernoulli Working Vacations

P. Rajadurai, M. Sundararaman, Sherif I. Ammar, and D. Narasimhan

Abstract In this investigation, a priority retrial queue with working vacations and negative customers is addressed. The priority clients don't shape any line and have an elite preemptive priority to get their services over normal customers. When the orbit is noticeably empty at the season of service consummation, the server takes for a working vacation. In working vacation period, the server works at a lower rate of service. Utilizing the supplementary variable technique (SVT), the probability generating function (PGF) of the system capacity is found. Some important special cases are discussed.

1 Introduction

In our day-to-day real life, we undergo many queueing situations in which service is not immediate and customer has to wait for receiving service. From these situations, retrial concepts act a unique role in computers and communication networks, communication protocols, manufacturing or production systems, flexible manufacturing systems, wireless communication systems, and email systems. For more general models of retrial, priority, and working vacation queueing models, the reader may refer to Artalejo and Gomez-Corral [1], Arivudainambi et al. [2], Gao [3], Gao et al. [4], and Rajadurai [5]. In this analysis, we have generalized the work of Gao [3] by consolidating the ideas of working vacations and vacation interruption in presence of G-queues. The scientific outcomes and theory of waiting lines of this model give to serve a particular and persuading application in the media

P. Rajadurai (✉) · M. Sundararaman · D. Narasimhan
Department of Mathematics, SRC, SASTRA Deemed University, Kumbakonam, India
e-mail: psdurai17@gmail.com; rajadurai@src.sastra.edu; msn_math@rediffmail.com; msn_math@src.sastra.edu; narasimhan@src.sastra.edu

S. I. Ammar
Department of Mathematics, Menoufia University, Shebin El Koum, Egypt

Department of Mathematics, Faculty of science, Taibah University, Medina, Saudi Arabia
e-mail: sherifamar2000@yahoo.com

V. Madhu et al. (eds.), *Advances in Algebra and Analysis*, Trends in Mathematics,
https://doi.org/10.1007/978-3-030-01120-8_43

transmission system and phone conference of medicinal administration system, manufacturing systems, stock control systems, and simulations.

2 Description of the Model

In this paper, we consider a priority retrial G-queue with working vacations and vacation interruption. The detailed model description is given as follows:

- ***The arrival process:*** Three types of customers arrive into the system: priority customers, ordinary customers, and negative customers. Priority customers have preemptive priorities over ordinary customers in service time of busy server. Assume that all three types of customers arrive from outside the system according to three independent Poisson processes with rates λ, δ, and α, respectively.
- ***The retrial process:*** We assume that there is no waiting space and therefore an arriving priority customer finds the server free; the customer begins its service immediately. At the arrival time of a priority customer, the server gives service for a priority customer or working vacations, and the newly arriving priority customer will depart the system directly without service. While the regular busy server is working with an ordinary customer, the arriving priority customer will interrupt the service of the ordinary customer, and the server begins its service immediately. We assume that when an ordinary customer is preempted by a priority customer, the ordinary customer who was just being served before starts the service of the priority customer and waits in the service area for the remaining service to complete. If an arriving ordinary customer finds the server is being busy or on working vacation, the arrivals join the pool of blocked customers called an orbit in accordance with FCFS discipline. That is, only one customer at the head of the orbit queue is allowed access to the server. From the above instant, the server becomes free, an external potential priority customer or ordinary customer and a retrial ordinary customer compete to reach the server. Inter-retrial times have an arbitrary distribution $R(t)$ with corresponding Laplace-Stieltjes transform (LST) $R^*(\vartheta)$. The retrial ordinary customer is required to give up the attempt for service if an external priority customer or ordinary customer arrives first. In that case, the retrial ordinary customer goes back to its position in the retrial queue.
- ***The regular service process:*** Whenever a new primary (priority) customer or retry customer arrives at the server idle state, then the server immediately starts normal service for the arrivals. The service time of priority customers follows a general distribution and is denoted by the random variable S_p with distribution function $S_p(t)$ having LST $S_p^*(\vartheta)$. The service time of ordinary customers follows a general distribution, and it is denoted by the random variable S_b with distribution function $S_b(t)$ having LST $S_b^*(\vartheta)$.
- ***The Bernoulli working vacation process:*** The server begins a working vacation each time when the orbit becomes empty, and the vacation time follows an exponential distribution with parameter θ. If there are any customers arrive in a vacation period, the server continues to work at a lower speed service rate.

The working vacation period is an operational period at a lower speed. If any customers in the orbit at a lower speed service completion instant in the vacation period, the server will stop the vacation and come back to the normal busy period which means vacation interruption happens. Otherwise, if no customers are in the system at the end of the vacation, the server either joins the system and remains idle to serve a new customer with probability p (single working vacation) to serve the arriving customers in regular mode or leaves for another working vacation with probability $q = 1 - p$ (multiple working vacation). When a vacation ends, if there are customers in the orbit, the server switches to the normal working level. During the working vacation period, the service time follows a general random variable S_v with distribution function $S_v(t)$ and LST $S_v^*(\vartheta)$.

- ***The removal rule:*** The negative customers (G-queue) arrive only at the regular service time of the positive customers. It can not cannot accumulate in a queue and do not receive service, will remove the positive customers (priority customer or ordinary customer) being in service from the system. These types of negative customers cause server breakdown, and the service channel will fail for a short interval of time. When the server fails, it will be sent for repair immediately. After completion of repair, the server will be as good as new. The repair time (denoted by G) of the server is assumed to be arbitrarily distributed with distribution function $G(t)$ having LST $G^*(\vartheta)$.
- We assume that inter-arrival times, retrial times, service times, and working vacation times are mutually independent of each other
- Throughout the rest of the paper, we denote by $\bar{F}(x) = 1 - F(x)$ the tail of distribution function $F(x)$. We also denote $F^*(x) = \int_0^\infty e^{-sx} dF(x)$ the LST of $F(x)$ and $\tilde{F(s)} = \int_0^\infty e^{-sx} F(x)dx$ to be the Laplace transform of $F(x)$, and we assume the notation $\bar{F^*}(s) = \frac{1-F^*(s)}{s}$.

We assume that $R(0) = 0, R(\infty) = 1, S_p(0) = 0, S_p(\infty) = 1, S_v(0) = 0,$ $S_v(\infty) = 1, G(0) = 0,$ *and* $G(\infty) = 1$ are continuous at $x = 0$, and $S_b(0) = 0$ *and* $S_b(\infty) = 1$ are continuous at $y = 0$. So that, the function $a(x)$, $\mu_p(x)$, $\mu_b(y)$, $\mu_v(x)$, and $\xi(x)$ are the conditional completion rates for retrial, regular service for priority customers and ordinary customers, lower rates service, and repair, respectively. That is, $a(x)dx = \frac{dR(x)}{1-R(x)}$, $\mu_p(x)dx = \frac{dS_p(x)}{1-S_p(x)}$, $\mu_b(x)dx = \frac{dS_b(x)}{1-S_b(x)}$, $\mu_v(x)dx = \frac{dS_v(x)}{1-S_v(x)}$, and $\xi(x)dx = \frac{dG(x)}{1-G(x)}$.
Let $R^0(t)$, $S_p^0(t)$, $S_b^0(t)$, $S_v^0(t)$, and $G^0(t)$ be the elapsed times of retrial, service of priority and ordinary customers, working vacation, and repair, respectively, at time t.

Further, we present the random variables, $C(t) = \{0, 1, 2, 3, 4, 5, 6\}$, for the following: the server is idle in working vacation period, the server is free in regular service period, the server is busy with a priority customer in regular service period, the server is busy with a priority customer with preempting an ordinary customer in regular service period, the server is busy with an ordinary customer in regular service period, the server is busy in working vacation period, and the server is under repair period.

The sequence of random vectors $Z_n = \{C(t_n+), N(t_n+)\}$ forms a embedded Markov chain which is ergodic if and only if $\rho < R^*(\lambda+\delta)$.

where $\rho = (R^*(\lambda+\delta)+\lambda\bar{R}^*(\lambda+\delta))X+\lambda\bar{S}_p^*(\alpha)(1+\alpha g^{(1)})$ and $X = A_b'\bar{S}_b^{*'}(A_b)+\alpha(\delta\bar{S}_p^*(\alpha)\bar{S}_p^{*'}(\alpha)+(1+\delta\bar{S}_p^*(\alpha))(\bar{S}_b^{*'}(A_b)+\lambda g^{(1)}\bar{S}_b^*(A_b)))$.

3 Steady-State Analysis

The steady-state difference-differential equations for the retrial queueing system are discussed in this section. The PGFs for the server states, mean system, and orbit sizes are derived. For the process $\{N(t), t \geq 0\}$, we define the probabilities $Q_0(t) = P\{C(t) = 0, N(t) = 0\}$ and $P_0(t) = P\{C(t) = 1, N(t) = 0\}$, and the probability densities are $P_{1,n}(x,t)$, $P_{2,n}(x,t)$, $P_{3,n}(x,y,t)$, $P_{4,n}(x,t)$, $P_{5,n}(x,t)$, and $P_{6,n}(x,t)$ for $t \geq 0, x \geq 0$.

If the stability condition is fulfilled in the sequel, we can set $P_0 = lim_{t\to\infty}P_0(t)$ and $Q_0 = lim_{t\to\infty}Q_0(t)$ and limiting densities for $t \geq 0, x \geq 0$, $n \geq 1$. $P_n(x) = lim_{t\to\infty}P_{1,n}(x,t)$; $P_{2,n}(x) = lim_{t\to\infty}P_{2,n}(x,t)$, $P_{3,n}(x,y) = lim_{t\to\infty}P_{3,n}(x,y,t)$, $P_{4,n}(y) = lim_{t\to\infty}P_{4,n}(y,t)$, $P_{5,n}(x) = lim_{t\to\infty}P_{5,n}(x,t)$, and $P_{6,n}(x) = lim_{t\to\infty}P_{6,n}(x,t)$.

By the method of SVT, we obtain the following system of equations,

$$(\lambda+\delta)P_0 = \theta p Q_0 \tag{1}$$

$$(\lambda+\delta+\theta)Q_0 = \theta q Q_0 + \int_0^\infty P_{2,0}(x)\mu_p(x)dx + \int_0^\infty P_{4,0}(y)\mu_b(y)dy + \int_0^\infty P_{5,0}(x)\mu_v(x)dx + \int_0^\infty P_{6,0}(x)\xi(x)dx \tag{2}$$

$$\frac{dP_{1,n}(x)}{dx} + (\lambda+\delta+a(x))P_{1,n}(x) = 0, n \geq 1 \tag{3}$$

$$\frac{dP_{2,n}(x)}{dx} + (\lambda+\delta+\mu_p(x))P_{2,n}(x) = \lambda P_{2,n-1}(x), n \geq 1 \tag{4}$$

$$\frac{\partial P_{3,n}(x,y)}{\partial x} + (\lambda+\alpha+\mu_p(x))P_{3,n}(x,y) = \lambda P_{3,n-1}(x,y), n \geq 1, \tag{5}$$

$$\frac{dP_{4,n}(y)}{dy} + (\lambda+\delta+\alpha+\mu_b(y))P_{4,n}(y) = \lambda P_{4,n-1}(y) + \int_0^\infty P_{3,n}(x,y)\mu_p(x)dx, n \geq 1 \tag{6}$$

$$\frac{dP_{5,n}(x)}{dx} + (\lambda+\delta+\mu_v(x))P_{5,n}(x) = \lambda P_{5,n-1}(x), n \geq 1 \tag{7}$$

$$\frac{dP_{6,n}(x)}{dx} + (\lambda+\xi(x))P_{6,n}(x) = \lambda P_{6,n-1}(x), n \geq 1 \tag{8}$$

The steady-state boundary conditions at $x = 0$ and $y = 0$ are

$$P_{1,n}(0) = \int_0^\infty P_{2,n}(x)\mu_p(x)dx + \int_0^\infty P_{4,n}(y)\mu_b(y)dx \tag{9}$$

$$+\int_0^\infty P_{5,n}(x)\mu_v(x)dx + \int_0^\infty P_{6,n}(x)\xi(x)dx, n \geq 1$$

$$P_{2,n}(0) = \delta \int_0^\infty P_{1,n}(x)dx, n \geq 1 \tag{10}$$

$$P_{3,n}(0, y) = \delta(P_{4,n}(y)), n \geq 0 \tag{11}$$

$$P_{4,n}(0) = \int_0^\infty P_{1,n+1}(x)a(x)dx + \lambda \int_0^\infty P_{1,n}(x)dx$$

$$+\theta \int_0^\infty P_{5,n}(x)dx, n \geq 1, \tag{12}$$

$$P_{5,n}(0) = (\lambda + \delta)Q_0, n = 0 \tag{13}$$

$$P_{6,n}(0) = \alpha\{\int_0^\infty P_{2,0}(x)dx + \int_0^\infty \int_0^\infty P_{3,n}(x, y)dxdy$$

$$+ \int_0^\infty P_{4,n}(y)dy\}, n \geq 1 \tag{14}$$

The normalizing condition is

$$P_0+Q_0+\sum_{n=1}^{\infty} \int_\infty^0 P_{1,n}(x)dx+\sum_{n=0}^{\infty}\left(\int_0^\infty P_{2,n}(x)dx+\int_0^\infty P_{5,n}(x)dx\right) \tag{15}$$

$$+\sum_{n=0}^{\infty}\left(\int_0^\infty \int_0^\infty P_{3,n}(x, y)dxdy+\int_0^\infty P_{4,n}(y)dy+\int_0^\infty P_{6,n}(x)dx\right)=1$$

To analyze the developed queueing model, we make use of SVP and PGF methods. We define the generating functions for $|z| = 1$ as follows:

$$P_1(x, z)=\sum_{n=1}^{\infty} P_{1,n}(x)z^n;\ P_2(x, z)=\sum_{n=0}^{\infty} P_{2,n}(x)z^n;\ P_3(x, y, z) = \sum_{n=0}^{\infty} P_{3,n}(x, y)z^n;$$

$$P_4(y, z)= \sum_{n=0}^{\infty} P_{4,n}(y)z^n;\ P_5(x, z) = \sum_{n=0}^{\infty} P_{5,n}(x)z^n;\ P_6(x, z) = \sum_{n=0}^{\infty} P_{6,n}(x)z^n;$$

From Eqs. (3)–(14) multiplying by z^n and summing over n, $(n = 0, 1, 2\ldots)$ and solving the partial differential equations, we get the limiting PGFs $P_1(x, z)$, $P_2(x, z)$, $P_3(x, y, z)$, $P_4(x, z)$, $P_5(x, z)$ and $P_6(x, z)$.

Integrating the limiting PGF's with respect to x and y, then we get the PGFs as $P_1(z) = \int_\infty^0 P_1(x,z)dx$, $P_2(z) = \int_\infty^0 P_2(x,z)dx$, $P_3(y,z) = \int_\infty^0 P_3(x,y,z)dx$, $P_4(z) = \int_\infty^0 P_4(x,z)dx$, $P_5(z) = \int_\infty^0 P_5(x,z)dx$, and $P_6(z) = \int_\infty^0 P_{6,n}(x,z)dx$

Theorem 1 *Under the stability condition $\rho < R^*(\lambda+\delta)$, the PGFs of number of customers in the orbit when server being idle, busy on priority, serivce and re-service in ordinary customer, on working vacation and repair is given by*

$$P_1(z) = \frac{Nr(z)}{Dr(z)} \tag{16}$$

$$Nr(z) = z\bar{R}^*(\lambda+\delta)Q_0\{((\lambda+\delta)V(z)+\theta p)X(z) \\ +((\lambda+\delta)(S_v^*(A_v(z))-1)-\theta p\}$$

$$Dr(z) = z-(R^*(\lambda+\delta)+\lambda z\bar{R}^*(\lambda+\delta))X(z)-z\delta\bar{R}^*(\lambda+\delta)B(z)$$

$$P_2(z) = \frac{\delta Q_0(1-S_p^*(A_p(z)))}{A_p(z)\times Dr(z)}\{z\bar{R}^*(\lambda+\delta)\{((\lambda+\delta)V(z)+\theta p)X(z) \\ +((\lambda+\delta)(S_v^*(A_v(z))-1)-\theta_p)\}\} \tag{17}$$

$$P_3(z) = \frac{\delta Q_0((1-S_b^*(A_b(z)))(1-S_p^*(A_p(z)))}{(A_p(z)*(A_b(z)*Dr(z))} \\ \times\{(R^*(\lambda+\delta)+\lambda z\bar{R}^*(\lambda+\delta))((\lambda+\delta)(S_v^*(A_v(z))-1)-\theta p) \\ +z((\lambda+\delta)V(z)+\theta p)(1-\delta\bar{R}^*(\lambda+\delta)B(z))\} \tag{18}$$

$$P_4(z) = \frac{Q_0(1-S_b^*(A_b(z)))}{(A_b(z)*Dr(z))} \\ \times\{(R^*(\lambda+\delta)+\lambda z\bar{R}^*(\lambda+\delta))((\lambda+\delta)(S_v^*(A_v(z))-1)-\theta p) \\ +z((\lambda+\delta)V(z)+\theta p)(1-\delta\bar{R}^*(\lambda+\delta)B(z))\} \tag{19}$$

$$P_5(z) = \left\{\frac{(\lambda+\delta)Q_0V(z)}{\theta}\right\} \tag{20}$$

$$P_6(z) = Q_0z\{\bar{R}^*(\lambda+\delta)\{((\lambda+\delta)V(z)+\theta p)X(z) \\ +((\lambda+\delta)(S_v^*(A_v(z))-1)-\theta p)\}\times\delta\bar{S}_p^*(A_p(z))\}+\bar{S}_b^*(A_b(z)) \\ (1+\delta\bar{S}_p^*(A_p(z)))((R^*(\lambda+\delta)+\lambda z\bar{R}^*(\lambda+\delta)((\lambda+\delta) \\ (S_v^*(A_v(z))-1)-\theta p)+z((\lambda+\delta)V(z)+\theta p) \\ (1-\delta\bar{R}^*(\lambda+\delta)B(z)))/Dr(z) \tag{21}$$

$$where\ Q_0 = \frac{R^*(\lambda+\delta)-\rho}{\eta}$$

$$P_0 = \frac{\theta p(R^*(\lambda+\delta)-\rho)}{(\lambda+\delta)\eta} \tag{22}$$

$$\eta = \{(R^*(\lambda+\delta)-\rho)+(\lambda+\delta)(1-S_v^*(\theta))(\bar{R}^*(\lambda+\delta)(1+X)(1+\bar{S}_p^*(\alpha))$$
$$\times(\alpha+\delta g^{(1)})+\bar{S}_b^*(A_b)(1+\delta\bar{S}_p^*(\alpha))(1+\delta g^{(1)})(R^*(\lambda+\delta)-\delta\bar{R}^*(\lambda+\delta)$$
$$(1-S_p^*(\alpha))((\lambda+\delta)/\alpha+g^{(1)})+((\lambda+\delta)(1-\delta\bar{R}^*(\lambda+\delta))/\theta p))\}$$

$$X(z)=(S_b^*(A_b(z))+\alpha G^*(b(z))\bar{S}_b^*(A_(z))(1+\delta\bar{S}_p^*(A_p(z))))$$

$$B(z)=(S_p^*(A_p(z))+\alpha G^*(b(z))\bar{S}_p^*(A_p(z)))$$

$$\rho=(R^*(\lambda+\delta)+\lambda\bar{R}^*(\lambda+\delta))X+\lambda\bar{S}_p^*(\alpha)(1+\alpha g^{(1)})$$

$$X=A_b'\bar{S}_b^{*'}(A_b)+\alpha(\delta\bar{S}_p^*(\alpha)\bar{S}_p^{*'}(\alpha)+(1+\delta\bar{S}_p^*(\alpha))(\bar{S}_b^{*'}(A_b)+\lambda g^{(1)}\bar{S}_b^*)(A_b)))$$

$$\bar{R}^*(\lambda+\delta)=\left(\frac{1-R^*(\lambda+\delta)}{\lambda+\delta}\right);\ \bar{S}_p^*(A_p(z))=\left(\frac{1-S_p^*(A_p(z))}{(A_p(z)}\right)$$

$$\bar{S}_b^*(A_b(z))=\left(\frac{1-S_b^*(A_b(z))}{(A_b(z))}\right);\ V(z)=\frac{\theta[1-S_v^*(A_v(z))]}{\theta+\lambda(1-z)};\ b(z)=(\lambda(1-z))$$

$$\bar{S}_p^*(\alpha)=\frac{1-\bar{S}_p^*(\alpha)}{\alpha};\qquad \bar{S}_b^*(A_b)=\frac{1-S_b^*(\alpha+\lambda(1-S_p^*(\alpha)))}{\alpha+\lambda(1-S_p^*(\alpha))};$$

$$\bar{S}_p^{*'}(\alpha)=\frac{\lambda}{\alpha^2}(1-S_p^*(\alpha)+\alpha S_p^{*'}(\alpha));\qquad A_p(z)=(\alpha+\lambda(1-z));$$

$$A_b(z)=(A_p(z)+\delta(1-S_p^*(A_p(z))));\qquad A_v(z)=(\theta+\lambda(1-z))$$

Proof Integrating the limiting PGF's with respect to 'x' and 'y', for finding P_0 and Q_0 by setting $z=1$ in (16)–(21) and applying L-Hospital's rule whenever necessary in $P_0+Q_0+P_1(1)+P_2(1)+P_3(1)+P_4(1)+P_5(1)+P_6(1)=1$, then we get the probability that the server is idle when no customer in the orbit (P_0). □

Corollary If the system in stability condition $\rho < R^*(\lambda+\delta)$,

1. The PGF of system size is

$$K_s(z)=P_0+Q_0+z(P_1(z)+P_2(z)+P_3(z)+P_4(z)+P_5(z))+P_6(z).$$

2. The PGF of orbit size is

$$K_s(z)=P_0+Q_0+P_1(z)+P_2(z)+P_3(z)+P_4(z)+P_5(z)+P_6(z).$$

4 Performance Measures

We discuss some system performance measures of this queueing model in this section.

1. The average number of customers in the system (L_s) and orbit (L_q) is found by differentiating (K_s) and (K_o) with respect to z and evaluating at $z = 1$,

$$L_s = K_s^{'}(1) = lim_{z\to 1}\frac{d}{dz}K_s(z) \text{ and } L_q = K_o^{'}(1) = lim_{z\to 1}\frac{d}{dz}K_o(z).$$

2. The average time a customer spends in the system (W_s) and the average time a customer spends in the queue (W_q) are found by using the Little's formula,

$$W_s = L_s/\lambda \quad and \quad W_q = L_q/\lambda$$

3. The steady-state availability of the server (A_v) is given by

$$A_v = 1 - lim_{z\to 1}P_6(z) = 1 - P_6(1).$$

4. The steady-state failure frequency is obtained as

$$F_f = \delta \times (P_2(1) \times P_3(1) \times P_4(1)).$$

5 Special Cases

In this section, we analyze briefly some special cases of our model, which are consistent with the existing literature.

- Case 1: **No priority arrival, no negative customer, no vacation interruption, and single working vacation**
  In this case, our model becomes an M/G/1 retrial queue with single working vacation. We assume that $(\delta, \alpha, \theta) \to (0, 0, 0)$ in the main results are coincided with the result of Arivudainambi et al. [1].
- Case 2: **No negative customer, no working vacation, and no vacation interruption**
  In this case, we put $\delta = \alpha = \theta = 0$, and our model can be reduced to a single server retrial queueing system with working vacations. The results coincided with the result of Gao [3].

6 Conclusion

In this work, we have considered a single server preemptive priority retrial G-queue with Bernoulli working vacations. The PGFs of the number of customers in the system are found. Important performance measures like mean system size and orbit size are obtained. Some important special cases are discussed. The novelty of this investigation is the introduction of preemptive priority retrial queueing system in presence of Bernoulli WVs with G-queues. This proposed work has unique practical application in phone conference of medicinal administration system, production and manufacturing systems, stock control systems, and simulations.

Acknowledgements The authors are thankful to the editor and reviewers for their valuable comments and suggestions for the improvement of the paper.

References

1. Arivudainambi, D., Godhandaraman, P., Rajadurai, P.: Performance analysis of a single server retrial queue with working vacation. OPSEARCH. **51**, 434–462 (2014).
2. Artalejo, J., Gomez-Corral, A.: Retrial Queueing Systems. Springer, Berlin, Germany (2008).
3. Gao, S.: A preemptive priority retrial queue with two classes of customers and general retrial times. Operational Research. **15**, 233–251 (2015).
4. Gao, S., Wang, J., Li, J.: An M/G/1 retrial queue with general retrial times, working vacations and vacation interruption. Asia-Pacific Journal Oper Res. **31**, 6–31 (2014).
5. Rajadurai, P.: A study on an M/G/1 retrial G-queue with unreliable server under variant working vacations policy and vacation interruption. Songklanakarin J. Sci. Technol. **40**, 231–242 (2018).

Time to Recruitment for Organisations Having n Types of Policy Decisions with Lag Period for Non-identical Wastages

Manju Ramalingam and B. Esther Clara

Abstract Announcements of policy decisions in organisations may result in loss of manpower (wastage) due to employee's dissatisfaction. Recruiting for each loss is not a good practice because of cost and time. Hence the recruitment has to be done at the time of threshold crossing using a suitable recruitment policy. Time to recruitment has to be predicted, to avoid the complete breakdown of the organisation. The intensity of attrition may not be the same for every policy decision, so in general there may be n types of policy decisions each with different intensity. Whenever policy decisions are taken, the wastages may not occur instantaneously so the lag period for wastages is introduced in this paper. Two stochastic models have been constructed to derive the performance measures of time to recruitment with non-identical wastages and the inter-policy decision times (IPDT) as independent and identically distributed (iid) random variables (rvs) or geometric process. The impact of the parameters on performance measures are found from numerical illustrations. A better model is suggested for the prediction of the time to recruitment. The advantages of introducing lag period for wastages and the way to control the faculty flow are discussed in the conclusion.

1 Introduction

The manpower planning process enables to estimate the time to recruitment by knowing the pattern of wastages due to policy decisions taken by the management and the number of personnel required for an organisation over a predefined period of time. Using statistical techniques the authors [2–4] have initiated the manpower planning studies. The lifetime distribution of a device which is subjected to shocks is studied using shock models and wear process in [5]. The problem of time to recruitment is studied by considering the IPDT as iid rvs in [13] and geometric

Manju Ramalingam (✉) · B. Esther Clara
PG and Research Department of Mathematics, Bishop Heber College, Trichy, Tamilnadu, India
e-mail: meetmemanju@yahoo.co.in; manjuramalingam@bhc.edu.in; jecigrace@gmail.com; estherclara@bhc.edu.in

V. Madhu et al. (eds.), *Advances in Algebra and Analysis*, Trends in Mathematics,
https://doi.org/10.1007/978-3-030-01120-8_44

process in [6, 10] by employing the reliability approach of [5] and univariate CUM policy of replacement in [11]. Since all the policy decisions may not produce same intensity of attrition of personnel, n types of policy decisions with different attrition rates have been studied by Ramalingam et al. [7, 8] assuming n-stage hyper-exponential distribution [1] for IPDT because the attritions due to policy decisions are independent with each other with certain probability of occurrences. In all the above cited papers, the wastages take place instantaneously whenever the policy decisions are announced, but in reality the wastages may not be instantaneous, due to the notification period before resignation, bonding period in the organisation, etc. So in this paper as in replacement model [12], the IPDT are taken as the sum of two subintervals of time. The first interval is the lag period for the wastages (wastages will not occur during this period), and the second is the time between the occurrence of first wastage and the subsequent policy decision (wastages occur during this period). Also the wastages may not be identical for all the policies, so in this work it is associated with the non-identical exponential distribution.

In this paper, two stochastic models are constructed to derive the expected time and variance of time to recruitment for a single graded manpower system based on shock model approach using univariate CUM policy of recruitment, with the assumption (a) the wastages are independent and non-identically distributed exponential random variables, (b) lag period for wastages follows exponential distribution, (c) the time between the occurrence of first wastage and the subsequent policy decision follows n-stage hyper-exponential distribution and (d) the IPDT is a sequence of iid rvs for Model-I and geometric process for Model-II.

2 Model Description and Analysis for Model-I

Consider the time to recruitment model for the single graded organisation, in which the management takes policy decisions at random epochs in $(0, \infty)$ where the occurrence of wastages is not instantaneous. So there exists lag period for each decision in which no wastages will take place. Let Y_i be the lag period for wastages after $(i-1)$th decision point and the first exit of personnel due to $(i-1)$th decision occur. Let us assume $\{Y_i\}_{i=1}^{\infty}$ forms iid exponential rvs with mean $\frac{1}{\lambda_L}$ where $\lambda_L > 0$, absence rate of lag period. Let $f_Y(.)$ be the probability density function (pdf) and $F_Y(.)$ be the cumulative distribution function (cdf) of Y_i, $i = 1, 2, \cdots$; let $f_Y^*(.)$ be the Laplace transform (LT) of $f_Y(.)$. Let $\lambda_1, \lambda_2, \cdots, \lambda_n$ be the attrition rates due to n types of policy decisions, each with the corresponding probability of occurrence as $p_1, p_2, \cdots, p_n$, where $0 < \lambda_1 < \lambda_2 < \cdots < \lambda_n$ and $0 < p_1, p_2, \cdots, p_n < 1$. Let Z_i be the time between the occurrence of first wastage due to $(i-1)$th decision and the ith policy decision point. Let us assume that $\{Z_i\}_{i=1}^{\infty}$ are iid n-stage hyper-

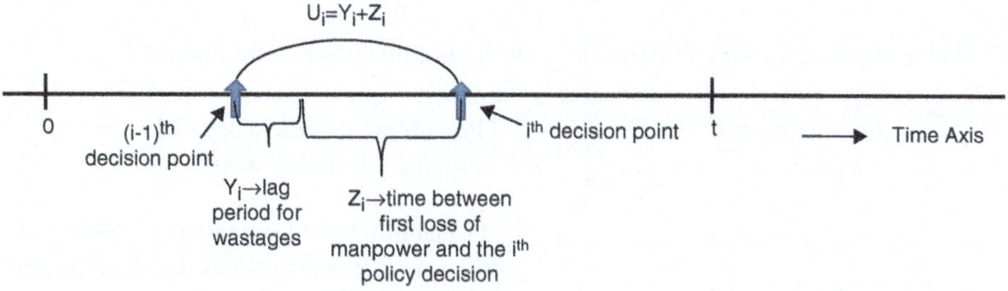

Fig. 1 Inter-policy decision times (IPDT) with Lag period for wastages

exponential rvs with mean $\sum_{j=1}^{n} \frac{p_j}{\lambda_j}$ where $\lambda_j, p_j > 0$ and $\sum_{j=1}^{n} p_j = 1$. Let $f_Z(.)$ be the pdf and $F_Z(.)$ be the cdf of Z_i, $i = 1, 2, \cdots$; let $f_Z^*(.)$ be the LT of $f_Z(.)$. Y_i and Z_i are independent. Let U_i be the IPDT between $(i-1)$th and ith decisions, for $i = 1, 2, \cdots$. Therefore, $U_i = Y_i + Z_i$, $i = 1, 2, \cdots$, as shown in Fig. 1. Let $f(.)$ be the pdf and $F(.)$ be the cdf of U_i, $i = 1, 2, \cdots$; let $f^*(.)$ be the LT of $f(.)$ and $f_k(.)$ be the pdf and $F_k(.)$ be the cdf of $\sum_{i=1}^{k} U_i$. It is assumed that the wastages due to each policy decision is linear and cumulative. Let X_i be the wastage due to ith policy decision, for $i = 1, 2, \cdots$. Let $\{X_i\}_{i=1}^{\infty}$ are independent and non-identically distributed exponential rvs with the respective absence rates of wastages $\alpha_i > 0$, mean $\frac{1}{\alpha_i}$ for $i = 1, 2, 3, \cdots$ and $\alpha_i \neq \alpha_j$ for $i \neq j$. Let $g(.)$ be the pdf and $G(.)$ be the cdf of X_i, for $i = 1, 2, \cdots$ and $g^*(.)$ be the LT of $g(.)$. Let S_k be the total wastages in the first k decisions, $S_k = \sum_{i=1}^{k} X_i$ for $k = 1, 2, \cdots$. Let $g_k(.)$ be the pdf and $G_k(.)$ be the cdf of S_k. Let T be the threshold for the cumulative wastages in the organisation which follows exponential distribution with mean $\frac{1}{\theta}$, where $\theta > 0$, and let $h(.)$ be the pdf of T. Let $N(t)$ be the number of policy decisions taken in the interval $(0, t]$. Let W be the time to recruitment for the organisation with mean $E(W)$ and variance $V(W)$. Let $l(.)$ be the pdf and $L(.)$ be the cdf of W and $l^*(.)$ be the LT of $l(.)$. It is assumed that the process of wastages, the process of IPDT and the breakdown threshold for wastages are statistically independent. The recruitment policy applied in this paper is univariate CUM policy and is defined as:

> Recruitment is done whenever the cumulative loss of manpower in the organisation exceeds the breakdown threshold T.

2.1 Main Result for Model-I

In this section, the performance measures are obtained for Model-I by deriving the survival function of the organisation using [5] and by applying the law of total probability for discrete case.

Thus, the `Survival function` of the organisation is obtained as

$$\begin{bmatrix}\text{Probability that the}\\ \text{organisation survives beyond t}\end{bmatrix} = \sum_{k=0}^{\infty}\begin{bmatrix}\text{Probability that there occurs}\\ \text{exactly k decisions in } (0,t]\end{bmatrix}$$
$$\times \begin{bmatrix}\text{Probability that the cumulative loss of}\\ \text{manpower does not cross its threshold level}\end{bmatrix}$$

By using the renewal theory result [9], $P(N(t)=k) = F_k(t) - F_{k+1}(t)$, $k = 0, 1, 2, \ldots$ with $F_0(t) = 1$, we get

$$L(t) = 1 - \sum_{k=0}^{\infty} \left(F_k(t) - F_{k+1}(t)\right) g_k^*(\theta) \tag{1}$$

The `newly derived pdf` of U_i for $i = 1, 2, \cdots$ is given by

$$f(u) = \lambda_L \sum_{j=1}^{n} \frac{p_j \lambda_j}{\lambda_L - \lambda_j}\left[e^{-\lambda_j u} - e^{-\lambda_L u}\right] \tag{2}$$

where $\lambda_L \neq \lambda_j$ for $j = 1, 2, \cdots, n$ with mean $\sum_{j=1}^{n} \frac{p_j}{\lambda_j} + \frac{1}{\lambda_L}$. Since $U_i = Y_i + Z_i$, $f(u)$ is the convolution of $f_Y(y)$ and $f_Z(z)$.

From the assumption of X_i, S_k is hypo-exponentially distributed with the pdf

$$g_k(x) = \sum_{i=1}^{k}\left(a_i \alpha_i e^{-\alpha_i x}\right) \text{ where } a_i = \prod_{\substack{l=1\\ l\neq i}}^{k}\left(\frac{\alpha_l}{\alpha_l - \alpha_i}\right), i=1,2,\ldots,k \text{ and } \sum_{i=1}^{k} a_i = 1. \tag{3}$$

Since $U_i's$ are iid rvs from (1), (2) and (3), we get the expected time to recruitment from $E(W) = -\frac{d}{ds}\left[l^*(s)\right]_{s=0}$ as

$$E(W) = \left(\sum_{j=1}^{n}\frac{p_j}{\lambda_j} + \frac{1}{\lambda_L}\right)\sum_{k=1}^{\infty} A_k \text{ where } A_k = \sum_{i=1}^{k}\left(\prod_{\substack{l=1\\ l\neq i}}^{k}\left(\frac{\alpha_l}{\alpha_l - \alpha_i}\right)\frac{\alpha_i}{\theta + \alpha_i}\right)$$

Now, the second moment of W is obtained from $E\left(W^2\right) = \frac{d^2}{ds^2}\left[l^*(s)\right]_{s=0}$ as

$$E\left(W^2\right)=2\left(\sum_{j=1}^{n}\frac{p_j}{\lambda_j}+\frac{1}{\lambda_L}\right)^2\sum_{k=1}^{\infty}(kA_k)+2\left[\sum_{j=1}^{n}\frac{p_j}{\lambda_j{}^2}+\frac{1}{\lambda_L^2}+\frac{1}{\lambda_L}\sum_{j=1}^{n}\frac{p_j}{\lambda_j}\right]\sum_{k=1}^{\infty}A_k$$

The variance of the time to recruitment is given by $V(W) = E\left(W^2\right) - (E(W))^2$.

Note When there is no lag period for wastages or $Y_i = 0$ (i.e. if instantaneous wastages are admitted in the organisation), for Model-I

$$E(W)=\left(\sum_{j=1}^{n}\frac{p_j}{\lambda_j}\right)\sum_{k=1}^{\infty}A_k \text{ and } E\left(W^2\right)=2\left(\sum_{j=1}^{n}\frac{p_j}{\lambda_j}\right)^2\sum_{k=1}^{\infty}A_k+2\left(\sum_{j=1}^{n}\frac{p_j}{\lambda_j{}^2}\right)\sum_{k=1}^{\infty}A_k$$

3 Model Description and Analysis for Model-II

Model description of Model-II is same as in Model-I, but the IPDT is assumed to form a geometric process with the first term of $\{U_i\}_{i=1}^{\infty}$ which follows the pdf as (2). Let $f(.)$ be the pdf ($F(.)$ be the cdf) of U_1, the 1st term of $\{U_i\}_{i=1}^{\infty}$ and $f^*(.)$ be the LT of $f(.)$ and $w_i(.)$ be the pdf ($W_i(.)$ be the cdf) of U_i, the ith term for $i = 2, 3, \cdots$ and $w_i{}^*(.)$ be the LT of $w_i(.)$.

3.1 Main Result for Model-II

The newly derived pdf of U_i for $i = 2, 3, \cdots$ is given by,

$$w_i(u) = a^{i-1}\lambda_L\sum_{j=1}^{n}p_j\lambda_j\left[\frac{e^{-a^{i-1}\lambda_j u}-e^{-a^{i-1}\lambda_L u}}{\lambda_L-\lambda_j}\right] \tag{4}$$

where $\lambda_L \neq \lambda_j$ for $j = 1, 2, \cdots, n$; $a > 0$. Since $\{U_i\}_{i=1}^{\infty}$ forms a geometric process, $w_i(u) = a^{i-1}f\left(a^{i-1}u\right)$ for $i = 2, 3, \cdots$.

Now from (4)

$$\frac{d}{ds}\left[\prod_{i=1}^{k}f^*\left(\frac{s}{a^{i-1}}\right)\right]_{s=0} = -\left(\sum_{j=1}^{n}\frac{p_j}{\lambda_j}+\frac{1}{\lambda_L}\right)\left(\frac{a^k-1}{a^{k-1}(a-1)}\right), \text{ for } a \neq 1 \tag{5}$$

Therefore, from (1), (5) and (3) we get

$$E(W) = \left(\sum_{j=1}^{n} \frac{p_j}{\lambda_j} + \frac{1}{\lambda_L}\right) \sum_{k=1}^{\infty} \left(\frac{A_k}{a^k}\right), \text{ for } a \neq 1$$

Therefore, from (5)

$$E\left(W^2\right) = 2\left(\sum_{j=1}^{n} \frac{p_j}{{\lambda_j}^2} + \frac{1}{\lambda_L^2} + \frac{1}{\lambda_L} \sum_{j=1}^{n} \frac{p_j}{\lambda_j}\right) \sum_{k=1}^{\infty} \left(\frac{A_k}{a^{2k}}\right)$$
$$-\left(\sum_{j=1}^{n} \left(\frac{p_j}{\lambda_j}\right) + \frac{1}{\lambda_L}\right)^2 \sum_{k=1}^{\infty} \left(\frac{A_k}{a^{2k}}\right) + \left(\sum_{j=1}^{n} \left(\frac{p_j}{\lambda_j}\right) + \frac{1}{\lambda_L}\right)^2$$
$$\times \sum_{k=1}^{\infty} \left(A_k \left(\frac{1 - 2a^{k+1} - a^2 + 2a^{k+2}}{a^{2k}(a-1)^2}\right)\right), \text{ for } a \neq 1.$$

Note When $Y_i = 0$ for Model-II, $E(W) = \left(\sum_{j=1}^{n} \frac{p_j}{\lambda_j}\right) \sum_{k=1}^{\infty} \left(\frac{A_k}{a^k}\right)$ where $a \neq 1$ and

$$E\left(W^2\right) = \left(2\sum_{j=1}^{n} \left(\frac{p_j}{{\lambda_j}^2}\right) - \left(\sum_{j=1}^{n} \left(\frac{p_j}{\lambda_j}\right)\right)^2\right) \sum_{k=1}^{\infty} \left(\frac{A_k}{a^{2k}}\right)$$
$$+\left(\sum_{j=1}^{n} \left(\frac{p_j}{\lambda_j}\right)\right)^2 \sum_{k=1}^{\infty} \left(\frac{1 - 2a^{k+1} - a^2 + 2a^{k+2}}{a^{2k}(a-1)^2}\right) A_k$$

3.2 Results

3.2.1 Result-1

When $\{X_i\}_{i=1}^{\infty}$ are iid exponential rvs with rate α (i.e. the wastages are identically distributed for all the policies), for Model-I $E(W) = \left(\sum_{j=1}^{n} \frac{p_j}{\lambda_j} + \frac{1}{\lambda_L}\right) \left(\frac{\theta+\alpha}{\theta}\right)$ where $|\alpha| < |\theta + \alpha|$. And for Model-II $E(W) = \left(\sum_{j=1}^{n} \frac{p_j}{\lambda_j} + \frac{1}{\lambda_L}\right) \left(\frac{a(\theta+\alpha)}{a(\theta+\alpha)-\alpha}\right)$ where $a \neq 1$.

Table 1 Comparison of models

Performance measures	Model-I	Model-II($a > 1$)	Model-II($a < 1$)
E(W)	77.23	38.67	834.48
V(W)	18,811.53	217.64	215,774.9

3.2.2 Result-2

When $\{Z_i\}_{i=1}^{\infty}$ follows iid exponential distribution with rate λ (i.e. the intensity of attrition is the same for all the policies), for Model-I $E(W) = \left(\frac{1}{\lambda} + \frac{1}{\lambda_L}\right) \sum_{k=1}^{\infty} A_k$, and for Model-II $E(W) = \left(\frac{1}{\lambda} + \frac{1}{\lambda_L}\right) \sum_{k=1}^{\infty} \left(\frac{A_k}{a^k}\right)$, where $a \neq 1$.

3.2.3 Result-3

When $\{X_i\}_{i=1}^{\infty}$ are iid exponential rvs with rate α and $\{Z_i\}_{i=1}^{\infty}$ follows iid exponential distribution with rate λ, for Model-I $E(W) = \left(\frac{1}{\lambda} + \frac{1}{\lambda_L}\right) \left(\frac{\theta+\alpha}{\theta}\right)$ and for Model-II $E(W) = \left(\frac{1}{\lambda} + \frac{1}{\lambda_L}\right) \left(\frac{a(\theta+\alpha)}{a(\theta+\alpha)-\alpha}\right)$ where $a \neq 1$.

4 Numerical Illustrations

To make a comparative study of these two models, numerical illustration is made and presented in Table 1 by fixing the values of the parameters as $\alpha_1 = 0.3$, $\alpha_2 = 0.5$, $\alpha_3 = 0.7$, $\alpha_4 = 0.9$, $\alpha_5 = 1.3$, $\lambda_1 = 0.04$, $\lambda_2 = 0.2$, $\lambda_3 = 0.4$, $p_1 = 0.2$, $p_2 = 0.3$, $p_3 = 0.5$, $\lambda_L = 0.2$, $\theta = 0.02$, $k = 5$ and $a = 1.5$ or 0.5. Here three types of policy decisions ($n = 3$) with low attrition rate (λ_1), medium attrition rate (λ_2) and high attrition rate (λ_3) are considered ($\lambda_1 < \lambda_2 < \lambda_3$). In Figs. 2, 3, 4, and 5, the effect of λ_L and αs on the performance measures are presented by varying them. In Figs. 6, 7, 8, 9, 10, and 11, by fixing $\lambda_L = 0.6$ the effect of attrition rates and αs are graphed.

Findings

1. It is observed from Table 1 that Model-II for $a < 1$ is preferable than Model-I and Model-I is preferable than Model-II for $a > 1$, since the time to recruitment is delayed.
2. It is observed from Figs. 2, 3, 4, 5, 6, 7, 8, 9, 10, and 11 that

   (a) As the absence rate of lag period, λ_A, decreases, then the average lag period increases, and hence the mean and variance for the time to recruitment increase.

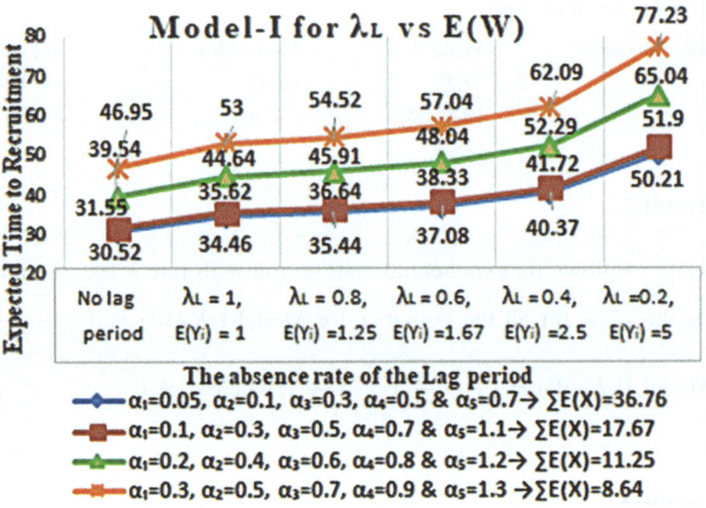

Fig. 2 Absence rate of lag period versus E(W) in Model-I

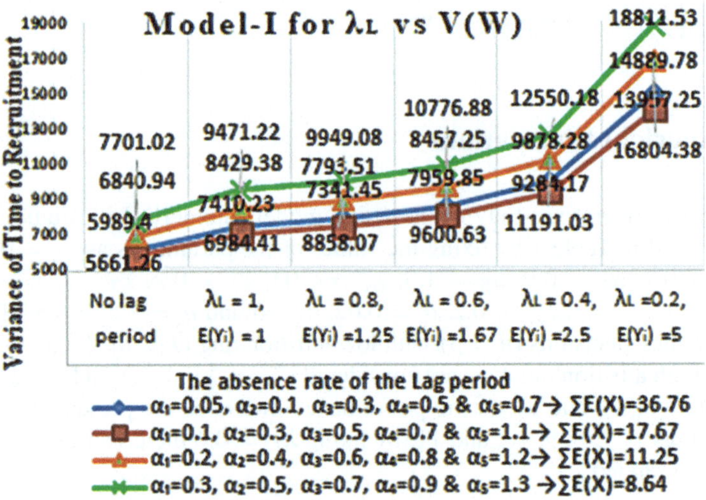

Fig. 3 Absence rate of lag period versus V(W) in Model-I

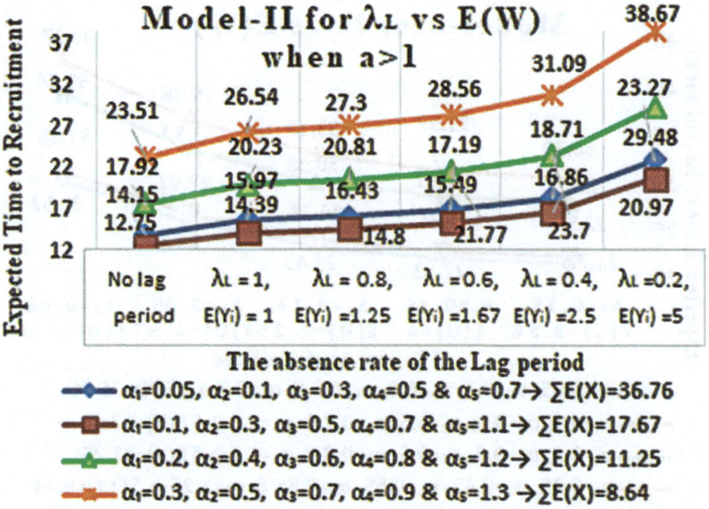

Fig. 4 Absence rate of lag period versus E(W) in Model-II ($a > 1$)

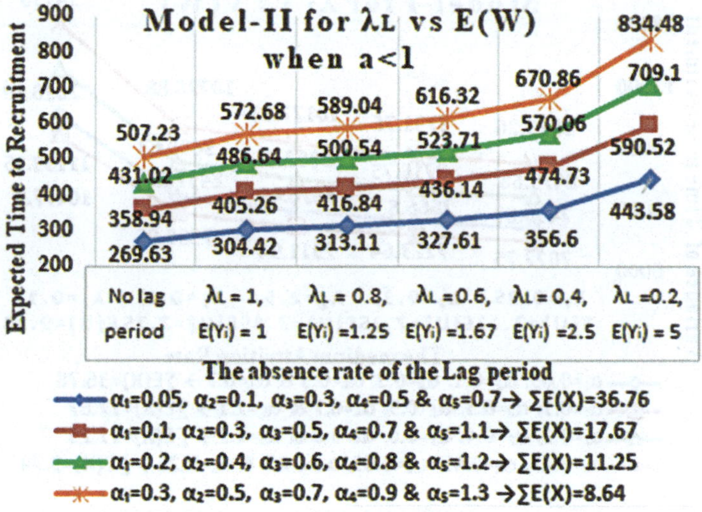

Fig. 5 Absence rate of lag period versus E(W) in Model-II ($a < 1$)

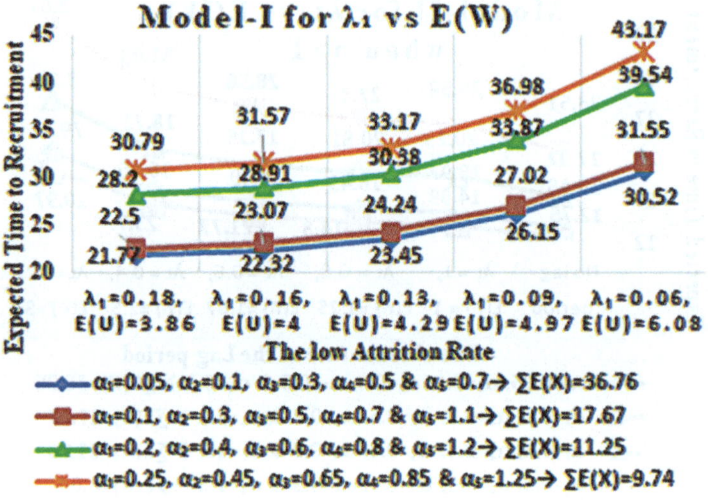

Fig. 6 Low attrition rate versus E(W) in Model-I

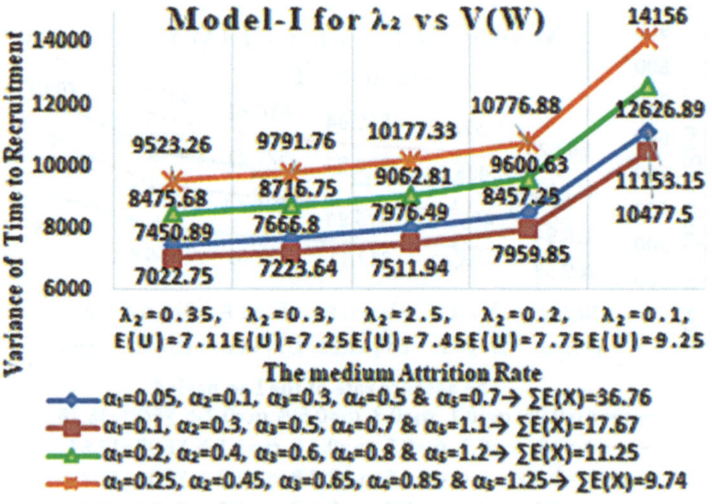

Fig. 7 Medium attrition rate versus V(W) in Model-I

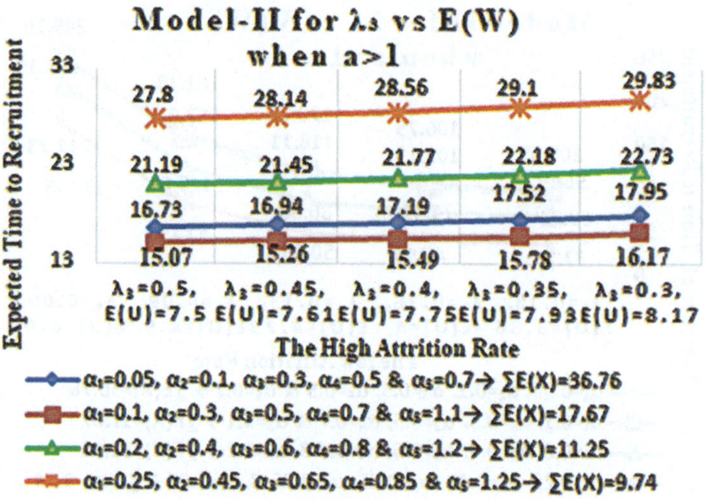

Fig. 8 High attrition rate versus E(W) in Model-II ($a > 1$)

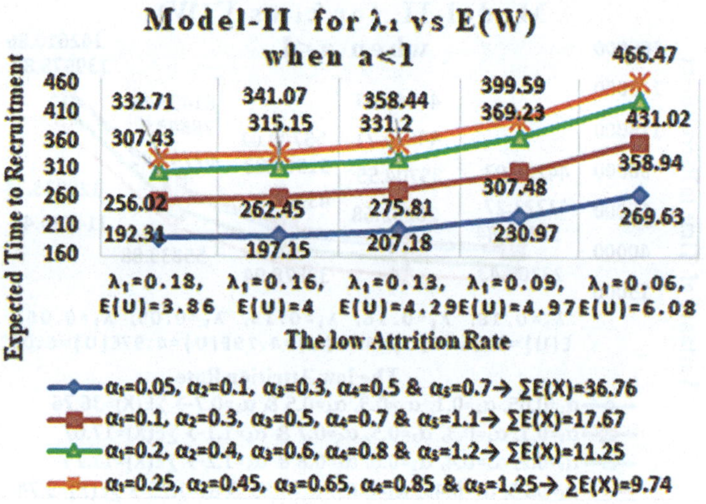

Fig. 9 Low attrition rate versus E(W) in Model-II ($a < 1$)

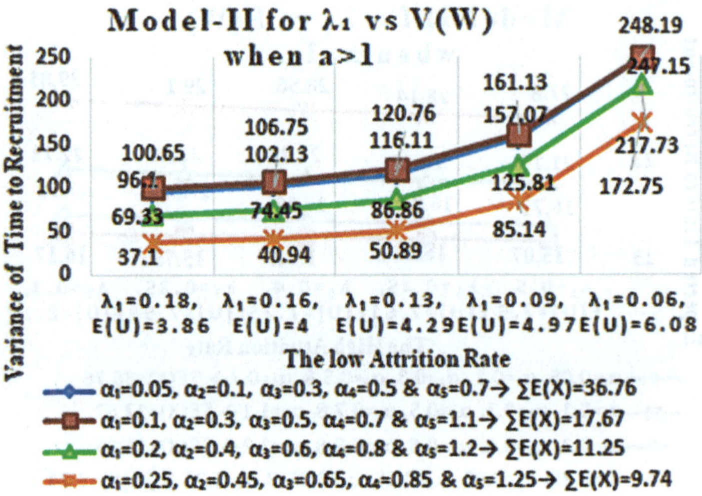

Fig. 10 Low attrition rate versus V(W) in Model-II ($a > 1$)

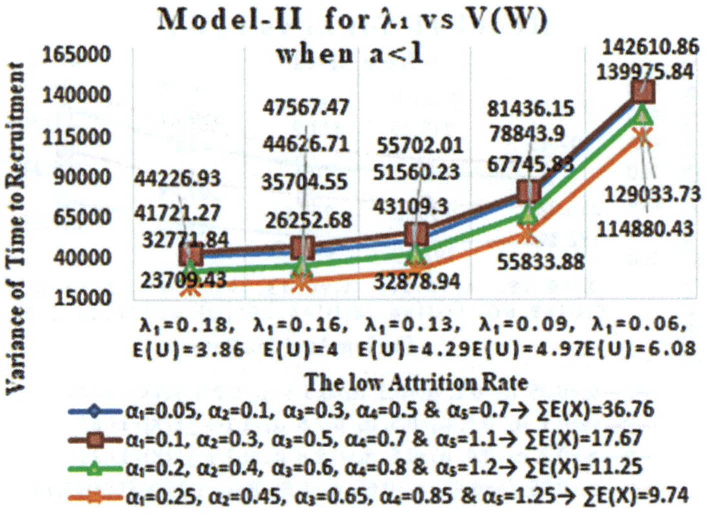

Fig. 11 Low attrition rate versus V(W) in Model-II ($a < 1$)

(b) When lag period exist, the expected time to recruitment is delayed than in the absence of lag period. So lag period is preferable.
(c) As any one of the attrition rate (λ_1, λ_2 or λ_3) decreases, then the average IPDT increases, and hence the expected time and variance for time to recruitment increases.
(d) As the absence rate of wastages increases, then the average wastages decrease, and hence the expected time to recruitment increases for both the models and the variance for time to recruitment increases for Model-I and decreases for Model-II. From these observations it is clear that Model-II is preferable.

5 Conclusion

In order to postpone the recruitment, the organisation may take policy decisions in such a way that the sequence of IPDT forms a stochastically increasing geometric process and also by introducing the practice of lag period for wastages.

References

1. Bartholomew, D. J.: Sufficient conditions for a mixture of exponentials to be a probability density function. Ann. Math. Statist. (1969). https://doi.org/10.1214/aoms/1177697296
2. Bartholomew, D. J.: The statistical approach to manpower planning model. The Statistician. (1971). https://doi.org/10.2307/2987003
3. Bartholomew, D. J.: Statistical Problems of Predication and Control in Manpower Planning. Mathematical Scientist. **1**, 133–144 (1976)
4. Bartholomew, D. J., Andrew Forbes, F.: Statistical Techniques for Manpower Planning. John Wiley and Sons, New York (1979)
5. Esary, J. D., Marshall A. W., Proschan, F.: Shock models and wear processes. Ann. Probab. (1973) https://doi.org/10.1214/aop/1176996891. MR350893
6. Esther Clara, B.: Contributions to the Study on Some Stochastic Models in Manpower Planning. Ph.D Thesis, Bharathidasan University, Tiruchirappalli (2012)
7. Manju Ramalingam, Esther Clara, B., Srinivasan, A.: A Stochastic Model on Time to Recruitment for a Single Grade Manpower System with n Types of Policy Decisions and Correlated Wastages using Univariate CUM Policy. Aryabhatta Journal of Mathematics and Informatics. **8(2)**, 67–74 (2016)
8. Manju Ramalingam, Esther Clara, B., Srinivasan, A.: Time to Recruitment for a Single Grade Manpower System with Two Types of Depletion Using Univariate CUM Policy of Recruitment. Annals of Management Science. (2017). https://doi.org/10.24048/ams5.no2.2017
9. Medhi, J. : Stochastic Processes. Third Edition, New Age International Publishers, India (2012)
10. Muthaiyan, A., Sathiyamoorthi, R.: A stochastic model using geometric process for inter-arrival time between wastage. Acta Ciencia Indica. **36(4)**, 479–486 (2010)
11. Revathy Sundarajan: Contributions to the study optimal replacement policies stochastic System. Ph.D Thesis. University of Madras, Chennai (1998)
12. Samuel Karlin, Howard M. Taylor: A first course in Stochastic Processes. Second Edition, Academic Press, New York (1975)
13. Sathyamoorthi, R., Elangovan, R.: Shock model approach to determine the expected time to recruitment. Journal of Decision and Mathematical Science. **3(1–3)**, 67–78 (1998)

A Novice's Application of Soft Expert Set: A Case Study on Students' Course Registration

Selva Rani B and Ananda Kumar S

Abstract A mathematical tool termed as soft set theory deals with uncertainty and was introduced by Molodtsov in 1999, which had been studied by many researchers, and some models were created to find a solution in decision-making. But, those models deal exactly with one expert in making a decision. There are situations in which more than one expert may get involved. S. Alkhazaleh and A.R. Salleh introduced a model with opinions from more than an expert which was coined as soft expert set in 2011. This method was found to be more effective compared with the traditional soft set theory. Now-a-days, educational institutions are relying on software tools and techniques in their academic processes. Applying Soft Expert Set in those processes would facilitate their decision making and yield better results. In this paper, the said concept would be applied for an institution's course registration process that would facilitate students to choose from the list of faculty members offering the same course based on the faculty's performance. The proposed approach may be generalized to a recommender system to accommodate institutions preferences over the set of deciding criteria.

Keywords Soft expert set · Decision-making · Agree set · Disagree set · Recommendation

1 Introduction

The uncertainty theories like soft expert sets find their applications in domains like business, economics, medical diagnosis, and engineering which deal with uncertainties. Innovative approaches using the aforementioned techniques were successfully employed for many real-world scenarios. Despite their applications and scope, several researchers are attempting to introduce innovative ideas over the

Selva Rani B (✉) · Ananda Kumar S
VIT, Vellore
e-mail: bselvarani@vit.ac.in

V. Madhu et al. (eds.), *Advances in Algebra and Analysis*, Trends in Mathematics,
https://doi.org/10.1007/978-3-030-01120-8_45

existing findings so as to improve one or the other parameters involved in decision-making process. Nowadays, all the institutions rely on automated processes that rely on one or many idealistic approaches in their decision-making stages. Most of the problems were found to be solved using several soft computing approaches. In this work, an attempt is made to idealize the impact of soft expert set utilized in decision-making process in place of traditional soft sets. In this paper, the concepts behind soft expert sets and some important operations on them are illustrated below. To explore the behavior of this soft expert set, a case study on student course registration system was considered. In any institution, more than one faculty member may offer the same course where in such scenarios, students may find it difficult to identify and enroll themselves with one among the faculty members offering that course based on their performance criteria. An attempt is made with soft expert set to assist students for identifying the faculty with desired performance characteristics and to enroll them. The same process is illustrated later in this work.

2 Literature Survey

To deal with uncertainties, Molodtsov formulated soft set theory [1]. Based on his work, some variations were presented by researchers as in Table 1.

Such theories were studied and applied to decision-making problems. Table 2 lists a brief summary of sample works in the domain.

Table 1 Soft set theories

Theory	Proposed by	Reference
Soft set	D. Molodtsov	[1]
Fuzzy soft set	P.K. Maji et al.	[2]
Bijective soft sets	K. Gong et al.	[3]
Exclusive disjunctive soft sets	Z. Xiao et al.	[4]

Table 2 Few applications of soft set theory

Application	Proposed by	Reference
An application of soft sets in a decision-making problem	P.K. Maji and A.R. Roy	[5]
A study of solving decision-making problem using soft set	R.K. Bhardwaj et al.	[6]
Soft set based association rule mining	Feng Feng et al.	[7]
A survey of decision-making methods based on two classes of hybrid soft set methods	Xueling Ma et al.	[8]
A survey of decision-making methods based on certain hybrid soft set methods	Xueling Ma et al.	[9]

Table 3 Multi-expert theories

Theory	Proposed by	Reference
Soft expert set	Alkhazaleh et al.	[10]
Fuzzy parameterized soft expert set	Maruah et al.	[11]
Fuzzy parameterized fuzzy soft expert set	Ayman et al.	[12]
Possibility fuzzy soft expert set	Maruah et al.	[13]
Vague soft expert set	N. Hassan et al.	[14]
Fuzzy soft expert set	Alkhazaleh et al.	[15]
Possibility intuitionistic fuzzy soft expert set	Ganeshsree et al.	[16]

The above contributions presented the idea behind the next level of soft sets and how it could be applied to other real-time problems like medical diagnosis and many more. But these models proposed exactly one expert to derive an opinion and forcing the users to perform union and other operations in case of multi-experts opinion needed. Alkhazaleh et al. proposed a model based on the idea of soft expert set, which seeks the opinion from more than one expert without any further operations [10]. This model could be more useful in almost all the decision-making problems, and the further extension principles are presented in Table 3.

3 Foundations

The fundamental ideas behind soft expert set is presented in this section. Some important operations on soft expert sets are also recalled here.

Let us consider the universe U, the set of parameters P, the set of experts E, and the set of opinions O. Here $Z = P \times E \times O$ and $X \subseteq Z$.

Definition 3.1 Soft Expert Set

A soft expert set is a pair (F, X) over the universe U, and F is a mapping of X to $P(U)$, which is the power set of U.

$$F : X \rightarrow P(U)$$

Definition 3.2 Soft Expert Subset and Soft Expert Superset

A soft expert set (F, X) is termed as a soft expert subset of (G, Y) over the common universe U, if:

- $X \subseteq Y$,
- $\forall \epsilon \in X, F(\epsilon) \subseteq G(\epsilon)$

which is denoted by $(F, X)\tilde{\subseteq}(G, Y)$.

Here, (G, Y) is known as the soft expert superset of (F, X) and is denoted by $(G, Y)\tilde{\supseteq}(F, Y)$.

Definition 3.3 Soft Expert Equal Sets

The soft expert sets (F, X) and (G, Y) over the universe U are termed as soft expert equal sets if (F, X) is the soft expert subset of (G, Y) and (G, Y) is the soft expert set of (F, X).

Definition 3.4 NOT set of Z

$Z = P \times E \times O$ where P is the set of parameters, E is the set of experts, and O is the set of opinions. The NOT of Z is denoted by $\neg Z$ and is defined as

$$\neg Z = \left\{(\neg p_i, e_j, o_k), \forall i, j, k\right\} \; where \; \neg p_i = not p_i$$

Definition 3.5 Complement Soft Expert Set

$(F, X)^c$ is termed as the complement of the soft expert set (F, X) over the universe U and defined by

$$(F, X)^c = (F^c, \neg X),$$

where F^c is a mapping of $\neg X$ to $P(U)$, and $\forall x \in \neg X, F^c(X) = U - F(\neg X)$

Definition 3.6 Agree-Soft Expert Set

Let (F, X) be the soft expert set over the universe U. The agree-soft expert set of (F, X) denoted as $(F, X)_1$ is defined by $(F, X)_1 = \{F_1(\epsilon) : \epsilon \in E \times X \times \{1\}\}$.

Definition 3.7 Disagree-Soft Expert Set

Let (F, X) be the soft expert set over the universe U. The disagree-soft expert set of (F, X) denoted as $(F, X)_0$ is defined by $(F, X)_0 = \{F_0(\epsilon) : \epsilon \in E \times X \times \{0\}\}$.

Definition 3.8 Union operation

Let (F, X) and (G, Y) be the two soft expert sets over the universe U. The union of these soft expert sets (H, Z) is the soft expert set denoted by $(F, X)\tilde{\cup}(G, Y)$, where $Z = X \cup Y \; \forall \epsilon \in Z$.

$$H(\epsilon) = \begin{cases} F(\epsilon) & if \epsilon \in X - Y \\ G(\epsilon) & if \epsilon \in Y - X \\ F(\epsilon) \cup G(\epsilon) & if \epsilon \in X \cup Y \end{cases}$$

Definition 3.9 Intersection operation

Let (F, X) and (G, Y) be the two soft expert sets over the universe U. The intersection of these soft expert sets (H, Z) is the soft expert set denoted by $(F, X)\tilde{\cap}(G, Y)$, where $Z = X \cap Y \; \forall \epsilon \in Z$.

$$H(\epsilon) = \begin{cases} F(\epsilon) & if \epsilon \in X - Y \\ G(\epsilon) & if \epsilon \in Y - X \\ F(\epsilon) \cap G(\epsilon) & if \epsilon \in X \cap Y \end{cases}$$

Definition 3.10 AND operation

Let (F, X) and (G, Y) be the two soft expert sets over the universe U. Now (F, X) AND (G, Y) is denoted as $(F, X) \wedge (G, Y)$ and defined by $(F, X) \wedge (G, Y) = (H, X \times Y)$, where $H(\alpha, \beta) = F(\alpha)\tilde{\cap}G(\beta)\ \forall(\alpha, \beta) \in X \times Y$.

Definition 3.11 OR operation

Let (F, X) and (G, Y) be the two soft expert sets over the universe U. Now (F, X) OR (G, Y) is denoted as $(F, X) \vee (G, Y)$ and defined by $(F, X) \vee (G, Y) = (I, X \times Y)$, where $I(\alpha, \beta) = F(\alpha)\tilde{\cup}G(\beta)\ \forall(\alpha, \beta) \in X \times Y$.

4 Soft Expert Set and Decision-Making

As aforementioned, the students registration case study considered here would enable them to identify one faculty member for enrolling into a course. Institutions rely on feedback from students' community as one of the criteria to evaluate the performance of their faculty members. So, the feedback collected thus would enable students to understand and analyze the performance based on the previous year's performance of an individual. The factors to evaluate an individual faculty member by students vary among institutions. The following factors are considered for evaluating ten faculty members by five students in this case so as to make illustration precise: subject knowledge, communication skill, encourage interaction, slow learners attention, and challenging assignments.

The universe is given by $U = \{f_1, f_2, f_3, f_4, f_5, f_6, f_7, f_8, f_9, f_{10}\}$. Let the evaluation parameters be represented as $P = \{e_1, e_2, e_3, e_4, e_5\}$, where,

e_1 : *subject knowledge*
e_2 : *communication skill*
e_3 : *encourage interaction*
e_4 : *slow − learner attention*
e_5 : *challenging assignments*

Let $E = \{p, q, r, s, t\}$ be the set of students providing feedback on the evaluation criteria. Assume that the following soft expert set (F, X) is obtained based on the students feedback. It can be represented as agree-soft expert set and disagree-soft expert set as in Tables 4 and 5, respectively.

To make the final decision on faculty, the following procedure is to be followed:

(a) Input the soft expert set (F, X).
(b) Find the corresponding agree-soft expert set and disagree-soft expert set.
(c) For the agree-soft expert set, compute $a_j = \sum_i f_{ij}$.
(d) For the disagree-soft expert set, compute $d_j = \sum_i f_{ij}$.
(e) Compute $x_j = a_j - d_j$.
(f) Find $\max(x_j)$ which is the optimal choice.

$$(F, X) =$$

$$\left\{\begin{array}{c}
((e_1, p, 1), \{f_1, f_2, f_4, f_6, f_8, f_{10}\}), ((e_1, q, 1), \{f_1, f_3, f_7, f_9\}), ((e_1, r, 1), \{f_2, f_6, f_8, f_{10}\}), \\
((e_1, s, 1), \{f_3, f_4, f_5, f_6, f_9\}), ((e_1, t, 1), \{f_1, f_2, f_4, f_5, f_6, f_8\}), ((e_2, p, 1), \{f_1, f_3, f_5, f_7, f_9, f_{10}\}), \\
((e_2, q, 1), \{f_2, f_3, f_6, f_7\}), ((e_2, r, 1), \{f_4, f_8, f_9, f_{10}\}), ((e_2, s, 1), \{f_3, f_4, f_5, f_9, f_{10}\}), ((e_2, t, 1), \{f_1, f_9, f_{10}\}), \\
((e_3, p, 1), \{f_3, f_6, f_8, f_9\}), ((e_3, q, 1), \{f_2, f_6, f_8\}), ((e_3, r, 1), \{f_9, f_{10}\}), ((e_3, s, 1), \{f_2, f_3, f_4, f_5, f_{10}\}), \\
((e_3, t, 1), \{f_6, f_7, f_8, f_{10}\}), ((e_4, p, 1), \{f_2, f_4, f_6, f_8, f_{10}\}), ((e_4, q, 1), \{f_1, f_9, f_{10}\}), ((e_4, r, 1), \{f_2, f_3, f_5, f_6\}), \\
((e_4, s, 1), \{f_3, f_7, f_8, f_9\}), ((e_4, t, 1), \{f_1, f_2, f_5, f_6\}), ((e_5, p, 1), \{f_2, f_3, f_6\}), ((e_5, q, 1), \{f_7, f_8, f_9\}), \\
((e_5, r, 1), \{f_2, f_4, f_6\}), ((e_5, s, 1), \{f_1, f_2, f_6, f_8, f_{10}\}), ((e_5, t, 1), \{f_2, f_3, f_5, f_6, f_8, f_{10}\}), ((e_1, p, 0), \{f_3, f_5, f_7, f_9\}), \\
((e_1, q, 0), \{f_2, f_4, f_5, f_6, f_8 f_{10}\}), ((e_1, r, 0), \{f_1, f_3, f_4, f_5, f_7, f_9\}), ((e_1, s, 0), \{f_1, f_2, f_7, f_8, f_{10}\}), \\
((e_1, t, 0), \{f_3, f_7, f_9, f_{10}\}), ((e_2, p, 0), \{f_2, f_4, f_6, f_8\}), ((e_2, q, 0), \{f_1, f_4, f_5, f_8, f_9, f_{10}\}), \\
((e_2, r, 0), \{f_1, f_2, f_3, f_5, f_6, f_7\}), ((e_2, s, 0), \{f_1, f_2, f_6, f_7, f_8\}), ((e_2, t, 0), \{f_2, f_3, f_4, f_5, f_6, f_7, f_8\}), \\
((e_3, p, 0), \{f_1, f_2, f_4, f_5, f_7, f_{10}\}), ((e_3, q, 0), \{f_1, f_3, f_4, f_5, f_7, f_9, f_{10}\}), \\
((e_3, r, 0), \{f_1, f_2, f_3, f_4, f_5, f_6, f_7, f_8\}), ((e_3, s, 0), \{f_1, f_6, f_7, f_8, f_9\}), ((e_3, t, 0), \{f_1, f_2, f_3, f_4, f_5, f_9\}), \\
((e_4, p, 0), \{f_1, f_3, f_5, f_7, f_9\}), ((e_4, q, 0), \{f_2, f_3, f_4, f_5, f_6, f_7, f_8\}), ((e_4, r, 0), \{f_1, f_4, f_7, f_8, f_9, f_{10}\}), \\
((e_4, s, 0), \{f_1, f_2, f_4, f_5, f_6, f_{10}\}), ((e_4, t, 0), \{f_3, f_4, f_7, f_8, f_9, f_{10}\}), ((e_5, p, 0), \{f_1, f_4, f_5, f_7, f_8, f_9, f_{10}\}), \\
((e_5, q, 0), \{f_1, f_2, f_3, f_4, f_5, f_6, f_{10}\}), ((e_5, r, 0), \{f_1, f_3, f_5, f_7, f_8, f_9, f_{10}\}), ((e_5, s, 0), \{f_3, f_4, f_5, f_7, f_9\}), \\
((e_5, t, 0), \{f_1, f_4, f_7, f_9\})
\end{array}\right\}$$

Table 4 Agree-soft expert set

U	f_1	f_2	f_3	f_4	f_5	f_6	f_7	f_8	f_9	f_{10}
(e_1,p)	1	1	0	1	0	1	0	1	0	1
(e_2,p)	1	0	1	0	1	0	1	0	1	1
(e_3,p)	0	0	1	0	0	1	0	1	1	0
(e_4,p)	0	1	0	1	0	1	0	1	0	1
(e_5,p)	0	1	1	0	0	1	0	0	0	0
(e_1,q)	1	0	1	0	0	0	1	0	1	0
(e_2,q)	0	1	1	0	0	1	1	0	0	0
(e_3,q)	0	1	0	0	0	1	0	1	0	0
(e_4,q)	1	0	0	0	0	0	0	0	1	1
(e_5,q)	0	0	0	0	0	0	1	1	1	0
(e_1,r)	0	1	0	0	0	1	0	1	0	1
(e_2,r)	0	0	0	1	0	0	0	1	1	1
(e_3,r)	0	0	0	0	0	0	0	0	1	1
(e_4,r)	0	1	1	0	1	1	0	0	0	0
(e_5,r)	0	1	0	1	0	1	0	0	0	0
(e_1,s)	0	0	1	1	1	1	0	0	1	0
(e_2,s)	0	0	1	1	1	0	0	0	1	1
(e_3,s)	0	1	1	1	1	0	0	0	0	1
(e_4,s)	0	0	1	0	0	0	1	1	1	0
(e_5,s)	1	1	0	0	0	1	0	1	0	1
(e_1,t)	1	1	0	1	1	1	0	1	0	0
(e_2,t)	1	0	0	0	0	0	0	0	1	1
(e_3,t)	0	0	0	0	0	1	1	1	0	1
(e_4,t)	1	1	0	0	1	1	0	0	0	0
(e_5,t)	0	1	1	0	1	1	0	1	0	1
$a_j = \sum_i u_{ij}$	$a_1 = 8$	$a_2 = 13$	$a_3 = 11$	$a_4 = 8$	$a_5 = 8$	$a_6 = 15$	$a_7 = 6$	$a_8 = 12$	$a_9 = 11$	$a_{10} = 13$

According to Table 6, $\max(x_j)$ is x_6, hence the students may opt for $faculty_4$ based on the previous performance feedback.

Table 5 Disagree-soft expert set

U	f_1	f_2	f_3	f_4	f_5	f_6	f_7	f_8	f_9	f_{10}
(e_1,p)	0	0	1	0	1	0	1	0	1	0
(e_2,p)	0	1	0	1	0	1	0	1	0	0
(e_3,p)	1	1	0	1	1	0	1	0	0	1
(e_4,p)	1	0	1	0	1	0	1	0	1	0
(e_5,p)	1	0	0	1	1	0	1	1	1	1
(e_1,q)	0	1	0	1	1	1	0	1	0	1
(e_2,q)	1	0	0	1	1	0	0	1	1	1
(e_3,q)	1	0	1	1	1	0	1	0	1	1
(e_4,q)	0	1	1	1	1	1	1	1	0	0
(e_5,q)	1	1	1	1	1	1	0	0	0	1
(e_1,r)	1	0	1	1	1	0	1	0	1	0
(e_2,r)	1	1	1	0	1	1	1	0	0	0
(e_3,r)	1	1	1	1	1	1	1	1	0	0
(e_4,r)	1	0	0	1	0	0	1	1	1	1
(e_5,r)	1	0	1	0	1	0	1	1	1	1
(e_1,s)	1	1	0	0	0	0	1	1	0	1
(e_2,s)	1	1	0	0	0	1	1	1	0	0
(e_3,s)	1	0	0	0	0	1	1	1	1	0
(e_4,s)	1	1	0	1	1	1	0	0	0	1
(e_5,s)	0	0	1	1	1	0	1	0	1	0
(e_1,t)	0	0	1	0	0	0	1	0	1	1
(e_2,t)	0	1	1	1	1	1	1	1	0	0
(e_3,t)	1	1	1	1	1	0	0	0	1	0
(e_4,t)	0	0	1	1	0	0	1	1	1	1
(e_5,t)	1	0	0	1	0	0	1	0	1	0
$d_j = \sum_i u_{ij}$	$d_1 = 17$	$d_2 = 12$	$d_3 = 14$	$d_4 = 17$	$d_5 = 17$	$d_6 = 10$	$d_7 = 19$	$d_8 = 13$	$d_9 = 14$	$d_{10} = 12$

Table 6 $x_j = a_j - d_j$

$a_j = \sum_i f_{ij}$	$d_j = \sum_i f_{ij}$	$x_j = a_j - d_j$
$a_1 = 8$	$d_1 = 17$	$x_1 = -9$
$a_2 = 13$	$d_2 = 12$	$x_2 = 1$
$a_3 = 11$	$d_3 = 14$	$x_3 = -3$
$a_4 = 8$	$d_4 = 17$	$x_4 = -9$
$a_5 = 8$	$d_5 = 17$	$x_5 = -9$
$a_6 = 15$	$d_6 = 10$	$x_6 = 5$
$a_7 = 6$	$d_7 = 19$	$x_7 = -13$
$a_8 = 12$	$d_8 = 13$	$x_8 = -1$
$a_9 = 11$	$d_9 = 14$	$x_9 = -3$
$a_{10} = 13$	$d_{10} = 12$	$x_{10} = 1$

5 Conclusion

The basic idea of soft expert set was presented in this work. Fundamental operations on the same set were also discussed. Finally, the same concept was applied for the students' course registration process of an institution and concluded with a simple idea. The same idea can also be applied to many problems involving decision-making. The number of factors and the number of students considered for faculty performance evaluation in this case were limited. This case can also be extended with fuzzy soft expert set.

References

1. Molodtsov, D.: Soft set theory-first results. Computers & Mathematics with Applications **37**, 19–31 (1999)
2. Maji, P.K., Biswas, R.K., Roy, A.: Fuzzy soft sets. Journal of Fuzzy Mathematics **9**, 589–602 (2001)
3. Gong, K., Xiao, Z., & Zhang, X.: The bijective soft set with its operations **60**, 2270–2278 (2010)
4. Xiao, Z., Gong, K., Xia, S., & Zou, Y. : Exclusive disjunctive soft sets. Computers & Mathematics with Applications **59**, 2128–2137 (2010)
5. Maji, P.K., Roy, A.R. and Biswas, R.: An application of soft sets in a decision making problem. Computers & Mathematics with Applications **44**, 1077–1083 (2002)
6. Bhardwaj, R.K., Tiwari, S.K. and Nayak, K.C.: A Study of Solving Decision Making Problem using soft set. IJLTEMAS **IV**, 26–32 (2015)
7. Feng, F., Cho, J., Pedrycz, W., Fujita, H. and Herawan, T.: Soft set based association rule mining. Knowledge-Based Systems **111**, 268–282 (2016)
8. Ma, X., Zhan, J., Ali, M.I. et al.: A survey of decision making methods based on two classes of hybrid soft set models. Artif Intell Rev (2018). https://doi.org/10.1007/s10462-016-9534-2
9. Ma, X., Liu, Q. & Zhan, J.: A survey of decision making methods based on certain hybrid soft set models. J. Artif Intell Rev (2017). https://doi.org/10.1007/s10462-016-9490-x
10. Alkhazaleh, S. and Salleh, A.R.: Soft expert sets. Adv. Decis. Sci. **15**, (2011)
11. Bashir, M. and Salleh, A.R.: Fuzzy parameterized soft expert set. Abstract and Applied Analysis **2012**, 1–15 (2012)
12. Hazaymeh, A., Abdullah, I.B., Balkhi, Z. and Ibrahim, R.: Fuzzy parameterized fuzzy soft expert set **6**, 5547–5564 (2012)
13. Bashir, M. and Salleh, A.R.: Possibility fuzzy soft expert set. Open Journal of Applied Sciences **12**, 208–211 (2012)
14. Hassan, N. and Alhazaymeh, K.: Vague soft expert set theory. AIP Conference Proceedings **1522**, 953–958 (2013)
15. Alkhazaleh, S. and Salleh, A.R.: Fuzzy soft expert set and its application. Applied Mathematics. **5**, 1349–1368 (2014)
16. Selvachandran, G. and Salleh, A.R.: Possibility intuitionistic fuzzy soft expert set theory and its application in decision making. International Journal of Mathematics and Mathematical Sciences, 1–11 (2015)

Dynamics of Stochastic SIRS Model

R. Rajaji

Abstract This article presents a SIRS epidemic model with stochastic effect. For the stochastic version, we prove the existence and uniqueness of the solution of this stochastic SIRS model. In addition, sufficient conditions for the stochastic stability of equilibrium solutions are provided. Finally, numerical visualization is presented to justify our results.

1 Introduction

Mathematical modeling is an important tool used in analyzing the spread of infectious diseases. One of the vital models in epidemiological patterns and disease control is SIR model. Kermack and McKendrick [5] initially suggested and analyzed the deterministic SIR model. After that, many authors have examined the deterministic SIRS model [2, 9].

The deterministic SIRS model can be written as

$$\begin{aligned} \frac{d\alpha}{dt} &= l - b\alpha\beta - m\alpha + c\gamma, \\ \frac{d\beta}{dt} &= b\alpha\beta - (k + m + a)\beta, \\ \frac{d\gamma}{dt} &= k\beta - (m + c)\gamma. \end{aligned} \tag{1}$$

where $\alpha(t)$, $\beta(t)$, and $\gamma(t)$ denote the number of susceptible, infective, and recovered individuals at time t, respectively, l is the recruitment rate of the population, m is the natural death rate, a is the death rate due to disease, b is the infection coefficient, k is the recovery rate of the infective individuals, and c is the rate at which recovered individuals lose immunity and return to the susceptible class.

R. Rajaji (✉)
Department of Mathematics, Patrician College of Arts and Science, Chennai, India
e-mail: rajajiranga@gmail.com

V. Madhu et al. (eds.), *Advances in Algebra and Analysis*, Trends in Mathematics,
https://doi.org/10.1007/978-3-030-01120-8_46

The model (1) may have at most two equilibrium solutions; they are an infection-free equilibrium solution $E_1 = (\alpha_1, \beta_1, \gamma_1)$, where $\alpha_1 = \frac{l}{m}$, $\beta_1 = 0$, *and* $\gamma_1 = 0$, and an endemic equilibrium solution $E_2 = (\alpha_2, \beta_2, \gamma_2)$, where

$$\alpha_2 = \frac{k+a+m}{b} = \frac{\alpha_1}{\mathscr{R}_0}, \; \beta_2 = \frac{m(m+c)(k+a+m)(\mathscr{R}_0 - 1)}{b(km+(m+c)(a+m))},$$

$$\gamma_2 = \frac{km(k+a+m)(\mathscr{R}_0 - 1)}{b(km+(m+c)(a+m))}.$$

The endemic equilibrium solution exists if the following condition holds

$$\mathscr{R}_0 = \frac{lb}{m(k+a+m)} > 1,$$

where $\mathscr{R}_0$ is the basic reproduction number. In [7], León studied the global stability of the model (1).

In real life, any system is irresistibly influenced by the environmental noise, which is an essential component in an ecosystem. So we introduce environmental noises into biological system by parameter perturbation, which is the procedure used in constructing SDE models [3, 10, 11, 14]. In the last few decades, several authors analyzed the effect of environmental noise on the transmission of disease dynamics by proposing epidemic SDE model [6, 8, 13, 14]. So, we perturbed the deterministic system (1) by a white noise and get a stochastic counterpart by replacing the rates b by $b + \sigma_1(\alpha - \alpha_2)\frac{dW_1}{dt}$ and k by $k + \sigma_2(\beta - \beta_2)\frac{dW_2}{dt}$, where σ_1, σ_2 are real constants and W_i $i = 1, 2$ are *i.i.d.* Wiener processes are defined on a filtered complete probability space $(\Omega, \mathscr{F}, \{\mathscr{F}_t\}_{t\geq 0}, \mathbb{P})$.

The stochastic SIRS model takes a form as:

$$\begin{aligned}
d\alpha &= (l - b\alpha\beta - m\alpha + c\gamma)\,dt - \sigma_1\alpha\beta(\alpha - \alpha_2)dW_1,\\
d\beta &= (b\alpha\beta - (k+m+a)\beta)\,dt + \sigma_1\alpha\beta(\alpha - \alpha_2)dW_1 - \sigma_2\beta(\beta - \beta_2)dW_2,\\
d\gamma &= (k\beta - (m+c)\gamma)\,dt + \sigma_2\beta(\beta - \beta_2)dW_2.
\end{aligned} \tag{2}$$

2 Existence of a Unique Global Solution

In this section, we prove the existence of a unique global solution of (2).

Define

$$\mathbb{D} = \left\{(\alpha, \beta, \gamma) \in \mathbb{R}^3 : \alpha > 0, \beta > 0, \gamma > 0, \alpha + \beta + \gamma \leq \frac{l}{m}\right\}.$$

Theorem 1 *Let* $(\alpha(t_0), \beta(t_0), \gamma(t_0)) = (\alpha_0, \beta_0, \gamma_0) \in \mathbb{D}$, *and* $(\alpha_0, \beta_0, \gamma_0)$ *is independent of* W. *Then the stochastic SIRS model* (2) *has a unique continuous time and global solution* $(\alpha(t), \beta(t), \gamma(t))$ *on* $t \geq t_0$, *and this solution is invariant (a.s) with respect to* $\mathbb{D}$.

Proof Since the argument is similar to that of Theorem 1 in [12], we here sketch the proof to point out the distinction with it.

Let

$$\mathbb{D}_n := \Big\{(\alpha, \beta, \gamma) : e^{-n} < \alpha < \frac{l}{m} - e^{-n}, e^{-n} < \beta < \frac{l}{m} - e^{-n}, \\ e^{-n} < \gamma < \frac{l}{m} - e^{-n}, \alpha + \beta + \gamma \leq \frac{l}{m}\Big\},$$

for $n \in \mathbb{N}$. The system (2) has a unique solution up to stopping time $\tau(\mathbb{D}_n)$.

Let

$$\begin{aligned} V(\alpha, \beta, \gamma) = \alpha - \ln\alpha + \left(\frac{l}{m} - \alpha\right) - \ln\left(\frac{l}{m} - \alpha\right) + \beta - \ln\beta \\ + \left(\frac{l}{m} - \gamma\right) - \ln\left(\frac{l}{m} - \gamma\right), \end{aligned} \tag{3}$$

defined on $\mathbb{D}$ and assume that $\mathbb{E}(V(\alpha, \beta, \gamma)) < \infty$. Note that $V(\alpha, \beta, \gamma) \geq 4$ for $(\alpha, \beta, \gamma) \in \mathbb{D}$. Let $W(\alpha, \beta, \gamma, t) = e^{-c(t-s)}V(\alpha, \beta, \gamma)$, defined on $\mathbb{D} \times [s, \infty)$, where

$$\begin{aligned} c = \frac{1}{4}\Bigg(3m + c + a + 2k + \frac{l}{m}\left(b + \frac{bl}{m} + m + c\right) \\ + \frac{3}{2}\left(\frac{l}{m}\right)^2\left(\sigma_1^2\left(\frac{l}{m}\right)^2 + \sigma_1^2\alpha_2^2 + \frac{2}{3}\sigma_2^2\right) + \sigma_2^2\beta_2^2\Bigg) \end{aligned} \tag{4}$$

Apply the infinitesimal generator $\mathscr{L}$ on Eq. (3), we have

$$\begin{aligned} \mathscr{L}V(\alpha, \beta, \gamma) = & m - b\alpha\beta\frac{1}{\left(\frac{l}{m} - \alpha\right)} + \frac{c\gamma}{\left(\frac{l}{m} - \alpha\right)} - \frac{l}{\alpha} + b\beta + m - \frac{c\gamma}{\alpha} \\ & + b\alpha\beta - b\alpha - (k + m + a)\beta + k + m + a + \frac{k\beta}{\left(\frac{l}{m} - \gamma\right)} - k\beta \\ & - \frac{(m + c)\gamma}{\left(\frac{l}{m} - \gamma\right)} + (m + c)\gamma + \frac{1}{2}\frac{\sigma_1^2\alpha^2\beta^2}{\left(\frac{l}{m} - \alpha\right)^2}(\alpha - \alpha_2)^2 \ldots \end{aligned}$$

$$\cdots + \frac{1}{2}\sigma_1^2\beta^2(\alpha-\alpha_2)^2 + \frac{1}{2}\sigma_1^2\alpha^2(\alpha-\alpha_2)^2$$
$$+ \frac{1}{2}\sigma_2^2(\beta-\beta_2)^2 + \frac{1}{2}\sigma_2^2\frac{\beta^2}{\left(\frac{l}{m}-\gamma\right)^2}(\beta-\beta_2)^2.$$

Since $\alpha+\beta+\gamma \le \frac{l}{m}$, we have,

$$\begin{aligned}\mathscr{L}V(\alpha,\beta,\gamma) \le & 3m + c + a + 2k + \frac{l}{m}\left(b + \frac{bl}{m} + m + c\right)\\ & + \frac{3}{2}\left(\frac{l}{m}\right)^2\left(\sigma_1^2\left(\frac{l}{m}\right)^2 + \sigma_1^2\alpha_2^2 + \frac{2}{3}\sigma_2^2\right) + \sigma_2^2\beta_2^2 = 4c.\end{aligned}$$

Since $V(\alpha,\beta,\gamma) \ge 4$, for $(\alpha,\beta,\gamma) \in \mathbb{D}$, $\mathscr{L}V(\alpha,\beta,\gamma) \le cV(\alpha,\beta,\gamma)$. Hence $\mathscr{L}W(\alpha,\beta,\gamma,t) = e^{-c(t-s)}(-cV(\alpha,\beta,\gamma) + \mathscr{L}V(\alpha,\beta,\gamma)) \le 0$. It is easy to see that, $\inf_{(\alpha,\beta,\gamma)\in\mathbb{D}\setminus\mathbb{D}_n} V(\alpha,\beta,\gamma) > n+1$, for $n \in \mathbb{N}$. Now we define $\tau_n := \min\{t, \tau(\mathbb{D}_n)\}$ and apply Dynkin's formula to obtain

$$\mathbb{E}\left[W(\alpha(\tau_n),\beta(\tau_n),\gamma(\tau_n),\tau_n)\right] \le \mathbb{E}\left[V(\alpha_0,\beta_0,\gamma_0)\right]. \tag{5}$$

We take the expected value of $e^{c(t-\tau_n)}V(\alpha(\tau_n),\beta(\tau_n),\gamma(\tau_n))$, and using the above inequality (5), we have

$$\mathbb{E}\left[e^{c(t-\tau_n)}V(\alpha(\tau_n),\beta(\tau_n),\gamma(\tau_n))\right] \le e^{c(t-s)}\mathbb{E}\left[V(\alpha_0,\beta_0,\gamma_0)\right], \tag{6}$$

Now, to show that $\mathbb{P}(\tau(\mathbb{D}_n) < t) = 0$, we use (6) and obtain

$$\begin{aligned} 0 \le & \mathbb{P}(\tau(\mathbb{D}) < t) \le \mathbb{P}(\tau(\mathbb{D}_n) < t), \qquad \text{since} \quad \mathbb{D}_n \subseteq \mathbb{D}\\ = & \mathbb{P}(\tau_n < t) = \mathbb{E}(\mathbf{1}_{\tau_n < t}), \qquad \text{where } \mathbf{1} \text{ is the indicator function}\\ \le & e^{c(t-s)}\frac{\mathbb{E}\left(V(\alpha_0,\beta_0,\gamma_0)\right)}{\inf_{(\alpha,\beta,\gamma)\in\mathbb{D}\setminus\mathbb{D}_n} V(\alpha,\beta,\gamma)} \le e^{c(t-s)}\frac{\mathbb{E}\left(V(\alpha_0,\beta_0,\gamma_0)\right)}{n+1}.\end{aligned}$$

Thus $\mathbb{P}(\tau(\mathbb{D}) < t) = \mathbb{P}(\tau(\mathbb{D}_n) < t) = 0$, for $(\alpha_0,\beta_0,\gamma_0) \in \mathbb{D}$ and $t \ge t_0$, that is, $\mathbb{P}(\tau(\mathbb{D}) = \infty) = 1$. This completes the proof.

3 Stochastic Stability of Infection-Free and Endemic Equilibrium States

In this section, we analyze stochastic stability of equilibrium solutions of (2). Generally global stability of infection-free equilibrium biologically means a disease always dies out and global stability of endemic equilibrium means disease continues to exist.

So, we now present stochastic stability of the model (2). The stochastic SIRS model (2) may have at most two equilibrium solutions; they are an infection-free equilibrium solution $E_1 = (\alpha_1, \beta_1, \gamma_1)$ and an endemic equilibrium solution $E_2 = (\alpha_2, \beta_2, \gamma_2)$.

Theorem 2 *The infection-free equilibrium solution* $E_1 = (\alpha_1, \beta_1, \gamma_1)$ *of* (2) *is globally stochastically asymptotically stable on* $\mathbb{D}$ *if* $am \geq lb$.

Proof We consider the function

$$V_1(\alpha, \beta, \gamma) = \frac{1}{2}\left(\alpha - \frac{l}{m} + \beta + \gamma\right)^2 + \frac{l}{m}\beta + \frac{l}{m}\gamma \tag{7}$$

Applying the infinitesimal generator $\mathscr{L}$, we have

$$\begin{aligned}\mathscr{L}V_1(\alpha, \beta, \gamma) = &(l - b\alpha\beta - m\alpha + c\gamma)\left(\alpha - \frac{l}{m} + \beta + \gamma\right) \\ &+ (b\alpha\beta - (k + m + a)\beta)\left(\alpha - \frac{l}{m} + \beta + \gamma + \frac{l}{m}\right) \\ &+ (k\beta - (m + c)\gamma)\left(\alpha - \frac{l}{m} + \beta + \gamma + \frac{l}{m}\right).\end{aligned}$$

$$\begin{aligned}\mathscr{L}V_1(\alpha, \beta, \gamma) = &- m\left(\alpha - \frac{l}{m} + \beta + \gamma\right)^2 - a\beta^2 - \left(a - \frac{l}{m}b\right)\alpha\beta \\ &- a\beta\gamma - l\beta - \frac{l}{m}(m + c)\gamma.\end{aligned}$$

If $am \geq lb$, then $\mathscr{L}V_1(\alpha, \beta, \gamma)$ is negative definite on $\mathbb{D}$. By Theorem 11.2.8 in [1], the stochastic SIRS model (2) is globally stochastically asymptotically stable on $\mathbb{D}$.

Theorem 3 *The endemic equilibrium solution* $E_2 = (\alpha_2, \beta_2, \gamma_2)$ *of the system* (2) *is stochastically asymptotically stable on* $\mathbb{D}$ *if* $\mathscr{R}_0 > 1$ *and satisfies the condition*

$$2(1 + 2a)m^3 \geq \sigma_1^2 l^2 \left(2al^2 + b\beta_2 m^2\right) \text{ and } 2(a + m)m^2 \geq \sigma_2^2 \left(bm^2\beta_2 + 2cl^2\right), \tag{8}$$

for $d_1 = \dfrac{m}{c}$, $d_2 = \dfrac{a + 2m}{b} + 2d_1\alpha_2$ *and* $d_3 = \dfrac{a + 2m}{2k}$.

Proof We consider a function

$$
\begin{aligned}
V_2(\alpha, \beta, \gamma) =& \frac{1}{2}(\alpha - \alpha_2 + \beta - \beta_2 + \gamma - \gamma_2)^2 + d_1(\alpha - \alpha_2)^2 \\
&+ d_2\left(\beta - \beta_2 - \beta_2 \ln\left(\frac{\beta}{\beta_2}\right)\right) + d_3(\gamma - \gamma_2)^2.
\end{aligned} \tag{9}
$$

Applying the infinitesimal generator $\mathscr{L}$ on V_2, we get

$$
\begin{aligned}
\mathscr{L}V_2(\alpha, \beta, \gamma) =& (\alpha - \alpha_2 + \beta - \beta_2 + \gamma - \gamma_2)\,[l - m\alpha - (m+a)\beta - m\gamma] \\
&+ 2d_1(\alpha - \alpha_2)(l - b\alpha\beta - m\alpha + c\gamma) \\
&+ d_2\,(\beta - \beta_2)\,(b\alpha - (k+m+a)) + 2d_3(\gamma - \gamma_2)\,(k\beta - (m+c)\gamma) \\
&+ \frac{1}{2}\sigma_1^2\alpha^2(\alpha - \alpha_2)^2\left(2d_1\beta^2 + d_2\beta_2\right) \\
&+ \frac{1}{2}\sigma_2^2(\beta - \beta_2)^2\left(d_2\beta_2 + 2d_3\beta^2\right).
\end{aligned} \tag{10}
$$

The following Eqs. (i) – (iv) help to simplify $\mathscr{L}V_2(\alpha, \beta, \gamma)$

(i) $l - m\alpha - (m+a)\beta - m\gamma = -m(\alpha - \alpha_2 + \beta - \beta_2 + \gamma - \gamma_2) - a(\beta - \beta_2)$
(ii) $l - b\alpha\beta - m\alpha + c\gamma = -b(\alpha - \alpha_2)\beta - b\alpha_2(\beta - \beta_2) - m(\alpha - \alpha_2) + c(\gamma - \gamma_2)$
(iii) $b\alpha - (k + m + a) = b(\alpha - \alpha_2)$
(iv) $k\beta - (m+c)\gamma = k(\beta - \beta_2) - (m+c)(\gamma - \gamma_2)$.

Using the above identities into (10), we obtain

$$
\begin{aligned}
\mathscr{L}V_2(\alpha, \beta, \gamma) =& -(m + 2d_1b\beta + 2d_1m)(\alpha - \alpha_2)^2 - (a+m)(\beta - \beta_2)^2 \\
&- (m + 2d_3(m+c))(\gamma - \gamma_2)^2 \\
&- (2m + a + 2d_1b\alpha_2 - d_2b)\,(\alpha - \alpha_2)(\beta - \beta_2) \\
&- (a + 2m - 2d_3k)(\beta - \beta_2)(\gamma - \gamma_2) - (2m - 2d_1c)(\alpha - \alpha_2)(\gamma - \gamma_2) \\
&+ \frac{1}{2}\sigma_1^2\alpha^2(\alpha - \alpha_2)^2\left(2d_1\beta^2 + d_2\beta_2\right) \\
&+ \frac{1}{2}\sigma_2^2(\beta - \beta_2)^2\left(d_2\beta_2 + 2d_3\beta^2\right).
\end{aligned}
$$

Choosing $d_1 = \frac{m}{c}$, $d_2 = \frac{a+2m}{b} + 2d_1\alpha_2$ and $d_3 = \frac{a+2m}{2k}$, we have

$$\mathscr{L}V_2(\alpha, \beta, \gamma) = -\left(m + 2d_1 b\beta + 2d_1 m - \frac{1}{2}\sigma_1^2\alpha^2\left(2d_1\beta^2 + d_2\beta_2\right)\right)(\alpha - \alpha_2)^2$$
$$- \left(a + m - \frac{1}{2}\sigma_2^2\left(d_2\beta_2 + 2d_3\beta^2\right)\right)(\beta - \beta_2)^2$$
$$- (m + 2d_3(m + c))(\gamma - \gamma_2)^2$$

Hence $\mathscr{L}V_2(\alpha, \beta, \gamma) = 0$ only at $(\alpha_2, \beta_2, \gamma_2)$. By the assumption (8), we have $\mathscr{L}V_2(\alpha, \beta, \gamma)$ is negative definite on $\mathbb{D}$.

By Theorem 11.2.8 in [1], the stochastic SIRS model (2) is stochastically asymptotically stable on $\mathbb{D}$.

4 Example

In this section we visualize our results numerically, we take parameters from [4], and some of the parameters are assumed. Figures 1A–C and 2A–C are plotted by expected values of Susceptible, Infected and Recovered versus time. They demonstrate that Susceptible, Infected and Recovered populations, in average, approach to the equilibrium solution. Figures 1D–F and 2D–F are plotted by variances of susceptible, infected, and recovered versus time. As it is seen, variance rapidly tends to zero. Hence the equilibrium solutions are approached.

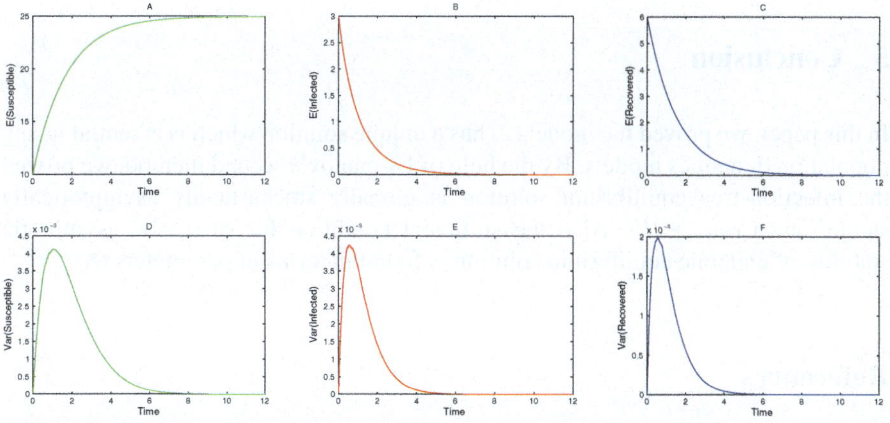

Fig. 1 The infection free equilibrium $(\alpha_1, \beta_1, \gamma_1) = (25, 0, 0)$ is globally asymptotically stochastically stable for the parameters: $l = 15, m = 0.6, a = 0.55, b = 0.02, k = 0.1$, $\sigma_1 = (0.6/15)^3$, $\sigma_2 = (0.6/15)^2$ and $c = 0.1$ ($\mathscr{R}_0 = 0.4000 < 1$)

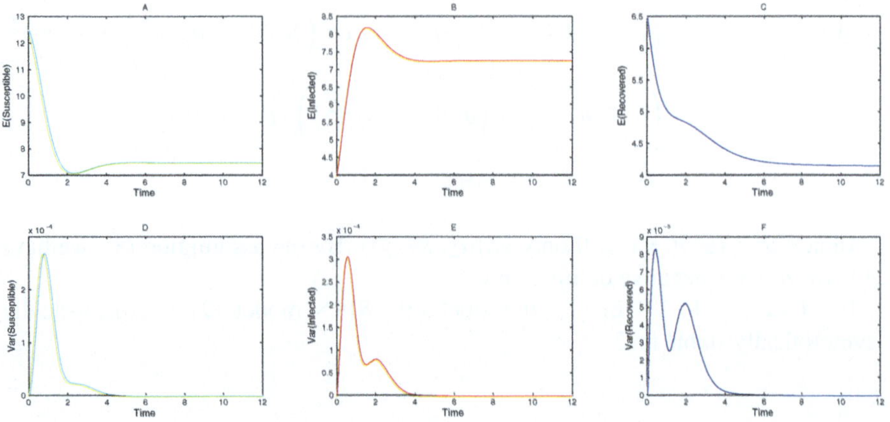

Fig. 2 The endemic equilibrium $E_2 = (\alpha_2, \beta_2, \gamma_2) = (7.5000, 7.2772, 4.1584)$ is asymptotically stochastically stable for the following parameters: $l = 15, m = 0.6, a = 0.5, b = 0.2, k = 0.4$ $\sigma_1 = (0.6/15)^3$, $\sigma_2 = (0.6/15)^2$, and $c = 0.1$ ($\mathscr{R}_0 = 3.3333 > 1$)

Figure 1 demonstrates that Theorem 2 is true, which says that, if $am - lb = 0.0300 \geq 0$, then the infection-free equilibrium solution $E_1 = (25, 0, 0)$ is globally stochastically asymptotically stable on $\mathbb{D}$.

Figure 2 shows that Theorem 3 is true, that is, the endemic equilibrium solution $E_2 = (7.5000, 7.2772, 4.1584)$ to the system (2) is stochastic asymptotic stability on $\mathbb{D}$.

5 Conclusion

In this paper, we proved the model (2) has a unique solution which is essential in any population dynamics models. By the help of Lyapunov's second method, we proved the infection-free equilibrium solution is globally stochastically asymptotically stable, when $am \geq lb$. Also the sufficient condition for stochastic asymptotic stability of endemic equilibrium solution is found in terms of parameters (8).

References

1. L. Arnold, Stochastic Differential Equations: Theory and Applications. Wiley, New York, 1974.
2. Y. Enatsu, Y. Nakata, Y. Muroya, Global stability of SIRS epidemic models with a class of nonlinear incidence rates and distributed delays, Acta Math. Sci. **32B** (3) (2012) 851–865.
3. C. Ji, D. Jiang, and N. Shi, Multigroup SIR epidemic model with stochastic perturbation. Phys A **390** (2011) 1747–1762.

4. M. E. Fatini, A. Lahrouz, R. Petterssonb, A. Settati, Regragui TakiStochastic stability and instability of an epidemic model with relapse. Appl. Math. Comput. **316** (2018) 326–341.
5. W. O. Kermack, A. G. McKendrick, Contributions to the mathematical theory of epidemics (Part I), Proc. Soc. London Ser. A **115** (1927) 700–721.
6. A. Lahrouz, L. Omari, D. Kiouach, A. Belmaâti, Deterministic and stochastic stability of a mathematical model of smoking, Statist. Probab. Lett., **81** (2011) 1276–1284.
7. De León, C. Vargas, Constructions of Lyapunov functions for classics SIS, SIR and SIRS epidemic model with variable population size, Foro-Red-Mat: Revista electrónica de contenido matemático **26**(5) (2009) 1–12.
8. M. Liu, K. Wang and Q. Wu, Survival analysis of stochastic competitive models in a polluted environment and stochastic competitive exclusion principle, Bull. Math. Biol., **73** (2011) 1969–2012.
9. Y. Muroya, Y. Enatsu, Y. Nakata, Global stability of a delayed SIRS epidemic model with a non-monotonic incidence rate, J. Math. Anal. Appl. **377** (2011) 1–14.
10. M. Pitchaimani, R. Rajaji, Stochastic Asymptotic Stability of Nowak-May Model with Variable Diffusion Rates, Methodol. Comput. Appl. Probab. **18** (2016) 901–910.
11. R. Rajaji, M. Pitchaimani, Analysis of Stochastic Viral Infection Model with Immune Impairment, Int. J. Appl. Comput. Math. **3** (2017) 3561–3574.
12. H. Schurz and K. Tosun, Stochastic asymptotic stability of SIR model with variable diffusion rates, J. Dyn. Diff. Equat., **27** (2014) 69–82.
13. Q. Yang, X. Mao, Stochastic dynamics of SIRS epidemic models with random perturbation. Math. Biosci. Eng. **11**(4) (2014) 1003–1025.
14. Y. Zhao, S. Yuan and J. Ma, Survival and stationary distribution analysis of a stochastic competitive model of three species in a polluted environment, Bull. Math. Biol., **77**(7) (2015) 1285–1326.

Steady-State Analysis of Unreliable Preemptive Priority Retrial Queue with Feedback and Two-Phase Service Under Bernoulli Vacation

S. Yuvarani and M. C. Saravanarajan

Abstract This article discusses the concepts of preemptive priority retrial queue with two-phase service, feedback, and Bernoulli vacation for an unreliable server, which consists of breakdown period. The queue involves two types of customers, known as priority and ordinary customers. The server provides first essential service and second essential service to the arriving customers or customers from the orbit. The server takes Bernoulli vacation, when an orbit becomes empty. The supplementary variable technique is used to obtain the steady-state probability generating functions for the system/orbit and some important system performance measures.

1 Introduction

Queueing theory has a dominant role in network communication, production areas, and operating systems. In recent times, queueing theory with retrial queues is recognized as an essential research area due to wide applications in many areas. In the earlier years, retrial queues with two classes of customers have been widely discussed by several researchers, Artalejo et al. [1] and Liu et al. [2]. Preemptive priority queue with general times was analyzed by Gao [3]. Yuvarani and Saravanarajan [5] developed the concepts of preemptive priority queue with bulk arrival, orbital search, and Bernoulli vacation. Rajadurai et al. [4] discussed the concepts of queue with single server and service in two phases, and the server may undergo to the breakdown while providing the service to the existing customers. To the author's best knowledge, there are several analyses related to retrial queue, but there is no work related to the concepts of preemptive priority retrial queue with service given in two phases, feedback, and Bernoulli vacation.

S. Yuvarani (✉) · M. C. Saravanarajan
Department of Mathematics, VIT, Vellore, India
e-mail: yuvasaran10@gmail.com; yuvarani.s2014@vit.ac.in; mcsaravanarajan@vit.ac.in

V. Madhu et al. (eds.), *Advances in Algebra and Analysis*, Trends in Mathematics,
https://doi.org/10.1007/978-3-030-01120-8_47

In this article, we analyzed the model with the concepts of a single-server preemptive priority retrial queue with feedback and two phases of service under Bernoulli vacation. The remaining part of this article is constructed in the following manner. The mathematical model is described in Sect. 2. In Sect. 3, the stability condition of the given model is analyzed. In Sect. 4, the steady-state joint distribution of the server in different states, the number of customers in the orbit and also in the system, and system performance measures were discussed. In Sect. 5, conclusion of the article is summarized.

2 Model Description

We considered two classes of customers like priority and ordinary customers arriving into the system. Assumption is made that both the customers arrive to the system with independent Poisson processes and with respective rates λ and δ. Retrial times considered to have an arbitrary distribution $R(t)$ with corresponding LST $R^*(\theta)$. After completion of the service, i.e., when the server is empty, the server may go for vacation with probability p with random length V. The server waits for providing service to the next customer with probability (1-p). The distribution function of vacation time of the server is V(t) and its LST is given as $V^*(\theta)$.

It is assumed that the priority customer service time follows a random variable $S_{(pi)}$ with distribution function $S_{(pi)}(t)$ and its LST is denoted as $S^*_{pi}(\theta)$. The first and second moments of priority customer are denoted as β^1_{pi} and β^2_{pi}, respectively. Also, the service time of ordinary customers follows general random variable S_{bi} with distribution function $S_{bi}(t)$ with its LST $S^*_{bi}(\theta)$. Moreover, the first and second moments of ordinary customers are denoted as $\beta^1_{(bi)}$ and β^2_{bi}, respectively. The repair time of the system is denoted by G with distribution function G(t) and its LST is denoted as $G^*_i(\theta)$. After service completion of each customer in all phases, the customer who is not satisfied about the service may rejoin into the orbit as feedback customer with probability r ($0 \leq r \leq 1$) and can receive another regular service or the customer may leave the system with probability (1-r).

3 Stability Condition

Let t_n; n = 1,2,... be the sequence of epochs at which regular service completion times for priority customer, ordinary customers, and ordinary feedback customers or completion of vacation period ends occurs. Then the bivariate Markov process (C(t),N(t),t$\leq$ 0) represents the different states of the system where C(t) denotes the various server states (0,1,2,3,4,5,6,7,8,9,10,11,12,13) depending if the server is free, busy with priority customers, preemptive priority customers, and ordinary customers, on vacation, and on repair. The number of customers in the orbit is denoted by N(t).

$C(t) =$

$$\begin{cases} 0, \text{ if the server is idle at time t,} \\ 1, \text{ if the server is busy with a priority customer in FES service at time t} \\ 2, \text{ if the server is busy with a preemptive priority customer in FES service} \\ \text{at time t} \\ 3, \text{ if the server is busy with an ordinary customer in FES service at time t} \\ 4, \text{ if the server is busy with a priority customer in SES service a time t} \\ 5, \text{ if the server is busy with an preemptive priority customer in SES service} \\ \text{a time t} \\ 6, \text{ if the server is busy with an ordinary customer in SES service a time t} \\ 7, \text{ if the server is on vacation at time t,} \\ 8, \text{ server is on repair at time t, when the priority customer is in FES service} \\ 9, \text{ server is on repair at time t, when the preemptive priority customer is} \\ \text{in FES service} \\ 10, \text{ server is on repair at time t, when the ordinary customer is in FES service} \\ 11, \text{ server is on repair at time t, when the priority customer is in SES service} \\ 12, \text{ server is on repair at time t, when the preemptive priority customer is} \\ \text{in SES service} \\ 13, \text{ server is on repair at time t, when the ordinary customer is in SES service} \end{cases}$$

If $\rho < R^*(\lambda + \delta)$, the embedded Markov chain is said to be ergodic. Also $\rho = \left(R^*(\lambda + \delta) + \bar{R}^*(\lambda + \delta)\right)\lambda + (\lambda + \delta) + \delta\bar{R}^*(\lambda + \delta)$

4 Steady-State Analysis of the System

We assume that R(0)=0, R(∞)=1, $S_p(0) = 0$, $S_p(\infty) = 1$, $S_b(0) = 0$, $S_b(\infty) = 1$, V(0) = 0, V(∞) = 1, G(0) = 0, and G(∞) = 1 are continuous at x = 0 in steady state. The conditional completion rates (hazard rate) for retrial, service of a priority customer, ordinary customer, vacation, and repair are given by a(x), μ_p(x), μ_b(x), γ(x), and ξ(x), respectively. For the above-defined process, we define the limiting probabilities as

$$P_0(t) = P\{X(t) = 0, N(t) = 0\}$$

$$P_n(x,t)dx = P\left\{C(t) = 0, N(t) = n, x < R^0(t) \le x + dx\right\}, \text{ for } t \ge 0, x \ge 0 \text{ and } n \ge 1.$$

$$\Pi^1_{1,n}(x,t)dx = P\left\{C(t) = 1, N(t) = n, x < S^0_{pi}(t) \le x + dx\right\}, \text{ for } t \ge 0, x \ge 0, n \ge 0.$$

$$\Pi^1_{2,n}(x,y,t)dx = P\left\{C(t) = 2, N(t) = n, x < S^0_{pi}(t) \le x + dx, y < S^0_{bi}(t) \le y + dy\right\}$$
$$\text{for } t, x, n \ge 0.$$

$$\Pi^1_{3,n}(x,t)dx = P\left\{C(t) = 3, N(t) = n, x < S^0_{bi}(t) \le x + dx\right\}, \text{ for } t \ge 0, x \ge 0, n \ge 0.$$

$$\Pi_{1,n}^{2}(x,t)dx = P\left\{C(t) = 4, N(t) = n, x < S_{pi}^{0}(t) \le x + dx\right\}, \text{for } t \ge 0, x \ge 0, n \ge 0.$$

$$\Pi_{2,n}^{2}(x,y,t)dx = P\left\{C(t) = 5, N(t) = n, x < S_{pi}^{0}(t) \le x + dx, y < S_{bi}^{0}(t) \le y + dy\right\}$$
$$\text{for } t, x, n \ge 0.$$

$$\Pi_{3,n}^{2}(x,t)dx = P\left\{C(t) = 6, N(t) = n, x < S_{bi}^{0}(t) \le x + dx\right\}, \text{for } t \ge 0, x \ge 0, n \ge 0.$$

$$\Omega_{n}(x,t)dx = P\left\{C(t) = 7, N(t) = n, x < v^{0}(t) \le x + dx\right\}, \text{for } t, x, n \ge 0.$$

$$R_{1,n}^{1}(u,x,t)dx = P\left\{C(t) = 8, N(t) = n, x < S_{pi}^{0}(t) \le x + dx, u < g_{1}^{0}(t) \le u + du\right\},$$
$$\text{for } t, u, x, n \ge 0.$$

$$R_{2,n}^{1}(u,x,y,t)dx = P\left\{C(t) = 9, N(t) = n, x < S_{pi}^{0}(t) \le x + dx, y < S_{bi}^{0}(t)\right.$$
$$\left.\le y + dy, u < g_{1}^{0}(t) \le u + du\right\}, \text{for } t, u, x, n \ge 0.$$

$$R_{3,n}^{1}(u,x,t)dx = P\left\{C(t) = 10, N(t) = n, x < S_{bi}^{0}(t) \le x + dx, u < g_{2}^{0}(t) \le u + du\right\},$$
$$\text{for } t, u, x, n \ge 0.$$

$$R_{1,n}^{2}(u,x,t)dx = P\left\{C(t) = 11, N(t) = n, x < S_{pi}^{0}(t) \le x + dx, u < g_{2}^{0}(t) \le u + du\right\},$$
$$\text{for } t, u, x, n \ge 0.$$

$$R_{2,n}^{2}(u,x,y,t)dx = P\left\{C(t) = 12, N(t) = n, x < S_{pi}^{0}(t) \le x + dx, y < S_{bi}^{0}(t)\right.$$
$$\left.\le y + dy, u < g_{2}^{0}(t) \le u + du\right\}, \text{for } t, u, x, n \ge 0.$$

$$R_{3,n}^{2}(u,x,t)dx = P\left\{C(t) = 13, N(t) = n, x < S_{bi}^{0}(t) \le x + dx, u < g_{2}^{0}(t) \le u + du\right\},$$
$$\text{for} t, u, x, n \ge 0.$$

4.1 The Steady-State Equations of the Model

The supplementary variable technique is used to formulate the system of governing equations for the given model which are given below.

$$(\lambda + \delta)\, P_0 = (1 - r)q \int_0^\infty \Pi_{1,0}^{2}(x)\mu_p^2(x)dx + (1 - r)$$

$$q \int_0^\infty \Pi_{3,0}^{2}(x)\mu_b^2(x)dx + \int_0^\infty \Omega_0(x)\gamma(x)dx \tag{1}$$

$$\frac{dP_n(x)}{dx} + (a(x) + \lambda + \delta)\, P_n(x) = 0, n \ge 1 \tag{2}$$

$$\frac{d\Pi_{1,n}^{i}(x)}{dx}+\left(\alpha+\lambda+\mu_{p}^{i}(x)\right)\Pi_{1,n}^{i}(x)$$
$$=\lambda\Pi_{1,n-1}^{i}(x)+\int_{0}^{\infty}R_{1,n}^{i}(x,u)\xi_{i}(u)du, n\geq 0, i=1,2 \tag{3}$$

$$\frac{\partial\Pi_{2,n}^{i}(x,y)}{\partial x}+\left(\alpha+\lambda+\mu_{p}^{i}(x)\right)\Pi_{2,n}^{i}(y,x)$$
$$=\lambda\Pi_{2,n-1}^{i}(x)+\int_{0}^{\infty}R_{2,n}^{i}(x,u)\xi_{i}(u)du, n\geq 0, i=1,2 \tag{4}$$

$$\frac{d\Pi_{3,n}^{i}(x)}{dx}+\left(\lambda+\delta+\alpha+\mu_{b}^{i}(x)\right)\Pi_{3,n}^{i}(x)$$
$$=\lambda\sum_{k=1}^{n}\chi_{k}\Pi_{3,n-k}^{i}(x)+\int_{0}^{\infty}R_{3,n}^{i}(x,u)\xi_{i}(u)du+\int_{0}^{\infty}\Pi_{2,n}^{i}(y,x)\mu_{p}^{i}(y)dy, i=1,2 \tag{5}$$

$$\frac{d\Omega_{n}(x)}{dx}+(\gamma(x)+\lambda)\,\Omega_{n}(x)=\lambda\Omega_{n-1}(x), n\geq 1 \tag{6}$$

$$\frac{dR_{1,n}^{i}(x,u)}{dx}+(\lambda+\xi_{i}(u))\,R_{1,n}^{i}(x,u)=\lambda R_{1,n-1}^{i}(x,u), n\geq 1, i=1,2 \tag{7}$$

$$\frac{dR_{2,n}^{i}(x,u,y)}{dx}+(\lambda+\xi_{i}(u))\,R_{2,n}^{i}(x,u,y)=\lambda R_{2,n-1}^{i}(x,y,u), n\geq 1, i=1,2 \tag{8}$$

$$\frac{dR_{3,n}^{i}(u,x)}{dx}+(\lambda+\xi_{i}(u))\,R_{3,n}^{i}(u,x)=\lambda R_{3,n-1}^{i}(x,u), n\geq 1, i=1,2 \tag{9}$$

The steady-state boundary conditions at x = 0 and y = 0 are

$$P_{n}(0)=(1-r)q\int_{0}^{\infty}\mu_{p}^{2}(x)\Pi_{1,0}^{2}(x)dx+q(1-r)\int_{0}^{\infty}\Pi_{3,0}^{2}(x)\mu_{b}^{2}(x)dx$$
$$+\int_{0}^{\infty}\gamma(x)\Omega_{0}(x)dx, n\geq 1 \tag{10}$$

$$\Pi^1_{1,n}(0) = \delta \int_0^\infty P_n(x)dx + \delta P_0, n \geq 0 \tag{11}$$

$$\Pi^2_{1,n}(0) = \int_0^\infty \Pi^1_{1,n}(x)\mu^1_p(x)dx \tag{12}$$

$$\Pi^1_{2,n}(0, x) = \delta \Pi_{3,n}(x), n \geq 0 \tag{13}$$

$$\Pi^2_{2,n}(0, x) = \int_0^\infty \Pi^1_{1,n}(x, 0)\mu^1_p(x)dx \tag{14}$$

$$\Pi^1_{3,n}(0) = \lambda P_0 + \int_0^\infty a(x)P_{n+1}(x)dx + \int_0^\infty \lambda P_n(x)dx \tag{15}$$

$$\Pi^2_{3,n}(0) = \int_0^\infty \Pi^1_{3,n}(x)\mu^1_b(x)dx \tag{16}$$

$$\Omega_n(0) = p(1-r) \int_0^\infty \Pi^2_{1,n}(x)\mu^2_p(x)dx + p(1-r) \int_0^\infty \Pi^2_{3,n}(x)\mu^2_b(x)dx$$

$$+pr \int_0^\infty \Pi^2_{1,n-1}(x)\mu^2_p(x)dx + pr \int_0^\infty \Pi^2_{3,n-1}(x)\mu^2_b(x)dx, n \geq 0 \tag{17}$$

$$R^i_{1,n}(0, x) = \alpha_i \Pi^i_{1,n}(x), n \geq 0, i = 1, 2 \tag{18}$$

$$R^i_{2,n}(0, x, y) = \alpha_i \Pi^i_{2,n}(x, y), n \geq 0, i = 1, 2 \tag{19}$$

$$R^i_{3,n}(0, x) = \alpha_i \Pi^i_{3,n}(x), n \geq 0i = 1, 2. \tag{20}$$

The normalizing condition is

$$P_0 + \sum_{n=1}^\infty \int_0^\infty P_n(x)dx + \sum_{n=0}^\infty (\int_0^\infty \Pi^1_{1,n}(x)dx + \int_0^\infty \int_0^\infty \Pi^1_{2,n}(x, y)dxdy + \int_0^\infty \Pi^1_{3,n}(x)dx$$

$$+\int_0^\infty \Pi_{1,n}^2(x)dx+\int_0^\infty\int_0^\infty \Pi_{2,n}^2(x,y)dxdy+\int_0^\infty \Pi_{3,n}^2(x)dx+\int_0^\infty \Omega_n(x)dx$$

$$+\sum_{n=0}^\infty(\int_0^\infty\int_0^\infty R_{1,n}^1(x,u)dxdu+\int_0^\infty\int_0^\infty R_{2,n}^1(x,y,u)dxdudy+\int_0^\infty\int_0^\infty R_{3,n}^1(x,u)dxdu$$

$$+\int_0^\infty\int_0^\infty R_{1,n}^2(x,u)dxdu+\int_0^\infty\int_0^\infty R_{2,n}^2(x,y,u)dxdudy+\int_0^\infty\int_0^\infty R_{3,n}^2(x,u)dxdu)=1$$

4.2 The Steady-State Solution

The steady-state solution of the given model is obtained by using the probability generating function technique. To solve the above equations, the PGFs are defined for $|z|\le 1$ as follows:

$$P(x,z)=\sum_{n=1}^\infty P_n(x)z^n;\ P(0,z)=\sum_{n=1}^\infty P_n(0)z^n;\ \Pi_1^i(x,z)=\sum_{n=0}^\infty \Pi_{1,n}^i(x)z^n;$$

$$\Pi_1^i(0,z)=\sum_{n=0}^\infty \Pi_{1,n}^i(0)z^n;\ \Pi_2^i(x,y,z)=\sum_{n=0}^\infty \Pi_{2,n}^i(x,y)z^n;\ \Pi_2^i(x,0,z)=\sum_{n=0}^\infty \Pi_{2,n}(x,0)z^n;$$

$$\Pi_3^i(x,z)=\sum_{n=0}^\infty \Pi_{3,n}^i(x)z^n;\ \Pi_3^i(0,z)=\sum_{n=0}^\infty \Pi_{3,n}^i(0)z^n;\ \Omega(x,z)=\sum_{n=0}^\infty \Omega_n(x)z^n;$$

$$\Omega(0,z)=\sum_{n=0}^\infty \Omega_n(0)z^n;\ R_i^1(x,z)=\sum_{n=0}^\infty R_{1,n}^i(x)z^n;\ R_1^i(0,z)=\sum_{n=0}^\infty R_{1,n}^i(0)z^n;$$

$$and\ R_3^i(x,z)=\sum_{n=0}^\infty R_{3,n}^i(x)z^n;\ R_3^i(0,z)=\sum_{n=0}^\infty R_{3,n}^i(0)z^{n}\text{"}\ for\ all\ i=1,2;$$

Equations (2)–(12) are multiplied by z^n and summed over n (n = 0,1,2...) and solved for the partial differential equations, to which we get the following:

$$P(x,z)=[1-R(x)]P(0,z)e^{-(\delta+\lambda)x};\qquad \Pi_1^i(x,z)=\Pi_1^i(0,z)[1-S_p^i(x)]e^{-A_p^i(z)x};$$

$$\Pi_2^i(x,y,z)=\Pi_2^i(0,y,z)[1-S_p^i(x)]e^{-A_p^i(z)x};\quad \Pi_3^i(x,z)=\Pi_3^i(0,z)[1-S_b^i(x)]e^{-A_b^i(z)x};$$

$$\Omega(x,z)=[1-V(x)]\Omega(0,z)e^{-b(z)x};\qquad R_1^i(x,z)=R_1^i(0,z)[1-G(x)]e^{-b(z)x};$$

$$R_1^i(x,y,z)=R_1^i(0,z)[1-G(x)]e^{-b(z)x};\qquad R_3^i(x,z)=R_3^i(0,z)[1-G(x)]e^{-b(z)x};$$

where $b_i(z) = (\lambda(1-z))\,, A_b^i(z) = \left(A_p^i(z) + \alpha\left(1 - S^{i*}_{p}\left(A_p^i(z)\right)\right)\right)$;

and $A_p^i(z) = \left(\lambda(1 - X(z)) + \alpha\left(1 - D^*\left(b_i(z)\right) G^*\left(b_i(z)\right)\right)\right)$;

4.3 System Performance Measures

1. The probability that the server is idle during the retrial is given by

$$P = P(1) = P_0 q \bar{R}^*(\lambda+\delta)\left\{\frac{\lambda(1+\beta_{bi}^1)+\delta\beta_{pi}^1}{\bar{R}^*(\lambda+\delta)-\rho}\right\}$$

2. The probability that the server is busy serving a priority customer without preempting an ordinary customer in first-phase service is given by

$$\Pi_1^1 = \Pi_1(1) = \delta P_0 q \beta_{p1}^1 \bar{R}^*(\lambda+\delta)\left\{\frac{(q(1+r)+pv_1^1+\left(\lambda(1+\beta_{b1}^1)+\delta(1+\beta_{p1}^1)\right)}{\bar{R}^*(\lambda+\delta)-\rho}\right\}$$

3. The probability that the server is busy serving a priority customer with preempting an ordinary customer in first-phase service is given by

$$\Pi_2^1 = \Pi_2(1) = \delta P_0 q \bar{q} \beta_{p1}^1 \beta_{b1}^1\left\{\frac{(\lambda-(\lambda+\delta))(pv_1^1+qr)+\delta\left(R^*(\lambda+\delta)\beta_p^1\right)}{\bar{R}^*(\lambda+\delta)-\rho}\right\}$$

4. The probability that the server is busy serving an ordinary customer in first-phase service is given by

$$\Pi_3^1 = \Pi_3(1) = \delta P_0 q \bar{q} \beta_{b1}^1\left\{\frac{(\lambda-(\lambda+\delta)\,(pv_1^1+qr)+\delta\left(R^*(\lambda+\delta)\beta_p^1\right)}{\bar{R}^*(\lambda+\delta)-\rho}\right\}$$

5. The probability that the server is busy serving a preemptive priority customer in a second-phase service is given by

$$\Pi_1^2 = \Pi_1(1) = \delta P_0 q \beta_{p1}^1 \beta_{p2}^1 \bar{R}^*(\lambda+\delta)\left\{\frac{(q(1+r)+pv_1^1+\left(\lambda(1+\beta_{b1}^1)+\delta(1+\beta_{p1}^1)\right)}{\bar{R}^*(\lambda+\delta)-\rho}\right\}$$

6. The probability that the server is busy serving a priority customer with preempting an ordinary customer is given by

$$\Pi_2^2 = \Pi_2(1) = \delta P_0 q \beta_{p1}^1 \beta_{b1}^1 \beta_{p2}^1\left\{\frac{(\lambda-(\lambda+\delta))(pv_1^1+qr)+\delta\left(R^*(\lambda+\delta)\beta_p^1\right)}{\bar{R}^*(\lambda+\delta)-\rho}\right\}$$

7. The probability that the server is busy serving an ordinary customer is given by

$$\Pi_3^2 = \Pi_3(1) = \delta P_0 q \beta_{b2}^1 \beta_{b1}^1 \left\{ \frac{(\lambda - (\lambda+\delta)(pv_1^1 + qr) + \delta\left(R^*(\lambda+\delta)\beta_p^1\right)}{\bar{R}^*(\lambda+\delta) - \rho} \right\}$$

8. The probability that the server is busy on vacation is given by

$$\Omega = \Omega(1) = P_0 prv^1 \left\{ \frac{\begin{array}{l}\beta_{p1}^1 \beta_{p2}^1 (\lambda+\delta)(pv_1^1+qr) + \delta\left(R^*(\lambda+\delta)\beta_p^1\right) \\ +\beta_{b2}^1 \beta_{b1}^1 (\lambda - (\lambda+\delta)(pv_1^1 + qr)) + \delta\left(R^*(\lambda+\delta)\beta_p^1\right)\end{array}}{\bar{R}^*(\lambda+\delta) - \rho} \right\}$$

9. The probability that the server is waiting for repair when priority customer on the first-phase service is given by

$$R_1^1 = R_1^1(1) = \delta P_0 q g_1^1 \beta_{p1}^1 \bar{R}^*(\lambda+\delta) \left\{ \frac{(q(1+r)+pv_1^1) + \left(\lambda(1+\beta_{b1}^1) + \delta(1+\beta_{p1}^1)\right)}{\bar{R}^*(\lambda+\delta) - \rho} \right\}$$

10. The probability that the server is waiting for repair when preemptive priority customer on first phase-service is given by

$$R_2^1 = R_2^1(1) = \delta P_0 q g_1^1 \beta_{p1}^1 \beta_{b1}^1 \beta_{p2}^1 \left\{ \frac{(\lambda - (\lambda+\delta))(pv_1^1+qr) + \delta\left(R^*(\lambda+\delta)\beta_p^1\right)}{\bar{R}^*(\lambda+\delta) - \rho} \right\}$$

11. The probability that the server is waiting for repair when ordinary customer on first-phase service is given by

$$R_3^1 = R_3^1(1) = \delta P_0 q g_1^1 \beta_{b1}^1 \beta_{b2}^1 \left\{ \frac{(\lambda - (\lambda+\delta))(pv_1^1 + qr) + \delta\left(R^*(\lambda+\delta)\beta_p^1\right)}{\bar{R}^*(\lambda+\delta) - \rho} \right\}$$

12. The probability that the server is waiting for repair when priority customer on second-phase service is given by

$$R_1^2 = R_1^2(1) = \delta P_0 q g_1^2 \beta_{p1}^1 \bar{R}^*(\lambda+\delta) \left\{ \frac{(q(1+r)+pv_1^1) + \left(\lambda(1+\beta_{b1}^1) + \delta(1+\beta_{p1}^1)\right)}{\bar{R}^*(\lambda+\delta) - \rho} \right\}$$

13. The probability that the server is waiting for repair when preemptive priority customer on second-phase service is given by

$$R_2^2 = R_2^2(1) = \delta P_0 q g_1^2 \beta_{p1}^1 \beta_{b1}^1 \beta_{p2}^1 \left\{ \frac{(\lambda - (\lambda+\delta))(pv_1^1 + qr) + \delta\left(R^*(\lambda+\delta)\beta_p^1\right)}{\bar{R}^*(\lambda+\delta) - \rho} \right\}$$

14. The probability that the server is waiting for repair when ordinary customer on second-phase service is given by

$$R_3^2 = R_3^2(1) = \delta P_0 q g_1^2 \beta_{b1}^1 \beta_{b2}^1 \left\{ \frac{(\lambda - (\lambda+\delta))(pv_1^1 + qr) + \delta\left(R^*(\lambda+\delta)\beta_p^1\right)}{\bar{R}^*(\lambda+\delta) - \rho} \right\}$$

4.4 Mean System Size and Orbit Size

The probability generating function of the number of customer in the orbit (Ko(z)) and the system (Ks(z)) is obtained by using

$$K_s(z) = P_0 + P(z) + z\left(\Pi_1^1(z) + \Pi_2^1(z) + \Pi_3^1(z) + \Pi_1^2(z) + \Pi_2^2(z) + \Pi_3^2(z) + R_1^1(z) + R_2^1(z) + R_3^1(z) + R_1^2(z) + R_2^2(z) + R_3^2(z)\right) + \Omega(z)$$

and $K_o(z) = P_0 + P(z) + \Pi_1^1(z) + \Pi_2^1(z) + \Pi_3^1(z) + \Pi_1^2(z) + \Pi_2^2(z) + \Pi_3^2(z) + R_1^1(z) + R_2^1(z) + R_3^1(z) + R_1^2(z) + R_2^2(z) + R_3^2(z) + \Omega(z)$
is as follows

$$K_s(z) = P_0(pV^*(b(z)) + q)(1 - r + rz) \times \left\{ \frac{\left[\left(\lambda S_{bi}^* A_b(z) + \delta z S_{p1}^* A_p(z)\right)\left[(\bar{R}^*(\lambda+\delta)\left[qV^*(b(z)) + (1 - V^*(b(z)))\right]\right]\right] + T}{D_r(z)A_p(z)b(z)} \right\}$$

where $T = [(b(z)^2 + \alpha(b(z)^2(1 - G^*(b(z))][z(1 - S_p^* A_p(z))[z - q(\lambda S_b^* A_b(z) + \delta z S_p^* A_p(z)] + z[1 - S_b^*(A_b(z))][\lambda(1 - ([pV^*(b(z)) + q](1 - r + rz) + \lambda S_b^*(A_b(z)) + [\lambda S_b^*(A_b(z)) + \delta z S_p^*(A_p(z))]([pV^*(b(z)) + q](1 - r + rz)R^*(\lambda+\delta) + q\lambda V^*(b(z)))))]$
and $K_0(z) = P_0 \left\{ \frac{\left[q\left(\lambda S_{bi}^* A_b(z) + \delta z S_{pi}^* A_p(z)\right)\left[(\bar{R}^*(\lambda+\delta)[pV^*(b(z)) + q(1 - r + rz) + (1 - V^*(b(z)))]\right]\right] + W}{D_r(z)A_p(z)b(z)} \right\}$
where $W = [(b(z)^2 + \alpha(b(z)^2(1 - G^*(b(z))][z(1 - S_p^* A_p(z))[z - q(\lambda S_b^* A_b(z) + \delta z S_p^* A_p(z)] + [1 - S_b^*(A_b(z))][\lambda(1 - ([pV^*(b(z)) + q](1 - r + rz) + \lambda S_b^*(A_b(z)) + [\lambda S_b^*(A_b(z)) + \delta z S_p^*(A_p(z))]([pV^*(b(z)) + q](1 - r + rz)R^*(\lambda+\delta) + q\lambda V^*(b(z)))))]$
If the system is in a steady-state condition,

1. The excepted number of customers in the orbit (Lq) is obtained by

$$L_q = K_o'(1) = \lim_{z\to 1} \frac{d}{dz} K_o(z) = P_0 \left[\frac{Nr'''_q(1)Dr''_q(1) - Dr'''_q(1)Nr''_q(1)}{3\left(Dr''_q(1)\right)^2} \right]$$

2. The excepted number of customers in the system (Ls) is obtained by

$$L_s = K_s'(1) = \lim_{z\to 1} \frac{d}{dz} K_s(z) = \mathrm{P}_0 \left[\frac{Nr'''_s(1)Dr''_q(1) - Dr'''_q(1)Nr''_q(1)}{3\left(Dr''_q(1)\right)^2} \right]$$

5 Conclusion

In this analysis, we have investigated a retrial queue with feedback and two phases of service with Bernoulli vacation. The probability generating functions for the numbers of customers in the system when it is free, busy on both the phases, on vacation, and under repair are found using the supplementary variable technique. Some important system performance measures, the explicit expressions for the average queue length of orbit and system, have been obtained. The present investigation includes features simultaneously such as

- Preemptive priority queue
- Two-phase service
- Feedback
- Bernoulli vacation

Our suggested model and its results have a specific and potential application in the field of telephone consultation service and in the area of computer processing system.

References

1. Artalejo, Jr., Gomez-Corral, A.: Retrial queueing systems: a computational approach. Springer, Berlin (2008)
2. Liu,Z., Gao,S.:Discrete-time Geo/Geo/1 retrial queue with two classes of customers and feedback. Math Comput Model. **53**: 1208–1220 (2011)
3. Gao,S.: A preemptive priority retrial queue with two classes of customers and general retrial times. Oper Res Int J (2015). https://doi.org/10.1007/s12351-015-0175-z
4. Rajadurai,P., Chandrasekaran,V.M., Saravanarajan,M.C.: Steady state analysis of batch arrival feedback retrial queue with two phases of service, negative customers, Bernoulli vacation and server breakdown. Int. J. Math. Oper. Res. **7**: 519–546(2015).
5. Yuvarani,S., Saravanarajan,M.C.: Analysis of a preemptive priority Retrial Queue with Batch arrival and orbital search Bernoulli vacation. International Journal of pure and applied Mathematics. **113**: 139–147(2017).

An Unreliable Optional Stage $M^X/G/1$ Retrial Queue with Immediate Feedbacks and at most J Vacations

M. Varalakshmi, P. Rajadurai, M. C. Saravanarajan, and V. M. Chandrasekaran

Abstract A repeated attempt queue with multistages of service and almost J vacations is investigated. Balking (or reneging) is applicable for customers. If the orbit is empty at service accomplishment time, the server takes at most J vacations. Busy server may fail for a short interval of time. Using supplementary variable technique, the steady results are deduced.

1 Introduction

In the history of queue, Kalidass and Kasturi [4] contributed different ways to the theory of feedback called immediate feedback. That is, if arrival wants service one more time, it will immediately get into service once again without joining the queue after receiving the first service. Wang and Li [7], Chen et al. [1], Choudhury et al. [2], Zhang and Zhu [8], Rajadurai et al. [5], Sumitha and Chandrika [6], and Jailaxmi et al. [3] are few authors who have studied queues under vacation policy.

This model has application in computer processing system, LAN, Simple Mail Transfer Protocol, etc. Whatever remains of the sections are given in short. The model portrayed numerically is given in Sect. 2. The steady-state results are talked about in Sect. 3. Performance measures and unique cases are talked about in Sects. 4 and 5. The conclusion and the use of the model considered are conveyed in Sect. 6.

M. Varalakshmi · M. C. Saravanarajan · V. M. Chandrasekaran (✉)
Department of Mathematics, School of Advanced Sciences, VIT, Vellore, India
e-mail: varalakshmi.m1989@gmail.com; varalakshmi.m2013@vit.ac.in; mcsaravanarajan@vit.ac.in; vmcsn@vit.ac.in

P. Rajadurai
Department of Mathematics, SRC, SASTRA Deemed University, Kumbakonam, India
e-mail: psdurai17@gmail.com; rajadurai@src.sastra.edu

V. Madhu et al. (eds.), *Advances in Algebra and Analysis*, Trends in Mathematics,
https://doi.org/10.1007/978-3-030-01120-8_48

2 System Description

The detailed description of the model is given as follows:

- ***The arrival process:*** Customers arrive in batches according to a compound Poisson process with rateλ. Let X_k denote the number of customers belonging to the kth arrival batch, where X_k, $k = 1, 2, 3, \ldots$ are with a common distribution $pr[X_k = n] = \chi_n$, $n = 1, 2, 3, \ldots$ and $X(z)$ denotes the probability generating function of X.
- ***The retrial process:*** We assume that there is no waiting space, and therefore if an arriving customer finds the server free, one of the customers from the batch begins his service and the rest of them join into orbit. If an arriving batch of customers find the server being busy, on vacation, or in breakdown, the arrivals either leave the service area with probability $1 - b$ or join pool of blocked customers called an orbit with probability b. If a primary customer arrives first, the retrial customer may cancel its attempt for service and either returns to its position in the retrial queue with probability r or quits the system with probability $1 - r$. Inter-retrial times have an arbitrary distribution $R(t)$with corresponding Laplace-Stieltjes transform (LST)$R^*(\vartheta)$.
- ***service process:*** The single server provides k phases of service in succession. The first-phase service (FPS) is followed by i $(i = 1, 2, \ldots, k)$ phases of service. The service time for all the phases has a general distribution. It is denoted by the random variable S_i with distribution function $S_i(x)$ having LST $S_i^*(\vartheta)$ and first and second moments are $E(Si)$ and $E(S_i^2)$.
- ***Immediate feedback rule:*** After completion of ith phase of service, the customer may go to the $(i + 1)$th phase with probability θ_i or may join into ith phase with probability p_i as immediate feedback customer or leaves the system with probability $1 - \theta_i - p_i = q_i$, for $i = 1, 2, \ldots, k - 1$. The customer in the last kth phase may join into kth phase with probability p_k or leaves the system with probability $q_k = 1 - p_k$.
- ***Vacation process:*** Whenever the orbit is empty, the server leaves for a vacation of random length V. If no customer appears in the orbit while returning from a vacation, it leaves again for another vacation of same length. Such pattern continues until it returns from a vacation to find at least one customer found in the orbit. If the orbit is empty at the end of the Jth vacation, the server remains idle for new arrivals in the system. At a vacation completion epoch if the orbit is nonempty, the server waits for the customers in the orbit or for a new arrival. The vacation time V has distribution function $V(t)$ and Laplace-Stieltjes transform $V^*(\vartheta)$, and first and second moments are $E(V)$ and $E(V^2)$.
- ***Breakdown process:*** While the server is working with any phase of service, it may break down any time and the service channel will fail for a short interval of time. The breakdowns, i.e., server's lifetimes, are generated by exogenous Poisson processes with rates α_i for ith phase, respectively, for $(i = 1, 2, \ldots, k)$.

- ***Repair process:*** As soon as breakdown occurs, the server is sent for repair; during that time it stops providing service to the primary customers till service channel is repaired. The repair time (denoted by G_i) distributions of the server for i phases are assumed to be arbitrarily distributed with d.f. $G_i(y)$ and LST $G_i^*(\vartheta)$ for$(i = 1, 2, \ldots, k)$.
- Various stochastic processes involved in the system are assumed to be independent of each other.
- In the steady state, we assume that $R(0) = 0$, $R(\infty) = 1$, $S_i(0) = 0$, $S_i(\infty) = 1$, $V(0) = 0$, $V(\infty) = 1$ are continuous at $x = 0$ and $G_i(0) = 0$, $G_i(\infty) = 1$ are continuous at $y = 0$. The state of system at time t can be described by the bivariate Markov process $\{C(t), N(t); t \geq 0\}$ where $C(t)$ denotes the server state $(0, 1, 2, 3, 4, \ldots, J+3)$ depending if the server is idle, busy on ith phase, on repair at ith phase, and on 1st vacation,... and Jth vacation, respectively. $N(t)$ corresponds to the number of customers in the orbit at time t. If $C(t) = 0$ and $N(t) > 0$, then $R^0(t)$ represent the elapsed retrial time. If $C(t) = 1$ and $N(t) \geq 0$, then $S_i^0(t)$ corresponds to the elapsed time of the customer being served on ith phase. If $C(t) = 2$ and $N(t) \geq 0$, then $S_i^0(t)$ corresponds to the elapsed time of the customer being served on ith feedback phase. If $C(t) = 3$ and $N(t) \geq 0$, then $G_i^0(t)$ corresponds to the elapsed time of the server being repaired on ith phase. If $C(t) = 4$ and $N(t) \geq 0$, then$V_1^0(t)$ corresponds to the elapsed 1st vacation time. If $C(t) = j+3$ and $N(t) \geq 0$, then $V_j^0(t)$ corresponds to the elapsed jth vacation time.

The conditional completion rates for repeated attempts, service on ith phase, vacation, and repair time of ith phase, respectively, $(for\ i = 1, 2, \ldots, k)$

$$a(x)dx = \frac{dR(x)}{1-R(x)}, \mu_i(x)dx = \frac{dS_i(x)}{1-S_i(x)}, \gamma(x)dx = \frac{dV(x)}{1-V(x)}, \varsigma(y)dy = \frac{dG_i(y)}{1-G_i(y)}$$

$Bi^* = S_1^* S_2^* \ldots S_i^*$ and $B_0^* = 1$. The first moment M_{1i} and the second moment M_{2i} of B_i^* are given by

$$M_{1i} = \lim_{z \longrightarrow 1} dB_i^*[A_i(Z)]/dz = \sum_{i=1}^{t} \lambda E(X)E(S_i)(1 + \alpha_i E(G_i))$$

$$M_{2i} = \lim_{z \longrightarrow 1} d^2 B_i^*[A_i(Z)]/dz^2 = \sum_{j=1}^{t} [\lambda E(X(X-1))E(S_j)(1 + \alpha_j E(G_j))$$

$$+ \alpha_j (\lambda E(X))^2 E(S_j) E(G_j^2) + (\lambda E(X))^2 E(S_j^2)(1 + \alpha_j E(G_j))^2]$$

The embedded Markov chain is $\{Z_n; n \in N\}$ ergodic if and only if $\rho < 1$ for our system is stable.

3 Analysis of Steady-State Probability Distributions

For the procedure $\{N(t), t \geq 0\}$, we characterize the probabilities $P_0(t) = P\{C(t) = 0, N(t) = 0\}$ and the probability densities (for $0 \leq j \leq m-1$ and $1 \leq i \leq k$)

$$\Psi_n(x,t)dx = P\{C(t) = 0, N(t) = n, x \leq R^0(t) < x + dx\}, \text{ for } t \geq 0, x \geq 0, n \geq 1,$$

$$Q_{i,n}(x,t)dx = P\{C(t) = 1, N(t) = n, x \leq S_i^0(t) < x + dx\}, \text{ for } t \geq 0, x \geq 0, n \geq 0,$$

$$P_{i,n}(x,t)dx = P\{C(t) = 2, N(t) = n, x \leq S_i^0(t) < x + dx\}, \text{ for } t \geq 0, x \geq 0, n \geq 0,$$

$$R_{i,n}(x,t)dx = P\{C(t) = 3, N(t) = n, x \leq G_i^0(t) < x + dx\}, \text{ for } t \geq 0, x \geq 0, n \geq 0,$$

$$\Omega_{j,n}(x,t)dx = P\{C(t) = j+3, N(t) = n, x \leq V_t^0(t) < x + dx\}, \text{ for } t \geq 0, x \geq 0, n \geq 0.$$

We expect that the stability condition is satisfied in the sequel and so that we would be able to set $t \geq 0, x \geq 0$,(for $0 \leq j \leq m-1$ and $1 \leq i \leq k$).

$$P_0 = lim_{t \to \infty} P_0(t),\ \psi_n(x) = lim_{t \to \infty} \psi_n(x,t) Q_n(x) = lim_{t \to \infty} Q_{i,n}(x,t);$$

$$P_{i,n}(x) = lim_{t \to \infty} P_{i,n}(x,t),\ \Omega_{j,n}(x) = lim_{t \to \infty} \Omega_{j,n}(x,t),$$

$$R_{i,n}(x) = lim_{t \to \infty} R_{i,n}(x,t).$$

Utilizing the strategy for supplementary variable technique, we get the accompanying system of equations that represent the behavior of the system.

$$\lambda b P_0 = \int_0^\infty \Omega_{J,0}(x)\gamma(x)dx \tag{1}$$

$$\frac{d\Psi_n(x)}{dx} + (\lambda + a(x))\Psi_n(x) = 0, n \geq 1 \tag{2}$$

$$\frac{dQ_{i,n}(x)}{dx} + (\lambda + \mu_i(x) + \alpha_i)Q_{i,n}(x) = \lambda b \sum_{k=1}^{n} \chi_k Q_{i,n-k}(x) + \lambda(1-b)Q_{i,n}(x) \tag{3}$$

$$+ \int_0^\infty R_{i,n}(x,y)\xi_i(y)dy, i = 1, 2, \ldots, k, n \geq 1$$

$$\frac{dP_{i,n}(x)}{dx} + (\lambda + \mu_i(x) + \alpha_i)P_{i,n}(x) = \lambda b \sum_{k=1}^{n} \chi_k P_{i,n-k}(x) + \lambda(1-b)P_{i,n}(x) \tag{4}$$

$$+ \int_0^\infty R_{i,n}(x,y)\xi_i(y)dy, i = 1, 2, \ldots, k, n \geq 1$$

$$\frac{d\Omega_{j,n}(x)}{dx} + (\lambda + \gamma(x))\Omega_{j,n}(x) = \lambda b \sum_{k=1}^{n} \chi_k \Omega_{j,n-k}(x)c \tag{5}$$

$$\cdot + \lambda(1-b)\Omega_{j,n}(x), j = 1, 2, \ldots, J, n \geq 1$$

$$\frac{dR_{i,n}(x,y)}{dx}+(\lambda+\xi_i(y))R_{i,n}(x,y)=\lambda b\sum_{k=1}^{n}\chi_k R_{i,n-k}(x,y) \tag{6}$$
$$+\lambda(1-b)R_{i,n}(x,y),\ n\geq 1$$

The steady-state boundary conditions at $x=0$ are the following:

$$\Psi_n(0)=\sum_{j=1}^{J}\int_0^\infty \Omega_{j,n}(x)\gamma(x)dx+\sum_{i=1}^{k-1}q_i\int_0^\infty Q_{i,n}(x)\mu_i(x)dx \tag{7}$$
$$+(1-p_k)\int_0^\infty Q_{k,n}(x)\mu_k(x)dx+\sum_{i=1}^{k}\int_0^\infty P_{i,n}(x)\mu_i(x)dx,\ n\geq 1$$

$$Q_{1,n}(0)=\int_0^\infty \Psi_{n+1}(x)a(x)dx+\lambda r\sum_{k=1}^{n}\chi_k\int_0^\infty \Psi_{n-k+1}(x)dx \tag{8}$$
$$+\lambda(1-r)\sum_{k=1}^{n}\chi_k\int_0^\infty \Psi_{n-k+2}(x)dx+\lambda b\chi_{n+1}P_0$$

$$Q_{i,n}(0)=\theta_{i-1}\int_0^\infty Q_{i-1,n}(x)\mu_{i-1}(x)dx,\ i=2,3,\ldots,k \tag{9}$$

$$P_{i,n}(0)=p_i\int_0^\infty Q_{i,n}(x)\mu_i(x)dx, \tag{10}$$

$$\Omega_{1,n}(0)=\sum_{i=1}^{k-1}q_i\int_0^\infty Q_{i,0}(x)\mu_i(x)dx+(1-p_k)\int_0^\infty Q_{k,0}(x)\mu_k(x)dx \tag{11}$$
$$+\int_0^\infty P_{i,0}(x)\mu_i(x)dx,\ n=0$$

$$\Omega_{j,n}(0)=\int_0^\infty \Omega_{j-1,0}(x)\gamma(x)dx,\ n=0 \tag{12}$$

$$R_{i,n}(x,0)=\alpha_i(Q_{i,n}(x)+P_{i,n}(x)) \tag{13}$$

The normalizing condition is

$$P_0+\sum_{n=1}^{\infty}\int_0^\infty \psi_n(x)dx+\sum_{n=0}^{\infty}\sum_{i=1}^{k}\left(\int_0^\infty Q_{i,n}(x)dx+\int_0^\infty P_{i,n}(x)dx\right) \tag{14}$$
$$+\sum_{n=0}^{\infty}\sum_{i=1}^{k}\int_0^\infty\int_0^\infty R_{i,n}(x,y)dxdy+\sum_{j=1}^{J}\int_0^\infty \Omega_{j,n}(x)dx=1$$

To solve the above equations, we define the generating functions for $|z| \leq 1$ (for $0 \leq j \leq m-1$ and $1 \leq i \leq k$).

$$\psi(x,z) = \sum_{n=1}^{\infty} \psi_n(x)z^n;\ Q_i(x,z) = \sum_{n=0}^{\infty} Q_{i,n}(x)z^n;$$

$$P_i(x,z) = \sum_{n=0}^{\infty} P_{i,n}(x)z^n;\ \Omega_i(x,z) = \sum_{n=0}^{\infty} \Omega_{i,n}(x)z^n;\ R_i(x,y,z) = \sum_{n=0}^{\infty} R_{i,n}(x,y)z^n;$$

Now multiplying Eqs. (2)–(13) by z^n and summing over n (for $1 \leq j \leq m-1$ and $1 \leq i \leq k$, $n = 0, 1, 2 \ldots$) and solving the PDE, we get the limiting PGFs $\psi(x,z)$, $Q_i(x,z)$, $P_i(x,z)$, $\Omega j(x,z)$, $R_i(x,z)$.

Theorem 1 *Under the stability condition $\rho < 1$, the joint probability distributions of the number of customers in the queue when the server is idle, busy on ith phase of service of the direct customer, busy on ith phase of service of the feedback customer, on vacation, and under repair are given by:*

$$\Psi(z) = bP_0(1 - R^*(\lambda))[z(N(z)-1) + X(z)\sum_{i-1}^{k} \Phi_i(z)]/Dr(z) \quad (15)$$

$$Q_i(z) = \lambda bP_0B(z)(1 - S^*(A_i(z)))\Theta_{i-1}B^*_{i-1}(A_{i-1}(z))/Dr(z) \quad (16)$$

$$P_i(z) = \lambda p_i bP_0B(z)(1 - S^*(A_i(z)))\Theta_{i-1}B^*_{i-1}(A_{i-1}(z))S^*(A_i(z))/Dr(z) \quad (17)$$

$$\Omega_j(z) = \frac{P_0(1 - V^*(b(z)))}{(1 - X(z))[V^*(\lambda b)]^{J-j+1}},\ (j = 1, 2, \ldots, J) \quad (18)$$

$$R_i(z) = \lambda\alpha_i bP_0[C(z)(1 - S^*(A_i(z)))\Theta_{i-1}B^*_{i-1}(A_{i-1}(z))]/Dr(z)b(z)A_i(z) \quad (19)$$

where $\rho = \omega + (1 - R^*(\lambda))(E(X) + r - 1)$

$$\sum_{i=1}^{k} \Theta_{i-1}(1-\theta_i) - 1 + \beta[E(X)R^*(\lambda) + (1-r)(1 - R*(\lambda))]$$

$$\beta = \sum_{i=1}^{k} \lambda b\Theta_{i-1}E(Si)(1 + \alpha_i E(G_i))(1 + p_i)$$

$$\omega = \sum_{i=1}^{k} \Theta_{i-1}[M_{1i}(1-\theta_i) + \lambda bE(X)E(S_i)(1 + \alpha_i E(G_i))(1 - \theta_i + P_i)]$$

$$b(z) = \lambda b(1 - X(z)) \quad \textit{and} \quad A_i(z) = b(z) + \alpha(1 - G^*_i(b(z)))$$

$$R(z) = R^*(\lambda) + \frac{X(z)}{z}(1 - R^*(\lambda))(1 - r + zr)$$

$\Theta_{i-1} = \theta_1\theta_2 \cdots \theta_{i-1}$ *and* $B^*_{i-1}(A_{i-1}(z)) = S^*_1(A_1(z))S^*_2(A_2(z)) \cdots S^*_{i-1}(A_{i-1}(z))$

$\Phi_i(z) = \Theta_{i-1}B^*_{i-1}(A_{i-1}(z))S^*_i(A_i(z))(q_i + P_i S^*_i(A_i(z)))$

$$Dr(z) = z - R(z)\sum_{i=1}^{k} \Phi_i(z)\ B(z) = [R(z)(N(z) - 1) + X(z)];$$

Proof Integrating the above partial generating functions w.r.to x and y, then define the partial PGFs as, for (i = 1,2,...,k and j = 1,2,...,J), $\Psi(z) = \int_0^\infty \Psi(x, z)dx$,

$$Q_i(z) = \int_0^\infty Q_i(x, z)dx,\ P_i(z) = \int_0^\infty P_i(x, z)dx,\ \Omega_j(z) = \int_0^\infty \Omega_j(x, z)dx,$$

$$R_i(x, z) = \int_0^\infty R_i(x, y, z)dy,\ R_i(z) = \int_0^\infty R_i(x, z)dx$$

Then, the probability that the server is idle can be determined using the normalizing condition and applying L'Hospital's rule, we get Eq. (20):

$$P_0 + \Psi(1) + \sum_{i-1}^{k}(Q_i(1) + P_i(1) + R_i(1)) + \sum_{j-1}^{J}\Omega_j(1) = 1 \quad (20)$$

Corollary Under the stability condition $\rho < 1$,

1. The probability distribution function of number of customers in the orbit is

$$K(z) = P_0 + \Psi(z) + \sum_{i=1}^{k}(Q_i(z) + p_i(z) + R_i(z)) + \sum_{j=1}^{J}\Omega_j(z) \quad (21)$$

2. The probability distribution function of number of customers in the system

$$H(z) = P_0 + \Psi(z) + \sum_{i=1}^{k}(Q_i(z) + p_i(z) + R_i(z)) + \sum_{j=1}^{J}\Omega_j(z). \quad (22)$$

4 Performance Measures

In this section, we obtain some interesting probabilities when the system is in different states.

1. Let ψ be the steady-state probability that the server is idle during the retrial time:

$$\psi = \psi(1) = \frac{(b(1 - R^*(\lambda))}{(Dr)}[N(1) - 1 + E(X)\sum_{i=1}^{k}\Theta_{i-1}(1 - \theta_i)$$

$$+\sum_{i=1}^{k}\Theta_{i-1}(M_{1i}(1 - \theta_i) + \lambda bE(X)E(S_i)(1 + \alpha_i E(G_i))(1 - \theta_i + p_i))]$$

2. Let Qi be the steady-state probability that the server is busy in ith phase of service:

$$Q_i = Q_i(1) = \frac{\Theta_{i-1}\lambda bE(S_i)}{Dr}[N^{`}(1) + E(X) - (1 - R^*(\lambda))(E(X) + r - 1)]$$

3. Let Pi be the steady-state probability that the server is feedback busy in ith phase of service:

$$P_i = P_i(1) = \frac{\Theta_{i-1}p_i\lambda bE(S_i)}{Dr}[N^{`}(1) + E(X) - (1 - R^*(\lambda))(E(X) + r - 1)]$$

4. Let Ω_j be the steady-state probability that the server is on jth vacation:

$$\Omega_j = \Omega_j(1) = \frac{P_0\lambda bE(V)}{V*(\lambda b)^{J-j+1}}$$

5. Let R be the steady-state probability that the server is under repair time:

$$R_i = R_i(1) = \frac{\Theta_{i-1}\alpha_i(1 + p_i)\lambda bE(S_i)E(G_i)}{Dr}$$
$$[N^{`}(1) + E(X) - (1 - R^*(\lambda))(E(X) + r - 1)]$$

6. The mean number of customers in the orbit (L_q) and in the system (L_s) under steady-state condition is obtained by differentiating $H(z)$ and $K(z)$ with respect to z and evaluating at $z = 1$,

$$L_q = \lim_{s\to 1}\frac{d}{dz}H(z) \text{ and } L_s = \lim_{s\to 1}\frac{d}{dz}K(z).$$

5 Special Cases

Case 1: Single arrival, no vacation, no balking, and no breakdown
From this case, the model reduces to an M/G/1 queue with immediate feedbacks and two-phase service. The results agree with Kalidass and Kasturi [4].

Case 2: Single arrival, single-phase service, no vacation, no feedback, no balking, and no breakdown
From this case, our model can be reduced to M/G/1 queue.

6 Conclusion

In this paper, we analyzed an unreliable retrial queueing system with optional service and immediate feedbacks under at most J vacation. The steady-state results are found by using supplementary variable technique. Some performance measures were deduced. The results discover applications in mailing system, ATMs, software design, production lines, and satellite communication.

References

1. Chen, P., Zhu, Y., Zhang, Y.: A retrial queue with modified vacations and server breakdowns. IEEE 978-1-4244-5540-9, 26–30 (2010)
2. Choudhury, G., Tadj, L., Deka, K.: A batch arrival retrial queueing system with two phases of service and service interruption. Comput Math Appl. 59, 437–50 (2012)
3. Jailaxmi, V., Arumuganathan, R., Senthil Kumar, M.: Performance Analysis of an M/G/1 Retrial Queue with General Retrial Time, Modified M-Vacations and Collisions. Operational Research An International Journal. 16, 1–19 (2016)
4. Kalidass, K., Kasturi, R.: A two phase service M/G/1 queue with a finite number of immediate Bernoulli feedbacks. OPSEARCH. 51(2), 201–218 (2014)
5. Rajadurai, P., Varalakshmi. M., Saravanarajan, M.C., Chandrasekaran, V.M.: Analysis of M[X]/G/1 retrial queue with two phase service under Bernoulli vacation schedule and random break-down. International Journal of Mathematics in Operations Research.7, 19–41 (2015)
6. Sumitha, D., Udaya Chandrika, K.: Two Phase Batch Arrival Retrial Queue with Impatient Customers, Server Breakdown and Modified Vacation. International Journal of Latest Trends in Engineering and Technology. 5(1), 260–269 (2015)
7. Wang, J., Li, J.: A single server retrial queue with general retrial times and two phase service. Jrl. Syst. Sci. Complex. 22(2), 291–302 (2009)
8. Zhang, F., Zhu, Z.: A discrete-time Geo/G/1 retrial queue with J vacations and two types of breakdowns. Journal of Applied Mathematics. 2013, Article ID 834731 (2013)

Weibull Estimates in Reliability: An Order Statistics Approach

V. Sujatha, S. Prasanna Devi, V. Dharanidharan, and Krishnamoorthy Venkatesan

Abstract The Exponential and Weibull distributions are well-known failure time distributions in reliability theory and survival analysis. Order statistics occur naturally in life testing and in survival analysis. The properties of order statistics and the results of order statistics are used to estimate the three-parameter Weibull distributions. This study ranges from order statistics to distribution theory and then to survival analysis. To know the survival distribution function S(t), many distributional forms have been used. Moments of order statistics help us to estimate the one-parameter and two-parameter forms, but to estimate the three-parameter Weibull distribution is a challenging one. Hence we make this study as a twofold study: firstly, to study the order statistics for Weibull distributions and their moments, and secondly, apply the computed moments of order statistics to estimate the location and scale parameters (with the shape parameter being fixed) based on complete as well as type II right-censored samples. To achieve the goal, data have been simulated for the failure time and their moments estimated based on the order statistics, and we explained how the conventional estimators play their role in reliability in order to verify the accuracy of the numerical computations.

V. Sujatha (✉)
School of Advanced Sciences, VIT University, Vellore, Tamilnadu, India
e-mail: sujatha.v@vit.ac.in

S. Prasanna Devi
Department of Computer Science, SRM University, Vadapalani, Chennai, Tamilnadu, India
e-mail: prasannasiva11@gmail.com; hod.cse@vdp.srmuniv.ac.in

V. Dharanidharan
Department of Computer Science, Apollo Engineering College, Chennai, Tamilnadu, India
e-mail: maildharaniv@gmail.com

K. Venkatesan
College of Natural Sciences, Arba Minch University, Arba Minch, Ethiopia
e-mail: krishnamoorthy.venkatesan@amu.et

V. Madhu et al. (eds.), *Advances in Algebra and Analysis*, Trends in Mathematics,
https://doi.org/10.1007/978-3-030-01120-8_49

1 Introduction

Weibull distribution is one of the distributions to model the failure time distribution; here an attempt has been made to estimate the parameter and their corresponding reliability functions. Based on the ordered observations from the Weibull distribution and functions, in terms of moments are derived. Data have been simulated to derive the empirical estimates and also a comparison has been made with their respective theoretical parameters [2].

Simulated data from three-parameter Weibull distribution with the parameters a=25, b=30, and c=2.5 for the sample size of n=100 has been taken and it has been replicated for 15 times, and with that of the same parameters, the corresponding order statistics are found with the respective moment estimators. To estimate the parameter "a," trial and error method is used, and to estimate the remaining two parameters "b" and "c," two different methods were used: first one being the "goodness of fit" and the second one being the "maximum likelihood" approaches. For our analysis we considered the complete dataset (N=100) and the censored datasets R=80 and R=70 for type II censoring, and for type I censoring, we fixed our time limits as T=60 and T=53.

2 Graphical Estimation for a Complete Sample

A complete sample includes all such samples in the entire population which satisfies a set of well-defined selection criteria. To estimate the parameters on Weibull distributions, graphical representation methods are used, for probability plotting. We simulate a dataset consisting of n=100 observations from Weibull distribution with parameters a=0, b=25, and c=2, is taken and it is ordered and then we plotted to their corresponding plotting positions.

2.1 Parameter Estimation Using Graphical Method and Moment Derivation

Parameters of three-parameter Weibull distributions have been identified using graphical method with near straight line strategy, and the corresponding parameters of "b" and "c" are estimated by using goodness of fit and MLE approaches. Figure 1 provides straight lines corresponding to six different "U" values leading to identification of the best one. This best option is used for further analysis.

For the above data, after identifying the parameter "a" value, the best combinations of "b" and "c" are identified using Table 1 with the criteria of least residual sum of squares. This procedure is repeated for other combinations of parameters "a," "b," and "c" and with different sample sizes (R=100, R=80, and R=70) and for different time periods (T=70, T=60, and T=53). The results indicate the estimated parameters as well as approximated parameters of the distribution [1].

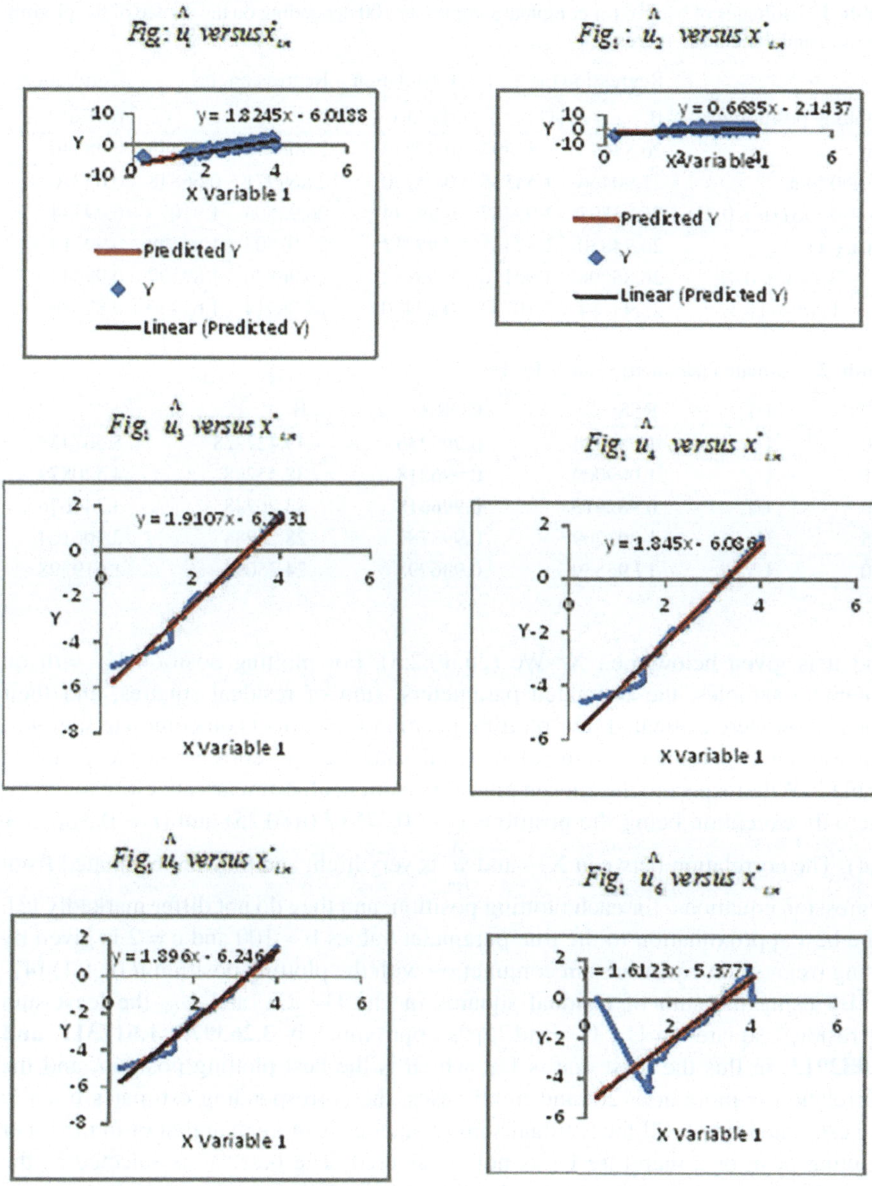

Fig. 1 Trial and error output for complete sample with regression model 1

The failure time for 75 elements is unknown. The correlation between $x^*_{i:n}$ and $\hat{u}_{i:n}$ is only 0.65770. To select the value of $\hat{a}$ by "trial and error approach," the residual sum of squares (RSS) of three different values of "a" was estimated,

Table 1 Estimates of b and c for complete samples T=100 depending on the choice of the plotting position and the kind of regression

	Regression (a)		Correlation	Regression (b)		Correlation
Plotting position $\overset{\Lambda}{u_i}$	B	C	r(x*,ũ)	b	c	r(x*,ũ)
i/n	26.84693	1.87893	0.98540	27.08448	1.82445	0.98540
(i − 0.5)/n	21.90466	1.54535	0.65770	24.69971	0.66848	0.65770
(i − 0.375)/(n + 0.25)	26.83795	1.93487	0.99374	26.93805	1.91073	0.99374
i/(n + 1)	26.94830	1.87441	**0.99212**	27.07703	1.84498	0.99212
(i − 0.3)/(n + 0.4)	26.86298	1.92116	0.99342	26.96870	1.89597	0.99342
(i − 1)/(n − 1)	25.93154	2.10733	0.87470	28.08914	1.61230	0.87470

Table 2 Estimated parametric values for U_4

A	U_4	RSS	CORR	B	c
10	U_4	1.383534	0.995236	43.427228	5.267433
15	U_4	1.099009	0.996218	38.35298	4.520879
20	U_4	**0.982913**	**0.996618**	33.26748	3.746676
25	U_4	1.803209	0.993787	28.20284	2.906164
30	U_4	17.95539	0.936295	24.25099	1.619398

and it is given below, i.e., X~We (25,30,2.5). For plotting position U_1 =i/n on complete samples, the estimated parameters, sum of residual squares, and their correlation were estimated. For plotting position U_4 =i/(n+1) on complete samples, the estimated parameters, sum of residual squares, and correlation are given in Table 2. With respect to the chosen plotting position, all estimates are close together, the only exception being the positions (i − 0.375) / (n+0.25) and (i − 0.3) / (n + 0.4). The correlation between $X^*_{i:n}$ and $\overset{\Lambda}{u}_{i:n}$ is very high, and they are estimated from regression equations, for each plotting position, and they do not differ markedly [3]. The best approximation to the true parameter values b = 100 and c = 2 is given by using (regression equations) in conjunction with the plotting position i/ (n + 1) [4].

By using the sum of residual squares in U_1, U_2, U_4, and U_6, the least sum of residual squares in U_1, U_2, and U_4 is approximately **3.263973, 1.612311, and 0.982913**; in this the least one is U_4, which is the best plotting position, and the correlation is more at a=20, and in all cases, the corresponding estimates b and c are estimated. In U_6 all the residual sum of squares is more than that of in the other plotting positions; therefore U_6 is not considered. The best "a" is selected as the plotted curve changes its convexity leading to straight line nature of the curve.

The curves display $\hat{u}_{i:n}$ as a function of $x^*_{i:n} = ln\left(x_{i:n} - \hat{a}\right)$ and are obviously concave, i.e., "a" must be greater than zero. Taking $\hat{a}$= 10, 15, 18, 19, 20, 21, 22, 25, 26 moves the curves to the left, reduces the concavity, and finally leads to convexity when a = 20. It is difficult to decide which of the nine curves is closest to linearity; to overcome the difficulty, calculate the residual sum of squares, and the least sum of residual squares is at $\hat{a}$=20, i.e., 20 moves the curves to the left, reduces the concavity, and finally leads to convexity when $\hat{a} = 20$; to decide which of the nine curves is closest to linearity, select the one with residual sum of squares value nearer to zero using trial and error. When censoring is involved in the data structure, it affects the value of the parameter estimators. The estimates of the reliability function differ by particular time because of censoring. The optimal parameter estimates corresponding to different sample sizes are consolidated and presented in Table 3 for regression approach. The corresponding solution for MLE approach is presented in Table 3 (Table 4).

From the above two tables, it is seen that as the sample size increases, the parameter estimates get close to the actual population parameters. This is true for both regression and MLE approaches. From the tables and Fig. 2, it is understood that the estimated value of "a" is 20 for complete data and it is 13 and 10 for

Table 3 Parameter estimates for different sample size and time limits regression approach

N=100	a	b	c
R=100	20	33.26	**3.75**
80	13	37.07	5.33
70	10	38.56	6.12
T=70	21	30.9	3.57
60	10	36.77	6.47
53	10	38.56	6.12

Table 4 Parameter estimates for different sample size and time limits-MLE approach

R	a	b	c
100	20	33.16	4.02
90	20	33.26	3.94
80	20	33.51	3.79
70	20	33.80	3.68
60	20	33.81	3.68
50	20	33.27	3.83

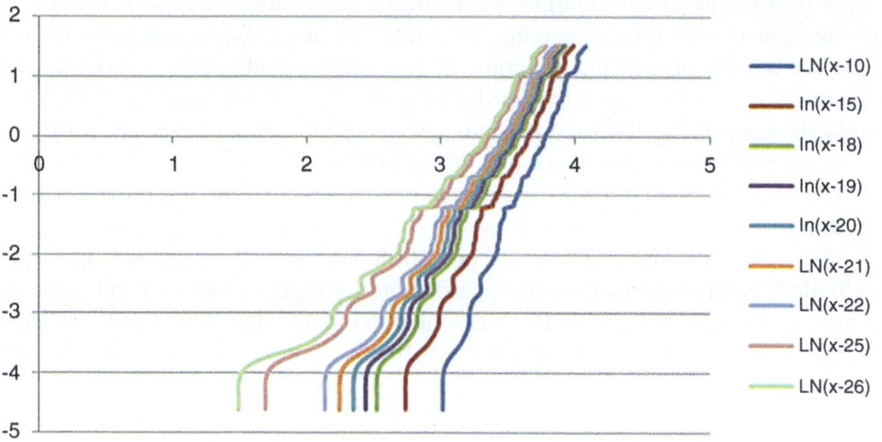

Fig. 2 Plotted lines for different values of a

Table 5 Estimation First Four moments of the ordered statistics and its transformed value

i	k	c	$E(U^k)$	$E(x^k)$	$Var(X_{i:n})$
1	1	3.7466	0.264185	**28.78677647**	6.839241
1	2	3.7466	0.075976	835.5177406	
1	3	3.7466	0.023322	17,245.02352	
1	4	3.7466	0.007547	720,771.8552	
2	1	3.7466	0.335148	**31.1470125**	5.003378
2	2	3.7466	0.116847	975.1397658	
2	3	3.7466	0.042166	23,483.40258	
2	4	3.7466	0.015689	970,240.2214	
3	1	3.7466	0.380390	**32.65176475**	4.13707
3	2	3.7466	0.148436	1070.274811	
3	3	3.7466	0.059289	28,015.79297	
3	4	3.7466	0.024195	1,163,068.373	
4	1	3.7466	0.414800	**33.79623137**	3.61623
4	2	3.7466	0.175328	1145.801485	
4	3	3.7466	0.075422	31,767.63468	
4	4	3.7466	0.032985	1,329,331.711	
5	1	3.7466	0.443088	**34.73711686**	3.260449
5	2	3.7466	0.199275	1209.927737	
5	3	3.7466	0.090896	35,055.46533	
5	4	3.7466	0.042020	1,479,620.274	

partial data which contains first 21 and 10 samples, respectively. It implies that when censoring is available, it affects the value of a. Changing “a” when the other parameters are held constant, it will result in a parallel movement of the density curve over the abscissa. Enlarging (reducing) a causes a movement of the density to the right (to the left). Changing “a” while “b” and “c” are held constant will alter the density at x in the direction of the ordinate. Enlarging “a” will cause a compression or reduction of the density and reducing “a” will magnify or stretch it, while the scale on the abscissa goes into the opposite direction. The shape parameter is responsible for the appearance of a Weibull density.

The moment estimates for the selected parameter c=3.7466 are provided in [5] Table 5.

The moments of ordered statistics up to fifth order statistics have been presented in Table 5. Similar works have been carried out for higher-order moments also. As a matter of verification, one can observe that the magnitude of first-order moment increases with the increasing order of the ordered statistics, as expected [6] (Tables 6 and 7).

Table 6 Theoretical parameter of Weibull distribution with parameters a=15, b=30, and c=2.5 for order samples of size 10

N	Mean	Variance	m3	m4	beta1	beta2
1	25.596000	20.564784	33.118457	1216.089777	0.126115	2.875526
2	30.159000	19.864719	19.545509	1078.055217	0.048736	2.731971
3	33.627000	19.604871	15.040746	1139.303799	0.030022	2.964228
4	36.675000	19.884375	12.481594	1113.773193	0.019815	2.816909
5	39.567000	20.522511	14.264925	1069.085227	0.023542	2.538349
6	42.468000	21.758976	14.113894	1289.860390	0.019337	2.724368
7	45.540000	23.738400	18.086328	1620.315796	0.024454	2.875390
8	49.011000	27.111879	33.133307	1857.506451	0.055087	2.527034
9	53.352000	34.284096	48.776380	3941.483884	0.059039	3.353315
10	60.186000	56.035404	179.325810	9939.448796	0.182767	3.165463

3 Conclusion

The study mainly focuses the optimum utilization of ordered statistics in applications concerned with reliability theory. This study particularly concentrates on two different failure time structures with Weibull failure time distribution. In particular, the study uses one-parameter exponential and three-parameter Weibull distributions. Trial and error method with graphical approximation is used to find the location parameter of the Weibull distribution. This is followed by the second-stage estimation of shape and scale parameters with two different approaches, one being the regression and the other being method of maximum likelihood. Different datasets with varying sample sizes and with varying time durations are employed and the process of estimation is tested in all these scenarios. With all these changing circumstances, it is found that with increasing sample size in type II censoring and with more time in type I censoring. The parameters are estimated with better accuracy. On the whole, the study uses effectively the concept of order statistics in the domain of reliability with focus on estimation and testing of its crucial parameters. As a by-product of the study, it is suggested that for future research, the method of moments and percentile estimation methods could be used and their results can be compared for the different combinations of parameters. Also all the three parameters of the Weibull Distribution could be simultaneously estimated with improved optimal criteria using global approaches such as "Genetic Algorithms" or "Neural Network".

Table 7 Theoretical parameter of Weibull distribution with parameters a=15, b=30, and c=2.5 for order samples of size 30

N	Mean	Variance	m3	m4	beta1	beta2
1	21.828000	8.548416	8.762071	217.874052	0.122901	2.981496
2	24.624000	8.088624	3.677125	149.649185	0.025550	2.287310
3	26.631000	7.639839	4.079545	94.369380	0.037323	1.616824
4	28.278000	7.384716	2.755662	162.301409	0.018856	2.976149
5	29.715000	7.208775	2.811652	121.185489	0.021103	2.331995
6	31.017000	7.065711	1.015644	193.950684	0.002924	3.884898
7	32.220000	7.041600	0.211896	130.106308	0.000129	2.623951
8	33.357000	6.910551	2.597301	99.854355	0.020441	2.090940
9	34.440000	6.926400	1.111968	149.359749	0.003721	3.113282
10	35.487000	6.882831	1.628683	203.976712	0.008135	4.305726
11	36.504000	6.927984	1.438864	181.199813	0.006226	3.775235
12	37.503000	6.974991	0.443475	253.119302	0.000580	5.202810
13	38.490000	7.029900	0.088398	225.661818	0.000022	4.566251
14	39.468000	7.196976	−0.163058	168.351409	0.000071	3.250249
15	40.449000	7.278399	1.930478	−2.311776	0.009665	−0.043639
16	41.436000	7.457904	−1.889820	215.329404	0.008610	3.871415
17	42.435000	7.550775	2.078926	165.974091	0.010039	2.911101
18	43.461000	7.351479	15.986932	−254.667130	0.643289	−4.712197
19	44.493000	8.202951	−8.992988	698.044785	0.146520	10.373924
20	45.582000	7.851276	21.194535	−632.412104	0.928169	−10.259346
21	46.695000	8.826975	−9.254945	712.333537	0.124541	9.142387
22	47.880000	8.985600	1.076544	443.620178	0.001597	5.494360
23	49.131000	9.754839	−7.787900	701.660148	0.065340	7.373719
24	50.487000	10.212831	1.488553	456.011049	0.002080	4.372029
25	51.972000	11.061216	2.143164	669.384002	0.003394	5.471036
26	53.640000	12.200400	8.093088	509.210220	0.036067	3.420968
27	55.578000	13.925916	13.359361	569.086991	0.066085	2.934480
28	57.951000	16.951599	12.229003	1154.084674	0.030701	4.016209
29	61.158000	22.599036	29.288113	2191.712316	0.074321	4.291447
30	66.624000	40.002624	148.324117	4835.519019	0.343683	3.021803

Acknowledgements I wish to thank my coauthors for their kind help and support in my research work throughout this paper. My heartfelt thanks to my institution for providing me infrastructural facilities and excellent resources to carry out my research in VIT University, Vellore.

References

1. Balakrishnan, N., Joshi, P.C.: Product moments of order statistics from doubly truncated exponential distribution. **31**, 27–31, (1984)
2. Harrel FE Jr.: Regression modelling strategies with applications to linear models, logistic regression and survival analysis. Springer, Berlin (2001)
3. Rinne, H.: The Weibull distribution the hand book. Chapman & Hall. New York, (2009)
4. Shea, B. L., Scallan, A. J.: AS R72. A remark on Algorithm AS128: approximating the covariance matrix of normal order statistics. Applied Statistics, **37**, 151–155 (1988)
5. Sujatha, V., Dharanidharan, V.: Moments of order statistics for exponential distribution. International conference on Advances in Mathematical Sciences. ICAMS (2017).
6. Sujatha, V., Dharanidharan, V.: Predictive Modelling and Analytics on Big Data for Diabetic Management. (2017) https://ieeexplore.ieee.org/document/8074497/.

Intuitionistic Fuzzy ANOVA and Its Application Using Different Techniques

D. Kalpanapriya and M. Mubashir Unnissa

Abstract A fuzzification is the conversion of an exact quantity to uncertainty. A more generalized fuzzy set called intuitionistic fuzzy set is defuzzified to propose its application in career development to find the homogeneity among students and their career using analysis of variance (ANOVA). The proposed test procedure is well illustrated using a numerical example. The main purpose of this paper is to give a view of intuitionistic fuzzy set with the application of ANOVA technique to emphasize the efficiency in career development.

Keywords Fuzzy set · Intuitionistic fuzzy set · Defuzzification · ANOVA

ANOVA is a unique technique in statistics which enables us to test the null hypothesis (more than two population means are equal) against the alternative hypothesis (they are not equal by using their sample data). This statistical technique is widely used in almost all areas of research. Devore [6] considered precise samples using conventional hypothesis which leads to test the significance in decision-making. However, the data values are vaguely specified in many applications; hence, imprecise data samples are used for testing the hypothesis. Fuzzy set theory was introduced by Zadeh [11], and intuitionistic fuzzy set (IFS) was introduced by Atanassov [3, 4] who is well known for the generalization of fuzzy set. The generalization consists of the degrees of membership, non-membership and hesitation margin which in turn gives more meaningful semantic representation about the data set. IFS is used as a tool in modelling the real-life problems like sales analysis, marketing and financial services, and also it is used in different fields of science. The fuzzy interval data by incorporating the concepts of fuzzy sets was proposed by Mikihiko Konishi et al. [8]. Wu introduced [10] one-factor ANOVA model for fuzzy data using the h-level set to solve optimization problems.

D. Kalpanapriya (✉) · M. Mubashir Unnissa
Department of Mathematics, School of Advanced Sciences, VIT University, Vellore, Tamilnadu, India
e-mail: dkalpanapriya@vit.ac.in; M.Mubashira@vit.ac.in

V. Madhu et al. (eds.), *Advances in Algebra and Analysis*, Trends in Mathematics,
https://doi.org/10.1007/978-3-030-01120-8_50

Kalpanapriya and Pandian [7] developed a new statistical fuzzy hypothesis testing of analysis of variance for testing the significant difference among more than two population means for an imprecise data. Ansari et al. [1] discussed a nonparametric method to test the statistical hypotheses of intuitionistic fuzzy data. Mohsen Arefi and Seyed Mahmoud Taheri [9] proposed a least-square regression model where explanatory variable(s) and the response variable and the parameters of the model are assumed to be Atanassov's intuitionistic fuzzy numbers. Gholamreza Hesamian and Mohamad Ghasem Akbari [13] proposed the statistical test based on intuitionistic fuzzy hypotheses. One-way analysis of variance in fuzzy environment was analysed by many authors like Alireza et al. [2] and Nourbakhsh et al. [14]. Hence, in this paper, the IFS is defuzzified to propose its application in career to find the homogeneity among students and their career using analysis of variance (ANOVA). The intuitionistic ANOVA problem has been defuzzified and solved using R-studio. Linguistic examples are provided to illustrate the approach.

1 Preliminaries

The following definitions of IFS, membership and non-membership function of an intuitionistic fuzzy set/number are obtained from Atanassov [3–5]. The need for the conversion of IFSs to fuzzy sets arises when we have intuitionistic fuzzy values (or data), and the available machine or computer package can only accept fuzzy values. This is due to the fact that IFS is relatively new and fuzzy set is industrially accepted for its efficiency and effective applications in computer programming and artificial intelligent and fuzzy control. Further, the technique to defuzzify the given IFS was motivated by Paul Augustine Ejegwa [12].

1.1 Definition

Let X be a nonempty set. An IFS A in X is given by a set of ordered triples, $A^{-1} = \{(x, \mu_A(x), \nu_A(x)); x \in X\}$, where $(\mu_A(x), \nu_A(x)) : X \to [0, 1]$ define, respectively, the degree of membership and degree of non-membership of the element $x \in X$ to the set A, which is a subset of X, and for every element, $0 \leq \mu_A(x) + \nu_A(x) \leq 1$. Furthermore, $\pi_A(x) = 1 - \mu_A(x) - \nu_A(x)$ is the IFS set index or hesitation margin and is the degree of indeterminacy concerning the membership of X in A and then $\pi_A(x) \in [0, 1]$. Whenever $\pi_A(x) = 0$, an IFS reduces automatically to fuzzy set.

1.2 Definition

Let X be a nonempty set. An intuitionistic fuzzy multiset (IFMS) A drawn from X is given as $A = \{\mu_A^1(x)\ldots\ldots.\mu_A^n(x), \nu_A^1(x)\ldots\ldots.\nu_A^n(x); x \in X\}$ where the functions $\mu_A^i(x), \nu_A^i(x) : X \rightarrow [0, 1]$ define the belongingness degrees and the non-belongingness degrees of A in X such that $\mu_A^1(x)\ldots\ldots.\mu_A^n(x)\ for\ i = 1$. If the sequence of the membership and non-membership (belongingness functions and $0 \leq \mu_A^i(x) + \nu_A^i(x)$ non-belongingness functions) functions has only n terms (i.e. finite), n is called the dimension of A. Consequently $A = \{\mu_A^1(x)\ldots\ldots.\mu_A^n(x), \nu_A^1(x)\ldots\ldots.\nu_A^n(x); x \in X\}\ for\ i = 1, \cdots n$. When no ambiguity arises, we write $\{\mu_A^i(x), \nu_A^i : x \in X\}$.

We henceforth denote the set of all IFMS over X as IFMS(X). Also, we denote an IFMS A as $(\mu_A^i(x), \nu_A^i(x))$ for simplicity.

1.3 Definition

Let $\{A_j\}_j \in J$ be an arbitrary family of IFMSs in X, where $A = (\mu_A^i(x), \nu_A^i(x)) \in IFMS(X)$. For each $j \in J$, we define

$$\bigcap_{j\in J} A = \left(\wedge\mu_{A_j}^i(x), \vee\nu_{A_j}^i(x)\right)$$

$$\bigcup_{j\in J} = \left(\wedge\mu_{A_j}^i(x), \vee\nu_{A_j}^i(x)\right)\forall \in X \tag{1}$$

1.4 Definition

For any two IFMSs A and B drawn from X, the following operations hold.

1. Inclusion: $A \subseteq B \Longrightarrow \mu_A^i(x) \leq and \mu_B^i(x) and \forall x\% \in X.$
2. Equality: $A = B \Longrightarrow \mu_A^i(x) = \mu_B^i(x) and \nu_A^i(x) = \nu_B^i(x) \forall x \in X$
3. Complement: $A^C = \left(\nu_A^i(x), \mu_A^i(x)\right) \forall x \in X.$
4. Union: $A \cup B = \mu_A^i(x) \vee \mu_B^i(x), \nu_A^i(x) \wedge \nu_B^i(x) \forall x \in X$
5. Intersection: $A \cap B = \mu_A^i(x) \wedge \mu_B^i(x), \nu_A^i(x) \wedge \nu_B^i(x) \forall x \in X$.
6. Addition: $A \oplus B = \left(\mu_A^i(x) + \mu_B^i(x) - \mu_A^i(x)\mu_B^i(x), \nu_A^i(x)\nu_B^i(x)\right) \forall x \in X$
7. Multiplication: $A \otimes B = \left(\mu_A^i(x)\mu_B^i(x), \nu_A^i(x) + \nu_B^i(x) - \nu_A^i(x)\nu_B^i(x)\right) \forall x \in X$
8. Difference: $A - B = \left(\mu_A^i(x) \wedge \nu_B^i(x), \nu_A^i(x) \vee \mu_B^i(x)\right) \forall x \in X$

1.5 Definition

Let IFMS(X), we define the support of A and the cross point of A as follows.

1. Supp(A)=$\{x \in X, \mu'^{i}_{A}(x) > 0, \nu'^{i}_{A}(x) < 1\}$, , $\forall x \in X$.
2. The crossover point of A is $\left\{x \in X, \mu'^{i}_{A}(x) = \frac{1}{2}, \nu'^{i}_{A}(x) = \frac{1}{2}\right\}$, $\forall x \in X$.

2 Two-Way Analysis of Variance

This analysis involves only two factors with more than two levels and different subjects in each of the experimental conditions. Using this model, it is more efficient to study two factors simultaneously rather than separately. Randomized block design (RBD) problems are solved by using a two-factor ANOVA technique. Consider a sample of size N of a given random variable X drawn from a normal population with variance σ^2. Let there be 'a' columns representing one factor and 'b' rows representing the other. The objective is to test the hypothesis of no difference against the alternative hypothesis.

Now, let μ_i be the column mean of the ith class and ν_j be the row mean of jth class. The testing hypotheses are given below: Null hypothesis for columns, $H_0^C : \mu_1 = \mu_2 = \ldots = \mu_s$ against the alternative H_A^C : $not\ all\ \mu_i's$ are equal , Null hypothesis for rows $H_0^R : \nu_1 = \nu_2 = \ldots = \nu_k$ against the alternative H_A^R : $not\ all\ \nu_j's$ are equal , hypothesis, Let x_{ij} be the value of the jth member of the ith class which contains n_i values and the number of the samples, N=ab.

Now, the two-factor ANOVA table is given below:

Source of variation	Sum of squares	Degrees of freedom	Mean square	F value
Between column classes	B_1	a−1	$MSC = \frac{B_1}{a-1}$	$F_1 = \frac{MSC}{MSE}$ or $F_1 = \frac{MSE}{MSC}$
Between row classes	B_2	b−1	$MSR = \frac{B_2}{b-1}$	
Residue or error	B_3	(a−1)(b−1)	$MSE = \frac{B_3}{(a-1)(b-1)}$	$F_2 = \frac{MSR}{MSE}$ or $F_2 = \frac{MSE}{MSR}$

The decision rules in F-test to accept or reject the null hypothesis and alternative hypothesis are given in Devore [6].

3 Intuitionistic Two-Way ANOVA

In this section, the two-factor analysis of variance is considered with respect to IFS such that each one of the factors occurs with its own number of levels. The effects of the two-way analysis of variance with respect to IFS were studied instead of two one-way ANOVAs. In converting IFSs to fuzzy sets, one could be tempted to just neglect the hesitation margin (not even minding its proportions), but doing this will definitely present a deceptive result. Some mathematical techniques which are effective and provide a pretty picture of IFS in terms of fuzzy set are introduced by Paul Augustine. The IFS problem reduced to fuzzy by taking the mean values of each of the parameters. Later a two-way ANOVA technique is used to compare the homogeneity among the groups. The proposed method is illustrated by the following example using R-studio.

3.1 Example

A career development is the significant part of human development. Every individual is interested to develop a good career in his lifetime. Most of them go through this individually, while there are career counsellors and trained specialists who lay suggestions to go for a successful upliftment in career. In this scenario, this paper aims in expressing the students about the influence of various factors affecting the career development using intuitionistic fuzzy set theory. The prime components that affect the development of the career among the five different students which are speed and efficiency, skills and interest, attitude, financial concerns, updated technology, family expectations, self-exploration and the decision-making are analysed using intuitionistic fuzzy set. Let S $= s_1, s_2, s_3, s_4$ and s_5 represent the five different students and Q = speed and efficiency, skills and interest, attitude, financial concerns, updated technology, family expectations, self-exploration and the decision-making be the set of all major influencing factors affecting the career development of an individual that are obtained as intuitionistic fuzzy data as shown in Table 1.

Applying Paul Augustine's techniques, we obtain the Tables 2, 3, 4.

Solving the above using the two-way ANOVA method in R-studio, we get the following result (Table 5).

Solving the above using the two-way ANOVA method in R-studio, we get the following result:

Table 1 Attributes influencing career development for students

Students	Speed and efficiency	Skills and interest	Attitude	Financial concerns	Updated technology	Family expectations	Self-exploration	Decision-making
s_1	(0.6, 0.4)	(0.675,0.325)	(0.65,0.35)	(0.55,0.45)	(0.65,0.35)	(0.67,0.325)	(0.775,0.225)	(0.725,0.275)
s_2	(0.775,0.225)	(0.75, 0.25)	(0.775,0.225)	(0.525,0.475)	(0.85,0.15)	(0.875,0.125)	(0.875, 0.125)	(0.9.0.1)
s_3	(0.75,0.25)	(0.725, 0.275)	(0.8,0.2)	(0.55,0.45)	(0.875,0.125)	(0.825,0.175)	(0.8,0.2)	(0.775,0.225)
s_4	(0.725,0.275)	(0.725,0.275)	(0.825,0.175)	(0.525,0.475)	(0.675,0.325)	(0.7,0.3)	(0.775,0.225)	(0.725, 0.275)
s_5	(0.65,0.35)	(0.5,0.5)	(0.725,0.275)	(0.7,0.3)	(0.65,0.35)	(0.5,0.5)	(0.65,0.35)	(0.625,0.375)

Table 2 Membership grades using Technique 1

Students	Speed and efficiency	Skills and interest	Attitude	Financial concerns	Updated technology	Family expectations	Self-exploration	Decision-making
s_1	0.600	0.675	0.650	0.550	0.650	0.675	0.775	0.725
s_2	0.725	0.705	0.775	0.525	0.850	0.875	0.750	0.900
s_3	0.750	0.750	0.800	0.550	0.875	0.825	0.800	0.775
s_4	0.725	0.725	0.825	0.525	0.675	0.700	0.775	0.725
s_5	0.650	0.500	0.725	0.700	0.650	0.500	0.650	0.625

Table 3 Membership grades using Technique 2

Students	Speed and efficiency	Skills and interest	Attitude	Financial concerns	Updated technology	Family expectations	Self-exploration	Decision-making
s_1	0.600	0.675	0.650	0.550	0.650	0.675	0.775	0.725
s_2	0.725	0.705	0.775	0.525	0.850	0.875	0.750	0.900
s_3	0.750	0.750	0.800	0.550	0.875	0.825	0.800	0.775
s_4	0.725	0.725	0.825	0.525	0.675	0.700	0.775	0.725
s_5	0.650	0.500	0.725	0.700	0.650	0.500	0.650	0.625

Table 4 Membership grades using Technique 3

Students	Speed and efficiency	Skills and interest	Attitude	Financial concerns	Updated technology	Family expectations	Self-exploration	Decision-making
s_1	0.061	0.000	0.206	0.056	0.138	0.141	0.082	0.074
s_2	0.000	0.078	0.260	0.106	0.889	0.094	0.094	0.094
s_3	0.078	0.000	0.083	0.113	0.000	0.000	0.000	0.079
s_4	0.153	0.325	0.089	0.053	0.147	0.072	0.082	0.153
s_5	0.067	0.050	0.076	0.000	0.067	0.100	0.067	0.133

Table 5 Membership grades using Technique 3

Students	Speed and efficiency	Skills and interest	Attitude	Financial concerns	Updated technology	Family expectations	Self-exploration	Decision-making
s_1	0.061	0.000	0.206	0.056	0.138	0.141	0.082	0.074
s_2	0.000	0.078	0.260	0.106	0.889	0.094	0.094	0.094
s_3	0.078	0.000	0.083	0.113	0.000	0.000	0.000	0.079
s_4	0.153	0.325	0.089	0.053	0.147	0.072	0.082	0.153
s_5	0.067	0.050	0.076	0.000	0.067	0.100	0.067	0.133

Techniques	Some of the squares	F ratio	Probability value
Technique 1	Between classes	2.287	0.0564
	Among classes	0.644	0.6357
Technique 2	Between classes	2.321	0.0532
	Among classes	0.620	0.6517
Technique 3	Between classes	1.534	0.197
	Among classes	0.730	0.579

Since the probability value is more than 5% level of significance in all the three techniques, we conclude that the students do not differ with these major influencing factors affecting the career development. But by comparing these three techniques, it is well cleared that Technique 3 has a better probability value among classes.

4 Conclusion

Intuitionistic fuzzy set is a vital tool in decision-making. In this paper, we proposed its application in career development to find the homogeneity among students and their career using the two-way analysis of variance (ANOVA). The proposed test procedure is well illustrated using a numerical example. The main purpose of this paper is to give a view of intuitionistic fuzzy set with the application of ANOVA technique to emphasize the efficiency in career development.

References

1. Q. Ansari, S. A. Siddiqui, J. A. Alvi :Mathematical techniques to convert intuitionistic fuzzy sets into fuzzy sets.Note on IFS **10**,13–17(2004)
2. Alireza Jiryaei, Abbas Parchami and Mashaalla Mashinchi: One-way Anova and least squares method based on fuzzy random variables, Turkish Journal of Fuzzy Systems, **4**, 18–33 (2013)
3. Atanassov.K.T :New operations defined over intuitionistic fuzzy sets, Fuzzy sets and Systems, **6**,137–142.(1994)
4. Atanassov: K. Intuitionistic fuzzy sets . In Proceedings of the VII ITKR's Session, Sofia (1983)
5. K.T. Atanassov: Intuitionistic fuzzy sets, Fuzzy Sets and Systems,**20**, 87–96(1986).
6. J.L.Devore, Probability and Statistics for Engineers, Cengage, 2008
7. D. Kalpanapriya and P. Pandian: Statistical Hypotheses testing for imprecise data, Applied Mathematical Sciences, **6**, 5285–5292 (2012)
8. Konishi,M., T. Okuda and K. Asai : Analysis of variance based on fuzzy interval data using moment correction method, International Journal of Innovative Computing, Information and Control, **2**, 83–99 (2006)
9. Arefi, M., and S.M. Taheri :Testing fuzzy hypotheses using fuzzy data based on fuzzy test statistic, Journal of Uncertain Systems, Vol. **5**, No.1, pp. 45–61 (2011)
10. Wu, H.C : Statistical confidence intervals for fuzzy data, Expert Systems with Applications, **36**, 2670–267(2006)

11. Zadeh, L.A: Probability measures of fuzzy events, Journal of Mathematical Analysis and Applications, **23**, 421–427 (1968)
12. Paul Augustine Ejegwa: Mathematical techniques to transform intuitionistic fuzzy multisets to fuzzy sets. Journal of Information and Computing Science.**2**,169–172(2015)
13. Akbari, M.G., and A. Rezaei: Bootstrap statistical inference for the variance based on fuzzy data, Austrian Journal of Statistics, **38**,121–130(2009)
14. Nourbakhsh M, M. Mashinchbi and A. Parchami: Analysis of Variance based on fuzzy observations, International Journal of Systems Science, **44**, 714–726 (2013).

A Resolution to Stock Price Prediction by Developing ANN-Based Models Using PCA

Jitendra Kumar Jaiswal and Raja Das

Abstract The application of artificial neural network (ANN) has become quite ubiquitous in numerous disciplines with different motivations and approaches. One of the most contemporary implementations accounts it for stock price behavior analysis and forecasting. The stochastic behavior of stock market follows numerous factors to determine the price vicissitudes such as GDP, supply and demand, political influences, finance, and many more. In this paper, we have considered two ANN techniques, viz., backpropagation-based neural network (BPNN) and radial basis function network (RBFN), first, without principal component analysis (PCA), and further modified the model with PCA, to execute financial time series forecasting for the next 5 days (which can also be extended for some other number of days) by accepting the input as historical data on the sliding window basis. Moreover, the empirical research is conducted to verify the forecasting impact on the stock prices for oil and natural gas sector in India with the developed model, and subsequently a comparison study has also been performed for the effectiveness of the two models without and with PCA, on the basis of mean square percentage error.

1 Introduction

The applications of ANN have been proliferated in multiple disciplines with their subtle influence in dealing with share market data to explore their dynamics. Neural networks can develop more effective models for a large class of data from different streams than which are developed from classical parametric prototypes [1] and [2]. Usually data from share market turn up in the time series format with noises, and different analysis approaches were considered along with ANN models to forecast share market prices [3]. ANN is highly applicable since it has

---

J. K. Jaiswal (✉) · R. Das
Vellore Institute of Technology, Vellore, Tamilnadu, India
e-mail: jit.frenz@gmail.com; jitendra.kjaiswal@vit.ac.in; rajadas@vit.ac.in

V. Madhu et al. (eds.), *Advances in Algebra and Analysis*, Trends in Mathematics,
https://doi.org/10.1007/978-3-030-01120-8_51

self-learning capability, a rigorous anti-jamming potential, and has been extensively utilized in share market price prediction, risk analysis, profits, and exchange rate estimations [4].

The backpropagation-based neural network technique, also called as backpropagation neural network (BPNN), performs the training of network to predict the stock market prices in a robust manner. Multilayer perceptron (MLP) is also considered as one of the most widely applied neural network techniques with the capability of establishing a relationship between input and output as a nonlinear function approximation [5] and [6]. Radial basis function network (RBFN) has been observed as one of the most effective forecasting approaches by researchers. It was first applied for multivariate interpolation of curve fitting starting with an initial set of points [7]. Versace et al. [8] implemented a hybrid approach of RBFN with genetic algorithm to forecast the exchange-traded capital (DIA). Further Wang et al. [9] implemented RBFN to attain close forecasting of stock market. Since RBFN has reflected a closed grip on stock market prediction, we have also approached for implementing PCA-BPNN and PCA-RBFN for the coming 5-day (next week) forecasting, based on historical data in a sliding-window manner in which the networks are being trained with new data discarding the old data. The application of PCA has been followed by many researchers, but they have performed prediction for the next day only; however, we have developed models to execute it for a number of forthcoming days.

2 Backpropagation-Based Neural Network (BPNN)

The application of artificial neural network has been increased extensively in numerous areas of researches and so in financial sector to analyze market behavior [10]. Being one of the most preliminary and robust approaches, backpropagation technique has been applied widely for learning in multilayer networks. A sketched diagram of backpropagation technique has been given in Fig. 1 with some description in the caption. Identity function has been applied at input layer, while hidden and output layers follow the sigmoid and linear function in the network.

Input-to-hidden node is calculated by the function

$$f_i(x) = \sum_{i=1}^{m} W1_{ih}x_i + b1_h \tag{1}$$

where $b1_h$ is the biased weight provided at the hidden layer. The resultant function at the hidden layer for the h_{th} node is given by

$$g_h(x) = \frac{1}{1 + e^{-f_h(x)}} \tag{2}$$

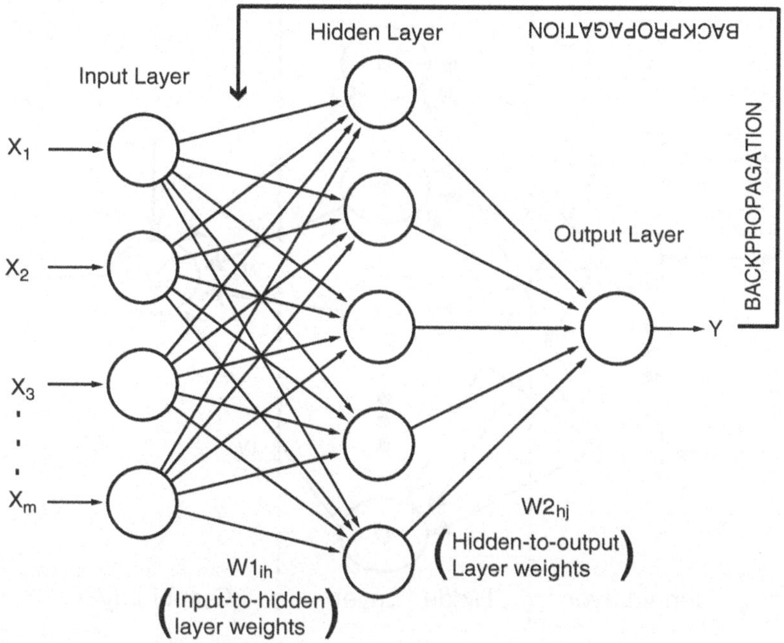

Fig. 1 Backpropagation-based technique again passes the output to the network to relearn by updating the weights from input-to-hidden layer as per the errors received from comparing it to the actual required output. It is an m×5×1 network where m is the number of nodes for m input variables, 5 is the number of nodes in the hidden layer, and there is one node in the output layer. Layers are also called as neurons.

At the output layer,

$$y_x = \sum W2_{hj} g_h(x) + b2 \tag{3}$$

where $b2$ is the biased weight at the output layer.

3 Radial Basis Function Network (RBFN)

RBFN is also a network of input, hidden, and output layers, while it can comprise only one neuron in the hidden layer, and it follows only a feed-forward technique; however, it may contain any number of nodes in the three layers. The basis stricture of RBFN is represented in Fig. 2 with some description in the caption.

The input-output pair $\mathbf{T} = \{X_i, d_i\}$ performs calculations with interpolation to acquire function f which takes input X_i and produce output close to the desired output d_i for n data sample.

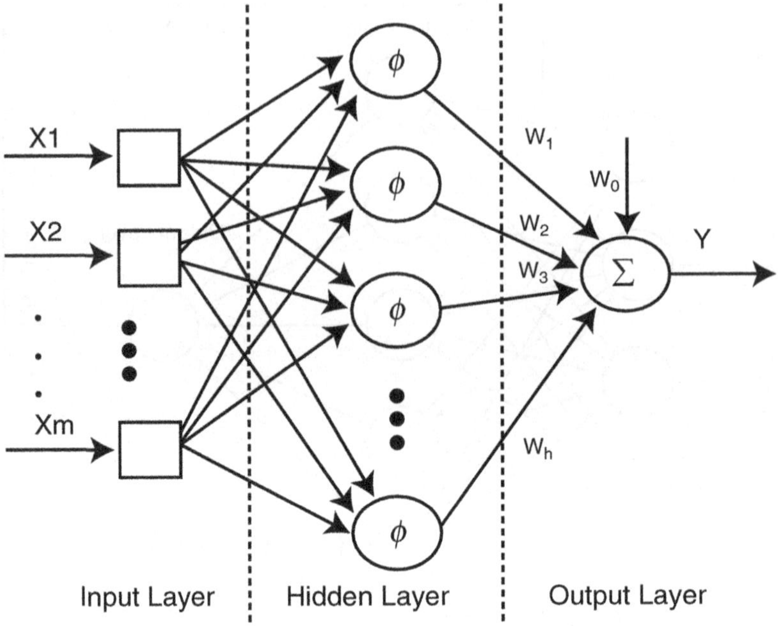

Fig. 2 RBFN with m nodes in input layer, h nodes in hidden layer, and one node in the output layer. The synaptic weights are not assigned from input to hidden layer, whereas the weights are charged with applied mathematical functions for hidden-to-output layer, and the output nodes trade on the heels of linear simulation

$$f(x_i) = d_i \quad i = 1, 2, \ldots, n \tag{4}$$

RBFN produces n function as basis $\phi(||X - X_i||), i = 1, 2, \ldots, n$ as nonlinear with Euclidean distance $(||X - X_i||)$, where X as applied input and X_i as points of training data. Then the mapping f is

$$f(x) = \sum_{i=0}^{n} w_i \phi(||X - X_i||) \tag{5}$$

From Eqs. (4) and (5),

$$\sum_{i=0}^{n} w_i \phi(||X - X_i||) = d_i \quad i = 1, 2, \ldots, n \tag{6}$$

For basis determination at the hidden layer, many functions can be applied. Some of them are given as follows:

1. *Gaussian function:*

$$\phi(x) = e^{\left(-\frac{(x-t)^2}{2\sigma^2}\right)} \quad \sigma > 0; \quad x, t \in \mathbf{R} \tag{7}$$

2. *Multiquadrics:*

$$\phi(x) = (x^2 + t^2)^{1/2} \tag{8}$$

The basis function ϕ is symmetric, and weights W can be estimated with the correct selection of ϕ as:

$$W = \phi^{-1} D \tag{9}$$

for $W = (w_1, \ldots, w_n)'$, and $D = (d_1, \ldots, d_n)'$.

4 Principal Component Analysis (PCA)

The approach of PCA has been widely applied mainly in signal and image processing. Basically it reduces the number of input variables on the basis of variance-covariance concepts. The procedure for PCA can be considered in the following steps:

1. Generate an $n \times m$ matrix **A** with m variables and n number of data sample.
2. Normalize the matrix with their mean so that for each column, $\sum_{i=0}^{m} x_i = 0$.
3. Find a symmetric matrix **C** of $m \times m$ dimension with factors of variance-covariance matrix from **A** for their individual and pairs of columns.
4. Calculate the eigenvalues $\lambda_1, \lambda_2, \ldots, \lambda_m$ with their respective eigenvalues $\nu_1, \nu_2, \ldots, \nu_m$.
5. Select the principal components as eigenvectors for some of the highest eigenvalues, and the important variables are determined as some of the highest values in considered principal components.

5 Experimentation and Result Analysis

5.1 Data Acquisition

Since financial data are readily available in the both forms, paid and free, we have downloaded free data from the NSE India website[1] for the stocks in Oil and Natural Gas Corporation Limited (ONGC), Oil India Limited (OIL), Selan Exploration

[1]Data can be downloaded from the website https://www.nseindia.com/products/content/equities/equities/eq_security.htm.

Technology Limited (SELAN), Aban Offshore Limited (ABAN), Hindustan Oil Exploration Company Limited (HINDOILEXP), and Cairn India (CAIRN) from the date April 29, 2016 to April 28, 2017 on the daily basis.

5.2 *Experimentation*

Each stock data consists of 12 variables, yesterday's closing, today's opening, high, low, last before closing, closing and day average price, total traded quantity, turnover, number of trades, deliverable quantity, and percentage of deliverable to traded quantity. On the basis of these variables, we have experimented to forecast 1-week (5 days)-ahead closing prices (Fig. 3).

We have followed Minitab® 14 software for PCA and performed neural network programming on the MATLAB software from MathWorks® for stock price prediction.

5.2.1 Error Estimation

The following error parameters may be followed to get the effectiveness of ANN algorithm by comparing actual ($y(i)$) and predicted values ($\hat{y}(i)$).

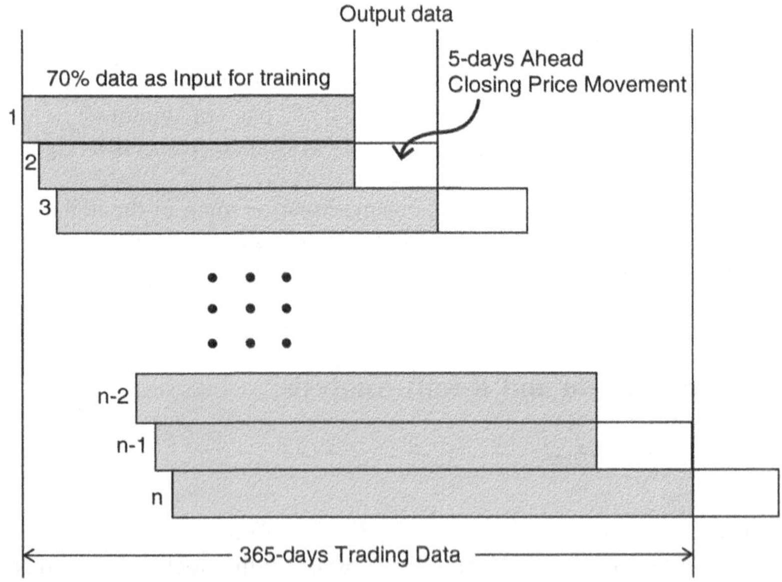

Fig. 3 Five-day-ahead closing price prediction. We have predicted for the next 5 days on the daily basis; that is, today we predicted for the next 5 days, tomorrow again we'll predict for the next 5 days, and so on. However prediction cannot be accurate, but it successfully works as signal for the trading

1. *Mean absolute percentage error (MAPE):*

$$MAPE = \frac{1}{n}\sum_{i=1}^{n}\frac{|y(i) - \hat{y}(i)|}{y(i)} \tag{10}$$

2. *Symmetric mean absolute percentage error (SMAPE):*

$$SMAPE = \frac{1}{n}\sum_{i=1}^{n}\frac{2|y(i) - \hat{y}(i)|}{|y(i) + \hat{y}(i)|} \tag{11}$$

3. *Root-mean-square error (RMSE):*

$$RMSE = \sqrt{\frac{1}{n}\sum_{i=1}^{n}(y(i) - \hat{y}(i))^2} \tag{12}$$

5.3 Result Analysis

PCA-based calculation can be observed in Fig. 4, where nearly 92% of variances of variables are being covered by the first four eigenvalues (arranged in decreasing order) only and selected variable can be observed in the respective four PCs.

We had conducted our experiment with different sets of variables proposed by PCA since calculations from PCA suggested us for some high and some equally impacted variables. It can be observed from the principal components PC1, PC2, PC3, and PC4 in Fig. 4. We found out that a set of equally impacted variables (obtained from PC3) are giving us more efficient and acceptable results. So we have presented here the results obtained from PCA-BPNN and PCA-RBFN only.

BPNN- and RBFN-based forecasting graph has been given in Figs. 5 and 6, and the error-estimation data can be observed in the given Table 1.

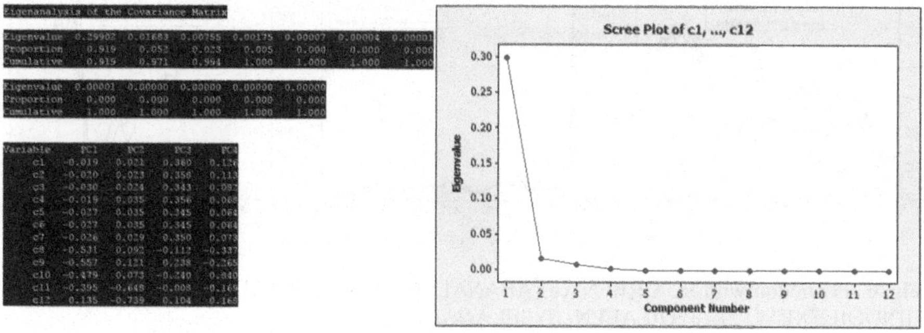

Fig. 4 PCA eigenvalues, principal components, and scree plot

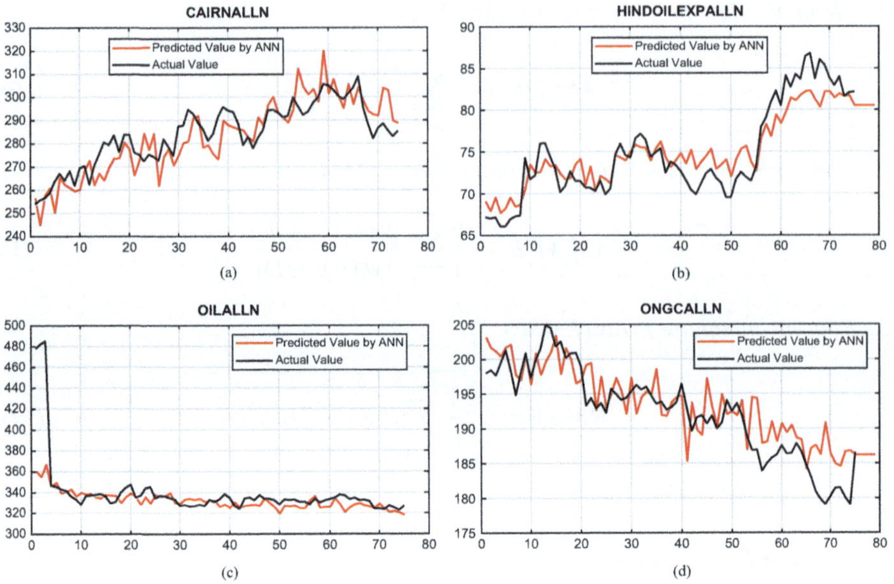

Fig. 5 Prediction with PCA-BPNN. (**a**) CAIRNALLN, (**b**) HINDOILEXPALLN, (**c**) OILALLN, (**d**) ONGCALLN

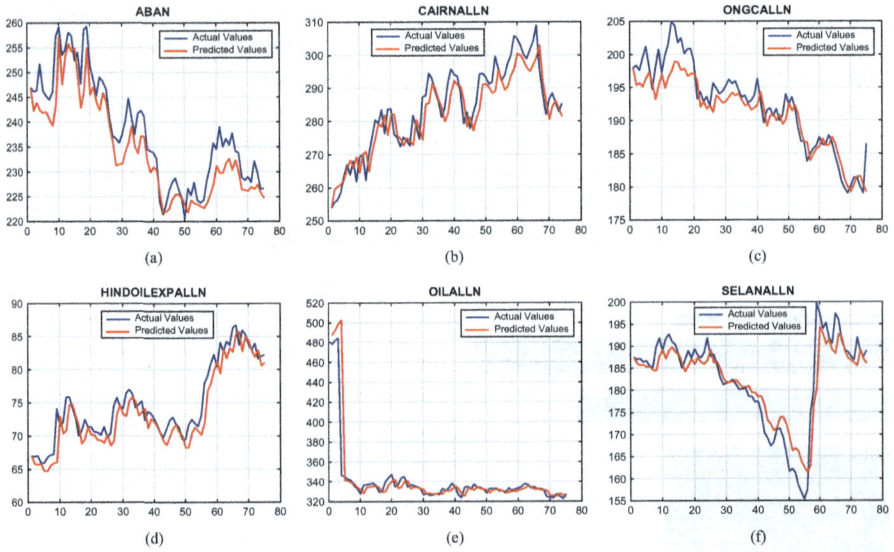

Fig. 6 Prediction with PCA-RBFN. (**a**) ABANALLN, (**b**) CAIRNALLN, (**c**) ONGCALLN, (**d**) HINDOILEXPALLN, (**e**) OILALLN, (**f**) SELANALLN

Table 1 Error estimation from different considered methods

	PCA-RBFN						PCA-BPNN			
	SELAN	ABAN	CAIRN	HIND	OIL	ONGC	CAIRN	HIND	OIL	ONGC
MAPE	160.28	127.53	145.98	225.22	32383.9	104.73	0.0234	0.0231	0.0243	0.0181
SMAPE	0.016	0.0128	0.0147	0.0236	0.0318	0.0106	0.0233	0.0230	0.0257	0.0172
RMSE	0.492	0.461	0.597	0.265	2.348	0.309	0.962	0.242	2.817	0.499

6 Conclusion and Scope

Different researches and empirical studies have been eventuated the fact that neural network techniques have been outstandingly performed in various disciplines, but no study has guaranteed that a particular technique will be performing best for all similar data. In this paper, we observed that both the techniques, viz., BPNN and RBFN, worked well and also performed better with those variables which are advised by PCA. We have presented graphs after PCA application along with BPNN and RBFN for the next 5 days in a sliding-window manner that can also be extended for some other numbers of days, but for a more longer period, it may not be equivalently effective. We also found out that PCA-RBFN has performed more efficiently than PCA-BPNN.

Since any particular ANN technique may not execute with all types of data in an efficient manner, there is a tremendous scope of modifying ANN techniques with hybridization and other numerous optimization approaches.

References

1. Enke, D., Thawornwong, D. S.: Thawornwong, D. S.: The use of data mining and neural networks for forecasting stock market returns. Expert Systems with Applications. **29**(4), 927–940 (2005) https://doi.org/10.1016/j.eswa.2005.06.024
2. Wang, F., Wang, J.: Statistical analysis and forecasting of return interval for SSE and model by lattice percolation system and neural network. Computers & Industrial Engineering. **62**(1), 198–205 (2012) https://doi.org/10.1016/j.cie.2011.09.007
3. Niu, H., Wang, J.: Volatility clustering and long memory of financial time series and financial price model. Digital Signal Processing. **23**(2), 489–498 (2013) https://doi.org/10.1016/j.dsp.2012.11.004
4. Pino, R., Parreno, J., Gomez, A., Priore, P.: Forecasting next-day price of electricity in the Spanish energy market using artificial neural networks. Engineering Applications of Artificial Intelligence. **21**(1), 53–62 (2008) https://doi.org/10.1016/j.engappai.2007.02.001
5. Balestrassi, P. P., Popova, E., Paiva, A. P., Lima, J. W. M.: Design of experiments on neural network's training for nonlinear time series forecasting. Neurocomputing. **72**(4–6), 1160–1178 (2009) https://doi.org/10.1016/j.neucom.2008.02.002
6. Liao, Z., Wang, J.: Forecasting model of global stock index by stochastic time effective neural network. Expert Systems with Applications. **37**(1), 834–841 (2010) https://doi.org/10.1016/j.eswa.2009.05.086

7. Powell, M. J. D.: Radial basis functions for multivariable interpolation: a review. Clarendon Press New York, USA (1987)
8. Versace, M., Bhatt, R., Hinds, O., Shiffer, M.: Predicting the exchange traded fund DIA with a combination of genetic algorithms and neural networks. Expert Systems with Applications. **27**(3), 417–425 (2004)
9. Wang, X. L., Sun, C. W.: Solve Fractal Dimension of Shanghai Stock Market by RBFNN. International Conference on Management Science & Engineering, Moscow. 1389–1394(2009)
10. Jaiswal, J. K., Das, R.: Application of artificial neural networks with backpropagation technique in the financial data. IOP Conference Series: Materials Science and Engineering. **263**(4), 042139(2017) http://stacks.iop.org/1757-899X/263/i=4/a=042139

A Novel Method of Solving a Quadratic Programming Problem Under Stochastic Conditions

S. Sathish and S. K. Khadar Babu

Abstract The optimization problem calculates an accurate solution, and the result of inequality constraints results in an approximate solution. The new method is developed under stochastic conditions to give an optimal expected value for the linear programming problem of downscaling data generated through an autoregressive integrated moving average (ARIMA) model. In this chapter, we predict future values using the ARIMA model for specified optimization problems. Wolfe's modified method is adopted to solve the linear programming problem under stochastic conditions.

1 Introduction

The stochastic process in the quadratic programming problem of generated downscaling data is a very novel approach in stochastic statistical analysis. The approximate value of the optimal solution under stochastic conditions is like a quadratic programming problem. Optimization provided results and generated downscaling data under a stochastic environment. The application of an optimization problem, which is obviously a nonlinear trend in stochastic hydrology, management science, financial engineering, and economics, can be predicted as an optimization that is useful in real life. The aim of this chapter is to show the result of the two applications of the stochastic process: optimization and generating downscaling data to carry out an idea for a quadratic problem and an autoregressive moving average stochastic optimization process, for which wide reading on the optimization method is expected based on a stochastic process.

S. Sathish (✉) · S. K. Khadar Babu
Department of Mathematics, School of Advanced Sciences, VIT University, Vellore, India
e-mail: sathish.s2016@vitstudent.ac.in; khadar.babu36@gmail.com; khadarbabu.sk@vit.ac.in

V. Madhu et al. (eds.), *Advances in Algebra and Analysis*, Trends in Mathematics,
https://doi.org/10.1007/978-3-030-01120-8_52

2 Review of Literature

Dantzig stated that the subject of mathematical optimization is derived from the stochastic process, the Markovian chain, operation research, and management science. In 1951, Kuhn and Tucker provided necessary and sufficient optimal conditions for the non-linear programming problem, now known as the Karush–Kuhn–Tucker (KKT) conditions. Earlier, in 1939, Karush had already developed conditions similar to those discovered by Kuhn and Tucker. The non-linear programming algorithms at this point in time, which constitute the basic theory of NLP, are not quite as exhaustive and systematic as the algorithms by Bazaraa et al. [1]. The optimality criteria for a class of non-differential non-linear programming were reported by Swarup et al. [2]. Prediction of the seasonal periods in the hydrological flow series of stochastic process using the Thomas–Fiering model was described by Sathish and Khadar Babu [3]. Vittal et al. [4] studied stochastic models for the amount of overflow in a finite dam with random inputs, random outputs, and exponential release policy. Linear programming for elementary introduction was described by Thompson [5]. A Markov decision model for economic production under stochastic demand and the factored Markov decision process to describe stochastic demand were observed by Mubiru [6] and Feng et al. [7]. The optimal control model in the production inventory system with stochastic demand and a comparative study of stochastic quadratic programming were published by Dhaiban [8].

3 Methods and Discussion

3.1 Statement of an Optimization Problem

The problem is as follows:

$$f(x) = \left\{ g_j(x) \leq 0,\ J = 1, 2, .., m,\ \ l_j(x) = 0,\ J = 1, 2, .p \right.$$

In the stochastic process, there is continuous optimization, i.e., $g_j(x) \leq 0$ and $l_j(x) = 0$ are called constraints. The stochastic process, which is deterministic or continuous, is designed to find the best feasible solution. The two objective functions are $f_1(x)$ and $f_2(x)$. Thus, t_1 and t_2 as f(x) = $t_1\ f_1(x) + t_2 f_2(x)$

3.2 Abstract in the Stochastic Process

The collection of random variables $x_t, t\ isT$ indexed on set T, denoted by S, which $x_t = x,\ \ forallt \in T$. The stochastic process is the collection of the function period t, f(X,T) verify x_1,$t_1 \geq 0$ similarly x_2,$t_2 \geq 0$.

3.3 Stochastic Process Linear Equality

In this concept, the best solution of the quadratic linear problem is stated as being continuous or deterministic. The fundamental concept of whether to identify is continuous or deterministic. Minimize $f(x) = c^T X = \sum_{k=1}^{n} c_j x_j$ subject to constraints $A_{ii}^T X = \sum_{j=1}^{n} a_{ij} x_j \leq b_i$,
$i = 1, 2, 3, \ldots\ldots m$
$j = 1, 2, 3, \ldots\ldots m,\ x_2 \geq 0$
$Var(f)$=$X^T V_X$
Mean: F(x)= $K_1 \sum_{j=1}^{n} \overline{cj} X_j$+$K_2 \sqrt{X^T V_X}$ $k_1 \geq 0, K_2 \geq 0$

3.4 Stochastic Process in Kuhn–Tucker Conditions of the Moving Average

Application 1: Downscaling data (Table 1, Fig. 1).

Application 2:

Solve quadratic programming using Wolfe's modified method

Max Z = $6x_1 + 3x_2 - 2x_1^2 - 3x_2{}^2 - 4x_1 x_2$

Subject to constraint $x_1 + x_2 \leq 1$, $2x_1 + 3x_2 \leq 4, x_1, x_2 \geq 0$ (Tables 2, 3, 4, 5 and 6)

Solution:

Table 1 Downscaling data

Downscaling data 30 days

YEAR 1985	1-15 DAYS	Water Level	YEAR 1985	16-30 DAYS	Water Level
	1/1/1985	21.9		1/1/1985	68.8
	2/1	21.8		2/1	71.1
	3/1	22.8		3/1	71
	4/1	22		4/1	68.1
	5/1	48.4		5/1	63.346
	6/1	62.374		6/1	47.9
	7/1	61.8		7/1	46.3
	8/1	57		8/1	45.7
	9/1	54.4		9/1	43.8
	10/1	57.5		10/1	56.5
	11/1	56.9		11/1	60.451
	12/1	58.9		12/1	57.623
	13/1	58.558		13/1	54.3
	14/1	60.451		14/1	56.9
	15/1	68.3		15/1	56.9..

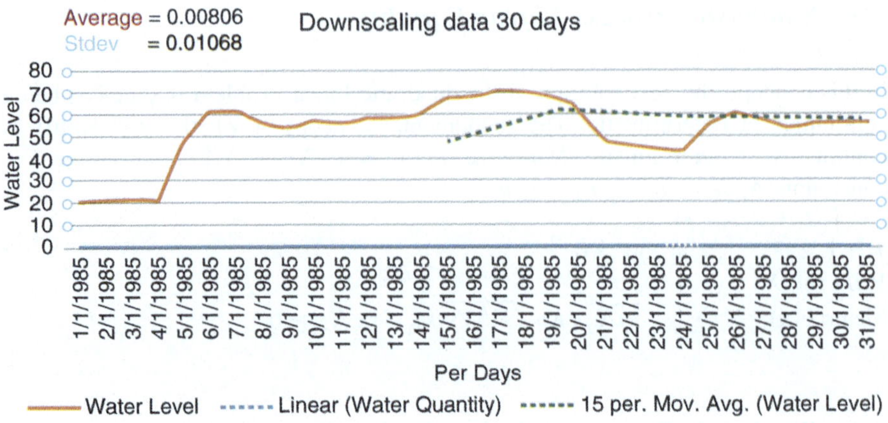

Fig. 1 Downscaling data

Table 2 Initial basic solution for quadratic programming

Basic	x_1	x_2	λ_1	λ_2	μ_1	μ_2	R_1	R_2	s_1	s_2	sol	Val
W	8	**10**	2	5	−1	−1	0	0	0	0	9	0.9
R_1	4	**4**	1	2	−1	0	1	0	0	0	6	1.5
R_2	**4**	**6**	**1**	**3**	**0**	**−1**	**0**	**1**	**0**	**0**	**3**	**0.5**
s_1	1	**1**	0	0	0	0	0	0	1	0	1	1
s_2	2	**3**	0	0	0	0	0	0	0	1	4	1.3

x_2 Entering variable
R_2 Leaving variable

Table 3 Iteration simplex method

Basic	x_1	x_2	λ_1	λ_2	μ_1	μ_2	R_1	R_2	s_1	s_2	sol	Val
W	**4/3**	0	1/3	0	−1	2/3	0	−5/3	0	0	4	3
R_1	**4/3**	0	1/3	0	−1	2/3	1	−2/3	0	0	4	3
x_2	**2/3**	**1**	**1/6**	**1/2**	**0**	**−1/6**	**0**	**1/6**	**0**	**0**	**1/2**	**0.75**
s_1	**1/3**	0	−1/6	−1/2	0	1/6	0	−1/6	1	0	1/2	1.5
s_2	**0**	0	−1/2	−3/2	0	1/2	0	−1/2	0	1	5/2	0

x_1 Entering variable
x_2 Leaving variable

Table 4 Iteration simplex method minimum values

Basic	x_1	x_2	λ_1	λ_2	μ_1	μ_2	R_1	R_2	s_1	s_2	sol	Val
W	0	−2	0	−1	−1	**1**	0	−2/3	0	0	3	3
R_1	0	−2	0	−1	−1	**1**	1	−1	0	0	3	3
x_1	1	3/2	1/4	3/4	0	**−1/4**	0	1/4	0	0	3/4	**−3**
s_1	**0**	**−1/2**	**−1/4**	**−3/4**	**0**	**1/4**	**0**	**−1/4**	**1**	**0**	**1/4**	**1**
s_2	0	0	−1/2	−3/2	0	**1/2**	0	−1/2	0	1	5/2	5

μ_2 Entering variable
s_1 Leaving variable

Table 5 Iteration for minimization

Basic	x_1	x_2	λ_1	λ_2	μ_1	μ_2	R_1	R_2	s_1	s_2	sol	Val
W	0	0	1	2	−1	0	0	−1	**4**	0	2	0.5
R_1	**0**	**0**	**1**	**2**	**−1**	**0**	**1**	**0**	**4**	**0**	**2**	**0.5**
x_1	1	1	0	0	0	0	0	0	**−1**	0	1	**−1**
μ_2	0	−2	−1	−3	0	1	0	−1	**4**	0	1/4	1
s_2	0	1	0	0	0	0	0	0	**2**	1	3/2	3

s_1 entering variable
R_1 Leaving variable

Table 6 Output result for minimization in feasible solution

Basic	x_1	x_2	λ_1	λ_2	μ_1	μ_2	R_1	R_2	s_1	s_2	Sol
W	0	0	0	0	0	0	−1	−1	0	0	0
s_1	0	0	1/4	1/2	−1/4	0	1/4	0	1	0	1/2
x_1	1	1	1/4	1/2	−1/4	0	1/4	0	0	0	3/2
μ_2	0	−2	−2	−5	1	1	−1	−1	0	0	−1
s_2	0	1	−1/2	−1	1/2	0	−1/2	0	0	1	1/2

Max Z = (6,3) $\begin{pmatrix} x_1 \\ x_2 \end{pmatrix} + (x_1, x_2) \begin{pmatrix} -2 & -2 \\ -2 & -3 \end{pmatrix}$

subject to $\begin{pmatrix} 1 & 1 \\ 2 & 3 \end{pmatrix} \begin{pmatrix} x_1 \\ x_2 \end{pmatrix} \leq \begin{pmatrix} 1 \\ 2 \end{pmatrix}$

$x_1, x_2 \geq 0$

Kuhn–Tucker conditions $\begin{pmatrix} -2D & A^T & -I & 0 \\ A & 0 & 0 & I \end{pmatrix} \begin{pmatrix} X \\ \lambda \\ U \\ S \end{pmatrix} = \begin{pmatrix} C^T \\ b \end{pmatrix}$

$$\begin{pmatrix} 4 & 4 & 1 & 2 & -1 & 0 & 0 & 0 \\ 4 & 6 & 1 & 3 & 0 & -1 & 0 & 0 \\ 1 & 1 & 0 & 0 & 0 & 0 & 1 & 0 \\ 2 & 3 & 1 & 3 & 0 & -1 & 0 & 0 \end{pmatrix} \begin{pmatrix} x_1 \\ x_2 \\ \lambda_1 \\ \lambda_2 \\ \mu_1 \\ \mu_2 \\ S_1 \\ S_2 \end{pmatrix} = \begin{pmatrix} 6 \\ 3 \\ 1 \\ 4 \end{pmatrix}$$

$$4x_1 + 4x_2 + \lambda_1 + 2\lambda_2 - \mu_1 = 6 \tag{1}$$

$$4x_1 + 6x_2 + \lambda_1 + 3\lambda_2 - \mu_2 = 3 \tag{2}$$

$$x_1 + x_2 + s_1 = 1 \tag{3}$$

$$2x_1 + 3x_2 + s_2 = 4 \tag{4}$$

$$4x_1 + 4x_2 + \lambda_1 + 2\lambda_2 - \mu_1 + R_1 = 6 \tag{5}$$

$$R_1 = 6 - 4x_1 - 4x_2 - \lambda_1 - 2\lambda_2 + \mu_1$$

$$4x_1 + 6x_2 + \lambda_1 + 3\lambda_2 - \mu_2 + R_2 = 3 \tag{6}$$

$$R_2 = 3 - 4x_1 - 6x_2 - \lambda_1 - 3\lambda_2 + \mu_2$$

$$x_1 + x_2 + s_1 = 1 \tag{7}$$

$$2x_1 + 3x_2 + s_2 = 4 \tag{8}$$

Min w = R_1 + R_2

$$\mathrm{W} = 9 - 8x_1 - 10x_2 - 2\lambda_1 - 5\lambda_2 + \mu_1 + \mu_2$$

$$\mathrm{W} + 8x_1 + 10x_2 + 2\lambda_1 + 5\lambda_2 - \mu_1 - \mu_2 = 9$$

x_1 = 3/2 x_2 = 0 Max z = 9/2 = 4.5

4 Stochastic Process in Quadratic Programming

Minimize f(x) = $\mathrm{C}^T\mathrm{X} = \sum_{k=1}^{n} C_j X_j$
Var (f)=$X^T V_X$.

Mean:
F(X) = $K_1 \sum_{j=1}^{n} \overline{Cj}\, X_j$ + $K_2\sqrt{X^T V_X}$ $K_1 \geq 0$, $K_2 \geq 0$
$h_i = \sum_{j=1}^{n} a_{ij} X_j - b_i$
$\bar{h}_1 = \bar{a}_{11}x_1 + \bar{a}_{12}x_2 - \bar{b}_1$
$6x_1 + 3x_2 - 2x_1^2 - 3x_2{}^2 - 4x_1x_2$
$\bar{h}_2 = \bar{a}_{21}x_1 + \bar{a}_{22}x_2 - \bar{b}_2$
$2x_1 + 3x_2 - 4$ {x(t), t≥0} by a stochastic process

K and t are independent variables according to the definition of a stochastic process.

The matrix of C_j is given by $V = \begin{pmatrix} b1 & 0 \\ 0 & b2 \end{pmatrix}$

$V = \begin{pmatrix} 0\,0 \\ 0\,4 \end{pmatrix} = 0$ is optimal, so that the variance of the objective function var (f) = $X^T V_X$
$= 4x_2^2$. The objective function can be taken as mean x_1=0 and x_2= 6
F = $k_1(x_1 + 6x_2)$ +k_2 $\sqrt{4x_2 2}$
The constrained can be taken in the probability conditions P (h_i $\leq$ 0) $\geq$ P_i. Verifying the above condition under stochastic process is optimal by changing to minimize the function.

F = $k_1(x_1 + 6)$ + k_2 $\sqrt{4x_2 2} \leq 0$ the constraints are $x_1 \geq 0$, $x_2 \geq 0$.

5 Conclusion

The stochastic downscaling of the predicted process is a novel procedure for solving the optimization problems using the linear programming problem approach. The currently developed solution is optimal and a predicted ARIMA is applicable. Finally, we conclude that the new procedure is the optimal approach to downscaling data under specified stochastic conditions.

References

1. M.S. Bazaraa, H.D. Sherali, C.N. Shetty, Nonlinear programming theory and algorithms, John Wiley and Sons, Third edition.**45**, 846 (1994)
2. K. Swarup, P.K. Gupta, M. Mohan, Operations Research, Sultan Chand&Sons.**22**, 69–72 (1978)
3. S. Sathish, S.K. Khadar Babu. Stochastic time series analysis of hydrology data for water resources., J. Iop Anal. Appl. **263**, (2017) doi:10.1088/1757-899X/263/4/042140
4. P.R. Vittal, V. Thangaraj, and V. Muralidhar, Stochastic models for the amount of overflow in a finite dam with random inputs, random outputs and exponential release policy, Stochastic Analysis and Applications. **29**, 473–485 (2011)
5. G.E. Thompson, Linear programming: An elementary introduction, Macmillan, New York, 1 May in ACM SIGMAP Bulletin New York, 26 (1973)
6. K. Mubiru, Optimal replenishment policies for two-Echelon inventory problems with stationary price and stochastic demand, International Journal of Innovation and Scientific Research **13**(1), 98–106 (2015)
7. H. Feng, Q. Wu, K. Muthuraman, and V Deshpande, Replenishment policies for multi-product stochastic inventory systems with correlated demand and joint-replenishment costs, Production and Operations Management **24**(4), 647–664 (2015)
8. A.K. Dhaiban, A comparative study of stochastic quadratic programming and optimal control model in production - inventory system with stochastic demand **37**, April (2017)

The manufacturer's authorised representative in the EU is Springer Nature Customer Service Centre GmbH, Europaplatz 3, 69115 Heidelberg, Germany. If you have any concerns regarding our products, please contact ProductSafety@springernature.com

Printed and bound by CPI Group (UK) Ltd, Croydon, CR0 4YY
30/09/2026
02225229-0017